Brazilian Marine Biodiversity

Series Editor

Alexander Turra, Unesco Chair for Ocean Sustainability
Oceanographic Institute and Advanced Studies Institute
University of São Paulo
São Paulo, Brazil

The book series Brazilian Marine Biodiversity was designed to communicate to a broad and international readership the diversified marine and coastal habitats along the large Brazilian coast.

***

The diversity of marine habitats found in Brazil is astonishing and includes estuaries, coral reefs, rocky shores, sandy beaches, rhodolith beds, mangroves, salt marshes, deep-sea habitats, vegetated bottoms, and continental shelf. These habitats are addressed from an ecosystem perspective across the series, and characterized in terms of distribution and peculiarities along the Brazilian coast, records of relevant species, and information on the prevailing structuring ecological and oceanographic processes governing biodiversity.

The series also presents an analysis of the role of biodiversity and the importance of ecosystem services, and discusses the threats to each habitat, such as pollution, habitat loss, invasive species, overfishing, and global environmental changes. Conservation efforts are also considered as well as gaps in scientific knowledge and science policy interface.

This series is an initiative of the UNESCO Chair for Ocean Sustainability, hosted by the Oceanographic Institute and the Advanced Studies Institute of the University of São Paulo, within the Brazilian Network for Monitoring Coastal Benthic Habitats (ReBentos; rebentos.org), which is supported by the Brazilian National Council for Scientific and Technological Development (CNPq), the Research Program on Biodiversity Characterization, Conservation, Restoration and Sustainable Use of the São Paulo Research Foundation (BIOTA-FAPESP), the Coordination for the Improvement of Higher Education Personnel (CAPES) and the Brazilian Innovation Agency (FINEP). ReBentos is part of the Brazilian Network on Global Climate Change Research (Rede Clima) and the Science and Technology National Institute on Climate Changes (INCT Mudanças Climáticas) at the Ministry of Science, Technology, and Innovation (MCTI).

Ruy K. P. Kikuchi
Zelinda M. A. N. Leão
Maria Elisabeth de Araújo • Tito M. C. Lotufo
Editors

Brazilian Coral Reefs

A Multidisciplinary Approach

Springer

Editors
Ruy K. P. Kikuchi
Laboratório de Estudo dos Recifes de Corais (RECOR)
Instituto de Geociências
Universidade Federal da Bahia
Salvador, Bahia, Brazil

Zelinda M. A. N. Leão
Laboratório de Estudo dos Recifes de Corais (RECOR)
Instituto de Geociências
Universidade Federal da Bahia
Salvador, Bahia, Brazil

Maria Elisabeth de Araújo
Departamento de Oceanografia
Universidade Federal de Pernambuco (UFPE) Grupo de Ictiologia Marinha Tropical (IMAT)
Pernambuco, Brazil

Tito M. C. Lotufo
Instituto Oceanografico
Universidade de São Paulo
São Paulo, SP, Brazil

Instituto Coral Vivo
Santa Cruz Cabralia
BA, Brazil

ISSN 2520-1077 ISSN 2520-1085 (electronic)
Brazilian Marine Biodiversity
ISBN 978-3-031-59151-8 ISBN 978-3-031-59152-5 (eBook)
https://doi.org/10.1007/978-3-031-59152-5

This Springer imprint is published by the registered company Springer Nature Switzerland AG
The registered company address is: Gewerbestrasse 11, 6330 Cham, Switzerland

Preface

Twenty years have passed since we published a comprehensive synthesis of the knowledge of corals and coral reefs of the Brazilian waters (Leão et al. 2003) after Laborel's seminal work of his thesis in 1967. During this period, an exponential increase in reef knowledge, in the number of investigators, and research groups occurred, bringing to light many approaches to reef ecosystem science, as shown in Chap. 2 of this book. It was high time we gathered the updated knowledge developed ever since in an extended volume. This project started a few years ago but gained momentum during the COVID-19 pandemic. While organizing the team that would elaborate this volume, we tried to gather people from different parts of the country and join experienced and young scientists to count on the most diverse team possible.

Reefs result from the gregariousness of many kinds of organisms, composing an ecosystem while they build their whereabouts. This communal strategy appeared early in the Earth's outer environment's history and is as old as 3400 million years (Ma) in the early Precambrian Eon. From the outset, the microorganisms that started this form of living passed it on to more complex organisms; they endured rough times when Earth's surface temperature was low and better ones when it got warmer.

The optimum conditions for reef growth are waters supersaturated in carbonate, where the temperature is higher than 20 °C and light is abundant, that is, shallow seas with little sediment in suspension. These conditions are not fully met in the Brazilian continental margin. Compared to tropical seas such as the Caribbean or the Indo-Pacific region, the river's runoff (small as it can be) and the water dynamics, mainly the wave regime, produce a turbid environment characteristic of marginal conditions for reef growth.

Reefs grow in Brazilian waters, though, mainly in the Tropical realm, but have been mapped to the sub-tropical domain recently. The fact that they grow in sub-optimal conditions makes them important locations to evaluate and understand the limits of their growth. On the other hand, being on the verge of survivorship also increases the risk of deteriorating because of global climate changes or local direct human influence. The year 2023 was especially warm, with strong heat waves spreading all over the globe, including Brazil. An active El Niño developed in the

Pacific threatens to cause an unprecedented bleaching event worldwide and can affect reefs of the Tropical South Atlantic.

Reefs are ecosystems that host a rich biodiversity that sustains human communities. They provide ecological benefits such as food through fisheries, coastal protection from marine and climatic hazards, culture, building materials, and biochemical substances that are highly valuable to human society. They are fundamental to traditional communities that depend on them for subsistence. In the Southwestern Atlantic, it is likewise.

This volume encompasses 12 chapters that cover three aspects of reefs' existence in the Southwestern Atlantic. The first three chapters present the regional approach and evolution of the research made on reefs and the biogeography of key groups of organisms (builders and dwellers).

Chapter 1 presents reef physical geography, geology, oceanography, the occurrence, morphology, and distribution of reefs on the continental shelf, with particular attention to the shallow water reefs from the coastline until 30 m of depth. We also present an ecogeographic compartmentalization of the Southwestern Atlantic, considering oceanographic and reef morphological characteristics.

Chapter 2 brings the history of research about the tropical Brazilian coral reef systems, which starts with the nineteenth-century naturalists through the twentieth century when a national community of researchers developed and reached the consolidation and internationalization of this community. It is a bibliometric analysis of over 700 articles obtained in the peer-reviewed literature that offer essential references concerning the studied articles' distribution in time and space. This analysis describes the main approaches and how they evolved, adopting new reef assessment and mapping technologies. It shows that contributions to management initiatives for minimizing the local and global impacts affecting the Brazilian reefs become significant only after the decade that began in 2010.

In Chap. 3, the authors present an overview of the origins, biogeography, and macroecology of the Southwestern Atlantic reef's biodiversity. This biodiversity is characterized by high endemism, driven by the isolation of the Atlantic from the world's centers of biodiversity (i.e., tropical Indo-Pacific) during the Paleogene Period, distinct ecological conditions, and the overlap of tropical and subtropical species. They also show that many evolutionary processes shape species' current distribution along the coast and offshore areas, such as seamounts and oceanic islands.

In Chaps. 4, 5, 6, and 7, biological and ecological aspects of invertebrates and vertebrates that build and dwell in reef environments of the Southwestern Atlantic are revisited.

As such, Chap. 4 is dedicated to synthesizing the ecological processes of herbivory and competition for space. The authors show that the morphology and distribution of reef buildups shelter an herbivore functional group that is low in diversity and highly endemic. Algal assemblages like reefs worldwide, coexisting and interacting with a unique collection of endemic species, provide an interesting scenario for interactions to be studied. This chapter compiles the results of coral reef studies on herbivory and competition for space over the past two decades and presents its implications for this singular reef system.

In Chap. 5, the authors address the general biological and ecological aspects of 12 benthic reef-dwelling taxa found in Brazil. They present each taxon's basic morphology, diversity, reproductive biology, and main functional roles. These organisms' roles include framework building and recycling (through bioerosion), sediment production, primary production, herbivory, predation (piscivory, omnivory), and cycling nutrients and organic matter.

The large vertebrates and their interactions are treated in Chap. 6. They are fundamental in maintaining the structure and dynamics of the trophic webs in the reefs due mainly to their capacity to perform different functions in the reefs. They are generally top predators, consuming other carnivores, but can also act as mid-ranking trophic levels in the reef food web. Reefs provide food, nursery areas, refuge from predation, shelter against currents, and the removal of parasites. Large vertebrates are threatened by global climate changes and local human impacts, such as overfishing, as are most species that build and dwell in reefs.

In Chap. 7, the authors use three approaches to discuss genetic and demographic connectivity in Brazilian reef environments. The first approach deals with the exchange of information through genetic connectivity within an evolutionary timescale. The authors discuss the relevance of biogeographical barriers and marine currents to genetically structured coral and fish species populations along the Brazilian coast. The second deals with demographic connectivity within the biophysical context on an ecological timescale. They use biophysical modeling tools to infer connectivity for fish and lobster species, shedding light on the most relevant processes involved. In the third approach, the focus shifts to the ecological connectivity between coral reefs and adjacent habitats, involving mainly the movement of individuals. It shows that mangroves and estuaries are vital for Brazilian reef fish species.

We approach human relationships with the reef ecosystem in the last four chapters (Chaps. 8, 9, 10, 11, and 12).

The first aspect is the benefit they provide as food reservoirs to humans, treated in Chap. 8. This chapter deals with artisanal fishers and fish as food for humans and the urgency to realize that healthy eating is the right of all of us and that these fishers are responsible for bringing fish to our tables. The authors show that difficulties in developing a sound conservation of coastal ecosystems, reefs in particular, cause socioeconomic and environmental imbalance and risk to food security, severely damaging human health. Furthermore, they also show that traditional communities' health is threatened by the contamination of fish and changes in dietary habits because of poverty and the consumption of industrialized foods.

A change in perspective appears in Chap. 9, where the authors discuss approaches to exploiting the biotechnological potential in an ecologically friendly or sustainable manner. They present an overview of marine biotechnology based on the myriad of organisms found in reefs. They also explore the origin and development of Brazilian marine biotechnology and the biotechnological potential of marine organisms from coral reefs.

Climate change and local impacts that threaten Brazilian coral reefs are described in Chap. 10. The authors discuss the Brazilian Reef Refugia Hypothesis (BRRH), based on the lower occurrence of coral bleaching events and mortality rates than other reef regions worldwide. They review the chronic and acute stressors that are currently affecting these reefs. They show an uneven incidence and frequency of stressors that impact these reefs. These stressors act in cumulative and synergistic ways and undermine BRRH. They conclude that the Brazilian reefs have limited potential to provide climate-change refugia and need urgent attention to reduce and mitigate impacts at local and global scales.

In Chap. 11, recreational and cultural relationships provided by the reefs are explored. Tourism, environmental education, and aquarium trade are discussed. The authors show that these are sources of employment, with foreign direct investment and gross domestic product. The economy of coastal states depends to some degree on these revenues, which are linked to recreational activities and the quality of the beaches. As poorly planned tourism causes severe damage to these delicate reef environments, it must be designed carefully, with educational activities contributing to the conservation and preservation of the reef environment. Among these activities, they suggest monitoring the capacity to support tourism, as well as incentives for ecotourism, practices of environmental education, and sustainable fishkeeping.

Last, Chap. 12 focuses on discussing the main strategies for the conservation and management of Brazilian reefs. The authors start from the principle that environmental protection areas are essential for conserving reef ecosystems. They show that although the existing system of MPAs covers 15.3% of the total reef area, this figure needs to be more accurate because only a fraction of reef habitats is covered by no-take MPAs (3.6%). Several reef types have no coverage by fully protected MPAs. This bias varies along the coastal regions and occurs together with a need for more effectiveness of the protection level of the MPAs. The authors also identified a general need for long-term monitoring programs in no-take MPAs and the absence of studies on coral reef restoration.

The COVID-19 pandemic compelled a pause in fieldwork, prompting us to analyze stored data and pending themes. Additionally, we observed a modest improvement in environmental quality stemming from reduced industrial production and social activities as previously practiced. As such, this study captures a retrospective moment of contemplation regarding our utilization of natural resources and the strain we impose on the ecological benefits of reefs and other coastal ecosystems. We anticipate that this comprehensive synthesis of different aspects of Brazilian coral reefs will shed more light on their distinctiveness and scientific, social, and ecological relevance. This collaborative effort among experts can guide us in improving our conservation strategies for these vital and sensitive ecosystems.

Salvador, Brazil Ruy K. P. Kikuchi
Salvador, Brazil Zelinda M. A. N. Leão
Recife, Brazil Maria Elisabeth de Araújo
São Paulo, Brazil Tito M. C. Lotufo

Reference

Leão ZMAN, Kikuchi RKP, Testa V (2003) Corals and coral reefs of Brazil. In: Cortés J (ed) Latin American coral reefs. Elsevier Science, Amsterdam, pp 9–52.

Acknowledgements

The editors thank all contributing authors for their efforts to make this book a valuable contribution to the reef users, research, and decision-making communities. This book is a contribution of the Nacional Institute of Science and Technology of the Tropical Marine Environment through grants by the National Council for Scientific Research and Technology (CNPq 465634/2014-1) and the Foundation for Research Support of the State of Bahia (FAPESB inc004/2019). We thank Dr. Alexander Turra for his trust and invitation to organize this book. At Springer, we thank Mr. João Pildervasser, Mr. Rahul Sharma, and Ms Kalai Shahethya for their patient guidance while preparing this work.

Contents

The original version of the book has been revised. A correction to this book can be found at https://doi.org/10.1007/978-3-031-59152-5_13

Contributors

Maria D. O. Alves Faculdade Frassinetti do Recife (FAFIRE), Recife, Pernambuco, Brazil

Thomás N. S. Banha Instituto Oceanográfico, Universidade de São Paulo, São Paulo, SP, Brazil

Leopoldo M. Barreto Universidade Federal do Recôncavo da Bahia (UFRB), Cruz das Almas, Bahia, Brazil

Eduardo L. Barros Instituto de Ciências do Mar-LABOMAR, Universidade Federal do Ceará (UFC), Fortaleza, Ceará, Brazil

Mariana G. Bender Laboratório de Macroecologia e Concervação, Departamento de Ecologia e Evolução, Universidade Federal de Santa Maria, Santa Maria, Rio Grande do Sul, Brazil

Adalto Bianchini Instituto de Ciências Biológicas, Universidade Federal do Rio Grande, Rio Grande, RS, Brazil

Janaina F. Calado Universidade Estadual do Amapá (UEAP), Macapa, Amapá, Brazil

Katia C. C. Capel Departamento de Genética e Evolução, Universidade Federal de São Carlos (UFSCAR), São Carlos, Brazil

Centro de Biologia Marinha (CEBIMAR), Universidade de São Paulo (USP), São Sebastião, Brazil

Clovis B. Castro Instituto Coral Vivo, Santa Cruz Cabrália, Bahia, Brazil

Museu Nacional, Universidade Federal do Rio de Janeiro, Rio de Janeiro, RJ, Brazil

Cesar A. M. M. Cordeiro Laboratory of Environmental Sciences, Universidade Estadual do Norte Fluminense, Campos dos Goytacazes, RJ, Brazil

Ralf T. S. Cordeiro Departamento de Biologia, Universidade Federal Rural de Pernambuco, Recife, Pernambuco, Brazil

Leticia V. Costa-Lotufo Laboratório de Farmacologia Marinha, Instituto de Ciências Biomédicas, Universidade de São Paulo, São Paulo, SP, Brazil

Luiz C. Cotovicz Jr Instituto de Ciências do Mar-LABOMAR, Universidade Federal do Ceará (UFC), Fortaleza, Ceará, Brazil

Ricardo Coutinho Instituto de Estudos do Mar Almirante Paulo Moreira, Arraial do Cabo, Rio de Janeiro, Brazil

Igor C. S. Cruz Instituto de Geociências, Universidade Federal da Bahia (UFBA), Salvador, Bahia, Brazil

Maria Elisabeth de Araújo Departamento de Oceanografia, Universidade Federal de Pernambuco, Recife, Pernambuco, Brazil

Beatriz Diniz Instituto de Ciências do Mar-LABOMAR, Universidade Federal do Ceará (UFC), Fortaleza, Ceará, Brazil

Linda Eggertsen Instituto Nacional de Pesquisas Espaciais (INPE), São José dos Campos, São Paulo, SP, Brazil

Hawai'i Institute of Marine Biology, University of Hawai'i at Manoa, Honolulu, HI, USA

Clarissa A. K. Endo Instituto Nacional de Pesquisas Espaciais (INPE), São José dos Campos, São Paulo, SP, Brazil

Institute of Marine Research, Bergen, Norway

Alice F. Feitosa Programa de Pós-Graduação em Ciências Marinhas Tropicais, Instituto VerdeLuz, Universidade Federal do Ceará (UFC), Fortaleza, Ceará, Brazil

Caroline V. Feitosa Universidade Federal do Ceará (UFC), Instituto de Ciências do Mar, LABOMAR-UFC, Fortaleza, Ceará, Brazil

João L. L. Feitosa Departamento de Zoologia, Universidade Federal de Pernambuco, Recife, Pernambuco, Brazil

Antonio V. Ferreira Jr Laboratório de Oceanografia Geológica, Departamento de Oceanografia, Universidade Federal de Pernambuco, Recife, Pernambuco, Brazil

Harildon M. Ferreira Instituto Chico Mendes de Conservação da Biodiversidade (ICMBIO), Federal District, Brasilia, Brazil

Sergio R. Floeter Laboratório de Macroecologia e Biogeografia, Departamento de Ecologia e Zoologia, Universidade Federal de Santa Catarina, Florianópolis, Santa Catarina, Brazil

João E. P. Freitas Instituto de Ciências do Mar-LABOMAR, Universidade Federal do Ceará (UFC), Fortaleza, Ceará, Brazil

Ana L. B. Gaspar Instituto Recifes Costeiros, Tamandaré, Pernambuco, Brazil

Douglas F. M. Gherardi Laboratorio de Estudos dos Oceanos e Atmosfera (LOA), Instituto Nacional de Pesquisas Espaciais (INPE), São José dos Campos, São Paulo, SP, Brazil

Leandro Godoy Universidade Federal do Rio Grande do Sul, Porto Alegre, Rio Grande do Sul, Brazil

Maria Tereza J. Gouveia Projeto Coral Vivo, Pesquisa e Educação Ambiental, Porto Seguro, Bahia, Brazil

Renan G. P. Guerra Instituto de Ciências do Mar-LABOMAR, Universidade Federal do Ceará (UFC), Fortaleza, Ceará, Brazil

Anne L. Gurgel Instituto de Ciências do Mar-LABOMAR, Universidade Federal do Ceará (UFC), Fortaleza, Ceará, Brazil

Livio M. Gurjão Instituto Brasileiro do Meio Ambiente e dos Recursos Naturais Renováveis (IBAMA), Fortaleza, Ceará, Brazil

Paula C. Jimenez Laboratório de Bioprospecção de Organismos Marinhos, Instituto do Mar, Universidade Federal de São Paulo, Santos, São Paulo, Brazil

Ruy K. P. Kikuchi Laboratório de Estudo dos Recifes de Corais (RECOR), Instituto de Geociências, Universidade Federal da Bahia, Salvador, Bahia, Brazil

Marcelo Kitahara Departamento de Ciências do Mar, Universidade Federal de São Paulo (UNIFESP), Santos, São Paulo, Brazil

Centro de Biologia Marinha da Universidade de São Paulo (CEBIMar), São Sebastião, São Paulo, Brazil

Zelinda M. A. N. Leão Laboratório de Estudo dos Recifes de Corais (RECOR), Instituto de Geociências, Universidade Federal da Bahia, Salvador, Bahia, Brazil

Luciana S. Lima Laboratorio de Estudos dos Oceanos e Atmosfera (LOA), Instituto Nacional de Pesquisas Espaciais (INPE), São José dos Campos, São Paulo, SP, Brazil

Bárbara P. M. M. Lippi Pesquisadora Independente, Recife, Pernambuco, Brazil

Guilherme O. Longo Marine Ecology Laboratory, Departamento de Oceanografia e Limnologia, Universidade Federal do Rio Grande do Norte, Natal, Rio Grande do Norte, Brazil

Tito M. C. Lotufo Instituto Coral Vivo, Santa Cruz Cabrália, Bahia, Brazil

Instituto Oceanográfico, Universidade de São Paulo, São Paulo, SP, Brazil

Caroline C. Lucas Instituto de Ciências do Mar-LABOMAR, Universidade Federal do Ceará (UFC), Fortaleza, Ceará, Brazil

Rafael A. Magris Instituto Chico Mendes de Conservação da Biodiversidade (ICMBIO), Federal District, Brasilia, Brazil

Eliana Matos Instituto de Ciências do Mar-LABOMAR, Universidade Federal do Ceará (UFC), Fortaleza, Ceará, Brazil

Sérgio M. G. Mattos Grupo de Ictiologia Marinha Tropical (IMAT), Independent Researcher, Recife, Pernambuco, Brazil

Melina F. Maxwell Laboratorio de Macroecologia e Concervação, Departamento de Ecologia e Evolução, Universidade Federal de Santa Maria, Santa Maria, Rio Grande do Sul, Brazil

Paulo W. Melo Grupo de Ictiologia Marinha Tropical (IMAT), Independent Researcher, Recife, Pernambuco, Brazil

Liana F. Mendes Departamento de Ecologia, Universidade Federal do Rio Grande do Norte (UFRN), Natal, Rio Grande do Norte, Brazil

Miguel Mies Instituto Coral Vivo, Santa Cruz Cabrália, Bahia, Brazil

Instituto Oceanográfico, Universidade de São Paulo, São Paulo, SP, Brazil

Ricardo J. Miranda Departamento de Zoologia, Universidade Federal de Pernambuco, Recife, Pernambuco, Brazil

Nemer E. Narchi Centro de Estudios en Geografia Humana, El Colegio de Michoacán, Zamora, Mexico

Marilia D. M. Oliveira Laboratório de Estudo dos Recifes de Corais (RECOR), Instituto de Geociências, Universidade Federal da Bahia, Salvador, Bahia, Brazil

Sandra V. Paiva Instituto de Ciências do Mar-LABOMAR, Universidade Federal do Ceará (UFC), Fortaleza, Ceará, Brazil

Renato C. Pereira Departamento de Biologia Marinha, Universidade Federal Fluminense e Instituto de Pesquisas Jardim Botânico do Rio de Janeiro, Rio de Janeiro, RJ, Brazil

Hudson T. Pinheiro Centro de Biologia Marinha, Universidade de São Paulo, São Sebastião, São Paulo, Brazil

Luisa V. M. V. Queiroz-Véras Departamento de Zoologia, Universidade Federal de Pernambuco, Recife, Pernambuco, Brazil

Juan P. Quimbayo Centro de Biologia Marinha (CEBIMAR), Universidade de São Paulo, São Sebastião, São Paulo, Brazil

Emanuelle F. Rabelo Universidade Federal Rural do Semiárido (UFERSA), Mossoró, Rio Grande do Norte, Brazil

Ligia M. Rocha Oceânica – Pesquisa, Conservação e Educação, Parnamirim, Rio Grande do Norte, Brazil

Luiz A. Rocha Departmento de Ictiologia, Academia de Ciências da California, San Francisco, CA, USA

Jones Santander-Neto Instituto Federal de Educação, Ciência e Tecnologia do Espírito Santo (IFES), Campus Piúma, Piúma, Espírito Santo, Brazil

Angelica R. Soares Grupo de Produtos Naturais de Organismos Aquáticos (GPNOA), Instituto de Biodiversidade e Sustentabilidade (NUPEM), Universidade Federal do Rio de Janeiro, Macaé, Rio de Janeiro, Brazil

Marcelo O. Soares Instituto de Ciências do Mar (LABOMAR), Universidade Federal do Ceará (UFC), Fortaleza, Ceará, Brazil

Dipartimento di Scienze e Tecnologie Biologiche e Ambientali (DISTEBA), Università del Salento, Lecce, Italy

Carlos E. P. Teixeira Instituto de Ciências do Mar-LABOMAR, Universidade Federal do Ceará (UFC), Fortaleza, Ceará, Brazil

Diego V. Wilke Laboratório de Bioprospecção e Biotecnologia Marinha, Núcleo de Pesquisa e Desenvolvimento de Medicamentos (NPDM), Departamento de Fisiologia e Farmacologia, Faculdade de Medicina, Universidade Federal do Ceará, Fortaleza, Ceará, Brazil

Carla Zilberberg Universidade Federal do Rio de Janeiro, Rio de Janeiro, RJ, Brazil

About the Editors

Ruy K. P. Kikuchi is Full Professor at the Department of Oceanography, Geosciences Institute of Universidade Federal da Bahia (UFBA) and is Vice Coordinator of the Geophysics and Geology Research Center (CPGG/UFBA). He took a degree in Geology at the University of São Paulo (USP, 1986) with a PhD in Geology from UFBA (2002). He is the Coordinator of the Reefs and Coralline Environments Working Group and the Oil Spill Working Group of the National Institute of Science and Technology for the Marine Tropical Environments (INCT AmbTropic). His research focuses on coral reef mapping, assessment, and monitoring, as well as coral sclerochronology and geochemistry.

Zelinda M. A. N. Leão is Professor Emeritus of the Institute of Geosciences of Universidade Federal da Bahia (UFBA). She is a Naturalist who graduated from UFBA (1958) with a PhD in Marine Geology from the University of Miami (Rosenstiel Marine and Atmospheric Sciences, RSMAS, 1982). She has carried out research on Brazilian reefs for over 40 years. She has numerous publications on coral reefs in Brazil. For many years, she has coordinated a team trained to, among other things, assess the conditions of Bahia's reefs, studying the biotic and abiotic processes operating in the reef environment and the condition of the fauna and flora that build the reef structures.

Maria Elisabeth de Araújo is Full Professor at the Department of Oceanography, Federal University of Pernambuco (UFPE). She is a biologist, master in zoology, and doctorate in genetics (UFRJ, 1997), worked with fishing and rural extension (SUDENE, 1982–1995), coordinated the Tropical Marine Ichthyology Group (IMAT/CNPq), and taught at the Department of Fisheries Engineering (UFC, 1995–2002). She trained more than 150 students (bachelor's and post-doctorate) and produced 80 scientific articles, three books, and 18 chapters. As a retired professor (2020), she seeks to apply her academic experience to provoke a more humanized interaction between people who experience ecological knowledge (scientific and popular) and desire nature conservation, including traditional communities.

Tito M. C. Lotufo is a biologist with an MSc in Ecology and a PhD in Zoology and Habilitation in Biological Oceanography, all from the University of São Paulo. He was formerly Professor at the Fisheries Engineering Department and the Institute of Marine Sciences at the Federal University of Ceará. Currently he is Associate Professor at the Oceanographic Institute, University of São Paulo, acting as the Head of the Biological Oceanography Department. He is also the Public Policies Coordinator of Projeto Coral Vivo, an NGO dedicated to the conservation of marine coralline environments. Prof. Lotufo serves as Associate Editor for the journals *Biota Neotropica*, *Zookeys*, *Revista de Investigaciones Marinas* (Cuba), and *Frontiers in Marine Science*.

Chapter 1
Morphology and Distribution of Reefs and Carbonate Buildups in the Southwestern South Atlantic

Ruy K. P. Kikuchi, Marilia D. M. Oliveira, and Zelinda M. A. N. Leão

Abstract Reefs are present along approximately 2000 km of the inner shelf and extend over long sections of the Brazilian coastline. Surveys made in the XXI century indicated a far more extensive occurrence of this carbonate environment, occupying the middle and outer regions of the Brazilian continental shelf. In this chapter, we intend to synthesize the present knowledge of the morphology and distribution of reefs on the continental shelf, with particular attention to the shallow water reefs that occur from the coastline until 30 m depth. We also present an ecogeographic compartmentalization of the Southwestern Atlantic, considering oceanographic and reef morphological characteristics. This chapter aims to contribute to understanding reef occurrence on the Brazilian continental shelf and to help improve environmental risk assessments related to global changes and anthropogenic impacts, such as energy production initiatives or carbonate mining projects that are underway and will affect reef ecosystems.

Keywords Reef morphology · Reef distribution · Reef habitats

1.1 Introduction

Reefs are a special kind of carbonate sedimentary environment. They result from the gregariousness of many different types of organisms, composing an ecosystem while they build their whereabouts. This association began in the Earth's outer environment's infancy, as we can see from the existence of stromatolites since the Precambrian eon. Reefs occur along the entire Brazilian Tropical continental margin. Recently, this range has been mapped to the sub-tropical realm. This margin is also a particular sedimentary environment known as a transition of carbonate to

R. K. P. Kikuchi (✉) · M. D. M. Oliveira · Z. M. A. N. Leão
Laboratório de Estudo dos Recifes de Corais (RECOR), Instituto de Geociências, Universidade Federal da Bahia, Salvador, Bahia, Brazil
e-mail: kikuchi@ufba.br

R. K. P. Kikuchi et al. (eds.), *Brazilian Coral Reefs*, Brazilian Marine Biodiversity,
https://doi.org/10.1007/978-3-031-59152-5_1

siliciclastic sediments, where carbonate sediments produced by marine organisms live within sediments that are brought to the basin from the continent, mainly by rivers, sometimes directly from landslides of highlands that reach the coast.

Reefs were instrumental to the Portuguese navigators in Brazil in the sixteenth century. They were responsible for the Portuguese choice for establishing the first settlement in the country, Porto Seguro (i.e., Safe Port). Reefs are intriguing natural structures. They cannot be defined in one single way. They are a self-building ecosystem. They are also obstacles and promote shelter to navigation. At the surface of the Earth, they provide food. After being buried by episodic loads of sediments along the planet's geological history, their stored organic matter was converted to hydrocarbons.

Reefs in the Southwestern Atlantic have been described and studied by Europeans, North Americans, and Brazilians since the nineteenth century (Chap. 2). The country's Eastern shelf reefs were the first and most investigated among the reefs known today. Attention and knowledge about reefs raised from the shallower waters and locations closer to the coast to deeper and farther from the coast. A new generation of coral reef scientists emerged in the 2010s when new technologies and new approaches allowed an expansion of the geographical extent and depth (in the water and the sedimentary rock) of the knowledge on reefs of the region (Chap. 2). The outline of the distribution and classes of the morphology of reefs in the Southwestern Atlantic was first set by Laborel (1969). This doctoral thesis was recently translated into Portuguese (Laborel-Deguem et al. 2019), which helped to increase its audience. Leão et al. (2003) built on this biogeographical model and a classification devised by Ginsburg and Schroeder (1969) to create a classification scheme and distribution map of these reefs.

During the beginning decades of the twenty-first century, a more precise description and mapping of reef sites expanded the geographic extension of the reef province to the equatorial region, such as the paper by Santos et al. (2007) that described the reefs deeper than 20 m on the northern shelf of the Brazilian Northeast region. On the other hand, new investigations widened the knowledge of local reefs to more marginal settings along the coast (reefs of the East Shelf, Carvalho and Kikuchi 2013) or greater depths (Abrolhos, Moura et al. 2013).

The shelf edge reefs were recognized in the 1990s (Kikuchi and Leão 1998, with descriptions, echosounding, and *in situ* photographs) and, more recently, the outcomes of research networks brought attention to the reefs on the shelf edge and associated canyons (Olavo et al. 2011; Moura et al. 2016; Silva et al. 2018).

In this chapter, we present a synthesis of the knowledge of the controlling factors of reef occurrence. We also describe reef types and their distribution using the habitat mapping scales approach defined by Greene et al. (1999). Reefs morphology was classified at the macrohabitat and mesohabitat scales of meters to kilometers, and their distribution at the megahabitat scale, along a modified eco-region approach of Spalding et al. (2007).

1.2 Environmental Settings

1.2.1 The Continental Shelf Morphology and Sedimentary Facies

The reefs and carbonate buildups along the tropical Brazilian middle and outer shelves, from north to south, are particular environments in a dominant carbonate sediment environment. They are the product of settlement of organisms that build massive carbonate skeletons on stable substrate, and organisms that bound the sediments, such as encrusting coralline algae and bryozoans. Bioclastic carbonate gravel and sands are also important constituents in the inner shelf in many areas (Dominguez and Leão 1994; Testa and Bosence 1998; Vieira et al. 2019). More commonly, the inner shelf constitutes a typical mixing zone of siliciclastic and carbonate sediments. The siliciclastic sediments originate from river discharges, coastal erosion, and reworked relict deposits of former lower sea-level stands. The carbonate sediments result from locally produced skeletons of calcareous organisms (such as coralline algae, mollusks, corals, among others) and their erosion.

This distribution occurs because the movement of coarse to fine siliciclastic sand or the suspension of siliciclastic mud in the water deter the large-scale production of carbonate skeletons and their cementation to build up a reef.

At the northernmost part of the Brazilian shelf, at the proximity of the mouth of the Amazon River, the shelf is wide (90 km to more than 250 km) and a high continental sediment load settles relatively quickly in the inner and mid-carbonate shelves, conditioning an unstable muddy benthic habitat for reef growth (Moura et al. 2016). The adjacent coastal zone is composed of Quaternary fluvial-marine and aeolian deposits of sand and mud, deposited under the influence of macrotides (on the western part of this region, as described in the next section) and a semi-arid climate east of São Luiz Bay (Fig. 1.1).

The northeastern coast is considered as starved of continental sediment supply, because of the low precipitation rates found in the highlands drained by rivers that run to the coast in this region (Dominguez et al. 1992; Fig. 1.1). The eastern coast, on the other hand, receives higher volumes of continental sediment because of the presence of larger rivers, such as the São Francisco, the Jequitinhonha, and the Doce rivers (Dominguez 2009). Neogene gravely sand deposits, known as Barreiras Group, occur commonly at, or near, the coastline of these regions, exposing cliffs at the coastline. Because of this trait, Knoppers et al. (1999) called them as the Tertiary (a geological period presently substituted for the Paleogene and the Neogene) coast (Fig. 1.1).

Near the São Francisco River mouth (10°30′S), the largest river draining to this region, the carbonate sediment production is reduced due to water turbidity caused by the river load plumes and occurs only on the outer shelf. In the inner shelf of the south part of this region, between the Jequitinhonha (15°00′S) and Doce (19°40′S) rivers, the shelf sediment is also influenced by river discharges. Plumes of fine sediments are seen to advance about 50 km offshore. In these areas, bioclasts occur only

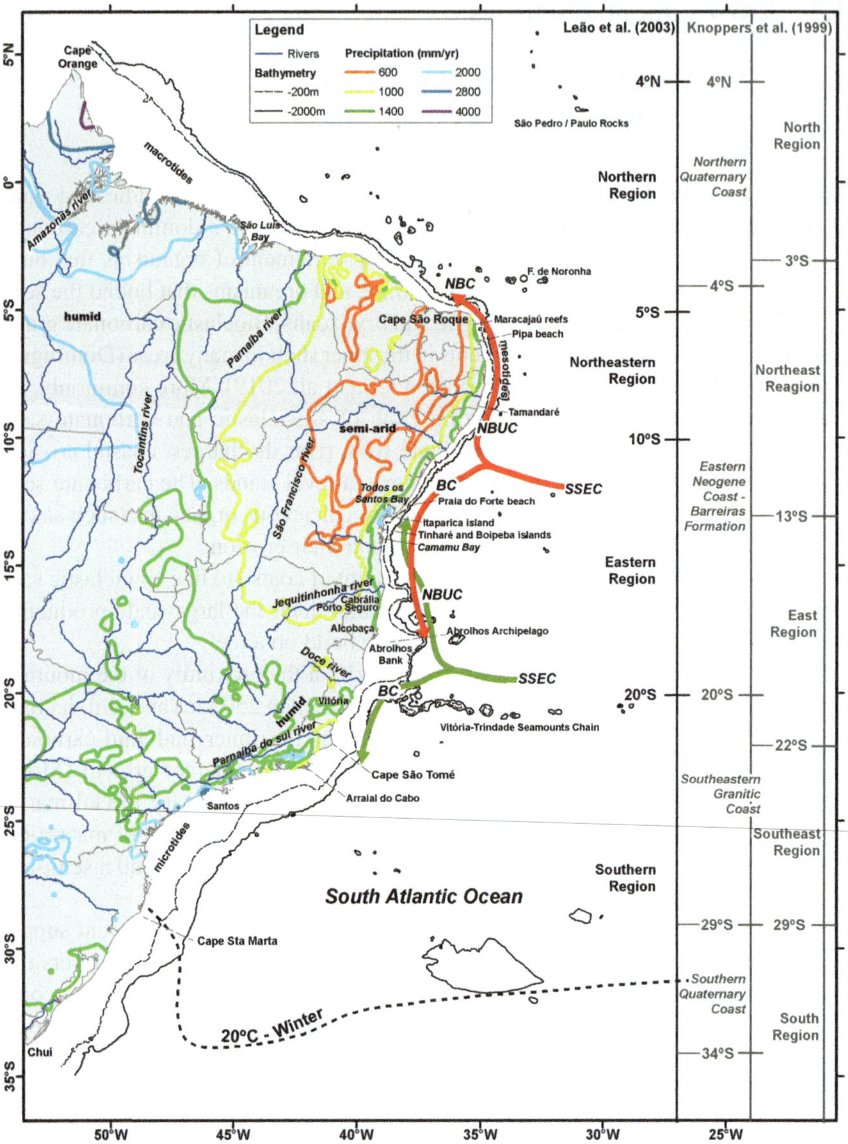

Fig. 1.1 Map that synthesizes the information of the controlling factors of the ecogeographical compartments of Southwestern Atlantic (Based on Knoppers et al. 1999; Stramma and England 1999; Dominguez 2009). NBC North Brazil Current, BC Brazil Current, NBUC North Brazil Under Current, SSEC Southern South Equatorial Current; Leão et al. (2003) based their zonation in Laborel (1967) that emphasized the occurrence of corals. The Northern region is the Laborel's "northern coral impoverishment region", the Northeast region is his stone reefs/beachrock coast, the Eastern Region was focused highly on the Abrolhos reefs region, and the Southern Region is his "southern coral impoverishment region"

on the middle and the outer shelves, and the main carbonate sediments are mollusk shells, benthic foraminifer's tests, remnants of calcareous algae, bryozoans, echinoids and, more rarely, coral gravel.

The fine sediment supply to the shelf and the ocean dynamics (mainly the action of the waves, described in the next section), result in a turbid environment that limits the light availability to carbonate-secreting organisms that depend on light to produce large amounts of carbonate particles, such as coralline algae and corals. This is a condition that controls the diversity and distribution of corals and other reef-building organisms (Freitas et al. 2019; Loiola et al. 2019) and, consequently, reef buildup (Freitas et al. 2019). See Chap. 4 for an extended approach to the biology of reef organisms.

The southeastern and the southern continental shelf does not receive significant fluvial discharges (except for the Doce and the Paraíba do Sul Rivers, Fig. 1.1) and carbonate sediment predominates. Despite the predominance of carbonate sediments, reefs are nearly absent south of Doce River. Some occurrences are described to the south of the city of Vitória (at about 20°45′S), corals form a special environment in Arraial do Cabo (close to 23°S) and a fringe of reef was described around an island in the vicinities of Santos city (around 24°30′S). The southeastern region, is characterized by the presence of Precambrian massifs at the coast, called as the Southeastern Granitic Coast by Knoppers et al. (1999) (Fig. 1.1) and the Southern coast is characterized by Quaternary marine deposits.

1.2.2 The Climatic and Oceanographic Settings

In the Brazilian tropical coast (eastern and northeastern regions; Fig. 1.1) the climate and associated oceanographic processes (waves and coastal currents) are controlled by frontal systems, trade winds and the Intertropical Convergence Zone[1] (ITCZ). The frontal systems (also named cold fronts) result from the northward displacement of polar air masses and are active all year around. The trade winds blow all year around originating in the South Atlantic high-pressure cell while the ITCZ varies its position throughout the year (Dominguez 2009).

The climate in the equatorial margin of Brazil, from Cape Orange to São Luiz Bay is classified as humid, and from São Luiz Bay, to west until Cape São Roque, and to south until the region of the São Francisco River mouth, it is mainly semi-arid (Table 1.1). The east coast, from the São Francisco River mouth to Arraial do Cabo it is classified as tropical humid, and south of Arraial do Cabo, it is classified as subtropical humid. Considering the distribution of annual rainfall (Nimer 1989),

[1] The Intertropical Convergence Zone (ITCZ) is a meteorological system that extend around the globe in the Equatorial region and is responsible for the equatorial rain. It consists of a cloud band that migrates from 14°S to 2°S in the Atlantic. It is formed by the convergence of the northeasterly trade winds of the northern hemisphere and the southeasterly trade winds of the southern hemisphere, that concentrate the humidity in this region.

Table 1.1 Synthesis of environmental conditions of the reef regions (as defined by Leão et al. 2003)

Region	Climate	Number of dry months	Surface current regime	Temperature Average	Waves	Tides
Northern (Cape Orange to Cape São Roque)	Humid do semi-arid	0 to >4	NBC	27–29 °C	Northeast	Macrotidal to Mesotidal
Northeastern (Cape São Roque to São Francisco River)	Semi-arid	>4	NBC and SEC	26–29 °C	Southeast and East	Mesotidal
Eastern (São Francisco River to Cape São Tomé)	Humid (tropical)	=<2	SEC and BC	24–28 °C	East, Northeast and Southeast	Mesotidal
Southern (Cape São Tomé do Chuí)	Humid (sub-tropical)	=<2	BC	21–25 °C		Microtidal

the northeast coast from 4°S to 6°S is characterized by an excess of four to five dry months, and from 6°S to 12°S the coastal segments with less than two dry months alternate with areas with four to five dry months. The east coast from 12°S to 20°S has less than two consecutive dry months.

In the northern region of the coast, northeastern trades prevail. Along the eastern coast a divergence zone of trade winds occurs in the southern part of the pressure cell, and northeastern winds blow to the south of this zone. A seasonal variation of this cell produces a north-south oscillation of the divergence zone between 10° and 20°S. This zone moves northward during summer and southward during winter. As a result, easterly and southeasterly winds dominate the coast far north of 13°S, year-round, with speeds ranging from 5.5 to 8.5 m s^{-1}. South of 13°S the easterly and southeasterly winds blow during fall and winter (April to September), and the northeasterly winds prevail during spring and summer (September to February); in this area the wind speed rarely surpass 5.5 m s^{-1}. This oscillation is reflected in the ocean surface circulation, as described in the next paragraph, and shown in Fig. 1.1. The polar front moves along the coast towards the Equator and can reach as far as 10°S during winter but, rarely, reaches latitudes lower than 15°S in summer (Dominguez et al. 1992). The advance of this polar system also generates additional south-southeasterly winds. Gale-force winds (25 m s^{-1}) have been measured with the advance of these polar fronts (Bandeira et al. 1975).

The Brazilian Current and the North Brazilian Current are the main surface currents on the Brazilian continental margin (Peterson and Stramma 1991; Silveira et al. 1994; Fig. 1.1). They originate from the southern branch of South Equatorial Current (SSEC) at about 10° (Nov-Dec, austral summer, Fig. 1.1 in red) and flow to the south (BC) with average velocities of 50 to 70 cm s^{-1}, and to the north and northwest (NBC) attaining velocities of 30 cm s^{-1}. This bifurcation migrates southward during the austral fall-winter and reach about 20°S in May-Jun (Fig. 1.1, in green). Data from the Atlas de Cartas Piloto (DHN 1993) show that between 10°S and 13°S, during July and August (austral winter) a reverse flow to the north can occur. North of 5°S, the North Brazilian Current becomes stronger because of combining with the South Equatorial Current. The surface currents distribution pattern is synthetized in Table 1.1.

Sea surface temperature is the most conservative parameter along the Brazilian coast. The SST maximum and minimum monthly mean calculated for the NOAA Regional Virtual Stations[2] of the Brazilian region varies in the northeast coast from an average of 29 °C during summer and fall (February to May) to 26 °C in the northeastern coast from the end of winter to the beginning of summer (August to December). In the east coast it varies from 28 °C (February to May) to 24 °C (July and August). The SST in the southeastern/southern region varies from 25 °C (February to May) to 21 °C (July to August) (Table 1.1). The isotherm of 20 °C, which is considered the limit for the thriving of carbonate sediments and reef buildups, is located around 30°S (Fig. 1.1).

The Brazilian coast is mainly dominated by sea waves (locally generated waves with periods lesser than 7s), and those with heights above 1 m account for more than 50% of the observations referred to in the U.S. Navy Marine Climatic Atlas (US Navy 1978). In the Brazilian north (equatorial coast) the waves moving from northeast dominate the year round. In the northeast coast, the waves moving from the southeast dominate the year round, and the waves from the east are important from January to May (summer-fall) and from September to November (spring). The southernmost part of the northeast coast and the east coast, on the other hand, are dominated by waves moving from the east during the whole year. In this region, waves from the northeast are only important from November to February (summer), while those from the southeast occur from March to August (winter) (Table 1.1).

Tides on the continental shelf are semi-diurnal (Table 1.1). Due to the large latitudinal extent of the shelf, three different areas are defined, according to the tidal range (Fig. 1.1): macro-tidal on the north coast (from Cape Orange to São Luiz Bay), meso-tidal on the northeast coast and on the east coast (from São Luiz Bay to Doce river), and micro-tidal on the entire south coast (south of the Doce river mouth). The tidal range increases from south (less than 1 m) to north (greater than 6 m) (Salles et al. 2000). The north macro-tides enhance the northwestern flow of the North Brazilian Current (BC), and periodically produce an intensification of this drift (Amorim et al. 2013; Teixeira et al. 2013).

[2] https://coralreefwatch.noaa.gov/product/vs/data.php

1.3 Reef Morphology

Originally, the term reef refers to an obstacle to navigation. Geologically, a reef is a rigid rocky structure that is exposed to and resists the action of waves and currents. Biologically, reefs are mainly composed of carbonate secreting skeletons, such as corals and calcareous algae, but other benthic organisms that produce rigid skeletons that can also contribute to build this cemented substrate. They may be called biogenic reefs. In addition to the biogenic reefs, two other kinds of stable substrates that support reef ecosystems are found along the coast of Brazil: the inorganic sedimentary rocks cemented by calcium carbonate, called beach-rocks, found all along the continental shelf of Brazil, and the submarine cemented terraces, generally found associated to coastal cliffs formed when the Neogene Barreiras Formation is at the coast. We refer to this set of rigid and stable structures as carbonate build-ups (Nolasco and Leão 1986). From an ecological perspective, we consider that reef morphology has a role in creating suitable habitats for organisms to attach or seek protection in. For this purpose, we reorganized and synthesized here (Fig. 1.2) the classification proposed in Leão et al. (2003) according to the habitat perspective of the reef morphotypes proposed in Araújo et al. (2020). Additionally, rocky shores also host reef-like communities. However, they will be object to a special volume of this series and are not a target of our study.

We recognize five forms of carbonate buildups along the Tropical and Subtropical coast of Brazil in terms of habitat complexity at range of macrohabitat, that were

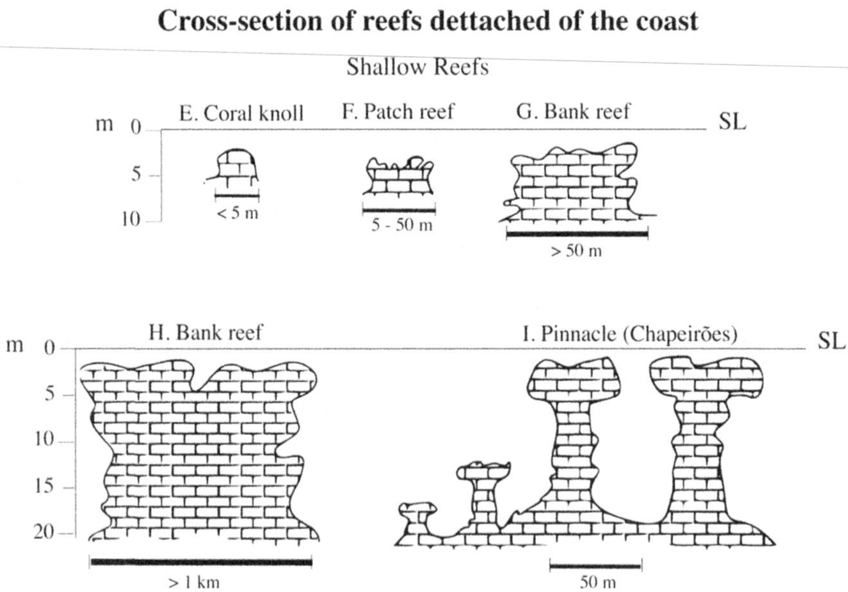

Fig. 1.2 Reef classification according to morphology and dimensions. (Adapted from Leão et al. (2003). SL means sea level)

used by Araújo et al. (2020) to analyze their influence on reef fish habitats and classified as habitat morphotypes. They are cemented terraces, fringing reefs and bank reefs attached to the coast, patch reefs and coral knolls, bank reefs, and pinnacle reefs. We also included two forms on the mesohabitat scale (*sensu* Greene et al. 1999) previously described by Leão et al. (2003), shelf edge reef and atoll. These reef forms are described, and examples are given below.

1.3.1 Cemented Terraces

Cemented terraces are formed by boulders and finer sediment as a result of cementation of cliff erosion products, found as plateaus that extend from tens to hundreds of meters along the coast (Morais 1969). These cliffs form on the Neogene Barreiras Formation, and are composed by reddish siliciclastic deposits of conglomerates, fine to coarse-grained sandstones and mudstones. Their base is marked by a surface of an iron oxide cemented conglomerate (Araújo et al. 2006; Rossetti and Góes 2009).

The terraces (named as rocky platforms by Albino et al. 2016) result from a combined process of marine abrasion of coastal cliffs, rock deposition, and reworking by the wave action, resulting in cemented plateaus. They are discontinuous shallow substrates in the intertidal zone that are exposed at low tide. Depressions on the surface result in small tidal pools that, along with some caves and small channels, are habitats for the reef biota. The reef at Pipa, at the northern coast of Brazil (about 6°S, Fig. 1.1), is an example of a cemented terrace in the Brazilian coast (Fig. 1.3a, b).

1.3.2 Fringing Reefs and Bank Reefs Attached to the Coast

Fringing reefs are found bordering the shores of landmasses (continents or islands). They are continuous structures growing up to several kilometers parallel to the coast (Fig. 1.3c). The reef front can be ca. 2–10 m deep, and the back reef zone may have a shallow lagoon (up to 2 m) where coral knolls or patch reefs can be found. They are submerged during high tide, and during low tide their tops (reef flats) are subaerially exposed. At this upper surface, tidal pools and shallow channels are habitats for a rich reef biota. The reef flat can be horizontal or slightly dipping toward the sea, following the slope of the beach.

This reef morphology can result from active growth of carbonate organisms in the bedrock of older (Cretaceous) sedimentary rocks, such as those that border the shores of the Itaparica Island in Todos os Santos Bay (Fig. 1.3d) or the islands of Tinharé and Boipeba, near the Camamu Bay, both in Eastern Coast, at about 13°S (Fig. 1.1). In this case they conform to shore and have a more irregular shape.

They can also be linear and straight along the coast with a variable thickness of biogenic carbonate that overgrows (or encrusts) ridges of cemented beach sands,

Fig. 1.3 Schematic Illustrations on the left of (**a**) cemented terrace, (**c**) fringing reef, and (**e**) and linear bank reef form. The photographs on the right depict (**b**) cemented terrace in Pipa beach, (**d**) the fringing reef of Itaparica island, and (**f**) linear bank reef of Arembepe beach. Locations of these sites are shown in the map of Fig. 1.1. (Illustrations (**a**) and (**e**) are modified from De Araújo et al. (2020). Photography (**b**) by Universos Paralelos de Can! Wikimedia Commons CC-BY-2.0 originally posted in https://www.flickr.com/photos/7845777@N07/3442742327)

known as beachrocks. Sometimes, seaward of fringing beachrocks one frequently finds linear bank reefs, that are also a result of carbonate organic deposition on a beachrock substrate (Dominguez et al. 1990) (Fig. 1.3e). They are separated from each other by depressions where sandy and muddy sediments accumulate. Some examples of fringing reefs are the Pinaúnas reef on Itaparica island (13°S, Fig. 1.1). Linear bank reefs are seen off Arembepe Beach (12°77′S, Fig. 1.3f), near Praia do Forte Beach, Eastern coast of Brazil. An example of a drowned fringing reef in the southernmost reef site described till now, close to Santos city (24°S, Fig. 1.1), was recently described by Pereira-Filho et al. (2021).

1.3.3 Patch Reefs and Coral Knolls

Patch reefs are structures of variable dimensions, detached from the coast, usually from five up to 50 m wide. If less than 5 m of width, they are known as coral knolls (as defined in Ginsburg and Schroeder 1969). They can also grow over larger reef surfaces, on deeper bank reefs. Along the Brazilian shelf, patch reefs occur as complexes in areas varying from tens of meters to some kilometers in geographical extension. They are usually found at less than 10 m deep and with variable distances from the coast. Some reefs reach sea level and may be exposed during low tide, whereas others stay submerged (Fig. 1.4a).

This reef morphology is very common in the Brazilian northeastern and Eastern ecoregions. In the Northeastern ecoregion patch reef concentrations may form mesohabitats such as those shown in Fig. 1.4b. Patch reefs can also be found at the vicinity of fringing and bank reefs.

1.3.4 Bank Reefs

Bank reefs are structures varying from 50 m to tens of kilometers wide. They occur attached to the shore or some kilometers further away from the shoreline. The reef flat may be in the intertidal zone. Differently to the fringing reef, no lagoon is formed, but tidal pools of reduced extension are common, generally shallower than 1 meter deep (Fig. 1.4c). The attached bank reefs differ from the fringing reefs by their smaller dimensions, and the absence of the typical back-reef zone on a lagoon. In this case, the reef front is less than 10 meters deep.

The bank reefs detached from coastline have heights from around 10 m to more than 20 m above the sea floor. Most are flat-topped reefs of varied shapes (irregular, circular, elongate, or arc-like). They may be subaerially exposed during low tide, where tidal pools and shallow channels can be found. Some reefs have abrupt walls, while others present lateral walls with several 'steps' created by the rigid organisms of the reef biota and may develop spur-and-grooves.

They occur in the whole tropical continental shelf, such as in the North Northeastern and Eastern ecoregions and are the main reef features of the Brazilian coast. Some examples are the shallow banks adjacent to the coast of Tinharé-Boipeba (Fig. 1.4d).

1.3.5 Pinnacle Reefs ("Chapeirão")

This distinctive reef type ranges in height above the sea floor from 5 to 25 m. The horizontal extension of their tops varies from 5 to 50 m. Chapeirão (or pl. in Portuguese, "chapeirões", meaning "big hats") is a term introduced in the scientific

Fig. 1.4 Schematic Illustrations on the left of (**a**) patch reefs, (**c**) bank reefs and (**e**) pinnacles. The aspect of the substrate where reefs grew are drawn on the sides of the diagrams. The photographs on the right depict (**b**) Maracajaú reefs, (**d**) bank reefs adjacent to fringing reefs off Tinharé island, and (**f**) pinnacle at Abrolhos complex, called locally as "chapeirão" meaning "big hat". In this last picture one sees the broad reef tops of several chapeirões. (Illustrations (**a**), (**c**) and (**e**) are modified from De Araújo et al. (2020). Photography (**b**) by Alvaro.maracajau Wikimedia Commons CC-BY-4.0 originally posted in https://en.wikipedia.org/wiki/File:Parracho.jpg)

literature by Charles F. Hartt in 1870 and coined by local fishermen to refer to this reef type. They are a reef growth form unique to the Brazilian reefs, which consists of isolated narrow pillars with tops expanded laterally, which resemble flat topped mushrooms (Fig. 1.4e). The flat top of the chapeirões can be sometimes over three times larger than its base. Seen from above, these structures generally have a rounded shape and are easily mapped from aerial photographs. At deeper sites, the pinnacles do not develop the mushroom shaped form.

They frequent between 15°S and 20°S on the Abrolhos Bank (Fig. 1.1). They can be found closer than 1 km from the coastline or as the Timbebas reefs at Alcobaça, and further off (more than 50 km) the coast (Figs. 1.1 and 1.4f), at the Abrolhos reef complex. They usually occur associated with bank reefs. This concentration of coral reefs extends outward to about 100 km from the coastline, forming two arches: (i) the internal arc is composed of bank reefs of various shapes and dimensions, commonly surrounded by shallower chapeirões; and (ii) the outer arc is formed by isolated and coalescent chapeirões at depths greater than 20 m.

1.3.6 Shelf Edge Reefs

Shelf edge reefs are reefs growing at the border of the continental shelf (Fig. 1.5a, b). They are features on the meso to mega-habitat scale. They were described for the first time near Praia do Forte beach (Fig. 1.1), at 13°S (Kikuchi and Leão 1998), where the continental shelf width varies from 8 to ca. 50 km. They occur at depths of 50 m, their relief at the back reef zone is about 5 m and the fore reef can achieve about 35 m. Presently, this structure is veneered by an extensive rodolith/maerl deposits, with sparse coral knolls and sponges (Kikuchi and Leão 1998), that can extend for up to 3 km across the shelf. Associated with this edge occurrence, there are also features of terraces in the upper part of the slope, at about 70 m and 90 m deep (Kikuchi and Leão 1998). These slope terraces might indicate former sea level stands. Olavo et al. (2011) studied the shelf break terraces along the Brazilian Eastern region (13°S to 22°S, Fig. 1.1) and show that these shelf break reefs bare important fishing spots to artisanal fishers. Equivalent morphology and sedimentary facies was described and extensively documented at the carbonate province on the Brazilian Northern region, adjacent to the Amazon River mouth (Fig. 1.1; Moura et al. 2016) and in Brazilian Northeastern region (5°S, Fig. 1.1, Gomes et al. 2020).

1.3.7 Atoll

The Rocas Atoll is the only Brazilian atoll. It is an oceanic ring-shaped reef that encloses a lagoon. It is roughly elliptical with diameters of 3.5 km and 2.5 km. Despite its small dimensions, the reef front, reef flat and a lagoon can be clearly distinguished and subdivided into discrete features (Fig. 1.5d). Its construction is mostly due to non-articulated coralline algal growth (Kikuchi and Leão 1997; Gherardi and Bosence 1999). Corals were estimated to account for 10% of an 11 m long core, cut at the western part of the reef flat (Kikuchi and Leão 1997).

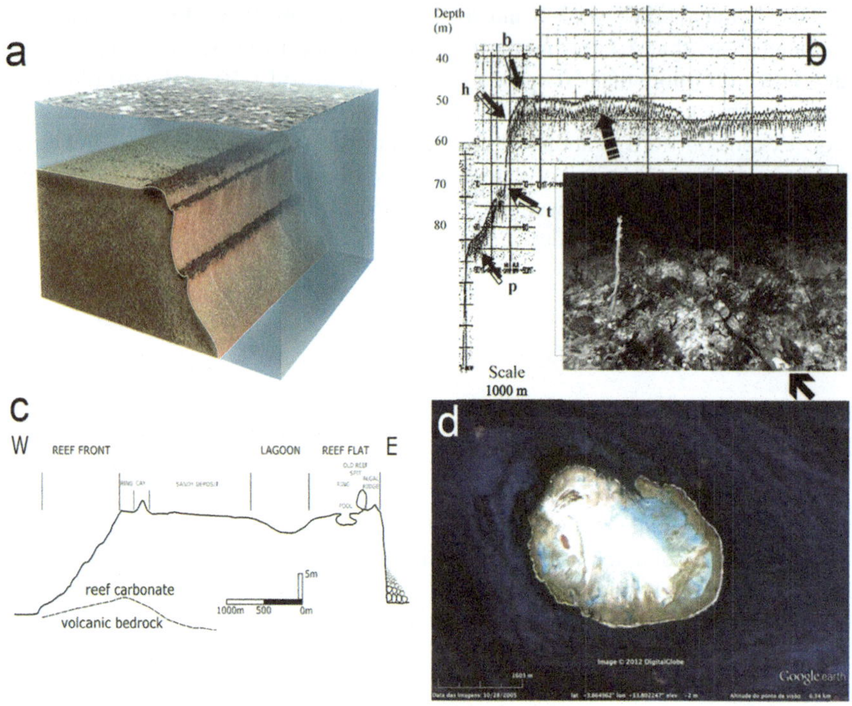

Fig. 1.5 Schematic Illustrations of (**a**) a shelf edge reef and (**c**) the Atoll das Rocas (Kikuchi and Leão 1997). At (**b**) an ecosounding profile at the shelf edge near Praia do Forte Beach, Brazilian Eastern region (13°S, Fig. 1.1), with a bottom surface photography showing the surface covered with rodoliths. In this profile, "**p**" stands for reef basement, "**t**" is the top of the reef talus deposit, "**h**" is the inflection point at −55 m and "**b**" is the shelf break, at -50 m (Kikuchi and Leão 1998). (**d**) a satellite image of Rocas Atoll (4°S, Fig. 1.1)

1.4 Reefs in the Brazilian Ecoregions

Here we gathered information of reef buildups available in the literature and in sites of state environmental institutions and present them within the megahabitats scales, namely, the shallow water reefs, mesophotic reefs and deep reefs (Fig. 1.6). Reefs ecosystems are found along the continental shelf in most of Tropical and Subtropical Western Atlantic, from shallow water (euphotic) down to the mesophotic realm. According to the traditional limits set to the mesophotic reefs (see Lesser et al. 2009 and references therein), the transition of the shallow water to mesophotic realm is set at the 30 m depth, reaching 150 m, from where we find the deep reefs.

According to Spalding et al. (2007), an ecoregion is the smallest scale unit in the marine system. An area of relatively homogenous species composition, clearly distinct from adjacent systems. Following this classification, Brazilian coral reefs occur in two marine provinces: The North Brazil Shelf and the Tropical Southwestern

Atlantic. But corals are found also in the Temperate South America province and in the Rio Grande province, in deep sites.

In the North Brazil Shelf province there is the Amazonian (AMZ) ecoregion, and in the Tropical Southwestern Atlantic province there are five ecoregions: the São Pedro and São Paulo islands (SPSP), the Fernando de Noronha islands and Atol das Rocas (FNAR), the Northeastern Brazil (NEB), the Eastern Brazil (EB) and Trindade and Martin Vaz islands (TMV), according to Spalding et al. (2007) classification.

Based on the characteristics of the reefs known up to now, mainly described in Leão et al. (2019), the occurrence of reef forms, differences in coral and calcareous hidroids distribution (see Chap. 3) and of the mapped shallow reef occurrence, we consider that the Northeastern Brazil region should be split in two ecoregions, the North Northeastern Brazil (NNEB) and the East Northeastern Brazil (ENEB). The seasonal variability of the South Equatorial Current (Fig. 1.1), the distribution of calcareous hidroids and fish biogeography (see Chap. 3) support the establishment of the São Francisco River delta as the limit between the East Northeastern Brazil and the Eastern Brazil (Fig. 1.6 and Table 1.2).

As a general trend, shallow reefs are rare in the Amazonia (AMZ) and in the North Northeastern (NNEB) ecoregions. The shallow reefs occupy larger areas in the East Northeastern Brazil (ENEB) ecoregion and are widely distributed in the East Brazil (EB) ecoregion. The mesophotic megahabitat was widely mapped in the Amazonian ecoregion and in the Abrolhos region, situated in the Eastern Brazil ecoregion, around 17°S (Moura et al. 2013). Last, deep reef habitats are widely mapped in the Southeastern and in the Rio Grande ecoregions. In this Chapter, we will mainly focus on the shallow reefs.

1.4.1 The North Brazil Shelf Province

The North Brazil Shelf Province (from 0°50′S to 5°00′S), is composed of the Amazonian ecoregion (Fig. 1.6 and Table 1.2), where an extensive, mostly drowned reef system occur at the Amazon River mouth and the reefs located at the coast of the state of Maranhão. This northernmost Brazilian region is the poorest reef area of the country continental shelf, compared to its northeastern and eastern areas, where shallow reefs are more frequent.

In the northern sector of the Amazonian ecoregion, structures are recorded near the shelf edge, comprising widely spaced (hundreds to thousands of meters) patches with lengths of up to 300 m and heights of up to 30 m. These irregularly shaped reefs tended to be elongated with a parallel shelf edge orientation, resembling erosive structures (Moura et al. 2016). In this north and central sector of the Amazonian ecoregion, reefs grow in mesophotic to deep habitats. Considering their situation relative to the sea level (Neumann and Macintyre 1985) they are give-up (drowned) reefs.

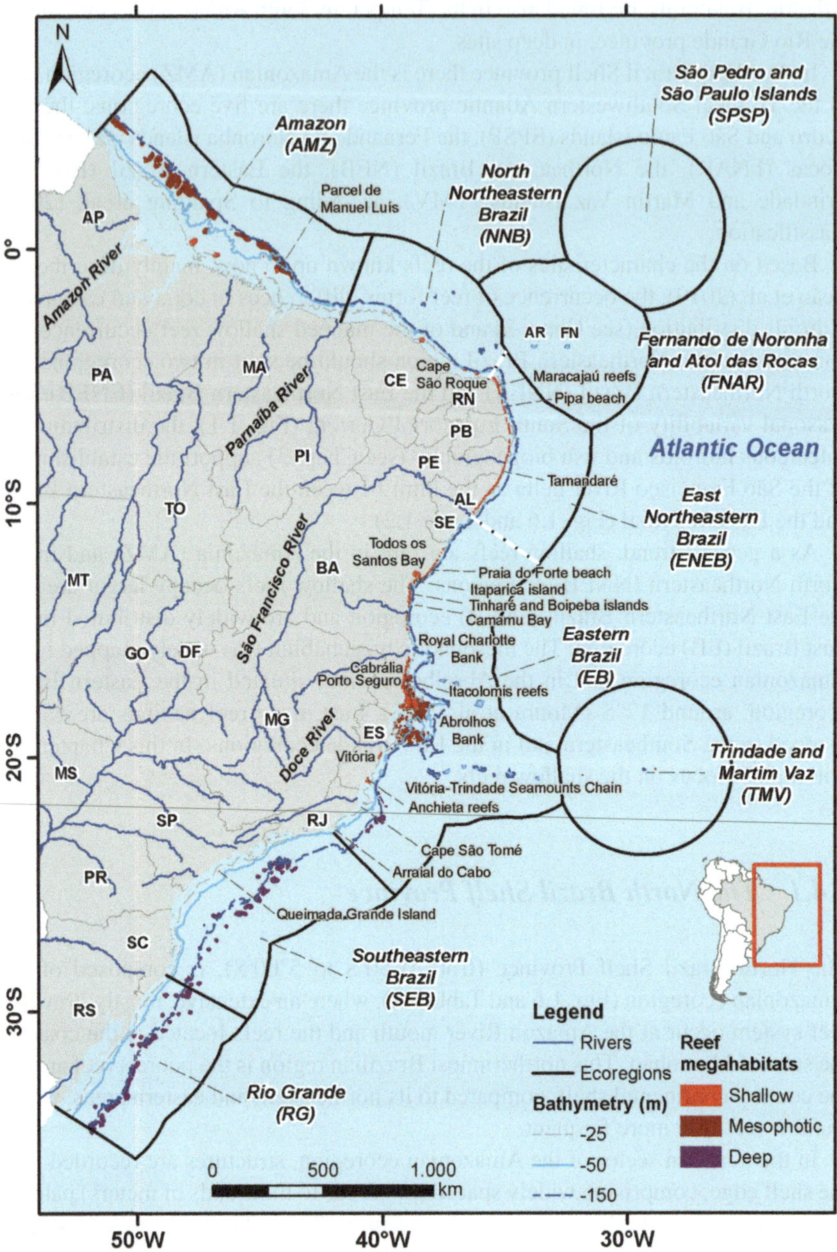

Fig. 1.6 Map that synthesizes the documented occurrence of reef megahabitats within a proposed division of Western South Atlantic ecoregion scheme, modified from the classification of Spalding et al. (2007). Reef spatial data composed by the authors' database of EB, ENEB and NNEB regions (Carvalho and Kikuchi 2013), added to Magris et al. (2021) (https://datadryad.org/stash/dataset/doi:10.5061/dryad.xsj3tx9d1) supplementary material

The southern sector of the Amazonian shelf, from the eastern part of Pará state shelf including the shelf of the state of Maranhão, approximately 90 km off its coastline, there are the Parcel de Manuel Luiz reefs which grow as giant pinnacles to depths of approximately 25 to 30 m, and their tops reach up to 2 m of water depth, but during spring low tides, some are at sea level (Coura 1994; Maida and Ferreira 1997; Amaral et al. 2006, 2007).

1.4.2 The Tropical Southwestern Atlantic Province

The Tropical Southwestern Atlantic Province extend from approximately 5°S to 19°S, and it includes six ecoregions: São Pedro and São Paulo islands, Fernando de Noronha islands and Atol das Rocas, North Northeastern Brazil, East Northeastern Brazil, Eastern Brazil and Trindade and Martin Vaz islands (Fig. 1.6 and Table 1.2).

The São Pedro and São Paulo islands ecoregion (00°56′S–29°22′W) are rocky islets located approximately 1100 km off the coast of Rio Grande do Norte state, which is composed of 15 islets, less than 100 m long and 50 m wide (Vaskes Júnior et al. 2007). Few coral species have been reported on the rocky shores of these islets (Amaral et al. 2006).

The Fernando de Noronha and Atol das Rocas ecoregion encompasses two very distinct locations. The Fernando de Noronha Archipelago (03°51′S–32°25′W) is composed of 21 islands and islets of volcanic origin and is located approximately 350 km off the coast of Rio Grande do Norte state (RN). These islands are part of the volcanic mountains from the Fernando de Noronha Chain (Almeida 1955). True reefs have not been found on the islands, but an abundant coral fauna grows on some of their rocky shores (Maida and Ferreira 1997; Castro and Pires 2001; Amaral et al. 2009). The Atol das Rocas (03°51′S–33°49′W) is an atoll that developed on a flat top of a seamount. The reef ring is open on its western and northern faces. Despite its small dimensions, all of the characteristic features of a reef can be distinguished, such as the reef front, reef flat and a shallow lagoon (Fig. 1.5). Atol das Rocas has been the subject of many studies over the last decades (Echeverría et al. 1997; Kikuchi and Leão 1997; Maida and Ferreira 1997; Gherardi and Bosence 2005; Pereira et al. 2010; Soares et al. 2011).

The North Northeastern Brazil (NNEB) ecoregion (from approximately 4°S to 5°S) is roughly east-west oriented from Parnaíba River mouth (Piauí State—PI) to Cape São Roque (Rio Grande do Norte State—RN, Fig. 1.6). Patch and bank reefs have been mapped on the inner shelf of these states (Laborel 1969; Santos et al. 2007), and more recently, Araujo et al. (2020) included in this ecoregion the cemented terraces, which are discontinuous shallow substrates found in the intertidal zone where they are exposed at low tide (Sect. 1.3.1; Fig. 1.3a).

The East Northeastern Brazil ecoregion (ENEB) is composed of the reefs along the coast extending from the cape São Roque to the São Francisco River mouth, along the eastern coast of the state of Rio Grande do Norte (RN), the coast of the states of Paraíba (PB), Pernambuco (PE) and Alagoas (AL). According to the description from Araujo et al. (2020), the major reef morphotypes found in this ecoregion are fringing reefs and patch reefs. Fringing and bank reefs originating from lines of beachrocks are common (Figs. 1.3 and 1.4). Some of these linear reefs are attached to and fringe the coastline while others are several kilometers offshore, generally aligned parallel to the coast at depths of approximately 5 m to 10 m, but can also be found at 20–25 m depth (Testa and Bosence 1999; Santos et al. 2007). Studies in the ENEB region have shown that these elongated reefs have developed over lines of beachrock (Delibrias and Laborel 1969; Laborel 1969; Dominguez et al. 1990; Testa and Bosence 1998; Maida and Ferreira 2003; Correia 2011) that constitute former coastlines (Gomes et al. 2020).

The Eastern Brazil ecoregion (from approximately 10°S to 19°S) extends between the São Francisco and the Doce rivers mouth over 1000 km along the coasts of the states of Sergipe (SE) and Bahia (BA) and the northern part of Espírito Santo (ES) state. This is the largest coral reef area along the entire Brazilian coast. The Sergipe shelf was considered depleted of coral reefs until the finding of submerged reef structures (Neves et al. 2006) that have not yet been mapped in detail. In this ecoregion most of the reef morphotypes described in this chapter are found. Along the coast of the state of Bahia, the reefs can be subdivided into the following four major reef areas: the North Bahia coast, the Todos os Santos-Camamu Bays area, the Cabrália/Porto Seguro area and the Abrolhos Bank reef complex. Shallow isolated bank reefs of various sizes occur in the North Bahia coast in the vicinities of Praia do Forte (Kikuchi and Leão 1998; Leão et al. 2003). A cored reef revealed Holocene carbonate buildups with a thickness of approximately 10 m lying on a pre-Cambrian rocky substrate (Nolasco and Leão 1986) (Fig. 1.4c) suggesting that deeper reef structures occur towards the continental shelf break (Kikuchi and Leão 1998). In the area from Todos os Santos Bay to Camamu Bay (Fig. 1.6), shallow fringing reefs that are relatively continuous border the islands shores while shallow bank reefs are observed in the interior of the bays (Leão et al. 2003; Cruz et al. 2009; Kikuchi et al. 2010; Loiola et al. 2014). The fringing reef that borders the shore of the Itaparica Island in Todos os Santos Bay was cored and exhibits a Holocene 8 m thick structure that grew above the island substrate (Araújo et al. 1984) (Fig. 1.3c, d). Near the Camamu Bay, a line of shallow fringing reefs borders the shore of the Tinharé and Boipeba Islands and a series of lines of elongated bank reefs (approximately north-south) occur down to depths of 25–30 m (Fig. 1.4c, d) (Kikuchi et al. 2008). The area of Cabralia/Porto-Seguro is characterized by the presence of bank reefs of various shapes and dimensions in locations no deeper than 20 m, running mostly parallel to the coastline (Costa Jr. et al. 2006). The most studied reef in this area is, an isolated bank reef a few kilometers off the coastline (Seoane et al. 2008, 2012). The elongated reefs may have grown on submerged strings of beachrocks (Laborel 1969) (Fig. 1.3e). Southward, there are the Itacolomis reefs, isolated bank reefs separated from one another by deep irregular channels (Castro et al. 2006; Cruz et al. 2008), which are the northernmost limit of the occurrence of the Brazilian giant "chapeirões". Patch reefs with varied dimensions are generally found at the vicinity of fringing and bank reefs in this region.

At approximately 17°S, the continental shelf widens and form the Abrolhos Bank in which the richest and the most well-known coral reefs of the Eastern Brazilian ecoregion are located (Hartt 1870; Laborel 1969; Leão 1996; Leão and Kikuchi 2001; Leão et al. 2003; Moura et al. 2013). These reefs form two arcs. The coastal arc is composed of bank reefs of various shapes and dimensions and the outer arc, located eastward of the islands of the Abrolhos Archipelago, is formed by isolated "chapeirões" in water deeper than 20 m. This distinctive reef type ranges in height above the sea floor from 5 to 25 m and the horizontal extension of their tops varies from 5 to 50 m. Incipient fringing reefs border the shores of the five islands that comprise the archipelago. A cored reef on the coastal arc revealed a Holocene coral reef structure, over 12 m thick, lying on a reefal carbonate rock of a probable Pleistocene age (Leão and Kikuchi 1999) (Fig. 1.4e). Southward in the Abrolhos Bank, in the northern part of the coast of Espírito Santo state, coral reefs have been described and classified as shallow "chapeirões" (Mazzei et al. 2017).

Furthermore, Milliman and Barretto (1975) have documented the occurrence of drowned reefs at the shelf break and mesophotic reefs are described across the mid and outer shelves on the Abrolhos Bank, at depths from 25 to 90 m. These reefs are structures described as submerged pinnacles, coalescent reefs 2–3 m high and sinkhole-like depressions known as "buracas" (these holes are 10–75 m in diameter and between 8 and 38 m in height, Bastos et al. 2013). These reef structures are nearly drowned reefs with low coral cover (Moura et al. 2013; Bastos et al. 2013).

The Trindade and Martin Vaz islands ecoregion (20°30′S–29°20′W) is formed by rocky islands in the Vitória—Trindade Seamounts Chain. They are located approximately 1200 km off the coast of Espírito Santo state. No true reefs exist, but coral species have been reported on these islands (Pereira-Filho et al. 2021).

1.4.3 The Warm Temperate South America Province

This province occurs in the southeastern region of the country (Fig. 1.6 and Table 1.2), from the mouth of the Doce River (19° 30′S) to the coast of the state of São Paulo. No reef was known there before the first decade of this century, although corals have been known in the region. The occurrence of cemented intertidal marine abrasion terraces (Albino et al. 2016) or shore platforms (Souza and Albino 2020) in association to the Neogene Barreiras Formation, which are similar to the cemented terraces, was more recently documented in this province. Dechnik et al. (2019) described and studied the evolution of Anchieta reefs, around 20°50′S, that were classified as fringing reefs. The buildups frequently align, forming linear reefs, generally attached to the shore but sometimes forming isolated patches.

Further south, at Arraial do Cabo in the state of Rio de Janeiro (RJ) (Fig. 1.6), in the so-called "Laborel's coral oasis" (Laborel 1969), few coral species have been found on rocky shores (Castro et al. 1995; Oigman-Pszczol and Creed 2004, 2006).

More recently, Pereira-Filho et al. (2019) described a reef fringing Queimada Grande Island, at 24°29′S (Fig. 1.6), in depths around 14 m, below a surface of

Table 1.2 Proposal for refinement of the ecoregion classification based on the features of reef habitats

Realm	Province		Ecoregion	
				Proposed in this paper
Tropical Atlantic	NB (13)	North Brazil Shelf	AMZ (72)	Amazonia
	TSA (14)	Tropical Southwestern Atlantic	SPSP (73)	São Pedro and São Paulo Islands
			FNAR (74)	Fernando de Noronha and Atol das Rocas
			NEB (75)	North Northeastern Brazil (NNEB)
				East Northeastern Brazil (ENEB)
			EB (76)	Eastern Brazil
			TMV (77)	Trindade and Martin Vaz Islands
Temperate South America	WTS (47)	Warm Temperate Southwerstern Atlantic	SEB (180)	Southeastern Brazil
			RG (181)	Rio Grande

Modified from Spalding et al. (2007); with the class number used in that paper, in parentheses

rocky shore that extends from the intertidal zone to ca. 12 m. According to the authors, only one species of coral (*Mussismilia hispida*) is sparsely found on the rocky shore, and the reef is formed by a coral framework (*Madracis decactis*) covered by living coralline algae (Pereira-Filho et al. 2021).

Following this pattern, up to about 27°18′S, at the coast of São Paulo (SP) and Santa Catarina (SC) states (Fig. 1.6), small patches of corals (*Madracis decactis*) are observed attached to rocky shores (Lindner 2016).

1.5 Final Remarks

Shallow and mesophotic reefs extend along nearly 3500 km of the Brazilian tropical and part of subtropical continental margin. In the last 10 years there was an expansion of mapping and a more detailed description of deeper reefs (mesophotic and deep reefs) in the Abrolhos Bank, in the North Brazil Shelf and in Southeastern Brazil. There still are new frontiers of discoveries or a more detailed description of occurrences known to fishermen on Eastern Brazil region, such as the Royal Charlotte Bank and in the vicinities of Todos os Santos Bay and Camamu Bay. Mapping efforts are crucial to support a better understanding the connectivity of reefs, along the shelf and across the shelf. This can also lead to an improvement in reef health assessment, in the design of monitoring programs and to better evaluations of the design of reef bearing protected areas and MPA networks. We also believe that a proper knowledge of reef distribution and morphological classification will help the improvement of environmental risk assessments related to global changes and anthropogenic impacts. Energy production projects, such as petroleum exploration and eolian mind power plants in shallow seas, or carbonate mining projects, for example, will affect reef ecosystems.

Acknowledgements RKPK benefits from a research productivity scholarship provided by the Brazilian National Council for Scientific and Technological Development (CNPq 314270/2023-0); ZMANL, MDMO, and RKPK are researchers participants of the National Institute of Science and Technology of Tropical Marine Environment (INCT AmbTropic – CNPq Grant 465634/2014-1); ZMANL, MDMO and RKPK integrate the ReBentos (The Brazilian Network for Study of Marine Benthic Organisms).

References

Albino J, Contti Neto N, Oliveira TCA (2016) The beaches of Espírito Santo. In: da Short AD, AHF K (eds) Brazilian beach systems. Springer International Publishing, Cham, pp 333–361

Almeida FFM (1955) Geologia e Petrologia do Arquipélago de Fernando de Noronha. DNPM/DGM, Rio de Janeiro

Amaral FMD, Silveira SRM, Vasconcelos SL, Ramos CAC (2006) Biodiversidade de cnidários bentônicos. In: Vasker Júnior T, Lessa RP, Nóbrega M et al (eds) Arquipélago de São Pedro e São Paulo: histórico e recursos naturais. Ed. Livro Rápido, pp 42–55

Amaral FMD, Hudson MM, Steiner AQ, Ramos CAC (2007) Corals and calcified hydroids of Manuel Luiz Marine Park (State of Maranhão, Northeast Brazil). Biota Neotropica 3:73–82

Amaral FMD, Ramos CAC, Leão ZMAN et al (2009) Checklist and morphometry of benthic cnidarians from the Fernando de Noronha Archipelago, Brazil. Cah Biol Mar 50:277–290

Amorim FN, Cirano M, Marta-Almeida M et al (2013) The seasonal circulation of the Eastern Brazilian shelf between 10°S and 16°S: a modelling approach. Cont Shelf Res 65:121–140. https://doi.org/10.1016/j.csr.2013.06.008

Araújo TMF, Leão ZMAN, Lima OAL (1984) Evolução do recife de coral da ilha de Itaparica determinada a partir de dados geológicos e geofísicos. In: Congr. Bras. de Geologia. SBG, Rio de Janeiro, pp 159–169

Araújo VD, Reyes-Peres YA, de Oliveira LR et al (2006) Fácies e sistema deposicional da Formação Barreiras na região da Barreira do Inferno, litoral oriental do Rio Grande do Norte. Geologia USP - Serie Cientifica 6:43–49

Araújo MED, Mattos FMGD, Melo FPLD et al (2020) Diversity patterns of reef fish along the Brazilian tropical coast. Mar Environ Res 160. https://doi.org/10.1016/j.marenvres.2020.105038

Bandeira AN Jr., Petri S, Suguio K (1975) Projeto Rio Doce. Petróleo Brasileiro S.A. Internal Report

Bastos AC, Moura RL, Amado-Filho GM et al (2013) Buracas: novel and unusual sinkhole-like features in the Abrolhos Bank. Cont Shelf Res 70:118–125. https://doi.org/10.1016/j.csr.2013.04.035

Carvalho RC, Kikuchi RKP (2013) ReefBahia, an integrated GIS approach for coral reef conservation in Bahia, Brazil. J Coast Conserv 17:239–252. https://doi.org/10.1007/s11852-013-0243-6

Castro CB, Pires DO (2001) Brazilian coral reefs: what we already know and what is still missing. Bull Mar Sci 69:357–371

Castro CB, Echeverría CA, Pires DO et al (1995) Distribuição de Cnidaria e Echinodermata no infra litoral dos costões rochosos de Arraial do Cabo, Rio de Janeiro, Brasil. Rev Bras Biol 55:471–480

Castro CB, Segal B, Pires DO, Medeiros MS (2006) Distribution and diversity of coral communities in the Abrolhos Reef Complex, Brazil. In: A biological assessment of Abrolhos Bank, Brazil, pp 19–39

Correia MD (2011) Scleractinian corals (Cnidaria: Anthozoa) from reef ecosystems on the Alagoas coast, Brazil. J Mar Biol Assoc U K 91:659–668. https://doi.org/10.1017/S0025315410000858

Costa OS Jr, Attrill MJ, Nimmo M (2006) Seasonal and spatial controls on the delivery of excess nutrients to nearshore and offshore coral reefs of Brazil. J Mar Syst 60:63–74. https://doi.org/10.1016/j.jmarsys.2005.11.006

Coura MF (1994) Contribuição do Plano de Manejo do Parque Estadual Marinho do Parcel de Manuel Luiz, MA – Brasil. Monografia de Especialização, Universidade Federal do Maranhão

Cruz ICS, Kikuchi RKP, Leão ZMAN (2008) Use of the video transect method for characterizing the Itacolomis Reefs, Eastern Brazil. Braz J Oceanogr 56:271–280

Cruz ICS, Kikuchi RKP, Leão ZMAN (2009) Caracterização dos Recifes de Corais da Unidade de Conservação Baía de Todos os Santos para fins de manejo, Bahia, Brasil. Revista de Gestão Costeira Integrada 9:3–23

Dechnik B, Bastos AC, Vieira LS et al (2019) Holocene reef growth in the tropical southwestern Atlantic: evidence for sea level and climate instability. Quat Sci Rev 218:365–377. https://doi.org/10.1016/j.quascirev.2019.06.039

Delibrias C, Laborel J (1969) Recent variations of sea-level along the brazilian coast. Quaternaria 14

DHN (1993) Atlas de Cartas Piloto. Ministério da Marinha, Diretoria de Hidrografia e Navegação

Dominguez JML (2009) The coastal zone of Brazil. In: Dillenburg SR, Hesp PA (eds) Geology and geomorphology of Holocene Coastal Barriers of Brazil. Springer, Berlin/Heidelberg/Berlin/Heidelberg, pp 17–51

Dominguez JML, Leão ZMAN (1994) Determinação das características sedimentológicas do fundo oceânico da área de influência dos emissários da CETREL e da TIBRÁS. CEPEMAR

Dominguez JML, Bittencourt ACSP, Leão ZMAN, Azevedo AEG (1990) Geologia do Quaternário costeiro do Estado de Pernambuco. Revista Brasileira de Geociências 20:208–215

Dominguez JML, Bittencourt ACSP, Martin L (1992) Controls on Quaternary coastal evolution of the east-northeasten coast of Brazil: roles of sea-level history, trade winds and climate. Sediment Gelogy 80:213–232

Echeverría CA, Pires DO, Medeiros MS, Castro CB (1997) Cnidarians of the Atol das Rocas. In: Lessios HA, Macintyre IG (eds) Int. Coral Reef Sym. ISRS, Panama, pp 443–446

Freitas LM, de Oliveira MDM, Leão ZMAN, Kikuchi RKP (2019) Effects of turbidity and depth on the bioconstruction of the Abrolhos reefs. Coral Reefs. https://doi.org/10.1007/s00338-019-01770-3

Gherardi DFM, Bosence DWJ (1999) Modeling of the ecological succession of encrusting organisms in recent coralline-algal frameworks from Atol das Rocas, Brazil. PALAIOS 14:145–158

Gherardi DFM, Bosence DWJ (2005) Late Holocene reef growth and relative sea-level changes in Atol das Rocas, equatorial South Atlantic. Coral Reefs 24:264–272

Ginsburg RN, Schroeder JH (1969) Introduction to the growth and diagenesis of Bermuda reefs. In: Bermuda conference on carbonates, p 13

Gomes MP, Vital H, Droxler AW (2020) Terraces, reefs, and valleys along the Brazil northeast outer shelf: deglacial sea-level archives? Geo-Mar Lett 40:699–711. https://doi.org/10.1007/s00367-020-00666-4

Greene HG, Yoklavich MM, Starr RM et al (1999) A classification scheme for deep seafloor habitats. Oceanol Acta 22:663–678. https://doi.org/10.1016/S0399-1784(00)88957-4

Hartt CF (1870) Geology and physical geography of Brazil. Fields, Osgood and Co., Boston

Kikuchi RKP, Leão ZMAN (1997) Rocas (Southwestern Equatorial Atlantic, Brazil): an atoll built primarily by coralline algae. In: Lessios HA, Macintyre IG (eds) Int. Coral Reef Sym. ISRS, Panama, pp 731–736

Kikuchi RKP, Leão ZMAN (1998) The effects of Holocene Sea level fluctuation on reef development and coral community structure, Northern Bahia, Brazil. An Acad Bras Cienc 70:159–171

Kikuchi RKP, Oliveira MDM, Leão ZMAN et al (2008) Os recifes de Tinharé-Boipeba-Camamu, Bahia. In: Rio Oil&Gas Conference. Instituto Brasileiro do Petróleo, Rio de Janeiro, pp 1–8

Kikuchi RKP, Leão ZMAN, Oliveira MDM (2010) Conservation status and spatial patterns of AGRRA vitality indices in Southwestern Atlantic reefs. Rev Biol Trop 58:1–31. https://doi.org/10.15517/rbt.v58i1.20021

Knoppers B, Ekau W, Figueiredo AG (1999) The coast and shelf of east and northeast Brazil and material transport. Geo-Mar Lett 19:171–178

Laborel J (1969) Les peuplements de madréporaires des côtes tropicales du Brésil. Abidjan

Laborel-Deguem F, Castro CB, Nunes FLD, Pires DO (eds) (2019) Recifes brasileiros: o legado de Laborel. Museu Nacional, Rio de Janeiro

Laborel J (1967) Madréporaires et hidrocoralliares récifaux des cotes brésiliennes - Systematique, Écologie, repartition verticale et géographique. Annales de l'Institut Océanographique 47:171–229

Leão ZMAN (1996) The Coral Reefs of Bahia: morphology, distribution and the major environmental impacts. Anais Acad Bras Cienc 68:439–452

Leão ZMAN, Kikuchi RKP (1999) The Bahian coral reefs - from 7000 years B.P. to 2000 years A.D. Ciência e Cultura J Brazil Assoc Adv Sci 51:262–273

Leão ZMAN, Kikuchi RKP (2001) The Abrolhos reefs of Brazil. In: Seeliger U, Kjerfve B (eds) Coastal marine ecosystems of Latin America. Springer, Berlin/Heidelberg/New York, pp 83–96

Leão ZMAN, Kikuchi RKP, Testa V (2003) Corals and coral reefs of Brazil. In: Cortés J (ed) Latin American coral reefs. Elsevier Science, Amsterdam, pp 9–52

Leão ZMAN, Kikuchi RKP, Oliveira MDM (2019) Chapter 35 - The Coral Reef Province of Brazil. In: Sheppard C (ed) World seas: an environmental evaluation, 2nd edn. Academic, pp 813–833

Lesser MP, Slattery M, Leichter JJ (2009) Ecology of mesophotic coral reefs. J Exp Mar Biol Ecol 375:1–8. https://doi.org/10.1016/j.jembe.2009.05.009

Lindner A (ed) (2016) A vida marinha de Santa Catarina, 2nd Revised edn. Editora UFSC, Florianópolis

Loiola M, Cruz ICS, Leão ZMAN, Kikuchi RKP (2014) Definition of priority areas for the conservation of a coastal reef complex in the eastern Brazilian coast. J Integ Coast Manag 14:611–624. https://doi.org/10.5894/rgci449

Loiola M, Cruz ICS, Lisboa DS et al (2019) Structure of marginal coral reef assemblages under different turbidity regime. Mar Environ Res. https://doi.org/10.1016/j.marenvres.2019.03.013

Magris RA, Costa MDP, Ferreira CEL et al (2021) A blueprint for securing Brazil's marine biodiversity and supporting the achievement of global conservation goals. Divers Distrib 27:198–215. https://doi.org/10.1111/ddi.13183

Maida M, Ferreira BP (1997) Coral reefs of Brazil: an overwiew. In: Proc. 8th Int. Coral Reef Symp. ISRS, Panama, pp 263–274

Maida M, Ferreira BP (2003) Área de Proteção Ambiental Costa dos Corais. In: Prates APL (ed) Atlas dos recifes de coral nas unidades de conservação brasileiras. Ministério do Meio Ambiente, Brasília, pp 86–90

Mazzei EF, Bertoncini AA, Pinheiro HT et al (2017) Newly discovered reefs in the southern Abrolhos Bank, Brazil: anthropogenic impacts and urgent conservation needs. Mar Pollut Bull 114:123–133. https://doi.org/10.1016/j.marpolbul.2016.08.059

Milliman JD, Barretto HT (1975) Part I. Background. In: Milliman JD, Summerhayes CP (eds) Upper continental margin sedimentation off Brazil. E. Schweizerbartsche Verlarsbuchhandlung, Sttutgart, pp 1–10

Morais J (1969) Aspectos correlativos de geologia litoral e submarina. Arquivos de Ciências do Mar 9:127–131

Moura RL, Secchin NA, Amado-Filho GM et al (2013) Spatial patterns of benthic megahabitats and conservation planning in the Abrolhos Bank. Cont Shelf Res 70:109–117. https://doi.org/10.1016/j.csr.2013.04.036

Moura RL, Amado-Filho GM, Moraes FC et al (2016) An extensive reef system at the Amazon River mouth. Sci Adv 2:e1501252–e1501252. https://doi.org/10.1126/sciadv.1501252

Neumann AC, Macintyre IG (1985) Reef response to sea level rise: keep-up, catch-up or give-up. In: Int. Coral Reef Sym. Antenne Museum EPHE, Tahiti, pp 105–110

Neves EG, Johnsson R, Sampaio C, Pichon M (2006) The occurrence of Scolymia cubensis in Brazil: revising the problem of the Caribbean solitary mussids. Zootaxa 1366:45–54. https://doi.org/10.11646/zootaxa.1366.1.2

Nimer E (1989) Climatologia do Brasil. Instituto Brasileiro de Geografia e Estatística, Rio de Janeiro

Nolasco MC, Leão ZMAN (1986) The carbonate buildups along the northern coasst of the State of Bahia, Brazil. In: Rabassa J (ed) Quaternary of South America and Antarctic Peninsula. A. A. Balkema, Rotterdam, pp 159–190

Oigman-Pszczol SS, Creed JC (2004) Size structure and spatial distribution of the corals Mussismilia hispida and Siderastrea stellata (Scleractinia) at Armação dos Búzios, Brazil. Bull Mar Sci 74:433–448

Oigman-Pszczol SS, Creed JC (2006) Distribution and abundance of fauna on living tissues of two Brazilian hermatypic corals (Mussismilia hispida (Verril 1902) and Siderastrea stellata Verril, 1868). Hydrobiologia 563:143–154. https://doi.org/10.1007/s10750-005-0002-2

Olavo G, Costa PAS, Martins AS, Ferreira BP (2011) Shelf-edge reefs as priority areas for conservation of reef fish diversity in the tropical Atlantic. Aquat Conserv Mar Freshwat Ecosyst 21:199–209. https://doi.org/10.1002/aqc.1174

Pereira NS, Manso VAV, Silva AMC, Silva MB (2010) Mapeamento Geomorfológico e Morfodinâmica do Atol das Rocas, Atlântico Sul. RGCI 10:331–345. https://doi.org/10.5894/rgci209

Pereira-Filho GH, Shintate GS, Kitahara MV et al (2019) The southernmost Atlantic coral reef is off the subtropical Island of Queimada Grande (24°S), Brazil. Bull Mar Sci 95:277–287. https://doi.org/10.5343/bms.2018.0056

Pereira-Filho GH, Mendes VR, Perry CT et al (2021) Growing at the limit: Reef growth sensitivity to climate and oceanographic changes in the South Western Atlantic. Glob Planet Chang 201:103479. https://doi.org/10.1016/j.gloplacha.2021.103479

Peterson RG, Stramma L (1991) Upper-level circulation in the South-Atlantic Ocean. Prog Oceanogr 26:1–73

Rossetti DF, Góes AM (2009) Marine influence in the Barreiras Formation, state of Alagoas, northeastern Brazil. An Acad Bras Cienc 81:741–755

Salles F, Bentes F, Santos J (eds) (2000) Catálogo de Estações Maregráficas. Fundação de Estudos do Mar, Rio de Janeiro

Santos CLA, Vital H, Amaro VE, Kikuchi RKP (2007) Mapping of the submerged reefs in the coast of the Rio Grande do Norte, near Brazil: Macau to Maracajau. Rev Bras Geofis 25:27–36. https://doi.org/10.1590/s0102-261x2007000500004

Seoane JCS, Arantes RCM, Castro CB, et al (2008) Atlas Físico do Parque Municipal Marinho do Recife de Fora, Porto Seguro, BA. Projeto Coral Vivo, Rio de Janeiro

Seoane JCS, Arantes RCM, Castro CB (2012) Benthic habitat mapping at Recife de Fora, Brazil: imagery and GIS. 5

Silva LLN, Gomes MP, Vital H (2018) The Açu Reef morphology, distribution, and inter reef sedimentation on the outer shelf of the NE Brazil equatorial margin. Cont Shelf Res 160:10–22. https://doi.org/10.1016/j.csr.2018.03.011

Silveira ICA, Miranda LB, Brown WS (1994) On the origins of the North Brazil Current. J Geophys Res 99:22501–22512

Soares MO, Lemos VB, Kikuchi RKP (2011) Aspectos biogeomorfológicos do Atol das Rocas, Atlântico Sul Equatorial. Revista Brasileira de Geociências 41:85–94

Souza BS, Albino J (2020) Development and morphodynamics of beaches on horizontal shore platforms on the microtidal southeast coast of Brazil. Geomorphology 363:107232. https://doi.org/10.1016/j.geomorph.2020.107232

Spalding MD, Fox HE, Allen GR et al (2007) Marine ecoregions of the world: a bioregionalization of coastal and shelf areas. Bioscience 57:573–583. https://doi.org/10.1641/B570707

Stramma L, England M (1999) On the water masses and mean circulation of the South Atlantic Ocean. J Geophys Res Oceans 104:20863–20883. https://doi.org/10.1029/1999JC900139
Teixeira CEP, Lessa GC, Cirano M, Lentini CAD (2013) The inner shelf circulation on the Abrolhos Bank, 18°S, Brazil. Cont Shelf Res 70:13–26. https://doi.org/10.1016/j.csr.2013.09.003
Testa V, Bosence DWJ (1998) Carbonate-siliciclastic sedimentation on a high-energy, ocean facing, tropical ramp, NE Brazil. In: Wright VP, Burchette TP (eds) Carbonate ramps. Geological Society, London, pp 55–71
Testa V, Bosence DWJ (1999) Physical and biological controls on the formation of carbonate and siliciclastic bedforms on the north-east Brazilian shelf. Sedimentology 46:279–301. https://doi.org/10.1046/j.1365-3091.1999.00213.x
US Navy (1978) South Atlantic Ocean. U.S. Navy, Washington, D. C
Vaskes Júnior T, LESSA R, Nóbrega M, et al (2007) Arquipélago de São Pedro e São Paulo: Histórico e Recursos Naturais
Vieira FV, Bastos AC, Quaresma VS et al (2019) Along-shelf changes in mixed carbonate-siliciclastic sedimentation patterns. Cont Shelf Res 187. https://doi.org/10.1016/j.csr.2019.103964

Chapter 2
History of Research About the Tropical Brazilian Coral Reef Systems

Zelinda M. A. N. Leão, Marília D. M. Oliveira, Antonio V. Ferreira-Junior, and Ruy K. P. Kikuchi

Abstract The history of research about the tropical Brazilian coral reef systems starts with the first foreign visitors to Brazil, in the 1800s and reaches the study of impacts of modern days in the ecosystem. A bibliometric analysis of over 700 articles, obtained in the peer-reviewed literature, offered important references concerning the distribution in time and space of the studied articles. The 1980 decade marks the beginning of the growth of the number of publications, and in the 1990s started the contribution among Brazilians and foreign researchers. This analysis has shown patterns of the morphology and distribution of reefs and that corals and reef fishes are the dominant targets of research. We also describe the main approaches and how they evolved in time, adopting new technologies for reef assessment and for reef mapping. We also show that contributions to management initiatives for minimizing the local and global impacts that are affecting the Brazilian reefs become significant after the decade that began in 2010. Future scope complemented with new approaches such as baseline surveys, integrated research, community involvement, policy, and continued management, will extend our knowledge contributing to compare the Brazilian coral reefs with worldwide counterpart.

Keywords Coral reefs · Reef research chronology · Bibliometric analysis · Literature review · Brazilian reefs distribution

---

Z. M. A. N. Leão (✉) · M. D. M. Oliveira · R. K. P. Kikuchi
Laboratório de Estudo dos Recifes de Corais (RECOR), Instituto de Geociências, Universidade Federal da Bahia, Salvador, Bahia, Brazil

A. V. Ferreira-Junior
Laboratório de Oceanografia Geológica, Departamento de Oceanografia, Universidade Federal de Pernambuco, Recife, Pernambuco, Brazil

R. K. P. Kikuchi et al. (eds.), *Brazilian Coral Reefs*, Brazilian Marine Biodiversity, https://doi.org/10.1007/978-3-031-59152-5_2

2.1 Introduction

This chapter focuses on the history of the research about the Tropical Brazilian coral reef systems, beginning with the visits of European naturalists to the New World, until the beginning of the implementation of the 2021–2030 United Nations Decade of the Ocean Science for Sustainable Development.[1] This historical narrative was complemented with a bibliometric analysis of the peer-reviewed literature produced in almost two centuries, which offers information about how this scientific literature grew along this time span.

This systematic analysis answered questions related to the distribution in time and in space of the reviewed literature. We aimed to evaluate the formation of a national research community followed by the insertion of this community in the international scene, and its participation in wider research networks. We also evaluated the diversification of approaches, which dominated over the time, and the evolution of the use of newer technology tools. At last, we also evaluated which reefs, along the continental shelf and their characteristics, were studied over the time span of scientific research.

This overview of the studies of Brazilian coral reef systems evaluated the number of scientific documents, the distribution of the publications of the peer-reviewed literature between 1832 and February 12nd, 2021, the participation of institutions abroad Brazil, the characteristics of the research, and what were the types and spatial distribution of the studied reefs.

The compilation of studies was undertaken based on the scientific documents, such as articles and book chapters, referred to in the Web of Science, Scopus, and databases provided by the authors search, using the terms "reef" and "Brazil" in Title, Abstract and Keywords. The searches resulted in over 1000 articles. From these first results, the scientific documents not related to the Brazilian Tropical Ecoregions (*sensu* Spalding et al. 2007; see Chap. 1) were eliminated, as for example the studies about rocky shores from the Brazilian Southeastern coast. However, we included the publications related to the Laborel's Coral Oasis of Rio de Janeiro state, located in Arraial do Cabo, southeastern coast of Brazil (Laborel 1970) (Fig. 2.1), due to its seminal importance. We excluded, also, papers dealing with artificial reefs, shipwrecks, the publications that were not peer-reviewed and few articles that were not accessible in the internet search. The final counting resulted in 755 documents analyzed.[2]

The data were plotted in electronic spreadsheets for identifying trends, and the following information was extracted: (a) journal title, (b) article title, (c) year of publication, (d) name of first author, (e) number of authors, (f) country of first author, (g) presence of foreign institution/author, (h) key words, (i) location of study area, (j) approaches, (k) methodology, (l) type of reefs, (m) depth range, (n) age of reefs and (o) geographical scale.

[1] https://www.oceandecade.org/

[2] The database can be found in the following address https://doi.org/10.5281/zenodo.10401848

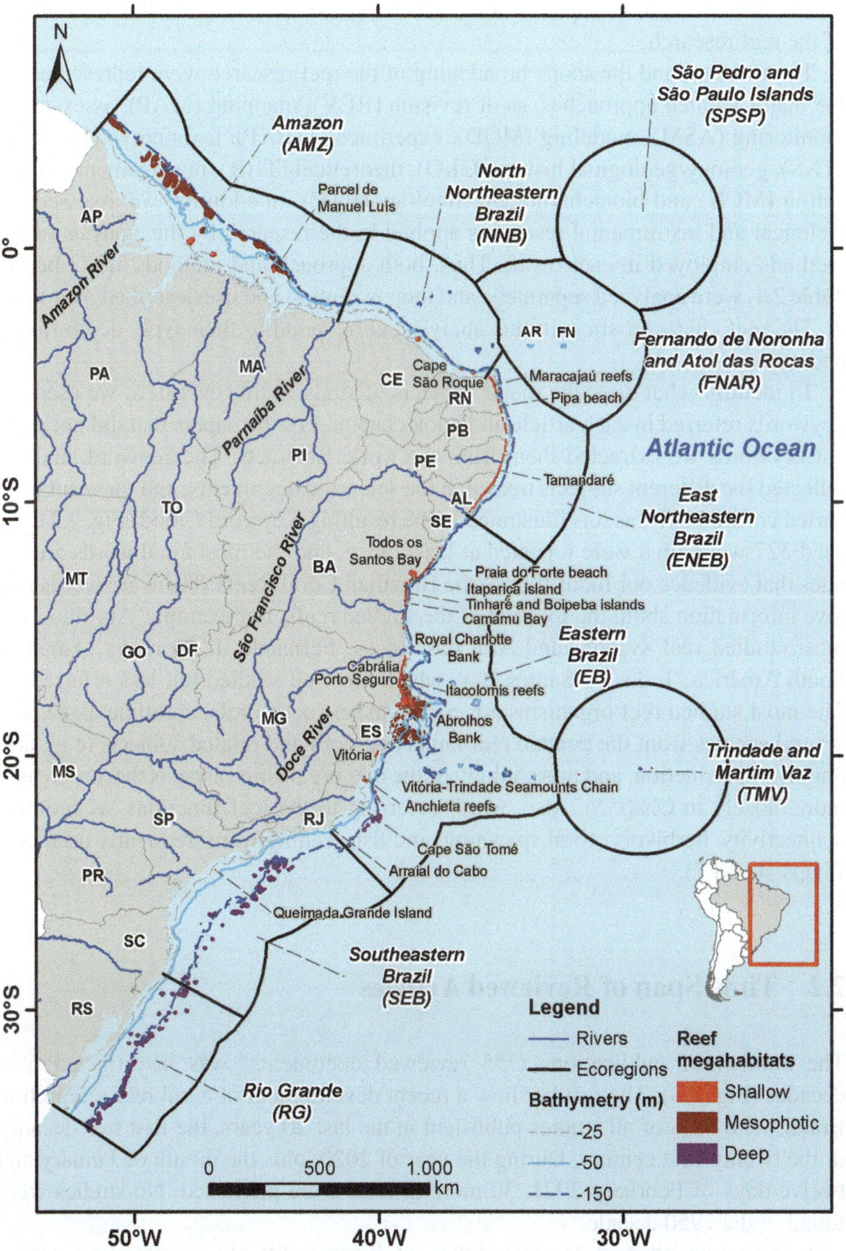

Fig. 2.1 Map of reef distribution within a proposed division of Western South Atlantic ecoregion scheme. See Chap. 1

The data was grouped per decade to better synthesize the growth and evolution of the reef research.

The interests and the scope broadening of the reef research were represented by the major applied approaches, as of revision (REV), mapping (MAP), assessment/monitoring (ASM), modeling (MOD), experimental (EXP), taxonomy/systematics (TXS), geology/geological history (GEO), theoretical (THR), management/conservation (MCN) and biotechnology/technology (BTC). In addition, we assessed the technical and instrumental resources applied in the research by the analysis of the methods employed in each work. Thus, both approach and methods, described in Table 2.1, were analyzed separately, and their evolution and use described over time.

The reef characteristics, in this analysis, correspond to their type, depth range, and age.

To identify what were the major subjects of study in this overview, we used the keywords referred in each article and book chapter. For the papers that did not present keywords we extracted them from the works abstracts. The keyword analysis reflected the different subjects treated in the scientific documents, and these subjects varied considerably, as it is illustrated in the resulting Keyword Cloud (Fig. 2.2). We used 327 words that were repeated at least twice, and the most cited words are the ones that evidence our focus theme, the Brazilian Coral Reefs. There are words that give information about the location of the studied reefs. For example, Abrolhos, the most studied reef system, and Atol das Rocas, Fernando de Noronha, Amazon, South America, Todos os Santos Bay, which are well studied but less referred to. The most studied reef organisms are corals, fishes, octocorals, coralline algae, and several species from the benthic reef fauna and flora and related subjects (e.g., taxonomy, reproduction, and morphology of the reef organisms; aspects that are treated more closely in Chap. 5). Also, words referring to the reef functions, as ecology, connectivity, herbivory, coral spawning, and fish feeding were frequently used (see Chaps. 4 and 7).

2.2 Time Span of Reviewed Articles

The number of publications (755 reviewed documents) was investigated over decades (Fig. 2.3). The results show a recent development of coral reef research in Brazil, with 70% of all studies published in the last 20 years, the first two decades of the twenty-first century. During the year of 2020, plus the month of January and twelve days of February 2021, 30 more articles were published. No studies were found in the 1950 decade.

In the nineteenth century and the first six decades of the twentieth century, there were only a few published studies (one to seven publications per decade). In the twentieth century, until the 1980, the number was still low, but in the 1980s, 23 articles and one book chapter were published, and in the 1990s the number of publications reached 56 articles and 5 book chapters (Fig. 2.3). The boom of publications about the Brazilian coral reefs occurred in the first two decades of the

Table 2.1 The main approaches and methods (techniques and/or instruments) identified in the peer-reviewed literature from 1832 to 2021 (February 12nd) and used to analyze coral reef research in Brazil

Approach	Code	Methods (techniques and/or instruments)	Code
Revision	REV	Narrative/best evidence	NBE
		Systematic	SYS
		Meta-analysis	MET
Mapping	MAP	Airborne and/or satellite remote sensing; GIS	ASR
		Marine geophysics (multibeam, side-scan-sonar)	MGE
		Video techniques	VTE
		Traditional cartography	TCA
		Secondary data (museums, bibliography)	SDT
Assessment/monitoring	ASM	Visual census	VSE
		Photo/video	PHV
		Biological sampling	BSA
		Water sampling (water quality)	WSA
		Sediment sampling	SSA
		Physical and chemical data sampling (field measurement instruments)	PCS
		Structured and free questionnaire	SFQ
Modeling	MOD	Ecological (e.g., Bayesian network)	ECO
		Physical	PHY
		Chemical	CHE
Experimental	EXP	Field	FLD
		Laboratory	LAB
Taxonomy/systematics	TXS	Morphology	MPH
		Molecular	MOL
Geology/geological history	GEO	Dating	DAT
		Geochemistry (isotope, element)	GCH
		X-radiation (X-radiography, computerized tomography)	XRY
		Petrography	PET
Theoretical	THR	Conceptual	CON
		Narrative/best evidence	NBE
Management/conservation	MCN	Narrative/best evidence	NBE
		Quantitative analysis	QAN
Biotechnology/technology	BTC	Organism detection/identification	OID
		Metabolites extraction/mass spectrometry	MEX
		Environmental parameter detection	ENV

twenty-first century, reaching 191 documents between 2000–2009 and 427 works in the last decade (2010–2019, including 2020 and beginning of 2021). During these periods the scientific articles dominate with circa 97% of the total publications.

Fig. 2.2 Keyword Cloud depicts the main subjects within the peer-reviewed literature (n = 755 scientific documents) from 1832 to 2021 (February 12nd). Three hundred and twenty-seven words that appeared at least twice were identified

2.2.1 *The New World of the Foreign Visitors*

The history of the research about the Brazilian coral reefs starts with the first reported scientific visit to the Brazilian reefs, when Charles Darwin, on the 28th of March 1832, aboard the H.M. Beagle, anchored near the Abrolhos Islands (Fig. 2.1), Darwin described the corals he observed around the islands as irregular masses which, although large, did not reach the water surface. We believe he was referring to the "chapeirões" that surround the islands of the Abrolhos Archipelago. Those descriptions are part of the Darwin Geological Observation on Coral Reefs, Volcanic Islands and South America, described in the Captain Fitzroy narratives of voyages of the HMS Adventure and HMS Beagle (Fitzroy, 1832). These descriptions were reprinted later in Darwin (1962).

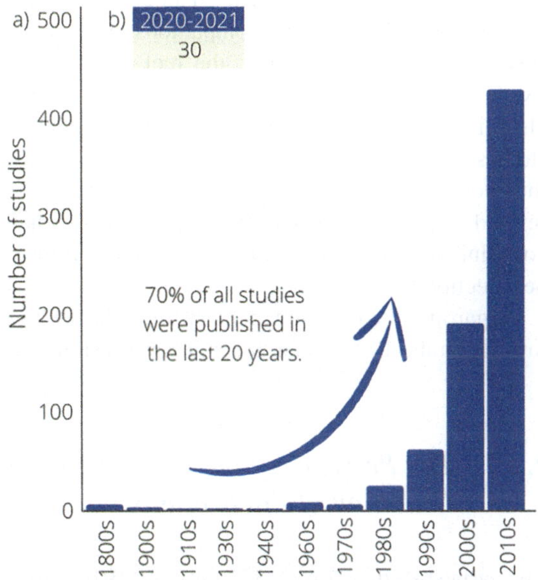

Fig. 2.3 Number of peer-reviewed publications (n = 755) about Brazilian coral reefs between 1832 and 2021 (February 12nd). References are arranged over decades, (**a**) decades between 1800 and 2010, (**b**) the 2020 year until February 12nd, 2021

Yet, in the nineteenth century, a description of the surface geology of the Brazilian coast as well as an account of the unusual morphology of the reefs and the coral zonation, particularly in the Abrolhos area, were given by the Canadian Scientist Charles F. Hartt. Hartt came to Brazil for the first time as a geologist of the Agassiz Expedition to South America in connection with the Thayer Expedition in the years of 1865 and 1866. He described that his first acquaintance with the coral reefs of Brazil was at Porto Seguro in South Bahia state in 1866, when he collected the first samples of the Brazilian coral species. In the following year, Hartt returned and spent several months examining the coral reefs between Pernambuco and Rio de Janeiro states, exploring more particularly the vicinity of Bahia and the islands and coral reefs of Abrolhos (Hartt 1868, 1869, 1870). Hartt studies mentioned that the reefs were composed by very few species of corals, and among them the massive forms were the most common. He also described the "chapeirões", for the first time in the literature, as coral structures, which grew in the shape of mushrooms, and are very common in the Abrolhos area. It is worth noting his illustration of a vertical transection along the wall of one of the Abrolhos shallow coastal reefs describing the distribution of the reef fauna until a depth of about 5 m (Hartt 1870, p. 210). One wonders what was the equipment used by Hartt to describe so precisely the coral depth distribution on a pinnacle. Although there is no mention to the type of diving equipment he used, he probably used a diving helmet.

The corals collected during Hartt's expeditions were identified by A. E. Verrill (1868, 1901, 1912). In his works, Verrill commented about the archaic character of several corals and pointed out that these coral species might be a small surviving remnant of an ancient coral fauna, dating back to the Early Tertiary Period. Other foreigner scientific researchers visited Brazil in the nineteenth century, such as R. Rathbun, one of Hartt's companions, who visited the reefs of Todos os Santos Bay and described, in detail, the reef that borders the coast of Itaparica Island (Rathbun 1876, 1878a, b, 1879), and J. C. Branner who studied the beachrocks and also visited shallow coral reefs along the northeast coast of Brazil (Fig. 2.1). The studies of J. C. Branner concerning the stone reefs were published late in 1904. In this work, he referred to Abrolhos as a barrier reef system, a description that was revised as aggregations of pinnacles and reef banks in two reef arcs parallel to the coastline in 1969, in a work by J. L. Laborel, whose contribution is described in the next section.

Regarding the twentieth century, until the 1980s the number of studies was still low, but in the 1990s, then, the number has shown a great increase (Fig. 2.3).

2.2.2 The Presence of the Brazilian Researchers—The Evolution of Publications

The earliest studies, from 1832 until the 1910–1919 decade had only the participation of foreign institutions. At the middle of the twentieth century, the contribution of foreigners was still significant, and the most comprehensive work about the Brazilian coral reefs was done at the Institute of Oceanographic Studies, from the Federal University of Pernambuco, by the French Biologist J. L. Laborel (1967, 1969, 1970). Laborel, who lived in Brazil for 3 years (1961–1964), made a detailed study of the corals and coral reefs, including coral taxonomy, geographic distribution, composition and evolution of the coral communities, and the morphology, origin, and development of the reefs. He compared the Brazilian and Caribbean corals, and made comments about the strong endemism of the Brazilian coral fauna. Up to this time, this was the most comprehensive study about the Brazilian coral reefs. In 2019, Laborel's work was commented and translated to the Portuguese language (Laborel-Deguen et al. 2019). Laborel's stay in Brazil coincides with the visit of other scientific researchers from abroad who were invited by the Federal University of Pernambuco, for participating in the foundation of the Institute of Oceanographic Studies. M. Kempf worked with the formations of vermetids and calcareous algae along the coast of NE Brazil (Kempf and Laborel 1968), and F. Ottmann presented the first hypothesis about the origin of beach rocks, as well as information on the origin of the Atol das Rocas (Ottmann 1960, 1963).

During and after Laborel's work, some investigations done by foreigner scientists were related to the study of the coral and sandstone reefs. Vallaux (1940) studied the geomorphologic characteristics of Atol das Rocas, while Mabesoone (1964) investigated the origin and age of the beach rocks from Northeastern Brazil. The taxonomy of the Brazilian milleporids was investigated by Boschma (1961, 1962) and later, Petuch (1979) described new gastropod species from the Abrolhos area.

The earliest studies with the participation of Brazilian researchers in the twentieth century date back to the 1930s. Duarte (1932) studied lagoon sediment and the islands soil composition of the Atol das Rocas. Later, Joly et al. (1969) wrote a paper claiming for the creation of a Marine National Park in Abrolhos. Two studies on Atol das Rocas, one about marine algae (Oliveira-Filho and Ugadim 1976) and another describing the mollusk fauna (Rios 1979), complement this list of early national efforts in studying coral reefs.

The presence of Brazilian researchers started to dominate from 1980 on, when a new time of coral research in Brazil marked the passage of the dominance of foreign scientific participants to the New Era of scientific studies dominated by Brazilian investigators. However, beside the dominance of Brazilian researchers, the foreign participation also grew over the years, showing the internationalization of Brazilian coral reef research community. For example, in the 1990s the participation of foreign authors was 12%, in the 2000s was 28% and in the 2010s it increased to 36%. In the whole studied period, about 30% of publications had the contribution from institutions outside Brazil (Fig. 2.4).

One additional aspect to consider is that, except for one, all the early publications (until the 1970s) had a single or two authors. Only after the 1980s papers with three or more co-authors began to be published (Fig. 2.5). Our data also show that articles involving two or more institutions were, similarly, more common after the 1990s, and increased up to date, as seen in Fig. 2.4. Thus, the Brazilian coral reefs studies started expanding, and the first major reef investigations mostly originated from scientific works that have their origin in student thesis and dissertations. After this period, the studies largely expanded in number of researchers and participant institutions, attaining more than six authors per article. This reflects a change in the way

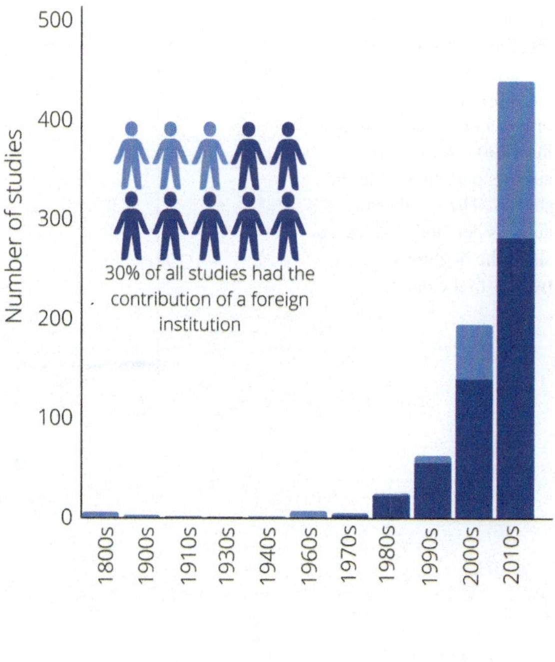

Fig. 2.4 Number of peer-reviewed publications (n = 755) about Brazilian coral reefs with the presence of foreign researchers between 1832 and 2021 (February 12th). The articles published in 2020 and January and February 2021 are not represented in the graphic. Brazilian researchers dominate from 1980 to the present days, being the contribution of foreign institutions, in the whole period, no more than 30%

science began to be done, with the organization of research networks along the country. For example, in the 2010s there were 39 studies with more than six authors (see Fig. 2.5).

The characterization of the reviewed literature involved the classification according to their approaches and the applied methods. About 98% of the analyzed studies had just one approach, and less than 2% had two or more approaches. The two most applied approaches are assessment/monitoring (40%) and taxonomy/systematic (22.1%), summing up 62.1% of all research papers. The use of experiments is the third most common approach, appearing in 9.3% of the papers examined. The other approaches occur at less than 7% of the publications (Fig. 2.6).

Regarding the methods, a great diversification of methods and technologies were observed in the studied decades. The most frequently used techniques were visual census (18.8%), morphology of organisms (13.1%), biological sampling (13.0%), collection for molecular studies (9.8%) and photo/video (6.8%), in works related to reef assessment and monitoring, and field procedures for data collection. Narrative/best evidence (7.0%) was used in revision papers. Experimental works were carried on in the laboratory (5.3%) in the field (4.6%), and among studies of the rocky structure of the reefs, geological dating (3.2%) was the most used. Less than 3% is concerned with various other techniques as shown in Fig. 2.7.

Fig. 2.5 Number authors in the peer-reviewed publications (n = 755) about Brazilian coral reefs between 1832 and 2021 (February 12th). The articles published in 2020 and January and February 2021 are not represented in the graphic. Early works were produced by a single or two investigators until the 1990s when more authors participated in the studies. The number of authors per study increased since the beginning of the twenty-first century

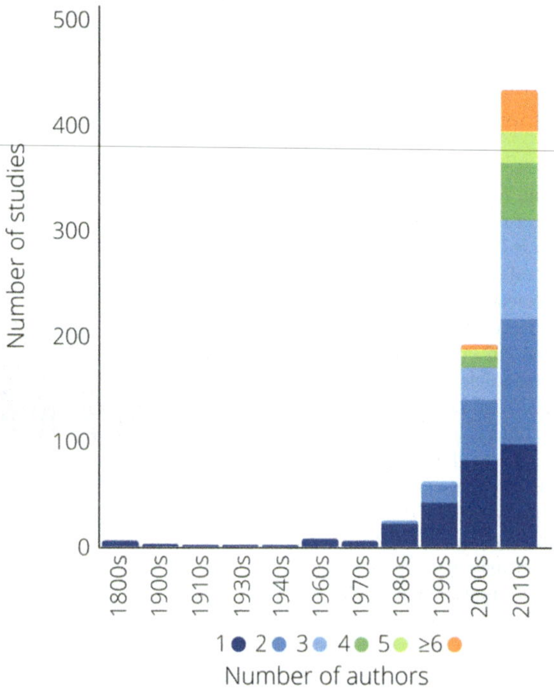

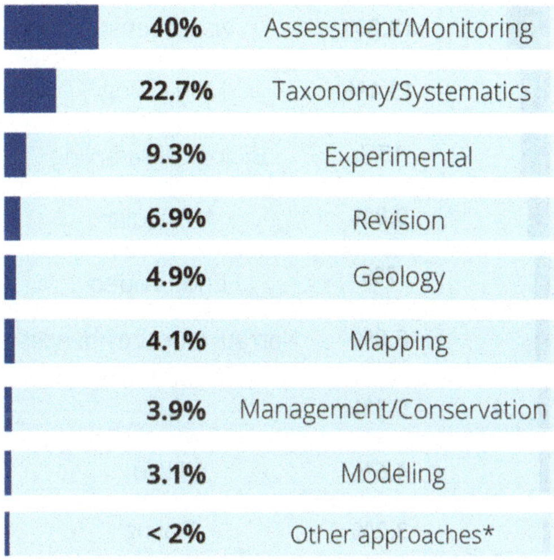

Fig. 2.6 Percentages of the different approaches in the peer-reviewed publications (n = 755) about Brazilian coral reefs between 1832 and 2021 (February 12th) The articles published after 2020 are not represented in this graphic *Assessment/Monitoring/Taxonomy/Systematic (1.6%); Theoretical (0.4%); Mapping/Management/Conservation (0.4%); Biotechnology/Technology (0.3%); Assessment/Monitoring/Experimental (0.3%); Modeling/Taxonomy/Systematic (0.3%); Biotechnology/Technology/Assessment/Monitoring/Taxonomy/Systematic (0.1%); Assessment/Monitoring/Modeling (0.1%); Assessment/Monitoring/Mapping (0.1%); Experiment/Taxonomy/Systematic (0.1%); Mapping/Geology/Geological history (0.1%); Mapping/Modeling (0.1%); Revision/Geology/Geological history (0.1%); Mapping/Taxonomy/Systematic (0.1%)

Until the 1970s, only five approaches were applied: theoretical (THR), revision (REV), taxonomy/systematic (TXS), mapping (MAP) and geology (GEO) (Fig. 2.8a, b).

In the 1800s the studies were mostly theoretical descriptions and broad revisions originated from the visits of foreign scientists to Brazil, as the publications of Fitzroy (1832), Hartt (1868, 1869, 1870) and Rathbun (1876, 1878a, b, 1879). From the beginning of the twentieth century the first studies applying the taxonomy/systematic approach based on the morphology of the reef organisms were the studies by Verrill (1868, 1901, 1912). From that time on these studies grew rapidly, and became the most common approaches in the 1960s and 1970s. Similarly, the first publications about mapping the reefs using traditional cartography occurred in the first decade of the twentieth century, and the first one is the study by Laborel (1970). On Geology, Vicalvi and Costa (1978) published their discovery of a paleo lagoon in the Abrolhos area, and called it the Abrolhos Depression.

Beginning in the 1980s, the number of the approaches more frequently applied increased, adding six more themes [theorical (THR), revision (REV), taxonomy/systematic (TXS), mapping (MAP), geology/geological history (GEO), assessment/monitoring (ASM), experiment (EXP), management/conservation

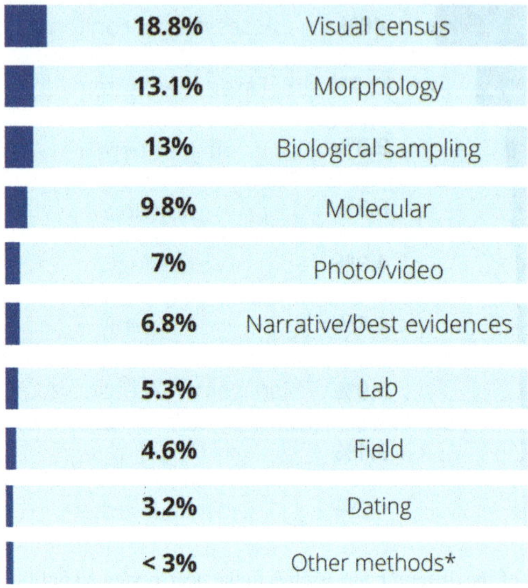

Fig. 2.7 Percentages of the different methods (techniques and/or instruments) in the peer-reviewed publications (n = 755; 850 refers to the total number of investigated words, including repetitions) about Brazilian coral reefs between 1832 and 2021 (February 12th). The articles published after 2020 are not represented in this graphic. *airborne and/or satellite remote sensing = 2.0%, ecological (Bayesian network = 2.0%, water sampling = 1.8%, marine geophysics = 1.4%, structured and free questionnaire = 1.3%, physical = 1.3%, geochemistry = 1.1%, sediment sampling = 1.1%, traditional cartography = 0.9%, meta-analysis = 0.7%, systematic = 0.7%, physical and chemical data sampling = 0.6%, petrography = 0.6%, video technique = 0.6%, secondary data = 0.4%, conceptual = 0.1%, organism detection = 0.1%, others = 0.8%

(MCN), modeling (MOD) and biotechnology/technology (BCT)] to the previous ones (Fig. 2.8a).

During this decade (the 1980s) six approaches were defined in 24 articles, which were produced regarding the so far most studied Brazilian reef areas, Abrolhos, and some reefs from the states of Pernambuco and Paraíba, in the Northeastern region. Still new revisions of the Abrolhos reefs described the beauty of the reefs, showing the ecological importance of the first National Marine Park, as well the need to call attention to the threats of predatory tourism (Castro and Sechinn 1981, 1982; Secchin 1986). A new work claimed against the exploitation of the Brazilian corals, as it had been observed that it was already common to sell coral skeletons as souvenirs, mainly of the hydrocoral *Millepora alcicornis* and secondary of brain corals, such as *Siderastrea* spp. and *Mussismilia hispida* (Maÿal 1986). Geological studies published ^{14}C dating of the calcareous reef structures from the coast of the state of Bahia (Leão 1983; Bernat et al. 1983; Araujo et al. 1984; Martin et al. 1985; Nolasco and Leão 1986; Leão et al. 1985, 1988) and from the Fernando de Noronha Archipelago (Jindrich 1983), presenting a developmental history of the reef structures. Those studies evidenced that Holocene reef growth, in Brazil, started after 8000 years BP, concomitantly with an ecological phenomenon widespread in the

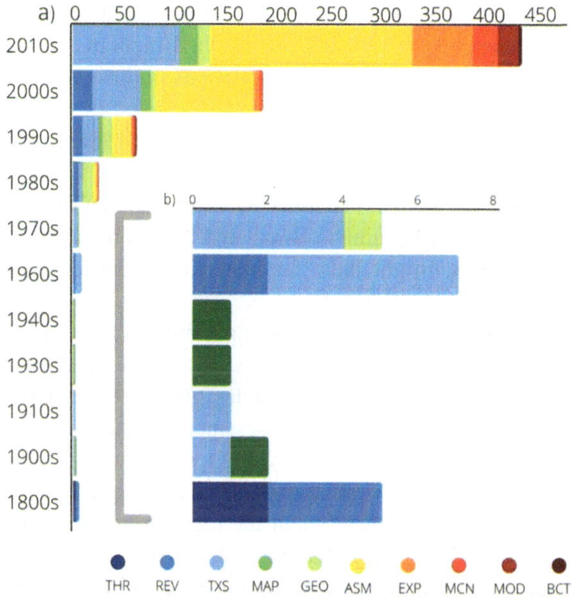

Fig. 2.8 Number of the different approaches in the peer-reviewed publications (n = 755) about Brazilian coral reefs between 1832 and 2021 (February 12th): (**a**) general view of the entire analyzed period; (**b**) detailed view of publications before the 1980s. The articles published after 2020 are not represented in the graphic. Approaches: Theoretical (THR), Revision (REV), Taxonomy/Systematic (TXS), Mapping (MAP), Geology/Geological history (GEO), Assessment/Monitoring (ASM), Experimental (EXP), Management/Conservation (MCN), Modeling (MOD), Biotechnology/Technology (BCT)

whole tropical world. However, the sea-level behavior along the coast of Brazil left distinct imprints on the development of the reefs, which allowed to recognize well marked phases during reef growth such as: reef initiation and establishment, rapid upward reef accretion, lateral growth of the reefs and reef degradation (Leão et al. 1988).

Using the method of visual census, new reef assessments produced more information about the status and composition of the Abrolhos reef fauna, mainly fishes and corals (Sazima 1984; Pitombo et al. 1988; Gonchorosky et al. 1989a). Based in the morphology of these reef organisms, new components of the reef fauna were identified, in Abrolhos (Rios and Barcellos 1980; Belém et al. 1982; Castro 1989) and in reefs located in the Brazilian Northeastern region (Young and Christoffersen 1984). At the end of this decade a pioneer experimental study, made in the field and laboratory, studied the effects of the decomposition of drift seaweeds in the area of coastal reefs from the state of Paraiba (Sassi et al. 1988), and a work claiming for the importance of the establishment of a management plan for the National Marine Park of Abrolhos was published (Gonchorosky et al. 1989b).

In the 1990s, the increase trend in the number of studies per decade occurred in the order of three times (61 studies in the 1900s, 191 studies in the 2000s, and 427 studies in the 2010), all of them are referred in the dataset shown in Supplementary Material. This increase in the publications about the Brazilian coral reefs reflected a growth of the Brazilian scientific community of reef studies and their participation in national and international coral reef scientific meetings, as well. As an example, we cite the International Coral Reef Symposium, when in 1988, during the Australia (Townsville) meeting, there were only two presentations by Brazilian researchers, while in the 1996 meeting, held in Panama City, more than ten Brazilian works were presented. The proximity of Panama to Brazil and the financial support received by the Brazilian researchers, might have facilitated their participation in the latter symposium.

At that time, the growing interest in knowing more about the Brazilian coral reef fauna and flora, the composition of the reefs structure and their distribution along the Brazilian coast was clear. In the 61 articles published during the 1990–1999 decade seven approaches were applied (assessment/monitoring ASM, taxonomy/systematic TXS, geology/geological history GEO, revision REV, management/conservation MNC, modeling MOD and mapping MAP) (Fig. 2.8a). Two early narrative revisions offered a general view about the reefs from the Northeastern region (Maida and Ferreira 1997) and from the reefs along the coast of the state of Bahia (Leão 1996), both reviews showed the importance of calling attention to the local impacts over the reefs. At the same time, there was an increase in reef assessments, producing more knowledge on the composition and distribution of the reef organisms, and on their conservation status. These studies were related, mainly, to the assessment of corals and fishes from the reefs of the Northeastern and Eastern regions, mostly using the visual censuses. Some of these works contributed to the identification of coral species, among them are the papers by Amaral (1992), Pires et al. (1992), and Marques and Castro (1995). Yet, the first assessment of the corals from the Fernando de Noronha islands (Maida et al. 1995) as well as one study about the taxonomy of a new species of octocorals in the country (Castro 1990), were published, both based in the organism morphology, and using the method of visual censuses.

In the context of geological studies an article published during the eighth International Coral Reef Symposium, showed the coexistence of coral reefs with muddy sediments in the region of Abrolhos (Leão and Ginsburg 1997). Furthermore, two pioneer articles using the method for dating the reef structure, described the geological history of the Atol das Rocas, and of the reefs located at the north coast of the state of Bahia (Kikuchi and Leão 1997, 1998). Those studies also showed the influence of Brazilian sea-level history on the development of the reef's structure. Two other pioneer studies that used coral the content of stable isotopes of oxygen and carbon in *Siderastrea* spp. skeletons as proxy of the water temperature and organic activity were done in the reefs of Bahia (Azevedo et al. 1992) and of the state of Pernambuco (Ferreira et al. 1998). The latter also suggested the possibility of interpreting the occurrence of bleaching based on the isotopic content in the coral skeleton.

As the knowledge about the Brazilian reefs, their distribution, evolution, and ecology evolved, the interest on the reefs status in order to protect these ecosystems gained momentum. This was represented by a proposition of a management plan for the reefs from the Parcel de Manoel Luis (Coura 1994), and by a study based in a 5 years survey of the effects of marine tourism in the coral reefs from the Abrolhos National Marine Park, which revealed that, although severely controlled, the visitor's activity in the park area had been hazardous to the reefs (Leão et al. 1993). Coral bleaching started to raise the awareness of the scientific community to the global change problems and the first report of a massive event of coral bleaching in the reefs of the Abrolhos region, in 1992, was assessed by visual census. It was observed that one species of hydrozoan, *Millepora alcicornis*, and 12 species of Scleractinia corals, including the three endemic species of the genus *Mussismilia*, were bleached (Castro and Pires 1999). In parallel, one pioneer study mapped, using satellite images, the shallow and mesophotic reefs at the Rio Grande do Norte continental shelf (Testa and Bosence 1998), which extended our knowledge about the geographic distribution of Brazilian reefs. A step further on the application of ecological theoretical and methodological tools was taken at the end of this decade, with the publication of the first ecological modeling for the Atol das Rocas proposed by Gherardi and Bosence (1999).

The beginning of the twenty-first century was marked by the boom of Brazilian coral reef studies, tripling every other decade. From this time on, there was an increase of the described working approaches (Fig. 2.6), as well of the applied methods and technologies (see Fig. 2.7).

Throughout the decades of 2000–2009 and 2010–2019, including the year of 2020, the month of January and twelve days of February, 2021, 648 articles were investigated, and nine approaches were identified: assessment/monitoring (ASM), taxonomy/systematic (TXS), revision (REV), mapping (MAP), geology/geological history (GEO), experiment (EXP), management/conservation (MCN), modeling (MOD) and biotechnology/technology (BCT), among which the most representative were taxonomy/systematic and assessment/monitoring (Fig. 2.7b). Among the 191 articles published during the first decade of this century (2000–2009), the study of the fish fauna grew, but corals were, still, the most frequent subject of study. Other organisms such as sponges, echinoderms, lobsters, octopuses, worms, foraminifera and plankton were also studied in these years.

At the beginning of the first decade of this century, three broad narrative revisions were published. One described the distribution, morphology and the geological history of the reefs of the Brazilian continental shelf, as well as the sedimentary characteristics around the coastal reefs, where the water turbidity was considered relatively high, a characteristic already referred to in early works (Leão et al. 2003). Another paper reported the distribution and characteristics of the reefs from Costa dos Corais, Pernambuco State, which include the reefs of Tamandaré (Maida and Ferreira 2003), where an environmental protection area was created in the previous decade as an outcome of the Brazilian scientific community initiative of the First International Year of the Reefs (1997). The third work presented a revision of the patterns of biodiversity and evolution of reef fishes from the Brazilian shelf (Rocha 2003). But during this period, circa 50% of the published research still referred to the assessment of the reef organisms, mostly using visual census. Photo/video sampling technique for benthic studies began to be also used by this time (Dutra et al. 2006; Fig. 2.9a). In this regard, two programs for monitoring the conservation status of Brazilian reefs were proposed. The first one was the Atlantic and Gulf Rapid Reef Assessment[3] (AGRRA) program, that aimed to assess the status of coral reefs from the Atlantic Ocean using the method of a Long-Term Monitoring Program site (PELD site 29) in the East Brazilian Province, especially in the Abrolhos reefs (Kikuchi et al. 2003a, b). The second one, applying the globally used Reef Check[4] protocol, resulted in a pilot project, which was presented during the tenth International Coral Reef Symposium, held in Japan (Ferreira et al. 2006). This protocol was further adopted by the Ministry of Environment as the method of the Brazilian National Reef Monitoring Program.

These reef monitoring programs gave important references for the knowledge about the condition of the Brazilian reefs, and environmental impacts over the reefs started to be reported on a global context as well (Garzón-Ferreira et al. 2002).

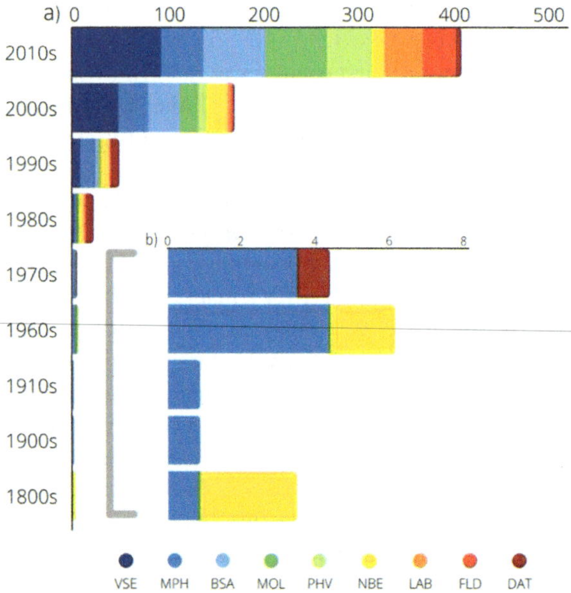

Fig. 2.9 Numbers of the different methods (techniques and/or instruments) in the peer-reviewed publications (n = 755; 850 refers to the total number of investigated words, including repetitions) about Brazilian coral reefs between 1832 and 2021 (February 12th). (**a**) general view of the entire analyzed period; (**b**) detailed view of publications before the 1980s. The articles published after 2020 are not represented in the graphic. VSE Visual Census, MPH Morphology, BSA Biological Sampling, MOL Molecular, NBE Narrative/Best Evidence, PHV Photo Video, LAB Laboratory, FLD Field work, DAT Dating. N in this figure is greater than the total number of studied papers because a same paper might use more than one technique

[3] https://www.agrra.org/

[4] https://www.reefcheck.org/country/brazil/

Several reef assessments showed different cases of reef impacts, as of the first occurrence of coral disease in the endemic coral *Mussismilia braziliensis* from the Abrolhos reefs (Francini-Filho et al. 2008), and of coral bleaching in Brazil. One work about the reefs from the state of Bahia reported the occurrence of coral bleaching in the reefs located along the coast, when over 10% of the coral colonies were affected, being the most severely impacted the species *Mussismilia hispida*, *Siderastrea* spp., *Montastraea cavernosa*, *Agaricia* spp. and *Porites astreoides* (Leão et al. 2008). Another research about coral bleaching compared zooxanthellae density from coral colonies affected by bleaching with healthy colonies, from a reef at the coast of the state of Pernambuco (Costa et al. 2001). Still, new research in the Abrolhos region showed the effects of sediment accumulation on measured parameters of the coral's condition, which seemed to influence the reef's vitality (Dutra et al. 2006).

Two other reef assessments referred to pioneering results about the coralline algae community from the Atol das Rocas, showing that the crustose coralline algae *Porolython* cf. *pachydermum* dominated the carbonate structure of the reefs (Gherardi and Bosence 2001), and the first report about the Itacolomis coral reefs, which was so far coral reef area still lacking proper investigation (Castro and Segal 2001).

Along with these reefs assessments, over 25% of the published research were related to the taxonomic identification and/or systematic revision of, mostly, fish and coral faunas using the organism morphology, such as the report by Echeverria (2002) about black corals, the presence of a new species of coral in eastern Brazil, the species *Scolymia cubensis*, by Neves et al. (2006), and a new article that reported the fish fauna from the Parcel de Manoel Luis State Park (Moura et al. 2009). By this time, pioneering studies using the molecular technique for genetic studies promoted an advance on these studies for the taxonomic identification of reef organisms, such as the paper by Bernandi et al. (2000) about the parrotfish genus *Sparisoma*, the one by Costa et al. (2008) regarding the coral *Siderastrea stellata*, and by Neves et al. (2008) discussing the presence of another species of the genus *Siderastrea* in Brazil, *Siderastrea radians*. Nunes et al. (2008) produced a paper that re-evaluated the systematics of the Brazilian endemic corals, proposing the change in the taxonomic classification of the coral species *Favia leptophylla* to the now denominated *Mussismilia leptophylla*.

Brazilian researchers started focusing on the impacts of climate warming, with two pioneer geological studies, one using X-Ray technique for linking the ENSO (El Niño Southern Oscillation) to coral growth in the region of Abrolhos, using samples from the coral species *Favia leptophylla* (now named *Mussismilia leptophylla*) (Evangelista et al. 2007), and another using carbon and oxygen isotopes of several coral species (*Siderastrea stellata*, *Porites astreoides*, *Porites branneri*, *Mussismilia hispida*, *Madracis decactis*, and *Montastraea cavernosa*) in order to assess thermal stress in Atol das Rocas (Mayal et al. 2009).

Less than 10% of the articles published during the 2000–2009 decade have as their objectives, studies related with reef mapping, influence of environmental conditions on organisms, conservation/management of environmental resources and ecological modeling. A mapping study, built on advances of the paper by Testa and Bosence (1998), used remote sensing for assessing submerged reefs at the coast of Rio Grande do Norte state (Santos et al. 2007). The use of laboratory experiments was widened. One example focused on responses to environmental light conditions of hydrocorals (Oliveira et al. 2008) evaluated the potential for cultivating *Millepora alcicornis* to use it as a tool to restore degraded reef areas. In addition, Riul et al. (2008) tested *Lithothaminion* spp. net photosynthetic rates to observe the environmental impacts of the presence of fine sediment plumes during the rodolith dredge activity, over the primary production of this coralline algae. Management and conservation programs, such as the narrative review of the Brazilian Marine Protected Areas done by Prates (2006), a work related to management of fisheries from an extractive reserve in south Bahia by Moura et al. (2009), ecological modeling of biogeography (Joyeux et al. 2001), trophic structure of fish fauna (Freire et al. 2008), and fishing resources (Fredou et al. 2009). The geographical range of these studies varied from a region (East and Northeast) to the entire tropical Atlantic.

In the decade that started in 2010, the end of the survey that included the year of 2020, the month of January and twelve days of February 2021, when 457 more articles were published, one sees a broadening of the scope of reef research, represented by a great diversification of approaches. Except for the theoretical study, all the other types of approaches are found. Assessment/monitoring is the most frequent, occurring in 45% of the searched articles. The taxonomy/systematic approach also largely expanded, occupying the second place with 24% of the studies. Experiments are seen in 13% of the articles, while the other approaches occur in less than 10% of the cited articles. Among these, less than 10% articles, the studies of management/conservation and modeling approaches had a great growth. The management/conservation approach grew from four articles in the previous decade to 24 articles in this decade, and the modeling approach, found in only one paper in the previous decade, summed up to 20 published papers in the 2010 decade. Biotechnology/technology studies appeared for the first time in the published research (see Fig. 2.8a).

The increase and predominance of assessment/monitoring studies is reflected also in the different methods employed in this approach, like visual census (18%), biological sampling (16%) and photo/video (9%), whereas the number of paper doing taxonomy and systematics result in high frequency in the methods like organism morphology (12%) and molecular/genetic (10%).

Based on the keywords expressed in each article, "coral" and "fish" were cited in most of the references, but several other marine organisms that are also part of the reef system or are just reef residents were also considered. Examples of this include coralline algae, mollusks, bryozoans, echinoderms, foraminifera, rodoliths, zoanthids, sponges, ascidians, as well the bacteria community, vibrio, zooxanthellae, microbes, and the reef dwellers, sharks, whales, sea turtles and many others. Subjects such as fishery, environmental protection, impacts to the reef environment, ethnobiology, reef conservation, ocean acidification, reef nutrification, reef carbonate sediments, thermal stress, were also part of the studied themes.

There was, also, an expansion of works focusing the impacts of human activities on the reefs. Among the discussed themes, there were the effects of the input of sediment from inland to the nearshore reef areas due to inadequate use of coastal zones. The theme of high sedimentation and high turbidity characteristic of the reefs in this region of the Atlantic, particularly in Abrolhos, has been treated since the beginning of the research on these reefs, especially in the 1990s, (as stated in Leão and Ginsburg 1997; Leão et al. 2003 and Dutra et al. 2006). Due to the high turbidity of the reef environment that makes the conditions for reef development less than optimal, these reefs were classified as marginal reefs (Sugget et al. 2012). Two assessments regarding the effects of sedimentation on the Abrolhos reefs suggested that the terrigenous sediment carried from inland may have been the major factor controlling coral communities and reef development (Segal and Castro 2011; Silva et al. 2013). Laboratory experiments with coral species (endemic, such as *Mussismilia braziliensis*, *M. hispida,* and *Siderastrea stellata*, or cosmopolitan, such as *Montastraea cavernosa*) from these reefs revealed the tolerance of the corals to the stress of turbid waters caused by sedimentation (Sugget et al. 2012; Loiola et al. 2013). Later, the influence of turbidity and sedimentation on the reef's bioconstruction (Freitas et al. 2019) and on the coral community structure (Loiola et al. 2019) were clearly shown.

Still, some subjects related to the reef impacts are considered in the reef assessments done by Costa et al. (2014) and by Miranda et al. (2016), about the expansion of the invasion of the sun coral *Tubastraea spp* in Brazil, and the occurrence of phase shift in the reefs of Todos os Santos Bay, where the dominance of the zoanthid *Palythoa* cf. *variabilis* over coral cover was observed (Cruz et al. 2015).

This decade was also marked by an expansion of documented reef sites and regions or more thorough descriptions of reef occurrences such as: (a) the reefs along the coast of the state of Ceará, reporting the occurrence of bleaching in the coral species *Siderastraea stellata* and *Favia gravida* (Soares and Rabelo (2014); (b) the mesophotic reefs in the Vitoria-Trindade seamounts (Meirelles et al. 2015); (c) the fish fauna that occurs in the St. Pedro and St. Paulo islands (Rosa et al. 2016); (d) the description of the Amazonian shelf reef system (Moura et al. 2016), and (e) the discovery of coral reefs at the coast of the state of Espirito Santo, in the southern part of the Abrolhos bank (Mazzei et al. 2017).

Studies using the molecular technique for taxonomy of the reef organisms started to be more common and were applied for identifying the fish community from the Trindade and Martin Vaz islands (Pinheiro et al. 2010, Rocha et al. 2010), as well the microorganisms of Abrolhos reefs (Bruce et al. 2012).

Ecological modeling approach expanded, giving information about the functions and relationship among the reef components, as the articles published by Magris et al. (2015), that advocated aspects of habitat quality and connectivity of species to help achieve conservation objectives of a Marine Protected Area, and by D'Agostini et al. (2015), which evaluated the effectivity and quality of connectivity among Marine Protected Areas in Eastern Brazilian Shelf. Krug et al. (2013) created a Bayesian network model to predict bleaching in the East Brazil Shelf, and the potential invasion of the coral *Tubastraea coccinea* in Brazil was modeled by Riul et al. (2013). A physical modeling by Costa et al. (2016) based on a field experiment at the coast of the municipality of Recife, focused on the influence of the reef morphology on the transformation of waves as they propagate into the adjacent shoreline and their role in shaping the morphology of the coast of the region.

The experimental studies increased and addressed mainly problems related to effects of global changes, either in the field and in the laboratory. On the calcification pattern and usefulness of corals as climate/environmental conditions proxy described the timing and rhythm of the formation of coral density bands of the species *Mussissmilia braziliensis* (Kikuchi et al. 2013). A relatively high resilience of calcified macroalgae and coralline algae, measured by its photosynthesis, was shown by Scherner et al. (2016). On the other hand, Hill et al. (2019) showed that the physiology, biochemistry and ultrastructure on the free-living form of *Symbiodinium* extracted from the endemic coral *Mussismilia braziliensis* was affected by acidification.

Among the articles that applied the management/conservation approach, new initiatives for reef management were presented: Rangely et al. (2010) described the strategies for management of artisanal fishermen in Alagoas state; Schiavetti et al. (2013) claimed for management and conservation of the Brazilian Marine Protected Areas, and Magris et al. (2017) presented a general plan for the Brazilian reef conservation. These articles showed that the new strategies for protection of marine areas had to focus on the expansion of the total protected area, once Brazil had not reached the minimum established of 10% protection of marine areas, as recommended by the Aichi target 11. They also claimed the need to engage a wider spectrum of stakeholders among the civil society in the discussion and in the implementation of the conservation efforts in Brazil.

Within the papers of this last decade, the biotechnology approach is better represented, among others, by the paper published by Sousa et al. (2015) about the extraction and isolation of secondary metabolites (amphidinolides, potentially anticancer agents) produced by the symbiosis of the octocoral *Stragulum bicolor* and the dinoflagellates that thrive in its tissue.

Among the 30 papers published during the years of 2020–2021 we recorded two articles still using the assessment approach, one regarding the artisanal fishing activity and its implication for reef conservation (Barbosa-Filho et al. 2020), and another describing the diversity patterns of the Brazilian reef fish fauna (Araújo et al. 2020). Two others articles used the modeling approach, one about the paleoclimatic distribution and phylogeography of the coral *Mussismilia braziliensis* that was done by Menezes et al. (2020), and a satellite model regarding the possible contamination of Abrolhos reefs by the dam tailings from the Samarco mining disaster in Mariana municipality through the mouth of Doce River (Coimbra et al. 2020; Freitas and Silva 2019). Yet in 2020, Mies et al. (2020) made a meta-analysis review regarding the Brazilian coral reefs, considering them as a warming refugium for reef coral fauna. The last paper of the studied period was published by Ledouc et al. (2021), which assessed the impacts of elevated sound pressure, produced during the Brazilian Carnival parade, in Salvador City, on the reef fish behavior, such as reduced feeding activity and on the flight-initiation distance under elevated noise levels.

2.3 Spatial Distribution of Studied Reefs

The published research referred to works done between the Amazon region and the coast of the Espirito Santo State, but we also included the region of the "Coral Oasis of Cabo Frio" in the state of Rio de Janeiro (see Arraial do Cabo, Fig. 2.1). We divided the extended area in ecoregions according to the classification suggested by Spalding et al. (2007) and revised in Chap. 1 (Fig. 2.1 and 2.10). They comprise the following ecoregions: the coral systems located at the eastern coast of the state of Amazon and the coast of Maranhão State (AMZ); the north part of the Northeastern

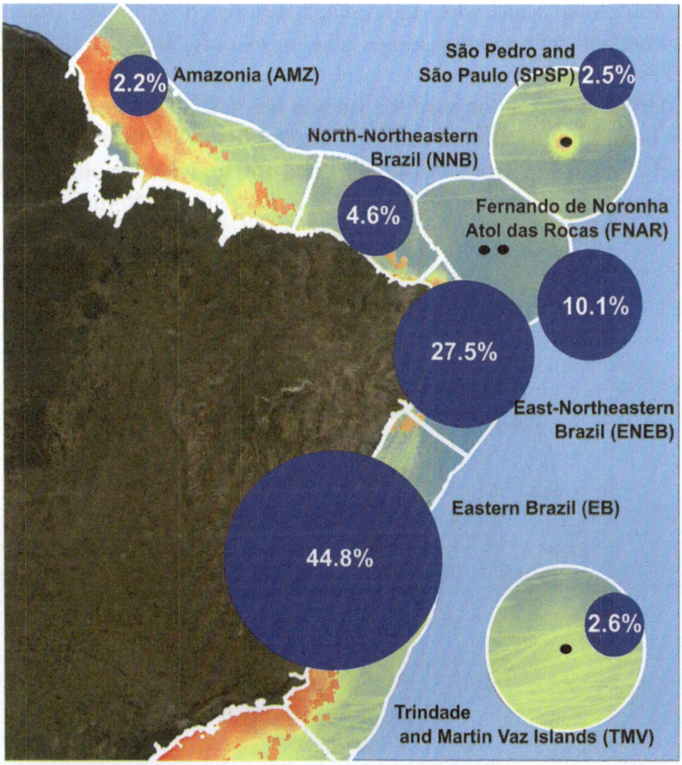

Fig. 2.10 Percentages of the peer-reviewed publications (n = 755) about Brazilian coral reefs between 1832 and 2021 (February 12th) within each of the Brazilian ecoregions (AMZ Amazonia ecoregion, SPSP São Pedro, São Paulo Archipelago, FNAR Fernando de Noronha and Atol das Rocas, NNEB North Northeastern ecoregion, ENEB East Northeastern ecoregion, EB Eastern Brazil ecoregion, TMV Trindade and Martin Vaz islands)

Region (NNEB) that include the coast of the state of Ceará and the northern part of the coast of the state of Rio Grande do Norte; the Fernando de Noronha Archipelago and Atol das Rocas (FNAR); the São Pedro and São Paulo Archipelago (SPSP); the east part of the Northeastern Region (ENEB) that include the east coast of the state of Rio Grande do Norte and the coast of the states of Paraiba, Pernambuco, Alagoas and part of Sergipe until the São Francisco River mouth; the Eastern Brazil Region (EB) extending from the southern part of the coast of Sergipe State, the entire coast of the state of Bahia and the coast of Espírito Santo state, and finally the Trindade and Martin Vaz islands (TMV). Most of the studied reefs are in the east part of the Northeastern region (27.5%) and in the Eastern region (44.8%) (Fig. 2.10).

Additionally, regarding the spatial scale of the studies, those conducted in one specific reef site were defined as local scale and dominated among all reviewed papers (67.9%). The ones classified as regional scale, e.g., a group of reefs located inside one Brazilian ecoregion, reached 12.9%. The national scale was considered when the study was done in reefs pertaining two or more Brazilian ecoregions (16.2%). The global scale, when the research compared/comprised Brazilian reefs and areas outside this regional range, add up to only 3.1% of all studied reefs (Fig. 2.11).

Considering the depth range of the studied reefs, 85.7% were done on the shallow reefs bordering the coastal zones, in depths shallower than 30 m. The mesophotic reefs, situated in depths between 31 and 150 m, were the object of 11.2% of the published studies, most of which located in Fernando de Noronha and Abrolhos

Fig. 2.11 Percentages of peer-reviewed publications (n = 755) about Brazilian coral reefs between 1832 and 2021 (February 12th) within classes of spatial scale (Local scale = reefs located inside a specific reef site; Regional scale = reefs located inside one Brazilian ecoregion; National scale = reefs located in two or more ecoregions; Global scale = International studies comparing Brazilian reefs with abroad reef areas)

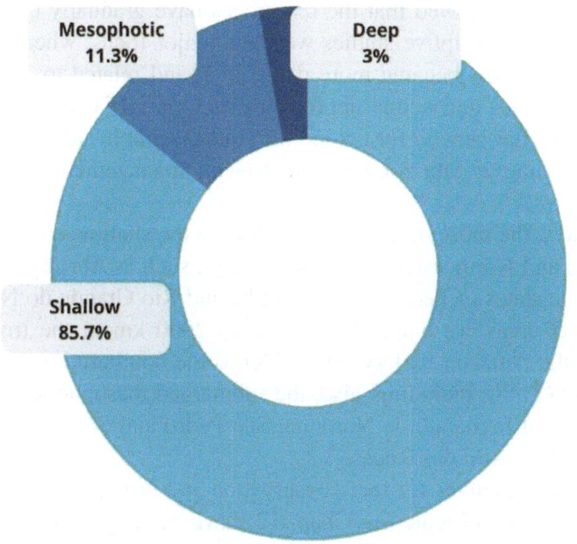

Fig. 2.12 Percentages of peer-reviewed publications (n = 792) about Brazilian coral reefs between 1832 and 2021 (February 12th) according to their depth range (Shallow reefs = depth less than 30 m; Mesophotic reefs = depths between 31 and 150 m; Deep reefs = depths greater than 150 m). N in this figure is greater than the total number of studied papers (n = 755) because different depth categories might be studied in a single paper

areas. The remaining 3% of papers refer to reefs located in depths greater than 150 m (Fig. 2.12).

Concerning the categories of geological time scales to which the reefs belong, most of the publications (circa 95%) refer to modern reefs, and only 5% of the studies reported to Holocene reef structures or sections.

2.4 Final Remarks

This chapter presents a systematic review of the literature about the coral reefs from the tropical coast of Brazil, describing the spatial extension and evolution of the themes, approaches, methods and techniques applied in the papers, over the studied period of time.

It is noteworthy to see that the great growth and diversification of the studies on Brazilian coral reefs occurred with the arrival of the second millennium and it increased markedly in the second decade of the twenty-first century. It is due to a new generation of researchers. The scientific production attained over 700 publications that dealt with several themes mostly related with the study of the modern shallow Brazilian coral reefs, that are presented in the next chapters of this book.

The main results showed that the researches have gradually changed from the early years, when descriptive studies were the major focus, whereas during more recent times the works became more diversified, and related to aspects that used modern technologies and actual approaches such as reef ecological and satellite modeling, molecular genetic for taxonomic studies, geophysics techniques for reef mapping, quantitative data for reef assessment, management and conservation, and so on.

In early days, the most studied reef areas were the shallow coastal reefs located in the Eastern and Northeastern regions of Brazil, such as Abrolhos and reefs from the coast of the states of Pernambuco, Paraíba and Rio Grande do Norte, whereas, nowadays, reef areas are mapped along about 2000 km of the tropical coast of Brazil, from the Amazon reef system (2°N) to the southern coast of the state of Espirito Santo (21°S), including, also, the submerged mesophotic reefs around the oceanic islands of Fernando de Noronha, São Pedro and São Paulo, Trindade and Martin Vaz and the Atol das Rocas.

Management initiatives for reef conservation and protection (an analysis of the present initiatives is presented in Chap. 12) have been growing upon new action towards acquiring more knowledge on the effects of global impacts on coral reefs, related to ocean warming and particularly coral bleaching events, as well as the effects of human activities (detailed in Chap. 10). Brazilian reefs protected areas are distributed along the entire region of reef occurrence, including the oceanic islands, with different management categories, federal, state and municipality. The involvement of coastal populations on evaluation of the existing MPAs, on the one hand, and the planning, designing, creating, and implementing new MPAs, on the other hand, in a co-production perspective is a process still to be pursued.

The generated dataset and information obtained through this work offered an objective picture of the evolution of the Brazilian coral reef systems studies. This bibliometric analysis was produced to comprise and analyze a comprehensive amount of information that will be beneficial for seeking new research paths. They will extend our knowledge contributing, for example, to compare the Brazilian coral reefs systems with counterpart all over the world. And these specific multidisciplinary and collaborative approaches involving scientists, local communities and policymakers are required. And they can include: (a) to continue baseline surveys for stablishing a benchmark of their current status and identifying more areas of concern; (b) to take an integrated approach to research, combining ecological, physical and social sciences to understand the interrelationships among these factors; (c) to engaging with local communities for understanding the human dimension of coral reefs management and conservation including efforts through participatory monitoring, ecotourism and sustainable fishing practices; and (d) to engaging with policymakers and stakeholders for developing policies that are grounded in scientific evidences and are socially and economically feasible. These approaches are essential for broadening our knowledge of the coral reefs of Brazil and developing effective strategies for their conservation and management.

Aknowledgements ZMAN and RKPK benefit research productivity scholarships provided by the Brazilian National Council for Scientific and Technological Development (CNPq); ZMANL, MDMO, AVFJr and RKPK are researchers participants of the National Institute of Science and Technology of Tropical Marine Environment (INCT AmbTropic—CNPq Grant 465,634/2014-1); ZMANL, MDMO and RKPK integrate the ReBentos (The Brazilian Network for Study of Marine Benthic Organisms).

References

Amaral FMD (1992) Sobre *Favia leptophylla* Verrill 1868 (Cnidaria, Scleractinia). Iheringia 3:117–118

Araujo TMF, Leão ZMAN, Lima OAL (1984) Evolução do recife de coral da ilha de Itaparica determinada a partir de dados geológicos e geofísicos Anais Congr. Bras Geol 33:159–169

Araújo ME, Mattos FMGD, Melo FPLD, Chaves LDCT, Feitosa, CV, Lippi DLF, Hackradt FC, Hackradt CW, Nunes JLS, Leão ZMAN, Kikuchi RKP, Ferreira Junior AV, Pereira PHC, Macedo CHR (2020) Diversity patterns of reef fish along the Brazilian tropical coast Mar Environ Res https://doi.org/10.1016/j.marevres.2020.105038

Azevedo AMR, Leão ZMAN, Kikuchi RKP, Azevedo AEG (1992) Composição isotópica do carbono e oxigênio em corais endêmicos do Brasil. In: Anais I Simpósio Regional de Geologia, Bahia – Sergipe, pp 11–14

Barbosa-Filho MLV, Seminara CI, Tavares DC, Siciliano S, Hauser-Davis RA, da Silva MJ (2020) Artisanal fisher perceptions on ghost nets in a tropical South Atlantic marine biodiversity hotspot: challenges to traditional fishing culture and implications for conservation strategies: ghost net impacts: artisanal fisher perceptions and challenges to fishing culture. Ocean Coast Manag. https://doi.org/10.1016/j.ocecoaman.2020/105189

Belém JC, Castro CB, Rohlfs C (1982) Notas sobre *Solanderia gracilis* Duchassaing & Michelin,1846 do Parcel de Abrolhos, Bahia, Brasil. Primeira ocorrência de Solanderidae (Cnidaria, Hidrozoa) no litoral brasileiro. Anais Acad Bras Cienc 54:585–588

Bernardi G, Robertson DR, Clifton KE, Azzurro E (2000) Molecular systematics, zoogeography, and evolutionary ecology of the Atlantic parrotfish genus *Sparisoma*. Mol Phylogenet Evol 15:292–300

Bernat M, Martin L, Bittencourt ACSP, Vilas Boas GS (1983) Io-U dates of a coral formation from the last interglacial age on the Brazilian coast. Use of Thorium229 as a tracer. C R Seances Acad Sci 296:197–200

Boschma H (1961) Notes on *Millepora braziliensis*, Verrill proceed. Royal Acad Sci Amsterdam Ser C 64:292–296

Boschma H (1962) On milleporine corals from Brazil proceed. Royal Acad Sci Amsterdam Ser C 65:302–312

Branner JC (1904) The stone reefs of Brazil, their geological and geographical relations with a chapter on the coral reefs. Bull Mus Comp Zool Harvard College Cambridge 44:207–275

Bruce T, Meirelles PM, Garcia G, Paranhos R, Rezende CE, Moura RL, Filho RF, Coni EOC, Vasconcelos AT, Filho G, Hatay M, Schmieder R, Edwards R, Dinsdale E, Thompson FL (2012) Abrolhos bank reef health evaluated by means of water quality, microbial diversity, benthic cover, and fish biomass data. PLoS One. https://doi.org/10.1371/journal.pone.0036687

Castro CB (1989) A new species of *Plexaurella* Valenciennes 1855 (Coelenterata, Octocorallia) from the Abrolhos reefs, Bahia, Brasil. Rev Bras Biol 49:597–603

Castro CB (1990) A new species of *Heterogorgia* Verrill, 1968 (Coelenterata, Octocorallia) from Brazil, with comments on the type species of the genus. Bull Mar Sci 47:411–420

Castro CB, Pires DO (1999) A bleaching event on a Brazilian coral reef. Rev Bras Ocean 47:87–90

Castro CB, Sechinn CA (1981) Abrolhos, santuário ecológico na costa da. Bahia Revista Geográfica Univ 81:26–37
Castro CB, Sechinn CA (1982) Um Parque Nacional Marinho para Abrolhos Ciência Hoje 1:38–43
Castro CB, Segal B (2001) The Itacolomis: large and unexplored reefs at the arrival point of the first Europeans in Brazil. Coral Reefs 20:18
Coimbra KTO, Alcântara E, Souza Filho CR (2020) Possible contamination of the Abrolhos reefs by Fundão dam tailings, Brazil—new constraints based on satellite data. Sci Total Environ. https://doi.org/10.1016/scitotenv.2020.138101
Costa CF, Sassi R, Amaral FMD (2001) Branqueamento em *Siderastraea stellata* (Cnidaria, Scleractinia) da Praia de Gaibu, Pernambuco, Brasil. Revista Nordestina Biol 15:15–22
Costa CF, Sassi R, Gorlach-Lira K (2008) Zooxanthellae genotypes in the coral *Siderastrea stellata* from coastal reefs in northeastern Brazil. J Exp Mar Biol Ecol 367:149–152
Costa TJF, Pinheiro HT, Teixeira JB, Mazzei EFBL, Hora MSC, Joyeux JC, Carvalho-Filho A, Amado-Filho G, Sampaio CLS, Rocha LA (2014) Expansion of an invasive coral species over Abrolhos Bank, Southwestern Atlantic. Mar Poll Bull 85:252–253. https://doi.org/10.1016/j.marpolbul.06.002
Costa MBSF, Araújo M, Araújo TCM, Siegle E (2016) Influence of reef geometry on wave attenuation on a Brazilian coral reef. Geomorphology 253:318–327
Coura MF (1994) Contribuição ao Plano de Manejo do Parque Estadual Marinho do Parcel de Manuel Luiz, Ma. Brasil. Departamento de Geociências Universidade Federal do Maranhão, 55p
Cruz ICS, Loiola M, Albuquerque T, Reis R, Nunes ACCJ, Reimer JD, Mizuyama M, Kikuchi RKP, Creed J (2015) Effect of phase shift from corals to zoantharia on reef fish assemblages. PLoS One. https://doi.org/10.1371/journal.pone.0139601
D'Agostini A, Gherardi DFM, Pezzi LP (2015) Connectivity of marine protected areas and its relation with total kinetic energy. PLoS One 10(10):e0139601. https://doi.org/10.1371/journal.pone.0139601
Darwin CR (1962) The structure and distribution of coral reefs. University of California Press, Berkeley/Los Angeles, 214p. Reprinted from Geological Observations on Coral reefs, volcanic Islands and South America, Elder and Company London 1851
Duarte PJ (1932) O Atol das Rocas. Arquivos de Pesquisas Agronômicas 1:61–70
Dutra LXC, Kikuchi RKP, Leão ZMAN (2006) Effects of sediment accumulation on coral reefs from Abrolhos, Bahia. Braz J Coastal Res 39:633–638
Echeverria CA (2002) Black corals (Cnidaria: Anthozoa: Antipatharia): first records and a new species from the Brazilian coast. Revista de Biologia Tropical 50:1067–1077
Evangelista H, Godiva D, Sifeddine A, Leão ZMAN, Rigozo NR, Segal B, Ambrizzi T, Kampel M, Kikuchi RKP, Cornec FL (2007) Evidences linking ENSO and coral growth in the Southwestern-South Atlantic. Clim Dyn 29:869–880
Ferreira VP, Maÿal EM, Sial NA, Exner MA (1998) Isotopic Variations (δ13C and δ18O) in *Siderastrea stellata* (Cnidaria-Anthozoa), Itamaracá Island, State of Pernambuco, Northeastern Brazil. Anais Acad Bras Cienc 70:849–855
Ferreira BP, Maida M, Castro CB, Pires DO, D'Amico TM, Prates APL, Marx D (2006) The status of coral reefs in Brazil. Proceeding of the 10th internatioal coral reef symposium 1:1011–1015
Fitzroy RN (1832) Extract of a letter from Cap. Fitzroy of H.M. Sloop Beagle on the subject of the Abrolhos Bank. Royal Geological Society, London, pp 315–316
Francini-Filho RB, Moura RL, Les K, Kikuchi RKP, Leão ZMAN, Thompson FL, Reis RM (2008) Diseases leading to accelerated decline of reef corals in the largest South Atlantic reef complex (Abrolhos Bank, eastern Brazil). Mar Poll Bull 56:1008–1014
Frédou T, Ferreira BP, Letourneur Y (2009) Assessing the stocks of the primary snappers caught in Northeastern Brazilian reef systems. 1: traditional modelling approaches. Fish Res 99:90–96
Freire KMF, Christensen V, Pauly D (2008) Description of the East Brazil large marine ecosystem using a trophic model. Sci Mar 72:477–491
Freitas CM, Silva MA (2019) Acidentes de trabalho que se tornam desastres: os casos dos rompimentos em barragens de mineração no Brasil. Rev Bras Med Trab 17:21–29

Freitas LM, Oliveira MDM, Leão ZMAN, Kikuchi RKP (2019) Effects of turbidity and depth on the bioconstruction of the Abrolhos reefs. Coral Reefs 38:1–13

Garzon-Ferreira J, Cortés J, Croquer A, Guzman H, Leão ZMAN, Rodriguez-Ramirez A (2002) Status of coral reefs in southern tropical America in 2000–2002: Brazil, Colombia, Costa Rica, Panama and Venezuela. Status of Coral Reefs of the World 2002:343–360

Gherardi DFM, Bosence DWJ (1999) Modeling of the ecological sucession of encrusting organisms in recent coralline-algal frameworks from Atol das Rocas. Brazil Palaios 14:145–158

Gherardi DFM, Bosence DWJ (2001) Composition and community structure of the coralline algal reefs from Atol das Rocas, South Atlantic, Brazil. Coral Reefs 19:205–219

Gonchorosky J, Sales G, Oliveira MCP (1989a) Environmental impact of the stranding of a freighter ship in the Parque Nacional Marinho de Abrolhos, Bahia, Brazil. Costline of Brazil. American Society of Civil Engeneering, p 304

Gonchorosky J, Sales G, Belém MJC, Castro CB (1989b) Importance, establishment and management plan of the Parque Nacional Marinho de Abrolhos. Coastlines of Brazil, pp 185–194

Hartt CF (1868) A naturalist in Brazil. Am Nat 2:1–13

Hartt CF (1869) The cruise of the Abrolhos. Am Nat 2:86–93

Hartt CF (1870) Geology and physical geography of Brazil Boston, fields. Osgood and Co., Boston, p 620

Hill LJ, Paradas WC, Willemes MJ, Pereira MG, Salomon OS, Mariath R, Moura RL, Atella GC, Farina M, Amado-Filho GM, Salgado LT (2019) Acidification-induced cellular changes in *Symbiodinium* isolated from *Mussismilia braziliensis*. PLoS One 14:e0220130. https://doi.org/10.1371/journal.pone.0220130

Jindrich V (1983) Structure and diagenesis of recent algal-foraminifer reefs, Fernando de Noronha, Brazil. J Sediment Petrol 53:449–459

Joly AB, Oliveira Filho EC, Narchi W (1969) Projeto de criação de um Parque Nacional Marinho na região de Abrolhos, Bahia. Anais Acad Bras Cienc 41:247–251

Joyeux JC, Floeter SR, Ferreira CEL, Gasparini JL (2001) Biogeography of tropical reef fishes: the South Atlantic puzzle. J Biogeogr 28:831–841

Kempf M, Laborel JL (1968) Formations de vermetset d'algues calcaires sur les côtes du Brésil. Recueil des Travaux de la Station Marine d1Encoume 59:325–435

Kikuchi RKP, Leão ZMAN (1997) Rocas (Southwestern Equatorial Atlantic, Brazil): an atoll built primarily by coralline algae. Proceedings of the 8th international Coral Reef Symposium Panamá 1:731–736

Kikuchi RKP, Leão ZMAN (1998) The effects of Holocene Sea level fluctuation on reef development and coral community structure, Northern Bahia, Brazil. Anais Acad Bras Cienc 70:159–171

Kikuchi RKP, Leão ZMAN, Testa V, Dutra LXC, Spanó S (2003a) Rapid assessment of the Abrolhos reefs, eastern Brazil (part 1: stony corals and algae). Atoll Res Bull 496:172–187

Kikuchi RKP, Leão ZMAN, Sampaio CLS, Telles MD (2003b) Rapid assessment of the Abrolhos reefs, eastern Brazil (part 2: fish communities). Atoll Res Bull 496:188–203

Kikuchi RKP, Oliveira MDM, Leão ZMAN (2013) Density banding pattern of the south western Atlantic coral *Mussismilia braziliensis*. J Exp Mar Biol Ecol 449:207–214

Krug LA, Gherardi DFM, Stech JL, Leão ZMAN, Kikuchi RKP, Suggett DJ (2013) The construction of causal networks to estimate coral bleaching intensity. Envir Modell Softw 42:157–167

Laborel JL (1967) A revised list of Brazilian scleractinians corals and descriptions of a new species. Postilla Mus Nat Hist Yale Univ 107:1–14

Laborel JL (1969) Madreporaires et hydrocoralliaires recifaux des côtes brésiliennes. Systematique, ecologie, repartition verticale et geographie. Ann Inst Oceanogr Paris 47:171–229

Laborel JL (1970) Les peuplements de madreporaires des côtes tropicales du Brésil. Ann Univ d'Abidjan, Ser E II, Fasc. 3:260p

Laborel-Deguen F, Castro CB, Nunes FLD, Pires DO (2019) Recifes Brasileiros: O legado de Laborel. Museu Nacional., Séries Livros 64 Rio de Janeir0:376p

Leão ZMAN (1983) Abrolhos: o refúgio pleistocênico de uma fauna terciária de corais Rev. Cienc Terra 8:22–24

Leão ZMAN (1996) The coral reefs of Bahia: morphology, distribution and the major environment impacts. Anais Acad Bras Cienc 68:439–452

Leão ZMAN, Ginsburg RN (1997) Living reefs surrounded by siliciclastic sediments: the Abrolhos coastal reefs, Bahia, Brazil. Proceedings of the 8th international Coral Reef Symposium 2:1767–1772

Leão ZMAN, Bittencourt ACSP, Dominguez JML, Nolasco MC, Martin L (1985) The effects of Holocene Sea level fluctuations on the morphology of the Brazilian coral reefs. Rev Bras Geoc 15:154–157

Leão ZMAN, Araujo TMF, Nolasco MC (1988) The coral reefs off the coast of eastern Brazil. Proceedings of the 6th international coral reef symposium Townsville 3: 339–347

Leão ZMAN, Telles MD, Sforza R, Bulhões HA, Kikuchi RKP (1993) Impacts of tourism development on the coral reefs of the Abrolhos area, Brazil. In: Proc Colloquium and Forum on Coral Reefs, pp 254–260

Leão ZMAN, Kikuchi RKP, Testa V (2003) Corals and coral reefs of Brazil, In Cortes J (Ed.) Latin American coral reefs Elsevier Science B.V., Amsterdam 9–52

Leão ZMAN, Kikuchi RKP, Oliveira MDM (2008) Branqueamento de corais nos recifes da Bahia e sua relação com eventos de anomalias térmicas nas águas superficiais do oceano Biota. Neotropica 8:69–82

Ledouc AOHCL, Nunes JACC, Araujo CB, Quadros ALS, Barros F, Oliveira HHQ, Simões CRMA, Winnandy GSM, Slabbekoorn H (2021) Land-based noise pollution impairs reef fish behavior: a case study with a Brazilian carnival. Biol Conserv 253:108910. https://doi.org/10.1016/j.biocon.2020.108910

Loiola M, Oliveira MDM, Kikuchi RKP (2013) Tolerance of Brazilian brain coral *Mussismilia braziliensis* to sediment and organic matter inputs. Mar Pollut Bull 7:55–62

Loiola M, Cruz ICS, Lisboa DS, Leão ZMAN, Oliveira MDM, Kikuchi RKP (2019) Structure of marginal coral reef assemblages under different turbidity regime. Mar Env Research 147:138–148

Mabesoone JM (1964) Origin and age of the sandstone reefs of Pernambuco (North-Eastern Brazil). J Sedim Petrol 34:715–726

Magris RA, Treml EA, Pressey RL, Weeks R (2015) Integrating multiple species connectivity and habitat quality into conservation planning for coral reefs. Ecography 38:001–016

Magris RA, Pressey RL, Mills M, Vila-Nova DA, Floeter S (2017) Integrated conservation planning for coral reefs: designing conservation zones for multiple conservation objectives in spatial prioritization. Global Ecol Conserv 11:53–68

Maida M, Ferreira BP (1997) Coral reefs of Brazil: an overwiew Proceedings of the 8th international coral reef symposium 1:263–274

Maida M, Ferreira BP (2003) Área de Proteção Ambiental Costa dos Corais. In: Prates APL (ed) Atlas dos recifes de coral nas unidades de conservação brasileiras, Brasília. Ministério do Meio Ambiente, pp 86–89

Maida M, Ferreira BP, Bellini C (1995) Avaliação preliminar do recife da Baía do Sueste, Fernando de Noronha, com ênfase nos corais escleractíneos. Bol Tec Cient CEPENE 3:37–47

Marques AC, Castro CB (1995) *Muricea* (Cnidaria, Octocorallia) from Brazil, with description of a new specie. Bull Mar Sci 56:161–172

Martin L, Flexor JM, Blitzkow D, Suguio K (1985) Geoid change indications along the Brazilian coast during the last 7,000 years. Proceedings of the 5th international Coral Reef Symposium 3:85–90

Maÿal EM (1986) Mercado de Corais. Rev Ciência Hoje 5:43

Maÿal EM, Sial NA, Ferreira VP, Fisner M, Pinheiro BR (2009) Thermal stress assessment using carbon and oxygen isotopes from Scleractinia, Rocas Atoll, northeastern Brazil. Int Geol Rev 51:166–188

Mazzei EF, Bertoncini AA, Pinheiro HT, Machado LF, Vilar CC, Guabiroba HC, Costa TJF, Bueno LS, Santos LN, Francini-Filho RB, Hostim-Silva M, Joyeux JC (2017) Newly discovered reefs in the southern Abrolhos Bank, Brazil: anthropogenic impacts and urgent conservation needs. Mar Poll Bull 1124:123–133

Meirelles PM, Amado-Filho GM, Pereira-Filho GH, Pinheiro HT, Moura RL, Joyeux JC, Mazzei EF, Bastos AC, Edwards RA, Dinsdale E, Paranhos R, Santos EO, IidaT GK, Nakamura S, Sawabe T, Rezende CE, Gadelha LMR, Francini-Filho RB, Thompson C, Thompson FL (2015) Baseline assessment of Mesophotic reefs of the Vitória-Trindade seamount chain based on water quality, microbial diversity, benthic cover and fish biomass data. PLoS One. https://doi.org/10.1371/journal.pone.0130084

Menezes N, Sobral-Souza T, Silva M, Solferini VN (2020) Paleoclimatic distribution and phylogeography of *Mussismilia braziliensis* (Anthozoa, Scleractinia), an endemic Brazilian reef coral. Mar Biodivers 50:1–12

Mies M, Francini-Filho RB, Zilberberg C, Garrido AG, Longo GO, Laurentino E, Güth AZ, Sumida PYG, Banha TNS (2020) South Atlantic coral reefs are major global warming refugia and less susceptible to bleaching. Front Mar Sci 7. https://doi.org/10.3389/fmars.2020.00514

Miranda RJ, Costa Y, Lorders FL, Nunes JACC, Barros F (2016) New records of the alien cup-corals (*Tubastraea* spp.) within estuarine and reef systems in Todos os Santos Bay, Southwestern Atlantic. Mar Biodivers Rec 35:10.1186/s41200-016-0053-2

Moura RL, Minte-Vera CV, Curado IB, Francini-Filho RB, Rodrigues HDCL, Dutra GF, Alves DC, Souto FJB (2009) Challenges and prospects of fisheries co-management under a marine extractive reserve framework in northeastern Brazil. Coast Manag 37:617–632

Moura RL, Amado-Filho G, Moraes FC, Brasileiro PS, Salomon PS, Mahiques M, Bastos SC, Almeida MG, Silva J, Araujo BS, Brito FP, Rangel TB, Oliveira J, Bahia R, Paranhos RP, Dias RJS, Siegle E, Figueiredo Jr, A, Pereira R, Leal C, Hajdu E, Gregor GB, Neuman-Leitão S, Yager PL, Francini-Filho R, Froes A, Campeão M, Silva B, Moreira AP, Oliveira L, Soares AC, Araujo L, Oliveira NL, Valle RAB, Thompson CC, Rezende CE, Thompson FL (2016) An extensive reef system at the Amazon River mouth. Sci Adv. https://doi.org/10.1126/sciadv.1501252

Neves EG, Johnsson R, Sampaio C, Pichon M (2006) The occurrence of *Scolymia cubensis* in Brazil: revising the problem of the Caribbean solitary mussids. Zootaxa 1366:45–54

Neves EG, Andrade SCS, Da Silveira FL, Solferini VN (2008) Genetic variation and population structuring in two brooding coral species (*Siderastrea stellata* and *Siderastrea radians*) from Brazil. Genetica 132:243–254

Nolasco MC, Leão ZMAN (1986) The carbonate buildups along the northern coast of the state of Bahia, Brazil. In: Rabassa A, J. (eds) Quaternary of South America and Antarctic Peninsula, vol 4. Balkema Publishing Company, Rotterdam, pp 159–190

Nunes F, Fukami H, Vollmer SV, Norris RD, Knowlton N (2008) Re-evaluation of the systematics of the endemic corals of Brazil by molecular data. Coral Reefs 27:423–432. https://doi.org/10.1007/s00338-007-0349-0

Oliveira MDM, Leão ZMAN, Kikuchi RKP (2008) Cultivo de *Millepora alcicornis* como ferramenta para restauração de manejo dos ecossistemas recifais do Nordeste do Brasil. Gestão Costeira Integrada 8:183–201

Oliveira-Filho EC, Ugadim Y (1976) A survey of the marine algae of Atol das Rocas (Brazil). Phycologia 15:41–44

Ottmann F (1960) Une hypotese sur l'origine des "arrecifes" du Nordeste brésilian. Compte Rendu Sommaire des Scéances de la Societé Geologique de France 7:175–176

Ottmann F (1963) L'Atol das Rocas das l'Atlantique Sud tropical. Revue de Géographie Physique et de Géologie Dinamique 2:101–107

Petuch EJ (1979) New gastropods from the Abrolhos Archipelago and reef complex, Brazil. Proc Biol Soc Washington 92:510–526

Pinheiro HT, Gasparini JL, Sazima I (2010) *Sparisoma rocha*, a new species of parrotfish (Actinopterygii: Labridae) from Trindade Island. South-western Atlantic Zootaxa 2493:59–65

Pires DO, Castro CB, Migotto A (1992) Benthic cnidaria of Fernando de Noronha archipelago. Bul Mus Nac Zool 354:1–21

Pitombo F, Ratto CC, Belém MJC (1988) Species diversity and zonation pattern of hermatypic corals at two fringing reefs of Abrolhos archipelago, Brazil. Proceedings of the 6th international coral reef symposium 2:817–820

Prates APL (2006) Unidades de Conservação, I: Prates APL (Org.) IBAMA – Atlas dos Recifes de Corais nas Unidades de Conservação Brasileiras, 2ª. Edição ampliada, Brasília, MMA, SBF 232 p

Rangely J, Fabré NN, Tiburtino C, Batista VS (2010) Strategies for artisanal fishing in Alagoas coast (Brazil). Bol Inst Pesca 36:263–275

Rathbun R (1876) Extinct coral reef at Bahia. Brazil Amer Naturalist 10:439–440

Rathbun R (1878a) Notes on the coral reefs on the Island of Itaparica, Bahia and of Parayba do Norte. Proc Boston Soc Nat Hist XX:34–41

Rathbun R (1878b) O recife de coral de Mar Grande. Arch Mus Nac Rio de Janeiro III:174

Rathbun R (1879) Brazilian corals and coral reefs. Am Nat 13:535–551

Rios EC (1979) Moluscos para o Atol das Rocas. V Encontro de Malacologistas Brasileiros. Fundação Zoobotânica do Rio Grande do Sul, Mossoró, p 75

Rios EC, Barcellos LP (1980) New occurrences of mollusks to Abrolhos Is., Bahia. Bull Amer Malac Union:109–112

Riul P, Targino CH, Farias JN, Vischer PT, Horta PA (2008) Decrease in *Lithothamnion* sp. (Rhodophyta) primary production due to the deposition of a thin sediment layer. J Mar Biol Assoc UK 88:17–19

Riul P, Targino CH, Junior LAC, Creed JC, Horta PA, Costa GC (2013) Invasive potential of the coral *Tubastraea coccinea* in the Southwest Atlantic. Mar Ecol Prog Ser 480:73–81

Rocha LA (2003) Patterns of distribution and processes of speciation in Brazilian reef fishes. J Biogeogr 30:1161–1171

Rocha LA, Pinheiro HT, Gasparini JL (2010) Description of *Halichoeres rubrovirens*, a new species of wrasse (labridae: perciformes) from the Trindade and Martin Vaz Island group, southeastern Brazil, with a preliminary mtDNA molecular phylogeny of new world *Halichoeres*. Zootaxa 2422:22–30

Rosa MR, Alves AC, Medeiros DV, Coni EOC, Ferreira CM, Ferreira BP, Rosa R, Amado-Filho GM, Pereira-Filho GH, Moura RL, Thompson FL, Sumida PYG, Francini-Filho RB (2016) Mesophotic reef fish assemblages of the remote St. Peter and St. Paul's Archipelago, Mid-Atlantic Ridge, Brazil. Coral Reefs 35:113–123

Santos CLA, Vital H, Amaro VE, Kikuchi RKP (2007) Mapping of the submerged reefs in the coast of the Rio Grande do Norte, near Brazil. Macau to Maracajau Revista Brasileira de Geofisica 25:27–36

Sassi R, Kutner MBB, Moura GF (1988) Studies on the decomposition of drift seaweed from the northeast Brazilian coastal reefs. Hydrobiologia 57:187–192

Sazima I (1984) Similarities in feeding behaviour between some marine and freshwater fishes in two tropical communities. J Fish Biol 29:53–65

Scherner F, Pereira CM, Duarte G, Horta PH, Castro CB, Barufi JB, Pereira SMB (2016) Effects of ocean acidification and temperature increases on the photosynthesis of tropical reef calcified macroalgae. PLoS One 11:e0154844. https://doi.org/10.1371/journal.pone.0154844

Schiavetti A, Manz J, Santos C.Z, Magro TC, Pagani MI (2013) Marine protected areas in Brazil: an ecological approach regarding the large marine ecosystems. Ocean Coastal Manage 76:96–104

Secchin C (1986) Abrolhos, Parque Nacional Marinho. Edit. Cor/Ação, Rio de Janeiro, p 128

Segal BS, Castro CB (2011) Coral community structure and sedimentation at different distances from the coast of the Abrolhos Bank, Brazil. Braz J Oceanogr 59:119–129

Silva AS, Leão ZMAN, Kikuchi RKP, Costa AB, Souza JRB (2013) Sedimentation in the coastal reefs of Abrolhos over the last decades. Cont Shelf Res 70:159–167

Soares MO, Rabelo EF (2014) Primeiro registro de branqueamento e corais no litoral do Ceará (NE, Brasil): indicador das mudanças climáticas? Geociência 33:1–10

Sousa TS, Nuzzo G, Torres MCM, Lopes NP, Cutignano A, Jimenez PC, Santos EA, Gomes BA, Sardo A, Pessoa ODL, Costa-Lotufo LV, Fontana A (2015) Amphidinolide P from the Brazilian octocoral *Stragulum bicolor*. Rev Bras Farmacognosia 5:600–604

Spalding MD, Fox HE, Allen GR, Davidson N, Ferdana ZA, Finlayson M, Halpern BS, Jorge MA, Lombana AL, Lourie SA, Martin KD, McManus MA, Molnar J, Rechia CA, Robertson J (2007) Marine ecoregions of the world: a bioregionalization of coastal and shelf areas. Bioscience 57:573–583

Sugget DJ, Kikuchi RKP, Oliveira MDM (2012) Photobiology of corals from Brazil's near-shore marginal reefs of Abrolhos. Mar Biol 159:1461–1473. https://doi.org/10.1007/s00227-012-1925-6

Testa V, Bosence DWJ (1998) Carbonate-siliciclastic sedimentation on a high-energy, ocean facing, tropical ramp, NE Brazil. Carbonate Ramps 149:55–71

Vallaux C (1940) La formation atollienne de Rocas (Brésil). Bulletin de L'Institut Océanographique 37:1–8

Verrill AE (1868) Notes of the radiates in the Museum of Yale College, with descriptions of new genera and species. 4. Notes of the corals and echinoderms collected by Prof. C.F. Hartt at the Abrolhos reefs, province of Bahia. Brazil Connecticut Acad Arts Sci Transact 1:351–371

Verrill AE (1901) Variations and nomenclature of Bermudian, West Indian and Brazilian reef corals with notes on various Indo-Pacific corals. Connecticut Acad Arts Sci Transact XI:66–116

Verrill AE (1912) The gorgonians of the Brazilian coast. J Acad Nat Sci Philadelphia 2:373–404

Vicalvi MA, Costa MPA, Kowsmann (1978) Depressão de Abrolhos: uma paleo-laguna Holocênica na plataforma continental leste brasileira Bol. Tec Petrobrás 21:279–286

Young PS, Christoffersen ML (1984) Recent coral barnacles of the genus *Ceratoconcha* (Cirripedia: Pyrgomatidae) from Northeast Brazil. Bull Mar Sci 5:239–252

Chapter 3
Origins, Biogeography and Macroecology of the Southwestern Atlantic Reef Biodiversity

Hudson T. Pinheiro, Mariana G. Bender, Guilherme O. Longo, Melina F. Maxwell, Luiz A. Rocha, and Sergio R. Floeter

Abstract The origin of most reef biodiversity of the southwestern Atlantic Ocean dates back to the Eocene, when the center of diversity of modern reef organisms was in the Tethys Sea, a vast extent of tropical shallow marine habitats situated between the Atlantic and Indian oceans. Following the closure of the Tethys Sea, the Atlantic reef biodiversity became increasingly isolated from the world's centers of biodiversity (*i.e.* tropical Indo-Pacific). Moderate rates of origination and extinction shaped endemic lineages and a lower diversity of species compared with the Indo-Pacific. Within the Atlantic, the Brazilian Province is characterized by high endemism and by the presence of a secondary center of biodiversity, driven by a combination of isolation, distinct ecological conditions, and the overlap of tropical and subtropical species. Many evolutionary processes shape the current distribution of species along the coast and in offshore areas such as seamounts and oceanic islands. The taxonomic and functional structure of reef fishes and benthic organisms in the Brazilian Province are related to both historical events and environmental factors, and changes considerably along its latitudinal extent. Today, increased and disordered human development, associated with inadequate conservation measures, is quickly

H. T. Pinheiro (✉)
Center for Marine Biology, University of São Paulo, São Sebastião, São Paulo, Brazil

Department of Ichthyology, California Academy of Sciences, San Francisco, CA, USA

M. G. Bender · M. F. Maxwell
Marine Macroecology and Conservation Laboratory, Departamento de Ecologia e Evolução, Universidade Federal de Santa Maria, Santa Maria, Rio Grande do Sul, Brazil

G. O. Longo
Marine Ecology Laboratory, Departamento de Oceanografia e Limnologia, Universidade Federal do Rio Grande do Norte, Natal, Rio Grande do Norte, Brazil

L. A. Rocha
Department of Ichthyology, California Academy of Sciences, San Francisco, CA, USA

S. R. Floeter
Marine Macroecology and Biogeography Lab, Departamento de Ecologia e Zoologia, Universidade Federal de Santa Catarina, Florianópolis, Santa Catarina, Brazil

R. K. P. Kikuchi et al. (eds.), *Brazilian Coral Reefs*, Brazilian Marine Biodiversity,
https://doi.org/10.1007/978-3-031-59152-5_3

threatening millions of years of eco-evolutionary processes responsible for the origin of this unique biodiversity.

Keywords Brazilian Province · Reef fish · Corals · Conservation · Distribution · Evolution

3.1 Introduction

The biodiversity associated with reef environments has a long and turbulent evolutionary history marked by shifts in global geological and climatic events. In our current geological time, shallow coral reefs exhibit the highest levels of species diversity among all marine ecosystems (Roberts et al. 2002; Tittensor et al. 2010), and biodiversity in marginal and mesophotic reefs is mostly derived from shallow coral ecosystems (Bellwood and Wainwright 2002; Tornabene et al. 2016). Most modern reef fish families originated during the Eocene, 56 to 33.9 million years ago (mya) (Bellwood et al. 2017). During that time, the center of marine diversity was the Tethys Sea, a shallow water marine seaway connecting the Atlantic and Indian oceans, where Europe and the Mediterranean Sea are today (Bellwood and Wainwright 2002; Stanley 2003; Floeter et al. 2008). However, analyses comparing the fossil record and extant species reveal many differences between the composition of the Eocene fauna and modern fish and coral assemblages (Bellwood et al. 2017). Many lineages that were widespread in the past went extinct, while others diversified and evolved into Atlantic and Indo-Pacific endemics (Stanley 2003; Siqueira et al. 2019).

The Oligocene (33.9 to 23 mya) and Miocene (23 to 5.3 mya) were marked by the origin and initial diversification of reef fish genera and coral groups that dominate modern reef assemblages (Stanley 2003; Bellwood et al. 2017). During this period, Africa collided with Eurasia, and the Tethys Sea closed. The center of diversity thus migrated from the Tethys to the Indo-west Pacific Region, where it is located today, further from the Atlantic Ocean (Renema et al. 2008; Leprieur et al. 2016), with secondary coral diversity centers found in the Red Sea and in Northern Madagascar (Veron et al. 2015). During that time, innovations in feeding strategies and changes in the trophic structure of fish communities prompted more complex fish–reef interactions, and adaptations to new habitats had a pivotal role in shaping fish diversity (Bellwood et al. 2017; Siqueira et al. 2020).

During the past five million years, diversification rates, strongly influenced by sea-level driven vicariant events, peaked both for fishes and corals, resulting in phylogenetic signatures as evident as strong variation in fish colors (Bellwood et al. 2017; Pinheiro et al. 2017; Hemingson et al. 2018), and as cryptic as molecular differences in morphologically identical coral species (Nunes et al. 2008, 2011). Today, reefs are marked by an astonishing diversity of conspicuous fish and coral families, many of which have global distributions, such as Labridae, Pomacentridae, and Chaetodontidae for fish, and Faviidae, Poritiidae, and Caryophyllidae for corals. These numerous reef organisms generate and occupy many niches, delivering a great diversity of ecosystem functions and services that directly and indirectly benefit humankind (Bellwood et al. 2017).

3.2 Historical Biogeography of the Reef Biodiversity in the Atlantic Ocean

Scleractinian corals started flourishing and building reef-like structures in the Eocene, after surviving through the K/T mass extinction (66 mya), likely using several refugia, and a Paleocene characterized by low coral diversity, paucity of tropical reefs and little carbonate accretion (Stanley 2003). Some modern reef fish lineages also originated in the Eocene Tethys, including ancestors of Labridae and damselfishes (Cowman and Bellwood 2013). These lineages showed little diversification along the Oligocene, and with the Atlantic becoming isolated by the movements of tectonic plates and the migration of the center of diversity towards the Indo-Pacific, few lineages were added by migration (Cowman and Bellwood 2013). During that time, the tropical Eastern Pacific was connected with the Atlantic in the Caribbean region, but reduced exchange of lineages between these provinces is reported (Cowman and Bellwood 2013; Cowman et al. 2017).

In the Miocene, a high diversification of modern lineages was observed worldwide, including a higher exchange of fish lineages between the Atlantic and tropical Eastern Pacific (TEP). The Miocene was characterized by the uplift of the Isthmus of Suez, which created a hard-biogeographic barrier separating tropical faunas of the Red Sea and Mediterranean. During this time, the Atlantic reef fish fauna experienced a strong reduction in diversity and turnover of its functional richness compared to the ancestral Tethyan trait space (Siqueira et al. 2019). Some lineages that adapted and diversified in the Atlantic and TEP include the genera *Anisotremus*, *Haemulon*, *Halichoeres*, and *Holacanthus*, many of them diverging between the Atlantic and TEP lineages after the closure of the Isthmus of Panama (Bernardi et al. 2008; Floeter et al. 2008; Rocha et al. 2008). The biogeographic barriers from the Miocene are linked to the origin of the coral genus *Mussismilia*, which was previously widely distributed in the Atlantic and is now a relict paleo-endemic, restricted to the Brazilian coast (Nunes et al. 2008). After the closure of the Tethys Sea, lineages that invaded the Atlantic in the Miocene came from the Indian Ocean, crossing through the southern African coast, and from the TEP, (Cowman and Bellwood 2013).

Overall species richness, nonetheless, has remained lower in the Atlantic and TEP compared to the Indo-Pacific, especially due to reduced immigration potential and high faunal turnover led by increased rates of lineage extinction (Budd 2000). Relict reef fishes and higher genetic diversity of endemic corals found in oceanic islands, identified as sources to coastal populations, are evidence of lineage extinctions along the continental margin of the Southwestern Atlantic (Pinheiro et al.

2017; Peluso et al. 2018). The closure of the Isthmus of Panama during the Pliocene completely isolated the Atlantic from the tropical Eastern Pacific, promoting vicariant events that split lineages of many families, such as Chaetodontidae, Gobiesocidae, Haemulidae, Labridae, Lutjanidae, Pomacanthidae, Pomacentridae, among others (Lessios 2008). Still in the Pliocene, and during the Pleistocene, rare events of migration from the Indian Ocean via southern Africa were capable of colonizing Atlantic provinces. These events are constrained due to the increasing strength of the biogeographic barrier formed by the cold-water upwelling off southwestern Africa (Rocha et al. 2005a). These colonization events occurred with contributions of the Agulhas Leakage, where Agulhas Rings containing warm Indian Ocean waters enter the Atlantic and join the South Atlantic Gyre. This process is suggested to have contributed to the colonization of the Eastern Atlantic by lineages of *Chromis* and *Lethrinus* (Floeter et al. 2008), and possibly also *Prionurus* and *Plectorhinchus. Centropyge* and *Gnatholepis* lineages found in the Western Atlantic and Mid-Atlantic Ridge, seem to have invaded the Atlantic even more recently (Rocha et al. 2005a; Bowen et al. 2006). Maps presenting the main routes of colonization over time and possible evolutionary scenarios are provided in Fig. 3.1.

3.3 Biogeographic Barriers and the Origin of Southwestern Atlantic Endemics

The modern lineages that evolved in the Atlantic diversified between and within biogeographical provinces. Allopatric speciation, one of the most common drivers of evolution in the terrestrial realm, is also driving diversification between biogeographic provinces in the Atlantic. A classic example is provided by the Amazon River discharge of freshwater and sediments, the greatest in the world, which has its origin around 10 mya, corresponding with the uplift of the Andes. The Amazon River discharge is considered a soft barrier, stronger during Ice Ages when the sea level is lower and the extensive continental shelf exposed, leaving little habitat available for migration between provinces (Rocha 2003; Nunes et al. 2011; Araujo et al. 2022) (Fig. 3.2). During high sea-level times, like today, an extensive mesophotic reef is present under the plume of the Amazon River, which can serve as a corridor for reef species able to live in these deeper habitats and conditions (Rocha et al. 2002; Rocha 2003; Nunes et al. 2009; Moura et al. 2016; Francini-Filho et al. 2018) (Fig. 3.2). Consequently, many species have crossed this barrier in both directions. However, speciation occurs mainly in small fishes and those that live in shallow waters (Floeter and Gasparini 2001; Rocha et al. 2002; Araujo et al. 2022; Pinheiro et al. 2018), as they cannot migrate under the Amazon plume and their colonization depends on stochastic events. Some examples of Brazilian fishes that share Caribbean sister species or clades include the genus *Acanthurus*, *Acyrtus*, *Bathygobius*, *Elacatinus*, *Gramma*, *Halichoeres*, *Malacoctenus*, *Opistognathus*, *Scarus*, *Sparisoma*, *Thalassoma*, among many others, in addition to corals within

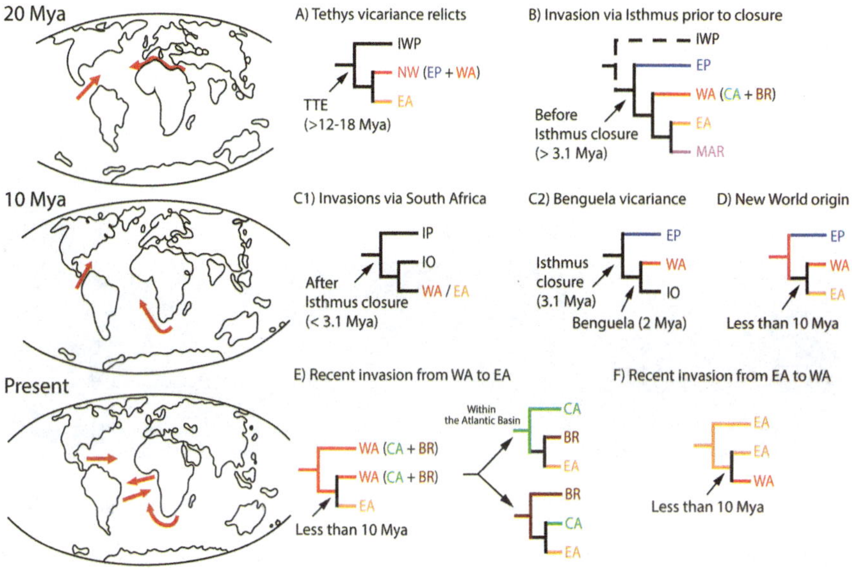

Fig. 3.1 Main routes (red arrows) of colonization over time (20 Million years ago—Mya, 10 Mya, and present) and possible evolutionary scenarios for Atlantic reef fish species. (A) Tethys vicariance relicts, (B) Invasion via Isthmus prior to closure, (C1) Invasions via South Africa, (C2) Benguela vicariance, (D) New World origin, (E) Recent invasion from Western Atlantic (WA) to Eastern Atlantic (EA), (F) Recent invasion from EA to WA. Different colours represent geographical marine areas: black: IWP Indo-West Pacific, IP Indo-Pacific, IO Indian Ocean, magenta: NW New World, blue: EP Eastern Pacific, red: WA Western Atlantic, orange: EA Eastern Atlantic, green: CA Caribbean, brown: BR Brazil, pink: MAR Middle Atlantic Ridge. (Maps adapted from Prof. Larry Braile (available at http://web.ics.purdue.edu/~braile). Evolutionary scenarios adapted from Floeter et al. (2008))

the genus *Favia* (*Favia gravida* in Brazil and *Favia fragum* in the Caribbean; see Nunes et al. 2008) (Fig. 3.3).

Species with great dispersal capacity are able to cross the Atlantic and have the chance to colonize new provinces (Nunes et al. 2008, 2009, 2011). Some ecological characteristics are good predictors of species that maintain their distribution in both Eastern and Western Atlantic, such as large body size and the ability to raft with flotsam for fish (Luiz et al. 2012), and the reproductive system in corals (*i.e.* brooders or spawners; Nunes et al. 2011). Other species cross barriers only very rarely, and the lack of connection causes them to evolve in different lineages. For instance, Brazilian species that share closely-related species with the Eastern Atlantic include *Clepticus brasiliensis* (Beldade et al. 2009), *Scartella* aff. *cristata* (Araújo et al. 2020) and *Stegastes fuscus* (Tang et al. 2021) (Fig. 3.4). Others, however, arrived in Brazilian waters more recently, being able to establish populations, as *Chromis limbata* (Anderson et al. 2017, 2020), or stay rare as vagrants, such as *Acanthurus monroviae* (Luiz et al. 2010) and *Heniochus acuminatus* (Luiz et al. 2014).

Species diversification (cladogenesis) within the southwestern Atlantic has been suggested to occur by a variety of speciation processes. The Brazilian Province

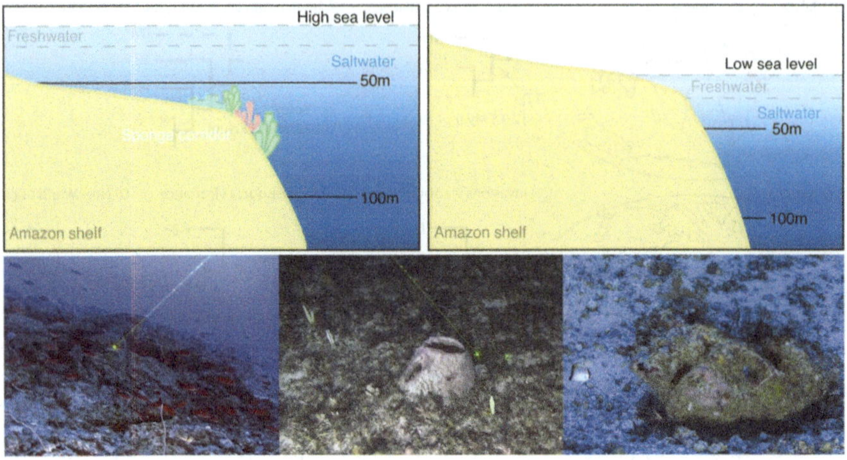

Fig. 3.2 The effect of sea-level fluctuations on the soft biogeographic barrier of the Amazon River discharge. During high sea-level periods (above-left), as today, a mesophotic ecosystem, known as Great Amazon Reefs (photos below), is extensive over the continental shelf, acting as a corridor for reef fish species and other marine organisms. However, this ecosystem is constrained in the sheer slopes and walls during periods of low-stand sea-levels (above-right), which decreases its corridor effect. (Schematic figure adapted from Rocha (2003). Photos by R.B. Francini-Filho)

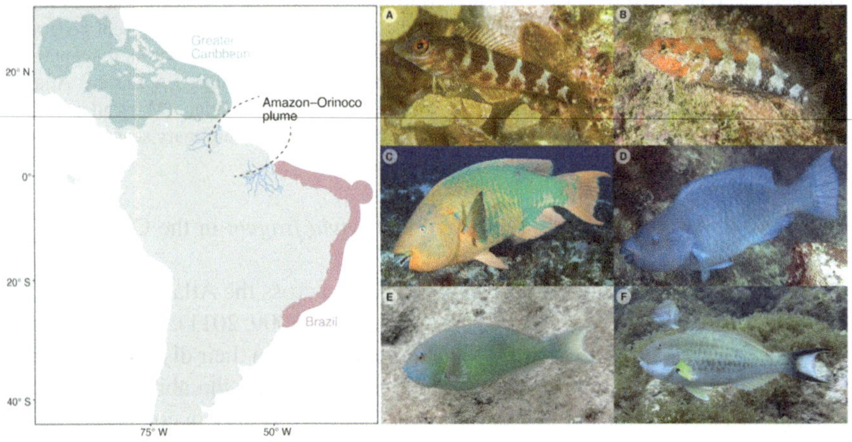

Fig. 3.3 Amazon-Orinoco biogeographical barrier between the Greater Caribbean and the Brazilian Province. In the right, examples of sister fish species found in Brazilian reefs (right bar) and Caribbean (left bar): (**a**) *Malacoctenus triangulatus*, (**b**) *Malacoctenus zaluari*, (**c**) *Scarus guacamaia*, (**d**) *Scarus trispinosus*, (**e**) *Sparisoma rubripinne*, and (**f**) *Sparisoma axillare*. (Photos by (**a**) F. Krasovec, (**b**) J.P. Krajewski, (**c**) J. Lyle, (**d**) C. Sampaio, (**e**) and (**f**) by S.R. Floeter)

harbors a center of diversity in its eastern coast, presenting a high number of endemic species that is not shared with the north-northeastern region (Pinheiro et al. 2018). The eastern coast presents a zoogeographical boundary between tropical and subtropical ecosystems (Pinheiro et al. 2015), and parapatric speciation has been suggested to explain the evolution of endemic species predominantly found in

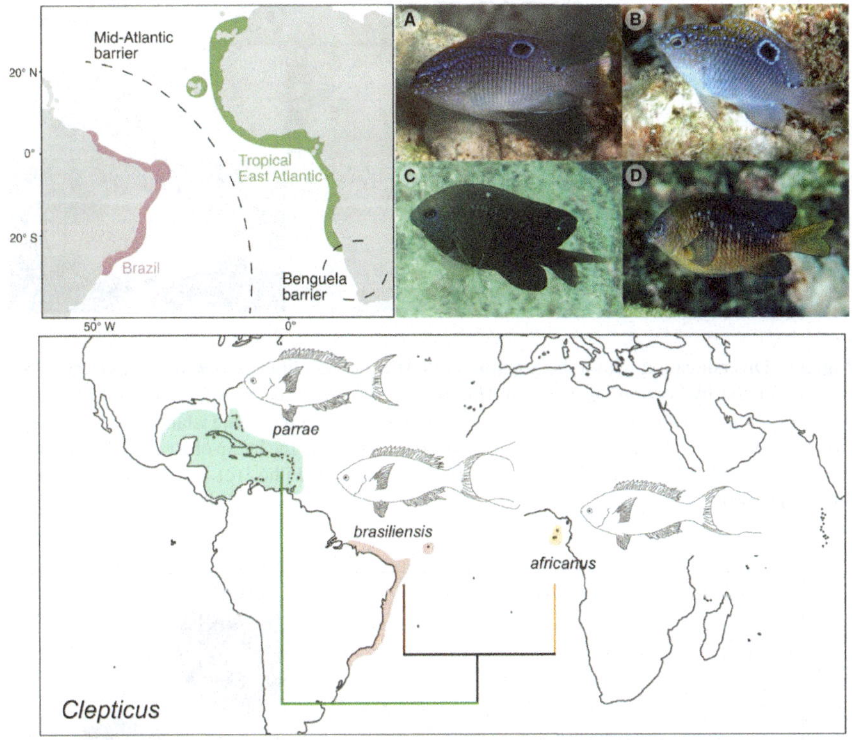

Fig. 3.4 Biogeographical barrier between Brazil and the Tropical Eastern Atlantic. A closely-related damselfish species found in Brazilian and Eastern Atlantic reefs: (**a**) *Stegastes fuscus* juvenile, (**b**) *Stegastes imbricatus* juvenile, (**c**) *Stegastes fuscus* adult, and (**d**) *Stegastes imbricatus* adult. Below, the phylogeny of the *Clepticus* genus, showing closely-related species occurring in the Caribbean region (*C. parrae*), Brazilian Coast (*C. brasiliensis*) and Sao Tome and Principe Island—Tropical East Atlantic (*C. africanus*). (Photos by (**a**) O.J. Luiz, (**c**) S.R. Floeter, (**b**) and (**d**) L.A. Rocha)

subtropical waters, potentially involving reef species within the genera *Gobiosoma*, *Opistognathus*, *Paraclinus*, *Serranus*, *Sparisoma* (Fig. 3.5), and the soft-bottom *Macrodon* (Santos et al. 2006; Pinheiro et al. 2018). However, vicariance is also important. The large São Francisco River discharge, associated with the split of the Brazil Current from the South Equatorial Current, is also suggested to influence reef fish biogeography and the genetic structure of marine populations in Brazil (Cunha et al. 2014; Pinheiro et al. 2018). Hydrocorals within the genus *Millepora* are an interesting example comprising four species in the Western Atlantic: *M. alcicornis*, broadly distributed in the tropical Atlantic, the Brazilian endemics *M. braziliensis* and *M. nitida,* which occur north and south of the São Francisco river, respectively, and *M. laborelli* restricted to a small area in northern Brazil (Souza et al. 2017) (Figs. 3.5 and 3.6); these patterns highlight the importance of diversification within the Southwestern Atlantic.

Antitropical species, such as *Mycteroperca microlepis, Chromis flavicauda* and *Halichoeres sazimai*, probably presented a wider distribution during Ice Ages,

Fig. 3.5 Diversification within the Southwestern Atlantic. São Francisco River as a vicariant barrier, as for the hydrocoral species: (**a**) *Millepora braziliensis* (occurring in the north) and (**b**) *Millepora nitida* (occurring in the south); Speciation in peripheral oceanic islands: (**c**) *Halichoeres rubrovirens*; and allopatric speciation with possible extinction in the hump of Brazil: (**d**) *Sparisoma tuiupiranga*. (Photos by (**a**) E.A. Vieira, (**b**) R.B. Francini-Filho, (**c**) H.T. Pinheiro, and (**d**) J.P. Krajewski)

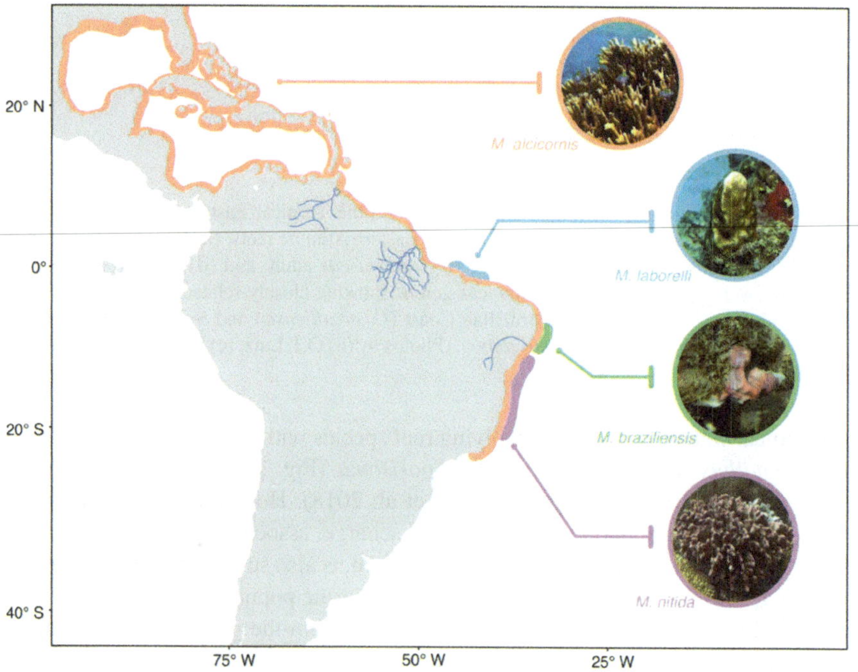

Fig. 3.6 Diversification and distribution of *Millepora* species in the Western Atlantic. *Millepora alcicornis* is widely distributed in the tropical Atlantic. The endemic *M. laborelli* is restricted to a small area in northern Brazil, and *M. braziliensis* and *M. nitida* which occur in the north and the south of the São Francisco River, respectively. (Photos by J. Bleuel, N. Roos, E.A. Vieira, and R.B. Francini-Filho)

evolving in isolation with the expansion of warm coastal waters during interglacial times. Moreover, vicariance is directly related to allopatric speciation in the Vitória-Trindade Chain (VTC) (Pinheiro et al. 2017). During the Pleistocene, lowstand sea-levels exposed deep seamounts, allowing weakly dispersing fishes to colonize remote islands via a stepping stones process. Sea level rise isolated these species, driving speciation (Pinheiro et al. 2017). Interestingly, low sea-level during the Last Glacial Maximum (~30,000–19,000 years ago) exposed most of the continental shelf, constraining coral populations to a much narrower area (*e.g. for Mussismilia braziliensis*; see Menezes et al. 2020) and seamounts (*e.g. Mussismilia hispida* see Peluso et al. 2018). During that time, oceanic locations like seamounts may have acted as refugia for coral populations, subsequently colonizing coastal areas as sea-level rose to current levels (Peluso et al. 2018).

Oceanic islands shelter 32% of the Brazilian endemic reef fishes, and the Saint Peter and Saint Paul's Archipelago (SPSPA), situated 1100 km off the northeastern Brazilian coast, presents the greatest relative endemism level of the province (Pinheiro et al. 2018). With one of the lowest biodiversity levels of shorefishes (86 species), SPSPA endemism reaches 9.3% (8 species), however, considering additional endemic species shared with other oceanic islands, its endemic fauna increases to 20.9% (Pinheiro et al. 2020). This high level of endemism is suggested to be related to the process of peripatric speciation, in which speciation of an isolated peripheral population occurs.

Ecological speciation may occur in any geographical context, driven by adaptations to different habitats, food, mating options, and among other biological factors. In Atlantic coral reefs, this process has been suggested to drive speciation in *Haemulon*, *Halichoeres*, *Hypoplectrus* and *Gramma* species (Bowen et al. 2013), most related to sexual and habitat selection. In Brazil, ecological speciation is suggested to drive the boundaries between *Halichoeres radiatus* and *Halichoeres brasiliensis*, found respectively in insular and mainland tropical waters of northeastern Brazil (Rocha et al. 2005b).

3.4 Composition and Distribution of Fishes and Corals

The Southwestern Atlantic shelters over 730 fish species associated with reef environments, of which 405 species are considered reef residents or strictly reef species (Pinheiro et al. 2018). Overall, when all species associated with reef environments are considered, Carangidae is the richest family recorded, with 35 species, followed by Gobiidae (31 species), Epinephelidae (25), Serranidae (25), Scorpaenidae (23), Labridae (21), Haemulidae (19), Muraenidae (19), and Ophichthidae (19). According to Morais et al. (2017), the families with highest biomass in the Brazilian Province are Haemulidae, Balistidae, Kyphosidae, Epinephelidae, Acanthuridae, Labridae, Holocentridae, Pomacentridae, Lutjanidae, Pomacanthidae, and Carangidae. Regarding the most species-rich genera, among all associated with reef environments, *Scorpaena* (14 species), *Carcharhinus* (11), *Haemulon* (10), *Lutjanus*

(9), *Serranus* (9), *Gymnothorax* (8), *Halichoeres* (8) and *Anchoa* (8) are the most important in the SWA (Pinheiro et al. 2018).

In terms of trophic structure of SWA reefs, mobile invertebrate feeders are the most important guild in richness (46%) (Pinheiro et al. 2018) and biomass (lower-level carnivores: 44%) (Morais et al. 2017). Macrocarnivores, although being the second richest trophic guild in the SWA with 27% of the reef fish fauna (Pinheiro et al. 2018), constituted only a small fraction of the standing biomass of most localities in the Brazilian Province (Morais et al. 2017). Species richness is evenly distributed through size classes (Pinheiro et al. 2018), however, most of the biomass is represented by fishes between 10 and 30 cm (Morais et al. 2017).

Concerning the geographic distribution of reef fishes occurring in the SWA, most species are widely distributed in the western Atlantic (46%), transatlantic and inter-oceanic species account for 20 and 6%, respectively, and SWA endemics constitute 27% of the resident reef fish fauna (Pinheiro et al. 2018) (Fig. 3.7). Another 49 endemic species are benthic-occasional and 14 are pelagic-occasional species, totaling 174 SWA endemic fishes (24% of the total species) associated with reef environments. Endemics are distributed in 45 families, in which Gobiidae and Labridae are the most species-rich families (15 species each), followed by Labrisomidae (10), Serranidae (10), Pomacentridae (8) and Blenniidae (8). Most endemics are small (0–10 cm, 42%) or medium-sized (10–25 cm, 28%) fishes, have sedentary (55%) and roving (39%) mobility, and none display rafting dispersal abilities (Pinheiro et al. 2018) (Fig. 3.7).

Coral fauna in the SWA is considered impoverished but highly endemic (~31%; Leão et al. 2016). The SWA coral biodiversity comprises 8 families of zooxanthellate scleractinian corals (Agariciidae, Astrocoeniidae, Faviidae, Meandrinidae, Montastraeidae, Pocilloporidae, Poritidae and Siderastreidae), 11 genera (*Agaricia, Stephanocoenia, Favia, Mussissmilia, Scolymia, Meandrina, Montastraea, Madracis, Porites* and *Siderastrea*) and 16 species of which 5 are endemic (*Favia leptophylla, Mussismilia braziliensis, M. harttii, M. hispida* and *Siderastrea stellata*; despite recent debate on the endemic status of the last one, see Garcia et al. 2017). In addition to scleractinian corals, SWA shallow reefs are also home to four species of hydrocorals (family Milleporidae), three of which are endemic (*Millepora braziliensis, M. nitida* and *M. laborelli*) (Fig. 3.6). This high endemism is related to large-scale barriers, such as the Amazon River discharge and the large distance that separates the eastern and western Atlantic, but also to meso-scale barriers, such as the São Francisco River (Nunes et al. 2008, 2009, 2011; Souza et al. 2017).

Within the SWA, a recent biogeographic analysis based on reef fish showed six sub-provinces (Fig. 8; Pinheiro et al. 2018): (1) St. Peter and St. Paul's Archipelago; (2) Fernando de Noronha Archipelago and Rocas Atoll; (3) Vitória-Trindade Chain (seamounts and islands) (VTC); (4) north and northeast Brazil; (5) east and southeast Brazil; and (6) Uruguay and Argentina (Argentinian Province). The greatest species richness is found in the east–southeastern Brazil (326 species), followed by north–northeastern Brazilian coast (259 species) and VTC (186 species) (Fig. 3.8). While the north–northeast subprovince shares most of its reef fish fauna (95%) with the east–southeast subprovince, over a quarter of the species found in the latter do

Fig. 3.7 Most of the endemic species in the Brazilian Province have small body size and sedentary mobility: (**a**) *Elacatinus figaro*, (**b**) *Gobiosoma alfie*, (**c**) *Lythrypnus brasiliensis*, (**d**) *Malacoctenus zaluari*, (**f**) *Paraclinus arcanus*, and (**e**) *Starksia brasiliensis*. (Photos by R.M. Macieira)

not occur in the former. Both sub-provinces in the mainland Brazilian coast share about 50% of their species with their adjacent oceanic islands.

When it comes to corals, there are two main diversity centers based on species composition: (1) eastern Brazil, encompassing the Abrolhos Bank and the states of Bahia and Espírito Santo; and (2) Northeastern Brazil, encompassing coastal areas between the states of Rio Grande do Norte and Alagoas, in addition to the oceanic islands of Fernando de Noronha and Rocas Atoll. Southern subtropical areas, like the state of São Paulo, and the remote locations, like St. Peter and St. Paul's Archipelago, have the lowest coral richness, and share widely distributed species such as *Madracis decactis*.

The Brazilian oceanic islands harbour 36 resident endemic reef fishes, while 40 are restricted to the continental shelf and slope, 28 occur in both Brazilian continental and oceanic subprovinces, and seven reach the Argentinian Province. The east–southeastern coast shelters the highest number of SWA endemic reef fish species (71 species), 26 of which are not recorded in the north–northeast coast, and 16 are exclusive to this subprovince. The north–northeastern coast shelters 48 SWA

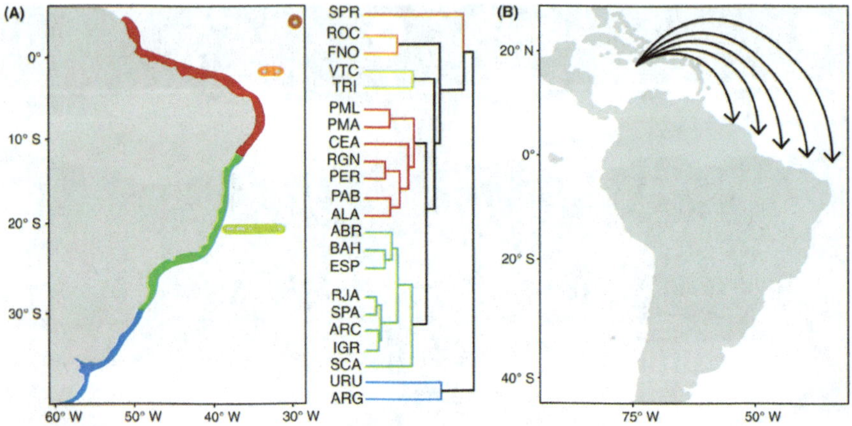

Fig. 3.8 (**a**) Southwestern Atlantic sub-provinces of reef fish species. For more details about the cluster and methods implemented, see Pinheiro et al. (2018). SPR St. Peter and St. Paul's Rocks, ROC Rocas Atoll, FNO Fernando de Noronha Archipelago, VTC Vitória-Trindade Chain, TRI Trindade-Martin Vaz insular complex, PML Parcel de Manuel Luís, PMA states of Pará and Maranhão, CEA state of Ceará, RGN state of Rio Grande do Norte, PER state of Pernambuco, PAB state of Paraíba, ALA state of Alagoas, ABR Abrolhos shelf, BAH state of Bahia, ESP state of Espírito Santo, RJA state of Rio de Janeiro, SPA state of São Paulo, ARC region of Arraial do Cabo, IGR Ilha Grande Bay, SCA state of Santa Catarina, URU Uruguay, ARG Argentina. (**b**) The reef fish composition in the Southwestern Atlantic is partially shaped by the target-area-distance model, also known as the propagule rain hypothesis, in which regions with similar coastline extensions and distance from the Caribbean (the biodiversity hotspot in the Atlantic) would have similar pool of colonist species, presenting low beta diversity. (Maps and cluster adapted from Pinheiro et al. (2018))

endemics, however, it displays one of the lowest local endemism among all SWA subprovinces (two species). St. Peter and St. Paul's Rocks has the highest local percent endemism level (9.3%, or 20.9% including oceanic island endemics; Pinheiro et al. 2020) in the SWA, followed by the VTC subprovince (6%) (Pinheiro et al. 2018). Differently from fishes, there are no coral species endemic to the oceanic islands. An interesting endemism case is that of the hydrocoral *Millepora laborelli* restricted to Parcel Manoel Luis and adjacent areas in the state of Maranhão, Northern Brazil (Souza et al. 2017).

3.5 Macroecological Patterns of Reef Fish Biodiversity

The taxonomic and functional structure of Atlantic fish assemblages are shaped both by historical events and environmental factors (Bender et al. 2013b). The proportion of species richness in fish families reveals a strong imprint of historical events that delineated reef fish biogeography. Nevertheless, proportions of species in different body-size classes and trophic groups respond to isolation from the center of biodiversity, and reef complexity. The number of species and relative abundance of fishes relying on relatively low-quality food (algae, detritus and sessile invertebrates) increase towards the tropics (Floeter et al. 2004). This pattern is likely related to algal productivity and rate of detritus decomposition. Moreover, these feeding strategies are a relatively recent evolutionary phenomenon, and a higher diversification of these groups in diversity hotspots helps explain their predominance in the tropics (Siqueira et al. 2020). Biogenic Caribbean reefs are also composed mainly by small species, a pattern driven largely by environmental factors, such as temperature and habitat complexity (Bender et al. 2013b). Peripheral provinces, which include Brazil, are dominated by larger species with diverse diets, which is related to dispersal abilities that allows them to migrate out of regional diversity hotspots.

Therefore, the Southwestern Atlantic is part of a biogeographic continuum that extends along the whole Western Atlantic. The biodiversity of the Brazilian Province is associated with the target-area-distance model, which is related to the propagule rain hypothesis (Pinheiro et al. 2018) (Fig. 3.8b). According to this hypothesis, the Caribbean as the center of diversity in the Atlantic, is the putative main source of propagules, and SWA regions with similar coastline extension and distance from the source (Caribbean) would have a similar chance to share the same pool of colonists, presenting low beta diversity. Indeed, many species widespread in the Caribbean occur in the north-northeastern Brazilian coast or in the northern oceanic islands, such as *Chromis scotii*, *Haemulon chrysargyreum*, *Haemulon melanurum*, *Haemulon vittata*, *Halichoeres radiatus* and *Lachnolaimus maximus* (Pinheiro et al. 2018). Conversely, the southeastern Brazilian coast presents many endemic fishes only shared among subtropical reefs, what increases dissimilarity from northeastern Brazil and the Caribbean, and decreases beta diversity among neighboring regions. The region where tropical and subtropical faunas overlap (~18–22 degrees latitude) hosts the highest richness of reef fishes and benthic organisms along the Brazilian coast (Aued et al. 2018; Pinheiro et al. 2018). Such a pattern does not conform with the classic latitudinal gradient of biodiversity.

This biogeographic gradient along the latitudinal extent also changes the trophic structure of fish assemblages (Fig. 3.9). While roving herbivores, such as parrotfishes (Scarinae) and surgeonfishes (Acanthuridae), are more abundant at low latitudes, browsing herbivores of the family Kyphosidae increase in relative importance with latitude (Ferreira et al. 2004). Omnivores, piscivores, and carnivores are more important in subtropical reefs (Fig. 3.9), while planktivores are more abundant in oceanic islands (Ferreira et al. 2004). However, snappers (Lutjanidae) present higher abundance and biomass in tropical waters (Ferreira et al. 2004; Morais et al. 2017). Such trophic structure is reflected in the intensity and composition of fish feeding pressure on the benthos, which is higher and dominated by roving herbivores (mostly Acanthurids) within the tropical region (from Parcel Manuel Luiz at 0° to Rio de Janeiro at 23°S). In comparison, subtropical reefs (*e.g.* Santa Catarina at 27°S) present about half the feeding pressure of tropical reefs, mostly dominated by omnivorous and invertivorous fishes (families Sparidae and Haemulidae, respectively; Longo et al. 2019). Algal turfs and macroalgae (including crustose corallines) are the dominant components of reef benthic communities throughout the

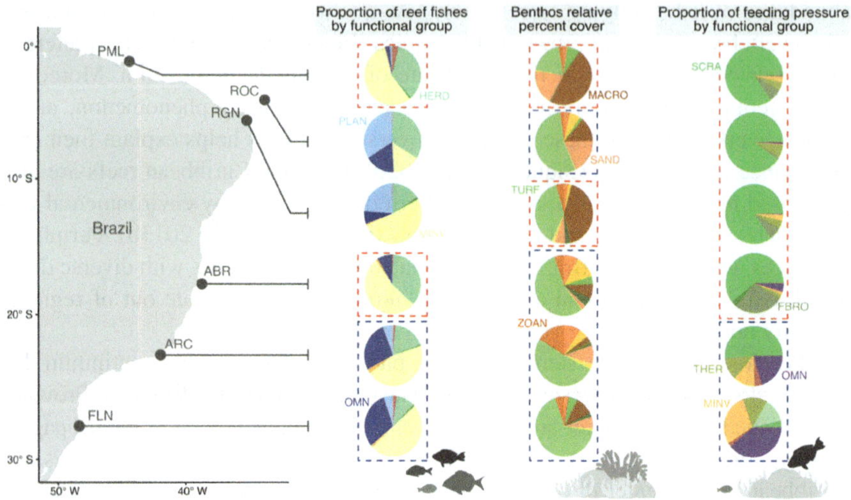

Fig. 3.9 Macroecological patterns of the Southwestern Atlantic reef biodiversity, considering the abundance-based proportion of reef fish functional groups, the relative percent cover of benthos, and the feeding pressure of reef fish functional groups on the benthos. Dashed lines indicate groups revealed by cluster analysis. Main reef fish trophic groups and categories of benthos cover are indicated—for more details please see Morais et al. (2017), Aued et al. (2018), and Longo et al. (2019), from which pie charts were adapted, respectively. Fish functional groups abbreviation: HERD herbivore detritivore, PLAN planktivore, OMN omnivore, SCRA scraper, FBRO fine browser, THER territorial herbivores, MIN mobile invertebrate feeder. Benthos cover abbreviation: MACRO macroalgae, SAND sand, TURF turf algae, ZOAN zoanthid. Abbreviation of the common sites between the three studies: PML Parcel de Manuel Luís, ROC Rocas Atoll, RGN Rio Grande do Norte, ABR Abrolhos shelf, ARC region of Arraial do Cabo, FLN Florianópolis

Brazilian province, but in the tropics, sponges can be abundant in deeper and corals and zoanthids in shallower reefs, while in the tropical-subtropical transition zones seaweeds, sponges, and corals share the benthic cover with gorgonids and octocorals (Aued et al. 2018; Roos et al. 2019).

Brazilian oceanic islands harbor the highest reef fish biomass of the province, followed by other remote and protected sites (Morais et al. 2017). In the oceanic islands, the great biomass is in part explained by the abundance of the black triggerfish *Melichthys niger* particularly in Trindade Island (Pinheiro et al. 2011) and St. Paul's Archipelago (Luiz et al. 2015). At the only atoll in South Atlantic, Rocas Atoll, reef fish biomass is dominated by large predators such as sharks (*Negaprion brevirostris*), snappers (*Lutjanus jocu*), and herbivorous fishes (*Acanthurus* spp.; Longo et al. 2015). Fish feeding pressure at these oceanic islands are mostly related to herbivory by Acanthuridae (Longo et al. 2015) or even by the omnivorous *M. niger* acting as a functional herbivore (Mendes et al. 2019) (Fig. 3.9).

Some ecological traits are associated with endemism and distributional patterns in Brazil (Pinheiro et al. 2018). Most of the Brazilian endemics have small sizes, occur only in shallow waters and are habitat specialists, occurring in few habitats. Conversely, species that occur across all Brazilian regions are mostly habitat generalists and present a wider depth range. Species restricted to the continental shelf have small sizes, most occur only in shallow waters and are dependent on brackish habitats. Species recorded in oceanic islands show better dispersal potential, as they are positively related to rafting use and large body sizes, but also are habitat generalists and are distributed along great depth ranges, characteristics that influence establishment (Pinheiro et al. 2018; Mazzei et al. 2021).

3.6 Final Remarks: Conservation of Brazilian Reef Biodiversity

Distributed along the tropical and subtropical coasts of a developing nation, Brazilian reefs are threatened by unsustainable human population growth near shore, which leads to increased pollution, habitat degradation, and overfishing (Floeter et al. 2006; Leão et al. 2010; Pinheiro et al. 2019) (Fig. 3.10). Human population density is a predictor of fish biomass and functional dispersion in coastal and oceanic reefs, with high biomass identified for remote locations (*e.g.*, Parcel do Manuel Luís, Maranhão State, reefs distant from the coast at Rio Grande do Norte state, and oceanic islands), or no-entry marine protected areas, such as Alcatrazes (São Paulo State). The impacts imposed by humans to reefs and associated habitats (Vila-Nova et al. 2011) prevent marine populations from recovering from frequent and intense disturbances. As a result, 12.4% (n = 78) of the fish species in Brazil are threatened with extinction (Bender et al. 2013a; Pinheiro et al. 2018; Ceretta et al. 2020). Among the threatened fishes, 23% (n = 18) are endemic to Brazilian reefs.

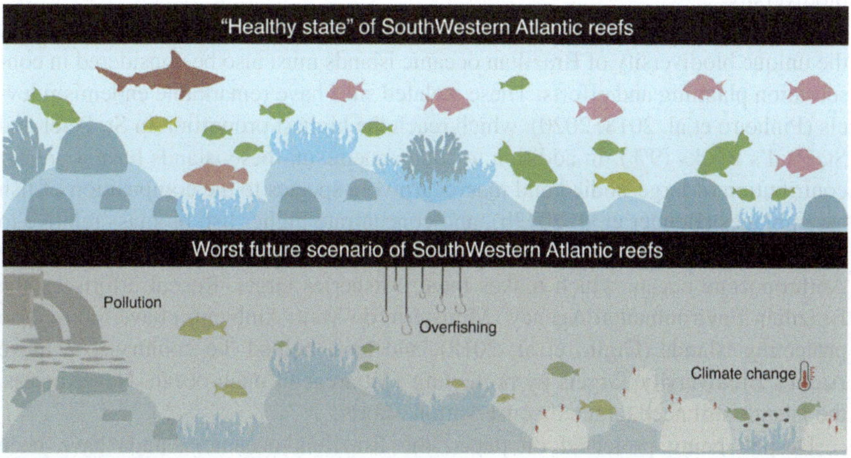

Fig. 3.10 Scenarios of the Brazilian reef biodiversity. Above, a "healthy state" of Brazilian reefs, which presents high endemism level, high abundance of large bodied species, mesopredators, and herbivores. However, currently, Brazilian reefs are threatened by pollution, overfishing and climate change (below), which cause habitat degradation and general biodiversity loss

Different biological and ecological attributes contribute to the extinction vulnerability of species. For Brazilian reef fishes, traits such as large body size, macro carnivorous diet and high mobility are good predictors of vulnerability to extinction (Bender et al. 2013a; Ceretta et al. 2020). Some species characteristics also interact with specific threats, enhancing their vulnerability. For instance, small-bodied species with complex reproductive strategies are targeted by the ornamental trade (Bender et al. 2013a), and the interaction of both—trait and threat—works as a double jeopardy (Hawkins et al. 2000). Traits of the coral fauna are highly conserved and present limited redundancy, particularly in areas of lower richness such as the Brazilian coast, enhancing their vulnerability to anthropogenic stressors (McWilliam et al. 2018). This is particularly true for functions such as carbonate accretion and structural complexity that can scale-up and affect the entire reef ecosystem.

Along the Brazilian coast, São Paulo and Espírito Santo states present the highest proportion of threatened reef fish species (Ceretta et al. 2020). Moreover, an astonishing marine biodiversity is found in the region that lies between Espírito Santo and Bahia. The coast of Espírito Santo can be considered a hotspot for endemic, threatened and targeted reef fish species (Vila-Nova et al. 2014). However, Vila-Nova et al. (2014) have also shown that this region is the least protected along the Brazilian coast. Unfortunately, the current network of Brazilian Marine Protected Areas does not protect hotspots of reef fish biodiversity (Vila-Nova et al. 2014). More recently, Magris et al. (2021) have combined the distribution of human-derived impacts, habitat types and threatened species, including reef habitats and species, into a large-scale spatial prioritization activity. This study identified coastal areas as high priorities for marine conservation in Brazil, and pointed to industrial fisheries, land-based activities, and climate change as the greatest threats to biodiversity.

Despite the great concentration of human threats and impacts on coastal reefs, the unique biodiversity of Brazilian oceanic islands must also be considered in conservation planning and efforts. These isolated sites have remarkable endemism levels (Pinheiro et al. 2018, 2020), which reach the highest proportion in St. Peter and St. Paul's Rocks (9%). In addition to such endemism, these islands have a greater contribution of large-bodied and macrocarnivore species to the composition of fish assemblages (Bender et al. 2013b), and concentrate higher fish biomass relative to the coast (Morais et al. 2017). Therefore, islands are havens of biodiversity in the Anthropocene ocean, which makes them a fisheries target. Recent efforts by the Brazilian Environmental Agency (Ministério do Meio Ambiente) have focused on protecting islands (Giglio et al. 2018), and have helped the country to achieve Aichi's Biodiversity Targets by protecting vast areas of open-ocean. Nevertheless, the islands still need targeted conservation efforts.

Despite being protected on paper, the Brazilian oceanic islands have been impacted by several types of fishing, from spearfishing on Trindade island (Pinheiro and Joyeux 2015; Guabiroba et al. 2020), to the industrial fishing vessels in the remote St. Paul's Rocks (Luiz and Edwards 2011; Giglio et al. 2018). Fishing efforts in these isolated islands have overexploited and threatened local populations of predators (Pinheiro et al. 2010; Luiz and Edwards 2011), which may cause disruptive ecological cascades and change insular community patterns (Guabiroba et al. 2020), influencing biogeographic inferences.

In addition to local stressors, global climate change can impose significant changes to Brazilian reef biodiversity. Although subject to 50–60% less thermal-stressed events in comparison to the Indo-Pacific and the Caribbean (Mies et al. 2020), coral bleaching and mortality affect the Brazilian reefs (Duarte et al. 2020), and ocean warming should also cause strong ecological changes along the Brazilian coast. Feeding pressure, for instance, can be severely reduced in tropical areas because predicted temperatures are likely to exceed the thermal limit of herbivorous fishes, and these species are likely to shift their distribution southwards (Inagaki et al. 2020), influencing community structure patterns. Similarly, the territorial behavior of the Brazilian endemic and ubiquitous damselfish *Stegastes fuscus* will likely be severely impaired by increasing temperatures (Silva-Pinto et al. 2020). This species has a critical role in shaping benthic communities (Ferreira et al. 1998) and structuring agonistic interactions (Fontoura et al. 2020), and the disruption of these processes can affect reef functioning. Therefore, climate change impacts on Brazilian reefs can go beyond coral bleaching, and although less evident than species extinction, still threaten low-redundant functions and impose severe changes to the reefs and the benefits they currently provide.

In conclusion, Brazilian reefs present a biodiversity that varies along the coastline and among oceanic and remote localities. Species distributions are driven by both ecological and evolutionary processes, shaping singular subprovinces, endemism hotspots and distinct macroecological patterns. However, the increased and disordered human development along the coast, associated with critical gaps in conservation and the global effects of climate change, is increasingly threatening these unique eco-evolutionary processes responsible for the origins of the biodiversity of the Brazilian Province.

Acknowledgements HTP and LAR thank Hope for Reefs Initiative of the California Academy of Sciences and Fundação de Amparo à Pesquisa do Estado de São Paulo (2019/24215-2; 2021/07039-6) for support and funding. GOL is grateful to a research productivity scholarship provided by the Brazilian National Council for Scientific and Technological Development (CNPq; 310517/2019-2) and Serrapilheira Institute for research support (Serra-1708-15364). SRF is grateful for continued CNPq grants over the years.

References

Anderson AB, Salas EM, Rocha LA, Floeter SR (2017) The recent colonization of South Brazil by the Azores chromis *Chromis limbata*. J Fish Biol 91:1–16

Anderson AB, da Silva JP, Sorvilo R, Francini CLB, Floeter SR, Barreiros JP (2020) Population expansion of the invasive Pomacentridae *Chromis limbata* (Valenciennes, 1833) in southern Brazilian coast: long-term monitoring, fundamental niche availability and new records. J Fish Biol 97:362–373

Araújo GS, Vilasboa A, Britto MR, Bernardi G, von der Heyden S, Levy A, Floeter SR (2020) Phylogeny of the comb-tooth blenny genus *Scartella* (Blenniiformes: Blenniidae) reveals several cryptic lineages and a trans-Atlantic relationship. Zool J Linnean Soc 190:54–64

Araujo GS, Rocha LA, Lastrucci NS, Luiz OJ, Di Dario F, Floeter SR (2022) The Amazon-Orinoco Barrier as a driver of reef-fish speciation in the Western Atlantic through time. Abstract Journal of Biogeography 49(8):1407–1419. https://doi.org/10.1111/jbi.v49.8. https://doi.org/10.1111/jbi.14398

Aued AW, Smith F, Quimbayo JP, Cândido DV, Longo GO, Ferreira CEL, Witman JD, Floeter SR, Segal B (2018) Large-scale patterns of benthic marine communities in the Brazilian Province. PLoS One 13:1–15

Beldade R, Heiser JB, Robertson DR, Gasparini JL, Floeter SR, Bernardi G (2009) Historical biogeography and speciation in the Creole wrasses (Labridae, *Clepticus*). Mar Biol 156:679–687

Bellwood DR, Wainwright PC (2002) The history and biogeography of fishes on Coral Reels. Coral Reef fishes. Academic, San Diego, pp 5–32

Bellwood DR, Goatley CHR, Bellwood O (2017) The evolution of fishes and corals on reefs: form, function and interdependence. Biol Rev 92:878–901

Bender MG, Floeter SR, Mayer FP, Vila-Nova DA, Longo GO, Hanazaki N, Carvalho-Filho, Ferreira CEL (2013a) Biological attributes and major threats as predictors of the vulnerability of species: a case study with Brazilian reef fishes. Oryx 47:259–265

Bender MG, Pie MR, Rezende EL, Mouillot D, Floeter SR (2013b) Biogeographic, historical and environmental influences on the taxonomic and functional structure of Atlantic reef fish assemblages. Glob Ecol Biogeogr 22:1173–1182

Bernardi G, Alva-Campbell YR, Gasparini JL, Floeter SR (2008) Molecular ecology, speciation, and evolution of the reef fish genus *Anisotremus*. Mol Phylogenet Evol 48:929–935

Bowen BW, Muss A, Rocha LA, Grant WS (2006) Shallow mtDNA coalescence in Atlantic pygmy angelfishes (genus *Centropyge*) indicates a recent invasion from the Indian Ocean. J Hered 97:1–12

Bowen BW, Rocha LA, Toonen RJ, Karl SA (2013) The origins of tropical marine biodiversity. Trends Ecol Evol 28:359–366

Budd AF (2000) Diversity and extinction in the Cenozoic history of Caribbean reefs. Coral Reefs 19:25–35

Ceretta BF, Fogliarini CO, Giglio VJ, Maxwell MF, Waechter LS, Bender MG (2020) Testing the accuracy of biological attributes in predicting extinction risk. Perspect Ecol Conserv 18:12–18

Cowman PF, Bellwood DR (2013) The historical biogeography of coral reef fishes: global patterns of origination and dispersal. J Biogeogr 40:209–224

Cowman PF, Parravicini V, Kulbicki M, Floeter SR (2017) The biogeography of tropical reef fishes: endemism and provinciality through time. Biol Rev 92:2112–2130

Cunha IMC, de Souza AS, Dias EA Jr, Amorim KDJ, Soares RX, Costa GWWF, García-Machado E, Galetti PM Jr, Molina WF (2014) Genetic multipartitions based on D-loop sequences and chromosomal patterns in Brown Chromis, *Chromis multilineata* (Pomacentridae), in the Western Atlantic. Hindawi Publ Corp 2014:1–11

Duarte GAS, Villela HDM, Silva DM, Barno A, Cardoso PM, Vilela CLS, Rosado P, Messias CSM, Chacon MA, Santoro EP, Olmedo DB, Szpilman M, Rocha LA, Sweet M, Peixoto RS (2020) Heat waves are a major threat to turbid coral reefs in Brazil. Front Mar Sci 7:179

Ferreira CEL, Gonçalves JEA, Coutinho R, Peret AC (1998) Herbivory by the Dusky Damselfish Stegastes fuscus (Cuvier, 1830) in a tropical rocky shore: effects on the benthic community. J Exp Mar Biol Ecol 229:241–264

Ferreira CEL, Floeter SR, Gasparini JL, Ferreira BP, Joyeux JC (2004) Trophic structure patterns of Brazilian reef fishes: a latitudinal comparison. J Biogeogr 31:1093–1106

Floeter SR, Gasparini JL (2001) Brazilian endemic reef fishes. Coral Reefs 19:292

Floeter SR, Ferreira CEL, Dominici-Arosemena A, Zalmon IR (2004) Latitudinal gradients in Atlantic reef fish communities: trophic structure and spatial use patterns. J Fish Biol 64:1680–1699

Floeter S, Halpern B, Ferreira C (2006) Effects of fishing and protection on Brazilian reef fishes. Biol Conserv 128:391–402

Floeter SR, Rocha LA, Robertson DR, Joyeux JC, Smith-Vaniz WF, Wirtz P, Edwards AJ, Barreiros JP, Ferreira CEL, Gasparini JL, Brito A, Falcón JM, Bowen BW, Bernardi G (2008) Atlantic reef fish biogeography and evolution. J Biogeogr 35:22–47

Fontoura L, Cantor M, Longo GO, Bender MG, Bonaldo RM, Floeter SR (2020) The macroecology of reef fish agonistic behaviour. Ecography 43:1278–1290

Francini-Filho R, Asp NE, Siegle E, Hocevar J, Lowyck K, D'Avila N, de Vasconcelos AA, Baitelo R, Rezende CE, Omachi CY, Thompson CC, Thompson FL (2018) Perspectives on the Great Amazon Reef: extension, biodiversity and threats. Front Mar Sci 5:1–5

García NAC, Campos JE, Musi JLT, Forsman ZH, Muñoz JM, Reyes AM, González JEA (2017) Comparative molecular and morphological variation analysis of *Siderastrea* (Anthozoa, Scleractinia) reveals the presence of *Siderastrea stellata* in the Gulf of Mexico. Biol Bull 232:58–70

Giglio VJ, Pinheiro HT, Bender MG, Bonaldo RM, Costa-lotufo LV, Ferreira CEL, Floeter SR, Freire A, Gasparini JL, Joyeux J, Paulo J, Lindner A, Longo GO, Lotufo TMC, Loyola R, Luiz OJ, Macieira RM, Magris RA, Mello TJ, Quimbayo JP, Rocha LA, Segal B, Teixeira JB, Vilanova DA, Vilar CC, Zilberberg C, Francini-Filho R (2018) Large and remote marine protected areas in the South Atlantic Ocean are flawed and raise concerns: comments on Soares and Lucas (2018). Mar Policy 96:13–17

Guabiroba HC, Santos MEA, Pinheiro HT, Simon T, Pimentel CR, Vilar CC, Joyeux J-C (2020) Trends in recreational fisheries and reef fish community structure indicate decline in target species population in an isolated tropical oceanic Island. Ocean Coast Manag 191:105194

Hawkins JP, Roberts CM, Clark V (2000) The threatened status of restricted-range coral reef fish species. Anim Conserv 3:81–88

Hemingson CR, Cowman PF, Hodge JR, Bellwood DR (2018) Colour pattern divergence in reef fish species is rapid and driven by both range overlap and symmetry. Ecol Lett 22:190

Inagaki KY, Pennino MG, Floeter SR, Hay ME, Longo GO (2020) Trophic interactions will expand geographically but be less intense as oceans warm. Glob Chang Biol 26:6805–6812

Leão ZMAN, Kikuchi R, Oliveira MDM, Vasconcellos V (2010) Status of Eastern Brazilian coral reefs in time of climate changes. Panam J Aquat Sci 5:224–235

Leão ZM, Kikuchi RK, Ferreira BP, Neves EG, Sovierzoski HH, Oliveira MDM, Johnsson R (2016) Brazilian coral reefs in a period of global change: a synthesis. Brazil J Oceanogr 64:97–116

Leprieur F, Descombes P, Gaboriau T, Cowman PF, Parravicini V (2016) Plate tectonics drive tropical reef biodiversity dynamics. Nat Commun 7:11461

Lessios HA (2008) The Great American Schism: divergence of marine organisms after the rise of the Central American Isthmus. Annu Rev Ecol Evol Syst 39:63–91

Longo GO, Morais RA, Martins CDL, Mendes TC, Aued AW, Cândido DV, de Oliveira JC, Nunes LT, Fontoura L, Sissini MN, Teschima MM, Silva MB, Ramlov F, Gouvea LP, Ferreira CEL, Segal B, Horta PA, Floeter SR (2015) Between-habitat variation of benthic cover, reef fish assemblage and feeding pressure on the benthos at the only Atoll in South Atlantic: Rocas Atoll, NE Brazil. PLoS One 10:e0127176

Longo GO, Hay ME, Ferreira CEL, Floeter SR (2019) Trophic interactions across 61 degrees of latitude in the Western Atlantic. Glob Ecol Biogeogr 28:107–117

Luiz OJ, Edwards AJ (2011) Extinction of a shark population in the Archipelago of Saint Paul's Rocks (equatorial Atlantic) inferred from the historical record. Biol Conserv 144:2873–2881

Luiz OJ, Sazima I, Waib LF, Ferreira CEL (2010) A honeymoon in Brazil: the spawning behavior of an exotic reef fish in the western South Atlantic. Neotrop Ichthyol 8:369–371

Luiz OJ, Madin JS, Robertson DR, Rocha LA, Wirtz P, Floeter SR (2012) Ecological traits influencing range expansion across large oceanic dispersal barriers: insights from tropical Atlantic reef fishes. Proc R Soc B Biol Sci 279:1033–1040

Luiz OJ, Comin EJ, Madin JS (2014) Far away from home: the occurrence of the *Heniochus acuminatus* (Pisces: Chaetodontidae) in the Atlantic. Bull Mar Sci 90:10–13

Luiz OJ, Mendes TC, Barneche DR, Ferreira CGW, Noguchi R, Villaça RC, Rangel CA, Gasparini JL, Ferreira CEL (2015) Community structure of reef fishes on a remote oceanic Island (St Peter and St Paul's Archipelago, equatorial Atlantic): the relative influence of abiotic and biotic variables. Mar Freshw Res 66:739–749

Magris RA, Costa MDP, Ferreira CEL, Vilar CC, Joyeux J-C, Creed JC, Copertino MS, Horta PA, Sumida PYG, Francini-Filho RB, Floeter SR (2021) A blueprint for securing Brazil's marine biodiversity and supporting the achievement of global conservation goals. Abstract Diversity and Distributions 27(2):198–215. https://doi.org/10.1111/ddi.v27.2 https://doi.org/10.1111/ddi.13183

Mazzei EF, Pinheiro HT, Simon T, Moura RL, Macieira RM, Pimentel CR, Teixeira JB, Floeter SR, Ferreira CEL, Ghisolfi RD, Francini-Filho RB, Quimbayo JP, Rocha LA, Gasparini JL, Joyeux J-C (2021) Mechanisms of dispersal and establishment drive a stepping stone community assembly on seamounts and oceanic islands. Marine Biology 168(109). https://doi.org/10.1007/s00227-021-03919-7

McWilliam M, Hoogenboom MO, Baird AH, Kuo CY, Madin JS, Hughes TP (2018) Biogeographical disparity in the functional diversity and redundancy of corals. Proc Nat Acad Sci 115:3084–3089

Mendes TC, Quimbayo JP, Bouth HF, Silva LPS, Ferreira CEL (2019) The omnivorous triggerfish *Melichthys Niger* is a functional herbivore on an isolated Atlantic oceanic Island. J Fish Biol 95:812–819

Menezes N, Sobral-Souza T, Silva M, Solferini VN (2020) Paleoclimatic distribution and phylogeography of *Mussismilia braziliensis* (Anthozoa, Scleractinia), an endemic Brazilian reef coral. Mar Biodivers 50:1–12

Mies M, Francini-Filho RB, Zilberberg C, Garrido AG, Longo GO, Laurentino E, Güth AZ, Sumida PYG, Banha TNS (2020) South Atlantic Coral Reefs are major global warming Refugia and less susceptible to bleaching. Front Mar Sci 7:1–13

Morais RA, Ferreira CEL, Floeter SR (2017) Spatial patterns of fish standing biomass across Brazilian reefs, Southwestern Atlantic. J Fish Biol 91:1642–1667

Moura RL, Amado-Filho GM, Moraes FC, Brasileiro PS, Salomon PS, Mahiques MM, Bastos AC, Almeida MG, Silva JM, Araújo BF, Brito FP, Rangel TP, Oliveira BCV, Bahia RG, Paranhos RP, Dias RJS, Siegle E, Figueiredo AG, Pereira RC, Leal CV, Hadju EASP, Gregoracci GB, Neumann-Leitão S, Yager PL (2016) An extensive reef system at the Amazon River mouth. Sci Adv 2:e1501252

Nunes FL, Fukami H, Vollmer SV, Norris RD, Knowlton N (2008) Re-evaluation of the systematics of the endemic corals of Brazil by molecular data. Coral Reefs 27:423–432

Nunes FL, Norris RD, Knowlton N (2009) Implications of isolation and low genetic diversity in peripheral populations of an amphi-Atlantic coral. Mol Ecol 18:4283–4297

Nunes FL, Norris RD, Knowlton N (2011) Long distance dispersal and connectivity in amphi-Atlantic corals at regional and basin scales. PLoS One 6:e22298

Peluso L, Tascheri V, Nunes F et al. (2018) Contemporary and historical oceanographic processes explain genetic connectivity in a Southwestern Atlantic coral. Sci Rep 8:2684. https://doi.org/10.1038/s41598-018-21010-y

Pinheiro HT, Joyeux J-C (2015) The role of recreational fishermen in the removal of target reef fishes. Ocean Coast Manag 112:12–17

Pinheiro HT, Martins AS, Gasparini JL (2010) Impact of commercial fishing on Trindade Island and Martin Vaz Archipelago, Brazil: characteristics, conservation status of the species involved and prospects for preservation. Brazilian Arch Biol Technol 53:1417–1423

Pinheiro HT, Ferreira CEL, Joyeux J-C, Santos RG, Horta PA (2011) Reef fish structure and distribution in a South-Western Atlantic Ocean tropical Island. J Fish Biol 79:1984–2006

Pinheiro HT, Madureira J, Joyeux J, Martins A (2015) Fish diversity of a southwestern Atlantic coastal Island: aspects of distribution and conservation in a marine zoogeographical boundary. Check List 11:1615

Pinheiro HT, Bernardi G, Simon T, Joyeux J-C, Macieira RM, Gasparini JL, Rocha C, Rocha LA (2017) Island biogeography of marine organisms. Nature 549:82–85

Pinheiro HT, Rocha LA, Macieira RM, Carvalho-Filho A, Anderson AB, Bender MG, DiDario F, Ferreira CEL, Francini-Filho RB, Gasparini JL, Joyeux J-C, Luiz OJ, Mincarone M, Moura RL, de Nunes JACC, Quimbayo JP, Rosa RS, CLSS S, Sazima I, Simon T, Vila-Nova DA, Floeter SR (2018) South-western Atlantic reef fishes: zoogeographic patterns and ecological drivers reveal a secondary biodiversity center in the Atlantic Ocean. Divers Distrib 24:951–965

Pinheiro HT, Teixeira JB, Francini-Filho RB, Soares-Gomes A, Ferreira CEL, Rocha LA (2019) Hope and doubt for the world's marine ecosystems. Perspect Ecol Conserv 17:19–25

Pinheiro HT, Macena BCL, Francini-Filho RB, Ferreira CEL, Albuquerque FV, Bezerra NPA, Carvalho-Filho A, Ferreira RCP, Luiz OJ, Mello TJ, Mendonça SA, Nunes DM, Pimentel CR, Pires AMA, Soares-Gomes A, Viana DL, Hazin FHV, Rocha LA (2020) Fish biodiversity of Saint Peter and Saint Paul's Archipelago, Mid-Atlantic Ridge, Brazil: new records and a species database. J Fish Biol 97:1143–1153

Renema W, Bellwood DR, Braga JC, Bromfield K, Hall R, Johnson KG, Lunt P, Meyer CP, McMonagle LB, Morley RJ, O'Dea A, Todd JA, Wesselingh FP, Wilson MEJ, Pandolfi JM (2008) Hopping hotspots: global shifts in marine biodiversity. Science (80-) 321:654–657

Roberts CM, Mcclean CJ, Veron JEN, Hawkins JP, Allen GR, Mcallister DE, Mittermeier CG, Schueler FW, Spalding M, Wells F, Vynne C, Werner TB (2002) Marine biodiversity hotspots and conservation priorities for tropical reefs. Science 295:1280–1284

Rocha LA (2003) Patterns of distribution and processes of speciation in Brazilian reef fishes. J Biogeogr 30:1161–1171

Rocha LA, Bass AL, Robertson DR, Bowen BW (2002) Adult habitat preferences, larval dispersal and the comparative phylogeography of three Atlantic *Acanthurus* (Teleostei: Acanthuridae). Mol Ecol 11:243–252

Rocha LA, Robertson DR, Rocha CR, Van Tassell JL, Craig MT, Bowen BW (2005a) Recent invasion of the tropical Atlantic by an Indo-Pacific coral reef fish. Mol Ecol 14:3921–3928

Rocha LA, Robertson DR, Roman J, Bowen BW (2005b) Ecological speciation in tropical reef fishes. Proc R Soc B Biol Sci 272:573–579

Rocha LA, Lindeman KC, Rocha CR, Lessios HA (2008) Historical biogeography and speciation in the reef fish genus *Haemulon* (Teleostei: Haemulidae). Mol Phylogenet Evol 48:918–928

Roos NC, Pennino MG, Carvalho AR, Longo GO (2019) Drivers of abundance and biomass of Brazilian parrotfishes. Mar Ecol Prog Ser 623:117–130. https://doi.org/10.3354/meps13005

Santos S, Hrbek T, Farias IP, Schneider H, Sampaio I (2006) Population genetic structuring of the king weakfish, *Macrodon ancylodon* (Sciaenidae), in Atlantic coastal waters of South America: deep genetic divergence without morphological change. Mol Ecol 15:4361–4373

Silva-Pinto T, Silveira MM, De Souza JF, Luisa A, Moreira P, Vieira EA, Longo GO, Luchiari AC (2020) Damselfish face climate change: impact of temperature and habitat structure on agonistic behavior. PLoS One 15:1–13

Siqueira AC, Bellwood DR, Cowman PF (2019) The evolution of traits and functions in herbivorous coral reef fishes through space and time. Proc R Soc B Biol Sci 286:20182672

Siqueira AC, Morais RA, Bellwood DR, Cowman PF (2020) Trophic innovations fuel reef fish diversification. Nat Commun 11:1–11

Souza JN, Nunes FL, Zilberberg C, Sanchez JA, Migotto AE, Hoeksema BW, Lindner A (2017) Contrasting patterns of connectivity among endemic and widespread fire coral species (*Millepora* spp.) in the tropical Southwestern Atlantic. Coral Reefs 36:701–716

Stanley GD Jr (2003) The evolution of modern corals and their early history. Earth Sci Rev 60:195–225

Tang KL, Stiassny MLJ, Mayden RL, DeSalle R (2021) Systematics of damselfishes. Ichthyol Herpetol 109:258–318

Tittensor DP, Mora C, Jetz W, Lotze HK, Ricard D, Vanden BE, Worm B (2010) Global patterns and predictors of marine biodiversity across taxa. Nature 466:1098–1101

Tornabene L, Van Tassell JL, Robertson DR, Baldwin CC (2016) Repeated invasions into the twilight zone: evolutionary origins of a novel assemblage of fishes from deep Caribbean reefs. Mol Ecol 25:3662–3682

Veron J, Stafford-Smith M, DeVantier L, Turak E (2015) Overview of distribution patterns of zooxanthellate Scleractinia. Front Mar Sci 1:81

Vila-Nova DA, Bender MG, Carvalho-Filho A, Ferreira CEL, Floeter SR (2011) The use of non-reef habitats by Brazilian reef fish species: considerations for the design of marine protected areas. Natureza & Conservação 9:79–86

Vila-Nova DA, Ferreira CEL, Barbosa FG, Floeter SR (2014) Reef fish hotspots as surrogates for marine conservation in the Brazilian coast. Ocean Coast Manag 102:88–93

Chapter 4
Biology of Brazilian Benthic Reef Builders and Dwellers

Miguel Mies, Thomás N. S. Banha, Clovis B. Castro, Ralf T. S. Cordeiro, Tito M. C. Lotufo, and Adalto Bianchini

Abstract Brazilian reef environments are composed of a unique fauna. This chapter addresses the general biological and ecological aspects of 12 benthic reef-dwelling taxa found in Brazil: (i) scleractinian corals, which are primary reef-builders that produce the tridimensional framework; (ii) calcified hydrozoans belonging to the *Millepora* genus; (iii) other cnidarians such as octocorals, zoanthids and anemones; (iv) crustose coralline algae, which are also major reef-builders that produce extensive rhodolith beds; (v) bryozoans and (vi) vermetid gastropods, which have contributed significantly to reef-building in the Abrolhos Bank and Rocas Atoll, respectively; (vii) echinoderms, with emphasis on sea urchins, that control algal overgrowth through herbivory but may also contribute to bioerosion; (viii) sponges, which are abundant in Brazilian reefs and both add structural complexity and recycle organic matter, but also contribute to reef erosion; (ix) polychaetes and (x) mollusks, both of which may occasionally contribute to reef accretion and also engage in burrowing and erosion; (xi) crustaceans, which interact with numerous taxa and

---

M. Mies (✉) · T. M. C. Lotufo
Instituto Coral Vivo, Santa Cruz Cabrália, BA, Brazil

Instituto Oceanográfico, Universidade de São Paulo, São Paulo, SP, Brazil
e-mail: miguel.mies@usp.br

T. N. S. Banha
Instituto Oceanográfico, Universidade de São Paulo, São Paulo, SP, Brazil

C. B. Castro
Instituto Coral Vivo, Santa Cruz Cabrália, BA, Brazil

Museu Nacional, Universidade Federal do Rio de Janeiro, Rio de Janeiro, RJ, Brazil

R. T. S. Cordeiro
Instituto Coral Vivo, Santa Cruz Cabrália, BA, Brazil

Departamento de Biologia, Universidade Federal Rural de Pernambuco, Recife, PE, Brazil

A. Bianchini
Instituto Coral Vivo, Santa Cruz Cabrália, BA, Brazil

Instituto de Ciências Biológicas, Universidade Federal do Rio Grande,
Rio Grande, RS, Brazil

R. K. P. Kikuchi et al. (eds.), *Brazilian Coral Reefs*, Brazilian Marine Biodiversity,
https://doi.org/10.1007/978-3-031-59152-5_4

form multiple trophic links; and (xii) ascidians, a diverse cryptic dweller of the reef matrix. The basic morphology is presented for each taxon, as well as their diversity, reproductive biology and main functional roles.

Keywords Coral · Crustose coralline algae · Invertebrate · Physiology · Sponge · Urchin

4.1 Introduction

Reefs are the most diverse aquatic environments in the planet, sustaining 25% of all marine life. Millions of prokaryotic and eukaryotic species are found within reefs, which harbor all but two animal phyla (Micrognathozoa and Onychophora). Corals are the main reef-builders and the most emblematic of reef inhabitants, but several other taxa play key roles as well. Mollusks, bryozoans and algae also engage in reef-building, while other groups such as echinoderms and sponges play dual roles by promoting reef diversity while at the same time contributing to the erosion of the reef framework. Therefore, the interaction of reef-dwelling organisms, through countless ecological and physiological processes, maintains this complex ecosystem integrity.

Brazilian reef environments are distinctive because of their turbid and nutrient-rich conditions Costa Jr et al. (2006), Santana et al. (2023). (refer to Chap. 1). Consequently, the Brazilian reef fauna is characterized by a comparatively lower diversity but high endemism. This chapter focuses on the biology and ecology of benthic organisms found in the unique Brazilian reef environments.

4.2 Scleractinian Corals and Milleporids

4.2.1 General Biology, Morphology and Functional Role

Scleractinian corals (Cnidaria: Anthozoa) are sessile animals that may form extensive colonies composed of numerous individual polyps. These organisms are part of a basal clade within Eumetazoa and first emerged approximately 250 million years ago, during the Triassic (Stanley and Fautin 2001). Most shallow-water scleractinians are light-dependent organisms because of their obligate symbiotic association with microscopic photosynthetic algae. However, the main characteristic that is diagnostic of scleractinian corals is the deposition of calcium carbonate skeletons that are crystallized in the form of aragonite. The colony skeleton architecture varies between species and there are many different morphological categories: branching, massive, encrusting, plate, columnar, among others (Veron 2000).

There are 837 extant photosymbiotic species described for Scleractinia (Veron et al. 2016; WoRMS 2021). Most of these (765) are found in the Indo-Pacific and associated with biodiversity hotspots such as the Coral Triangle and Great Barrier Reef. Only 72 species are found in the Atlantic Ocean, none of which are shared with the Indo-Pacific. Seventeen photosymbiotic scleractinian species are found within Brazilian reefs (Table 4.1), four of which are endemic—all belonging to the *Mussismilia* genus (Leão et al. 2003, 2016; Mies et al. 2020). The majority of these species display massive growth forms, such as the dominant *Mussismilia* spp., *Siderastrea* spp. and *Montastraea cavernosa*. Branching scleractinians, which are dominant in the Indo-Pacific, are entirely absent from Brazilian reefs. However, there are four species of photosymbiotic milleporid hydrocorals (*Millepora* spp.) in Brazil, three of which may display branching morphology and occupy this niche (Luza et al. (2023); Amaral et al. 2008). Hydrocorals are calcified hydrozoans and they are morphologically different from scleractinians, especially in having two main different types of polyps: the gastrozooids, used for feeding and digestion, and the dactylozooids, used for prey capture and protection.

Corals perform several key functions in the reef ecosystem. Most importantly, they are responsible for building the reef framework (*i.e.* hermatypic). The tridimensional structure and numerous topographic and hydrodynamic features allow for a high degree of habitat complexity, thus serving as shelter, feeding and reproduction site for millions of marine eukaryotic and prokaryotic species (Knowlton et al. 2010; Graham and Nash 2013). Corals and their photosynthetic symbionts also engage in intensive primary production, ranking coral reefs among the most productive ecosystems in the planet (Atkinson 2011). In addition, corals and other typical reef inhabitants (*e.g.* sponges and bivalves) perform nutrient recycling and reintroduce carbon, nitrogen and phosphorus into the food web by feeding on dissolved and particulate organic matter (Sorokin 1973; Rix et al. 2017).

4.2.2 The Symbiotic Relationship

As is the case for virtually all animals, corals are holobiontic organisms. They harbor multiple symbiont communities composed of bacteria, viruses, archaeans, fungi and endolithic algae. However, their main symbionts are photosynthetic dinoflagellates that belong to the Symbiodiniaceae family. Corals first engaged in a symbiotic association with symbiodiniaceans during the Jurassic (LaJeunesse et al. 2018). This association allowed for multiple adaptive radiation episodes and the widespread colonization of tropical shallow areas (Stanley 2003).

The dinoflagellate symbionts are kept in an intracellular compartment called symbiosome, where they have access to nitrogen, phosphorus and CO_2 derived from the host metabolism (Davy et al. 2012). In addition, the host provides the symbionts with a safe environment—within the host tissue, the dinoflagellates avoid predation

Table 4.1 Morphology, reproductive characteristics and conservation status for the 21 photosymbiotic scleractinian and milleporid species found in Brazilian coral reefs

Species	Primary growth morphology	Sexual pattern	Mode of development	Conservation status
Agaricia fragilis	Plate	?	Brooding	LC
Agaricia humilis	Encrusting	Hermaphroditism	Brooding	LC
Favia gravida	Massive	Hermaphroditism	Brooding	LC
Madracis decactis	Massive	Hermaphroditism	Brooding	LC
Meandrina brasiliensis	Flabelloid	?	?	DD
Millepora alcicornis	Branching	Gonochorism	Broadcast spawning	LC
Millepora braziliensis[E]	Branching	Gonochorism[*]	Broadcast spawning[*]	DD
Millepora laboreli[E]	Columnar	Gonochorism[*]	Broadcast spawning[*]	VU
Millepora nitida[E]	Branching	Gonochorism[*]	Broadcast spawning[*]	LC
Montastraea cavernosa	Massive	Gonochorism	Broadcast spawning	LC
Mussismilia braziliensis[E]	Massive	Hermaphroditism	Broadcast spawning	VU
Mussismilia harttii[E]	Phacelloid	Hermaphroditism	Broadcast spawning	EN
Mussismilia hispida[E]	Massive	Hermaphroditism	Broadcast spawning	LC
Mussismilia leptophylla[E]	Massive	?	?	DD
Porites astreoides	Massive	Hermaphroditism	Brooding	LC
Porites branneri	Massive	Hermaphroditism	Brooding	LC
Scolymia wellsii	Solitary	Hermaphroditism	Brooding	LC
Siderastrea radians	Massive	Gonochorism	Brooding	DD
Siderastrea siderea	Massive	Gonochorism	Broadcast spawning	n/a
Siderastrea stellata	Massive	Gonochorism	Brooding	DD
Stephanocoenia intersepta	Massive	Gonochorism	Broadcast spawning	DD

Morphology data were taken from Veron et al. (2016) and Mies et al. (2020); reproductive data were retrieved from Kerr et al. (2011), Madin et al. (2016) and Pires et al. (2016); conservation status is given according to ICMBio (2018)—LC, VU, EN and DD correspond to "least concern", "vulnerable", "endangered" and "data deficient", respectively

[E]Endemic to Brazil

[*]Further evidence required for confirmation

?: unknown

n/a: not available; occurrence in Brazil disputed

in the plankton by grazers while still having access to light. In exchange, symbionts translocate photosynthates to their hosts, mainly in the form of glucose and glycerol, which may account to up to 100% of the host energetic need (Muscatine et al. 1981; Muscatine 1990). Furthermore, the symbiont facilitates the deposition of the coral skeleton through light-enhanced calcification (Moya et al. 2006). Therefore, it is a mutualistic association as both sides benefit. However, for the vast majority of photosymbiotic corals, this relationship is obligate and they cannot survive without their dinoflagellate symbionts.

Symbiodiniaceae dinoflagellates are also remarkably diverse, with more than one thousand phylotypes described. Because symbiodiniaceans are morphologically indistinguishable, identification is performed through molecular markers such as the ITS2 (internal transcribed spacer 2—see LaJeunesse 2001; Hume et al. 2018). Currently, there are 11 genera and dozens of species formally described within the group (LaJeunesse et al. 2018; Nitschke et al. 2020; Pochon and LaJeunesse 2021). There is significant functional diversity within Symbiodiniaceae. Phylotypes differ in their response and tolerance to disturbances such as increased temperatures and irradiance (Ulstrup and van Oppen 2003; Robison and Warner 2006; Swain et al. 2017). They also differ in their effectiveness as symbionts, with some phylotypes translocating a higher quantity of photosynthates to their hosts (Stat et al. 2008). Symbiodiniaceans may be also described as either generalists or specialists depending on the number of host taxa they associate with (Thornhill et al. 2014; Mies et al. 2020).

Investigations on Symbiodiniaceae diversity and ecology in Brazilian reefs are still scarce. So far, 24 phylotypes have been described for Brazilian reefs, 6 of which are endemic (Garrido et al. 2022). These phylotypes are distributed among five genera (*Symbiodinium*, *Breviolum*, *Cladocopium*, *Fugacium* and *Gerakladium*—Mies et al. 2020). Functional and physiological assessments of Brazilian symbiodiniaceans are still pending.

4.2.3 Reproduction

Corals reproduce both sexually and asexually. Asexual reproduction typically occurs through budding of polyps, which is the process that will eventually lead to the formation of large colonies. Budding can take place in either intratentacular or extratentacular fashion (internal division of existing polyps, or development of new polyps from adjacent tissues, respectively—see Veron 2000). Additional forms of asexual reproduction such as fragmentation can also take place. However, although asexual reproduction may offer advantages such as rapid colonization, critical processes including genetic variability and dispersal are only achieved through sexual reproduction (Harrison and Wallace 1990).

Scleractinians species may be either gonochoric or hermaphroditic (reviewed in Fadlallah 1983; Harrison and Wallace 1990). The polyps and colonies of gonochoric species belong to a single sex. Hermaphroditic species are usually simultaneous, and a single polyp develops both female and male gametes at the same time. In both cases, gonads are not present and reproductive cells develop in the mesenteries. The reproductive mode of development (reviewed in Harrison 2011) can be distinguished between broadcast spawning and brooding. Broadcast-spawning species will release their gametes in the water column and both fertilization and the embryonic development take place externally. Brooders undergo internal fertilization, and embryonic and early larval development take place within the parental polyp before the release of an advanced-stage planula larva. Brooders transmit their dinoflagellate symbionts vertically, as they are passed on from the parental colony to the oocytes. Broadcast spawners, however, usually produce aposymbiotic planulae that must acquire symbionts from the plankton (Baird et al. 2009). Spawning and larval release are regulated by several abiotic factors, which include temperature, lunar cycle and wind pattern (Shlesinger and Loya 1985; van Woesik 2010). Lecithotrophic larvae typically spend between 3 and 15 days in the plankton before settling and metamorphosing into a founding polyp (Babcock and Heyward 1986).

The reproduction of Brazilian corals has been investigated since the 1990s. Later, in 2006, the Coral Vivo Research Station was established in Porto Seguro (Bahia State) and since then keep monitoring natural and induced reproductive events for most Brazilian shallow-water scleractinian species (Pires et al. 2016). The large and massive *Mussismilia* species are broadcast-spawning hermaphrodites that will spawn two or three times a year (Table 4.1; Pires et al. 1999; Neves and Pires 2002). Smaller species (*e. g., Agaricia humilis*, *Favia gravida* and *Porites branneri*) show a tendency towards brooding (Pires et al. 2016). Although investigations are still lacking for Brazilian milleporids, they are all likely broadcast-spawning gonochorists that release their gametes attached to small medusae (Lewis 2006).

4.2.4 Adaptations to Local Environmental Conditions

Brazilian reefs are found under a unique oceanographic setting because the Southwestern Atlantic continental shelf is influenced by the discharge of several large rivers (*e.g.* Amazon, São Francisco, Jequitinhonha, Doce, Paraíba do Sul and Plata see Leão et al. 2003; Omachi et al. 2019, Chap. 1). The river plumes carry nutrients and sediments to reef areas, prompting exceptionally high levels of nitrate (often >3.0 μM), phosphate (> 0.5 μM), total organic carbon (> 5.0 mg L^{-1}) and sedimentation rates (> 50 mg cm^{-2} day^{-1}) (Costa Jr et al. 2000; Segal and Castro 2011; Bruce et al. 2012). Compared to the Indo-Pacific and Caribbean, which are usually associated with clear and oligotrophic waters, the high level of nutrients and suspended sediments found in the Brazilian shelf produce an adverse scenario for reef development. This likely led to the establishment of a resistant coral fauna that possesses several mechanisms for coping with such conditions (Leão et al. 2003).

One of the main features that confer higher resistance to Brazilian scleractinian communities is that most species display massive skeletal architecture. Massive corals are more tolerant to environmental disturbances because of their thicker tissues, higher content of carbon and soluble proteins, and elevated metabolic rates (Gates and Edmunds 1999; Schlöder and D'Croz 2004). An additional feature that promotes higher tolerance to environmental disturbances is symbiotic flexibility. Generalist coral species, *i.e.*, corals that associate with a wide variety of Symbiodiniaceae phylotypes, are able to shift their symbiont communities and therefore acclimate and adapt more rapidly (Baker 2003; Little et al. 2004). Brazilian corals are known generalists, with 75% of species associating with multiple Symbiodiniaceae genera, compared to 45% for the Caribbean and 28% for the Indo-Pacific (Mies et al. 2020).

Brazilian corals also display adaptations for living under turbid conditions. The larger and deeper polyps found in most massive species such as *Mussismilia* spp. allow for a more effective removal of sediment deposited on their surface (Logan 1988; Leão et al. 2003). Furthermore, whenever under higher turbidity and reduced photosynthetic capacity, *Mussismilia* spp. display an intense heterotrophic behavior as a compensation mechanism (Mies et al. 2018; Marangoni et al. 2019b). In fact, it has been reported that *Mussismilia braziliensis* (Fig. 4.1) increases its reproductive output under higher turbidity and nutrient concentrations (Pires et al. 2011). Some Brazilian species also display tolerance to reduced salinity. Pereira et al. (2020) report that *Favia gravida* larvae settle and metamorphose normally at 25 ppt, without significant change in their metabolic activity. However, the physiological processes associated with the higher tolerance to elevated nutrient concentrations are not well-understood.

4.2.5 Physiological Responses to Stressors

Coral reefs and their organisms currently face the combined effects of multiple stressors, of both local and global scales. These stressors include global warming, ocean acidification, pollution, overfishing, competition with invasive species and tourism, among others. In this section we focus specifically on the coral response to global warming, ocean acidification and inorganic pollution, to which they display a wide array of physiological responses.

Global warming is responsible for the bleaching phenomenon. Under elevated temperatures, both Symbiodiniaceae dinoflagellates and the host coral engage in the production of reactive oxygen species (ROS) (Lesser 2006; Weis 2008). ROS accumulation leads to membrane, lipid and DNA damage (Lesser 2006; Weis 2008; Yakovleva et al. 2009). After antioxidant mechanisms become exhausted, the coral's adaptive response is to release its photosymbionts, which, together with host pigment breakdown, leaves the white calcareous skeleton visible underneath the translucent soft tissue (Glynn 1993)—*i.e.*, "bleached". However, excessive symbiont exocytosis leads to host starvation and death. Bleaching also leads to a destabilization in the holobiont nutrient cycling—corals under thermal stress release more ammonia, which causes for the dinoflagellates to become carbon-limited instead of

Fig. 4.1 Some of the main reef-building and reef-dwelling taxa in Brazil: (**a**) The endemic scleractinian coral *Mussismilia braziliensis*, (**b**) the milleporid hydrocoral *Millepora alcicornis*, (**c**) the anemone *Condylactis gigantea*, (**d**) crustose coralline algae belonging to *Litophyllum* and *Amphiroa* genera, (**e**) the bryozoan *Celleporaria mordax*, (**f**) the vermetids *Petaloconchus varians*. (**g**) the green sea urchin *Lytechinus variegatus*, (**h**) the haplosclerid sponge *Callyspongia* sp., (**i**) the lobster *Panulirus echinatus*, (**j**) the serpulid polychaete *Spirobranchus giganteus*, (**k**) the sea snail *Cyphoma macumba*, and (**l**) the ascidian *Didemnum aurantium*. Photo credits: Fábio Negrão/Coral Vivo (**a**), Athila Bertoncini/Coral Vivo (**b, h, k, l**), Miguel Mies (**c, g**), Thomás Banha (**d**), Tito Lotufo (**e, i**), Athila Bertoncini/Ilhas do Rio (**f**), Leones Lopes/Coral Vivo (**j**)

nitrogen-limited, thus ceasing carbon translocation to their host (Rädecker et al. 2021). Therefore, corals are already starving prior to the release of their symbionts.

Fig. 4.1 (continued)

The scleractinian *Mussismilia harttii* and the hydrocoral *Millepora alcicornis* have been widely used as model organisms for the investigations of the physiological impacts associated with thermal stress. Investigations show that photosynthetic efficiency, measured through the Fv/Fm metric, is significantly reduced under elevated temperature (Fonseca et al. 2017). Thermal stress is also well-correlated with lipid peroxidation (LPO) and total antioxidant capacity (TAC) (Marangoni et al. 2019a). However, short-term exposure (up to 8 days) to increasing temperature was shown to reduce LPO in *M. harttii*, thus suggesting the occurrence of a remodeling process in the lipid composition of biological membranes (Fonseca et al. 2017).

Landmark studies have also reported that peroxynitrite (a reactive nitrogen species) generation was linked to the onset and intensity of bleaching for both species under thermal stress (Marangoni et al. 2019b).

Bleaching impacts are on the rise in Brazil, with mortality increasing over time. Branching milleporids underwent mortality higher than 90% in several areas (Duarte et al. 2020; Ferreira et al. 2021). However, nearly all other species have suffered little mortality despite being exposed to intense heatwaves (Banha et al. 2020). This increased resilience is attributed to ecological and physiological aspects, including (i) massive growth morphology; (ii) symbiosis flexibility; (iii) increased heterotrophic input; (iv) protection from irradiance and heat by thriving in turbid areas; and (v) extended bathymetric distribution to higher (and less warm) depths (Mies et al. 2020). However, albeit in better shape than their Indo-Pacific and Caribbean counterparts, Brazilian reefs are facing an alarming advance of bleaching-associated mortality Braz et al. (2022), Pereira et al. (2022), Corazza et al. (2024).

Besides global warming, ocean acidification is a major threat to corals because it compromises the calcification process. The introduction of CO_2 in the seawater releases protons that will associate with carbonate, thus reducing its availability for coral calcification (see Hoegh-Guldberg et al. 2007). Investigations using Brazilian species have shown that *Millepora alcicornis* can cope, to some extent, with long-term (30 days) CO_2-driven acidification of seawater (pH 7.5) (Marangoni et al. 2017a). In this case, it was reported that an increased Ca-ATPase activity played a key role in the maintenance of calcification rate under scenarios of moderate (pH 7.8) and intermediate (pH 7.5) levels of seawater acidification. However, despite observing increases in Ca-ATPase and carbonic anhydrase activities, they were not enough to compensate for the effects of CO_2-driven reduction in seawater pH on the net calcification rate under a scenario of severe ocean acidification (pH 7.2).

Brazilian reefs have also been exposed to pollutants such as nutrients from continental runoff, chemical contamination associated with metals, pesticides and oil spills. Exposure to inorganic aquatic contaminants such as copper (Cu) are shown to negatively affect Brazilian corals in the field (Dal Pizzol et al. 2021), as well as under laboratory and mesocosm conditions (Marangoni et al. 2017b; Fonseca et al. 2019, 2021a). Field studies performed by Dal Pizzol et al. (2021) evaluated the influence of metal accumulation on the oxidative status (LPO and TAC) and carbonic anhydrase activity for *Mussismilia harttii* and *Millepora alcicornis* collected at Abrolhos Bank, potentially impacted by the rupture of the ore mining tailings dam occurred in Mariana municipality (Minas Gerais State) in November 2015. Findings show that these species had higher iron (Fe) and manganese (Mn) levels than those expected for preserved offshore reefs. Furthermore, increased concentrations of arsenic (As), chromium (Cr) and Mn were shown to induce carbonic anhydrase activity inhibition, which may be related to the oxidative stress condition observed in the holobiont.

Under experimental (laboratory and mesocosm) conditions, metal (Cu) exposure was shown to induce an oxidative stress condition (increased TAC and DNA damage) and a possible reduced calcification ability for *M. harttii* (Marangoni et al.

2017b). More recently, Fonseca et al. (2021a) demonstrated that exposure to dissolved Cu for 12 days in a marine mesocosm system induced an increased LPO level in *Mussismilia harttii*. Under *in vitro* conditions, it was shown that Cu induced inhibition of both lactate dehydrogenase (LDH) and electron transport system activity (ETS) by direct binding, thus negatively affecting the energy metabolism in Cu-exposed corals (Fonseca et al. 2021b).

Brazilian corals are currently threatened by both global and local impacts. It is important to note that these environmental stressors are prone to influence corals not as isolated factors, but likely in combination. Several investigations have shown that thermal stress, acidification and pollution often produce synergistic effects on Brazilian corals (Fonseca et al. 2017, 2021a; Zebral et al. 2019).

4.2.6 General Aspects of Other Reef Cnidarians

Non-calcifying cnidarians are also abundant in reef environments. Octocorals, black corals, sea anemones and zoanthids (all anthozoans) are typical reef inhabitants that both add structural complexity and contribute to nutrient recycling (Roberts et al. 2006). Some of these organisms, especially octocorals and zoanthids, may be the dominant taxa in some reef environments.

Octocorals (Anthozoa: Octocorallia) constitute the most diverse group within Cnidaria, with almost 3500 extant species described. However, only 25% are exclusive to shallow reef environments (Cairns 2007; Perez et al. 2016). Seventeen shallow reef species are known from Brazil, most of which are gorgonians (Castro et al. 2010). The main species belong to genera *Plexaurella*, *Muriceopsis* and *Phyllogorgia*, and similar to scleractinians, octocorals can function as mixotrophic organisms that both feed heterotrophically and rely on Symbiodiniaceae for nutrition (Pupier et al. 2021). Most Southwestern Atlantic gorgonians possess an inner horny axis and a discontinuous skeleton formed by microstructures called sclerites, which play multiple roles in the colony structure. Externally, they are adaptively related to abrasion, flexibility and predation (Lewis and Wallis 1991; Clavico et al. 2007), whereas internally they form channels responsible for nutrient flow, growth and regeneration (Pinzón et al. 2014). Feeding assays revealed that the concentration and length of sclerites in octocoral tissues are inversely proportional to palatability by predators (West 1998).

Black corals (Antipatharia) form colonies that may live for millennia and are usually found at mesophotic depths, including at deeper Abrolhos Bank reefs (Loiola 2007; Wagner et al. 2012). Their main characteristic is having long, spinous and dark-coloured axial skeletons (Bo et al. 2011; Brugler et al. 2013). Twenty-five black coral species occur in Brazil, three of which are likely endemic (Lima et al., 2019). Even though most antipatharians associate with symbiodiniaceans, they are suspension feeders and do not rely on their symbionts for meeting the energetic demand (Wagner et al. 2012).

There are nearly 1200 anemone (Actiniaria) species described, 50 of which are found in Brazil (WoRMS 2021). Anemones are solitary predators that use their long tentacles to catch prey items that may be considerably larger than their mouth, including fish. Smaller species such as those belonging to the *Bunodosoma* genus are very common in Brazil, especially in intertidal zones and rocky reefs. Larger species, such as *Condylactis gigantea* (Fig. 4.1), may be found throughout the coast, but its abundance has decreased severely because of collection associated with the aquarium trade, rendering it an endangered species status (Gasparini et al. 2005).

Zoanthids (Zoantharia) are colonial cnidarians that may form extensive mats. Approximately 360 species have been described, 13 of which are found in Brazil (WoRMS 2021). Zoanthid zones are very common, especially in shallow rocky reefs, where populations of *Palythoa caribaeorum* often become dominant (Laborel 1970). Additional zoanthid species typical of Brazilian reef environments are *P. variabilis, P. grandiflora, Parazoanthus swiftii, Zoanthus pulchellus* and *Z. sociatus*. Zoanthids have smaller tentacles and their nutrition relies mostly on the carbon translocated by Symbiodiniaceae dinoflagellates, and possibly on the acquisition of dissolved organic matter as well (Leal et al. 2017).

4.3 Other Reef-Building Organisms

4.3.1 Crustose Coralline Algae

Crustose coralline algae (CCA) are calcifying red algae (Rhodophyta) (Fig. 4.1). This group displays fossil records dating back to the Permian, 260 million years ago (Chen et al. 2020). CCA are marine multicellular plants with a calcified cell wall, largely composed of magnesium calcite (and occasionally, aragonite) crystals. They are composed of three thallii, which produce heavily calcified plates: the epithallus, the photosynthetic perithallus and the basal hypothallus (Littler and Littler 2013). Their red or pinkish coloration is due to the presence of pigments such as phycocyanin and phycoerythrin.

CCA are usually found in either encrusting or rhodolith form. The latter is formed by non-geniculate species that produce detached nodules, called "rhodoliths" (Bosellini and Ginsburg 1971). CCA are gonochorists and female gametes are held in the perithallus and internally-fertilized by flagellum-less sperm cells called spermatia. Eggs are then brooded within the plant (Santelices 2002). Species generally display a triphasic life cycle composed of a haploid gametophyte stage, a diploid sporophyte and a second spore-forming stage (Littler and Littler 2013).

Seventy-nine CCA species have been recorded for Brazil (Sissini et al. 2021). Several form extensive and flat rhodolith beds, which are found throughout the Brazilian continental shelf. This country displays the largest rhodolith beds in the world, most notably those found in the Great Amazon Reef System and the Abrolhos Bank (Amado-Filho et al. 2012; Cordeiro et al. 2015; Moura et al. 2016). Amongst the most relevant reef-building species are *Porolithon antillarum*, *Lithothamnion crispatum* and *Lithophyllum congestum*. (Figueiredo et al. 2008; Brasileiro et al. 2016). CCA are also abundant at mesophotic reefs, aided by the efficiency of light-harvesting by phycoerythrin even at higher depths (Goldberg 2013).

CCA are the most effective reef-builders and the primary biogenic element found in the framework of most Brazilian reefs, including those at Abrolhos Bank (Leão et al. 2003; Bastos et al. 2018). Small CCA and/or rhodolith mounds are also an early successional stage that leads to the formation of reefs. The action of sponges, fish and other organisms brings individual rhodoliths together, forming a small vertical reef that will serve as the platform for a larger reef structure (Pereira-Filho et al. 2015). In addition, the calcareous substrate associated with CCA is a relevant cue for the settlement and metamorphosis of coral larvae (Negri et al. 2001). In general, the higher presence of CCA is an indicator of good reef health.

4.3.2 *Bryozoans*

Organisms belonging to the Bryozoa phylum are small and sessile animals that possess a lophophore, which is a gill-like organ used for feeding and gas exchange. This group has its origins in the early Ordovician and became dominant reef-builders in the Devonian (Wood 1999). Bryozoans are coelomate and triploblastic organisms that typically produce encrusting or branching colonies. These colonies are composed of individuals called zooids, although many reef bryozoans also produce heterozooids, which are non-feeding individuals that may display additional functions such as protection or attachment (Goldberg 2013). The bryozoan external skeleton (called "zoecium") is usually heavily calcified and may be composed of calcite and/or aragonite (Smith et al. 2006).

The lophophore is composed of the mouth and its surrounding ciliated and hollow tentacles, in similar shape to a crown. Feeding takes place with an upstream collection system, in which the water current generated by the cilia movements directs food towards the mouth (Goldberg 2013). The bryozoan digestive system is arranged in the form of "U", positioning the protruded anus very close to the mouth. Some species also display an operculum-like structure that closes in order to protect the retracted lophophore (Todd and Havenhand 1989).

Similar to corals, bryozoans may reproduce both sexually and asexually. Asexual reproduction occurs through zooid budding. As for sexual reproduction, the zooids of most species are hermaphroditic, and both simultaneous and sequential hermaphroditism have been reported (Buchsbaum et al. 1987). In most cases, sperm is released in the water column before being captured by the tentacles of conspecific colonies. Eggs are then fertilized and brooded within the zoecium or specific chambers called ovicells (Buchsbaum et al. 1987). After the completion of the embryonic development, a planktonic cyphonautes larva is released in the water column, until settling and metamorphosing into a founding zooid called ancestrula (Brusca and Brusca 2003).

Approximately 5500 extant bryozoan species have been described, most of which are marine (WoRMS 2021). For Brazil, about 460 marine species have been reported (Vieira et al. 2016). The most important functional role performed by bryozoans is reef-building. Although there are notable bryozoan-built reefs in locations such as Florida and New Zealand (Bradstock and Gordon 1983; Lombardi et al. 2020), bryozoans are more often secondary frame builders (Wood 1999). It is the case for Brazilian reefs, where bryozoans such as *Celleporaria atlantica*, *Parasmittina* sp. and *Stylopoma* sp. were critical in the formation of the unique "chapeirões" found in the Abrolhos Bank (Bastos et al. 2018).

4.3.3 Vermetids

Also known as "worm snails", animals belonging to Vermetidae family are sessile gastropods that produce rigid and uncoiled shells. Originally established during the Mesozoic (Vermeij 1977; Bandel 1993), they inhabit both intertidal and subtidal habits, and are often present in reef crests along with CCA. Their thick shell (formed by aragonite and/or calcite) and operculum allow for them to withstand high wave energy and desiccation associated with reef crests. Vermetid communities may reach exceptionally high densities superior to 20,000 individuals m^{-2} (Hadfield et al. 1972; Miloslavich et al. 2010).

Vermetids are suspension feeders that capture food particles through mucous traps that extend beyond their shells, and also by filtering through their gill cilia. Like other gastropod species, a radula is present and is used both for feeding and also enlarging the tube aperture. Worm snails are typically gonochorists and fertilization occurs internally (Scheuwimmer 1979; Calvo et al. 1998). Male individuals produce spermatophores in the mantle cavity and expel them until they reach female individuals, where the offspring is brooded. Similar to nudibranchs, some vermetid species are poecilogonous and hatchlings include both veliger larvae and juvenile individuals (Hadfield et al. 1972). Eventually, juveniles will permanently cement themselves to hard substrata and initiate shell production.

More than 100 vermetid species have been described (WoRMS 2021). Five genera have been reported in Brazil, including the non-indigenous *Eualetes tulipa*, but diversity assessments at species level are still undergoing further investigations (Spotorno et al. 2012; Spotorno-Oliveira et al. 2018). However, the dominant species seem to be *Vermetus irregularis* and *Petaloconchus varians*. Vermetid reefs are well-known throughout the globe, especially the conspicuous "cup reefs" found in Bermuda. Within Brazil, vermetids are dominant reef-builders at the Rocas Atoll together with CCA (Gherardi and Bosence 2001). However, they also make an important contribution to the reef matrix in most Brazilian reefs, including rocky reefs at the southeastern coast (Leão et al. 2003; Breves et al. 2016).

4.4 Echinoderms

4.4.1 General Biology, Morphology and Reproduction

Echinodermata is an animal phylum comprised exclusively of marine species whose body plan is arranged in pentaradial symmetry. The oldest echinoderms in the fossil record date back from the Cambrian (Zamora et al. 2012). They display a multitude of unique morphological characteristics (synapomorphies). This includes a water vascular system that controls the action of podia (tube feet) and pedicellariae, which are structures used mainly for locomotion and food capture, respectively. Additional typical features include a body wall skeleton, the presence of mutable connective tissue and the madreporite, which is a ciliated porous plate from where water and oxygen enter the vascular system (Wessel 2018). The endoskeleton is formed by magnesium calcite ossicles, with its architecture varying significantly according to each class. These organisms are usually oriented in oral-aboral fashion, with most taxa positioning the mouth towards the substrate.

Crinoids are filter-feeding organisms that are usually attached to the substrate by a stalk and holdfast (Goldberg 2013). Particles are captured by a crown of arms containing lateral appendages called pinnules. Holothuroids are usually associated with sandy substrata as most species are deposit-feeders that sift through sediment for food. Their body plan is cylindrical and radially symmetric along a longitudinal axis. Echinoids are mainly represented by sea urchins. They are spherically-shaped and display a test made of ten fused plates with multiple spines (Lawrence 2020). Tube feet extend from additional plates called ambulacral plates. A feeding structure that is unique to echinoids is the Aristotle's lantern, which is composed of triangular calcareous plates with multiple teeth and is used for grabbing and scraping. In general, echinoids are vagile animals and function as detritivores, omnivores or herbivores. Ophiuroids display five arms that are connected to a central body disk; these arms may regenerate if damaged. Brittle stars are free-living detritivores that typically inhabit crevices in consolidated substrata. Asteroids also have five arms in most cases, but this number can vary greatly among species (Lawrence 2013). Like brittle stars, the arms may regenerate. Most asteroids are free-living omnivores that feed by projecting the stomach from the mouth and performing extracellular digestion.

Several echinoderm species may reproduce asexually through fissiparity, especially within Holothuroidea, Ophiuroidea and Asteroidea. As for sexual reproduction, most species are gonochorists and hermaphroditism is reported only for some holothuroid, ophiuroid and asteroid species (Wessel 2018). Although brooding is common for all classes, most echinoderm species are broadcast spawners (Wessel 2018). Therefore, fertilization is mostly external and larvae develop in the plankton. All echinoderm larval forms are supposedly derived from an ancestral larval stage called dipleurula (Byrne et al. 2007). Some classes display a complex life cycle, with multiple larval stages (McEdward and Miner 2001; Carrier et al. 2018).

Currently, there are approximately 7000 extant echinoderm species, of which *ca.* 1500 are associated with reef environments (Clark 1973; WoRMS 2021). In general, Brazilian echinoderm diversity is similar to that found in the Caribbean. More than 300 species have been detected, with an endemism level superior to 10% (Gondim et al. 2018). There are several ubiquitous species such as *Echinometra lucunter* (Echinoidea) (Fig. 4.1), *Echinaster brasiliensis* (Asteroidea), *Isostichopus badionotus* (Holothuroidea) and *Tropiometra carinata* (Crinoidea).

4.4.2 Functional Role

Echinoderms perform multiple functional roles in reef environments. Numerous species are detritivores, especially of Ophiuroids and Holothuroids, thus aiding in carbon remineralization and recycling. Many species also add habitat complexity to reef environments. It is the case of crinoids, whose arms harbor multiple epizoic organisms including crustaceans, mollusks and fishes among others (Fishelson 1974). Sea urchins perform a similar role—their long spines provide shelter for juvenile and adult fishes, while some are also capable of excavating the reef framework creating holes that add sheltered spaces in the reef. Within Brazilian reef environments, the echinoids *Diadema antillarum* and *Echinometra lucunter* are among species that contribute to increased habitat complexity (Giglio et al. 2018; Grande et al. 2020).

The most notable functional role performed by sea urchins is herbivory (see Chap. 5). These organisms are effective grazers that consume a considerable volume of algae (including cyanobacteria and turf algae—see McClanahan et al. 1996; Carpenter 1997), which are opportunistic competitors that outgrow corals (McManus and Polsenberg 2004). *In situ* experiments have shown that areas stocked with sea urchins have less algae and higher coral cover than the control area, which had received no urchins (Chiappone et al. 2003). Further evidence is that a massive *D. antillarum* die-off episode in the Caribbean during the 1980s promoted algal overgrowth, reduced structural complexity and a decline in overall reef diversity (Sammarco 1982; Lessios et al. 1984). In Brazilian reefs, among the main species that reportedly exert control over algal growth are *E. lucunter*, *Lytechinus variegatus*, *Paracentrotus gaimardi* and *Arbacia lixula* (Cordeiro et al. 2020).

Although sea urchins are effective in controlling algal overgrowth, their excavating and scraping activities through the use of spines and the Aristotle's lantern may incur in bioerosion, thus weakening the reef framework (Bak 1994). However, net erosion of the reef tends to take place only when sea urchin populations are found at abnormally high densities, often when their predators are overfished (Carreiro-Silva and McClanahan 2001; Dang et al. 2020). Therefore, sea urchins are keystone organisms that play pivotal roles in maintaining reef health and in its carbonate budget.

4.5 Sponges

4.5.1 General Biology, Morphology and Reproduction

Sponges (phylum Porifera) are the simplest extant metazoans. They are sessile suspension-feeders whose body plan contains multiple pores, channels and chambers where water is pumped through. Sponges usually have no symmetry and lack true tissues. They are likely among the most ancient multicellular organisms, present in the fossil record even before the Cambrian, approximately 700 million years ago (Love et al. 2009; Van Soest et al. 2012).

The morphology of sponges is considerably variable. Although controversial, sponges are classified into 10–12 different morphological forms including encrusting, massive, globular, pedunculate and flabellate types (Bell and Barnes 2001). Their evolutionary heritage naturally plays a role in defining their form, but so does the environment. Irradiance, wave action and turbidity are some of the abiotic factors that may influence sponge morphology (Palumbi 1986; McDonald et al. 2002). The sponge skeleton is formed by spicules, which are located in the mesohyl (a middle layer formed by a gelatinous matrix). Spicule shape and size are major characteristics used in poriferan taxonomic classification (Uriz et al. 2003). Species with siliceous spicules are called "glass sponges" and placed within the Hexactinellida class, while species with calcitic spicules are called "calcareous sponges" and placed in the Calcarea class. Species having siliceous spicules coated with and connected by spongin fibers belonging to the Demospongiae class are the most abundant sponges in reef environments.

Sponges reproduce both sexually and asexually as well. Asexual reproduction is attained through fission, fragmentation or gemmulation. Fission occurs spontaneously, while fragmentation is often a consequence of mechanical impact. Gemmulation, however, is the production of an enclosure (gemmule) consisted of a protective layer and spicules. Inside it are archeocytes, which are stem cells that can morph into any sponge cell type (reviewed in Goldberg 2013). Most sponges are hermaphrodites (including both simultaneous and sequential hermaphroditism). Gonads are entirely absent—sperm and eggs are generated in choanocytes and archeocytes (Müller 2006). Most reef species are broadcast spawners with external fertilization. Larval development is lecithotrophic and may range from a few minutes to several weeks (Carrier et al. 2018). Larval morphology can be quite diverse with amphiblastula and parenchymella forms most common.

Approximately 9500 sponge species have been described, of which about 700 are found in Brazil (Hooper et al. 2021). The Guyana Shelf and the Amazon-Orinoco plume mark transition zones in sponge biodiversity. Although several species are shared between Brazil and the Caribbean, this area is the southernmost distribution of multiple Caribbean species, and also the northernmost limit for several Brazilian species (Van Soest 2017). Similar to corals, the level of endemism for Brazilian poriferans is elevated, at 35% (Muricy et al. 2011). Sponges are particularly abundant and diverse at Brazilian mesophotic reefs and rhodolith beds, including the Great Amazon Reef System (Pereira-Filho et al. 2015; Moura et al. 2016).

4.5.2 *Functional Role*

In very similar fashion to corals, sponges may be considered holobionts. They host a diverse bacterial community (including cyanobacteria) that plays a relevant role in both carbon and nitrogen fixation and may even surpass the sponge biomass. In addition, sponges perform the "sponge-loop", a process in which they consume and thus recycle large volumes of dissolves organic matter (De Goeij et al. 2013). They then release particulate organic carbon, which is consumed by benthic detritivores and thus reintroduced to the trophic web (Rix et al. 2017; Pawlik and McMurray 2020). These processes are critical for sustaining productivity in oligotrophic systems such as reef environments.

Sponges also contribute to the reef structural complexity, especially in Brazil where they are ubiquitous reef inhabitants (Francini-Filho et al. 2018). They also promote inquilinism and provide shelter for numerous sponge-dwelling organisms that include crustaceans, polychaetes, and mollusks among several others (Padua et al. 2013). Although sponges may have been relevant reef-building organisms in the past, currently their contribution to reef accretion is negligible. However, demosponge spicules may be a relevant source of siliceous sediment to the reef surroundings. Despite not forming reefs, sponges may agglutinate and bind sediments and rubble, thus forming small mounds that can serve as a site for reef establishment and development (Sheppard et al. 2018).

Despite their several functional roles in sustaining reef health and diversity, sponges are also among the most relevant bioeroders. Through the use of pseudopodia, sponges bore into coral skeleton and contribute significantly to the degradation of the reef matrix (Schönberg 2008). Bioeroding demosponges, especially those belonging to the *Cliona* genus, are widespread in the Abrolhos Bank reefs, where they have primarily eroded structures formed by CCA (Moraes et al. 2019). With the advances of ocean acidification, it is likely that sponge bioerosion increases in the coming decades (Webb et al. 2017).

4.6 Other Invertebrates

4.6.1 *Polychaetes*

Polychaetes (phylum Annelida) are very common in reef environments. They are segmented worms with multiple bristles located on the surface of parapodia. In general, reef polychaetes are either free-moving or sessile and tubiculous.

Free-moving polychaetes are often detritivores, but some larger species such as *Hermodice carunculata* are carnivores whose jaws may consume a significant quantity of coral tissue (Sussman et al. 2003). Tubiculous polychaetes are suspension-feeders that use a branchial crown for both particle collection and gas exchange. These animals are typical reef inhabitants and their tubes may be composed of calcium carbonate, protein or sand grains. Among the most representative reef families are Serpulidae, Spionidae, Sabellariidae and Sabellidae (Goldberg 2013). Most species are gonochoric and their mode of development may be either broadcast spawning or brooding. The planktotrophic larva is called a trochophore.

There are approximately 8000 polychaete species described, of which nearly 1500 are found in Brazil (WoRMS 2021). Notable reef species include the sabellid *Branchiomma luctuosum* and the serpulid *Spirobranchus giganteus* (Fig. 4.1). The latter builds calcareous tubes that are very often associated with large colonies of scleractinians such as *Montastraea cavernosa*. Although marginally when compared to corals, serpulids may contribute to reef-building. Some other species such as free-moving eunicids, however, may use their jaws to burrow and excavate into the reef framework and thus also contribute to bioerosion (Cardona-Gutiérrez and Londoño-Cruz 2020).

4.6.2 Mollusks

Mollusks (phylum Mollusca) are soft-bodied animals that, in most cases, produce a carbonate calcium shell that is secreted by a thick mantle tissue. Although there are seven extant mollusk classes (Aplacophora, Monoplacophora, Polyplacophora, Gastropoda, Cephalopoda, Bivalvia and Scaphopoda), this section will focus largely on gastropods (with the exception of vermetids—see Sect. 4.3.3) and bivalves due to their more prominent role in reef environments.

Gastropods are typically represented by snails and slugs, which account for nearly 80,000 species, with about 1100 of them found in Brazil (Simone 1999). These organisms are vagile (in most cases) and feeding habits vary significantly among species, including herbivory, carnivory and suspension-feeding (Goldberg 2013). Gastropods feed through the use of a radula, a chitinous structure with multiple denticles that is used for scraping the hard surface. Among the notable Brazilian gastropods is *Aplysia dactylomela*, an herbivorous sea hare (Bezerra et al. 2004). Gastropods may be either gonochoric or hermaphroditic, with most species conducting internal fertilization and brooding of the veliger larvae before release into the plankton (Conn 2000).

Bivalves are easily distinguishable by the presence of two shells (valves) connected by a hinge. Within this group are oysters, scallops and clams, among others. Bivalves are sessile and usually epibenthic or infaunal. They are filter-feeders that collect water through an incurrent siphon that leads food to a gill-like organ called ctenidium. There are nearly 7000 bivalve species described, with 400 reported in Brazil (Simone 1999). Bivalves are mostly gonochoric, although hermaphroditism is very common. Most species are broadcast spawners, and larval development undergoes three stages before metamorphosis: a non-feeding trochophore, a planktotrophic veliger, and a benthic pediveliger larva (Conn 2000).

Rudist bivalves were the main tropical reef-building organisms approximately 100 million years ago (Wood 1999). With the notable exception of oyster reefs in Chesapeake Bay and giant clam reefs in the Indo-Pacific (Hargis Jr and Haven 1999; Mies 2019), bivalve contribution to reef-building is nowadays much reduced if compared to the past. Still, both bivalves and gastropods do contribute to reef accretion because of the high density that populations may reach (Wood 1999). Dead and broken shells contribute to the formation of calcareous sediment deposits. However, bivalves may burrow into the reef matrix, thus promoting reef erosion (Tribollet and Golubic 2011). Likewise, although many gastropod species are herbivorous and control turf algae overgrowth, others may consume CCA and degrade the reef surface through scraping (Steneck and Watling 1982). However, some gastropod species are corallivorous—the Brazilian species belonging to the *Cyphoma* genus (Fig. 4.1) are notorious octocoral predators (Pinto et al. 2016). Although to a lesser extent than sea urchins, mollusks may also be keystone organisms in reef environments.

4.6.3 Crustaceans

Crustaceans belong to the phylum Arthropoda and possess multiple articulated appendages. These structures are protected by a cuticle that is made primarily of chitin and protein. When enriched with calcium, the cuticle becomes a fortified exoskeleton. As crustaceans grow, their cuticle becomes too small and they molt (*i.e.*, discard the older and smaller exoskeleton before quickly forming a new one).

There is a large diversity of reef-dwelling crustaceans. The reef epifauna is riddled with minute crustaceans including copepods, amphipods, isopods, ostracods and several others. Although these organisms are morphologically very different, they are in general omnivores that feed on detritus, algae and other small organisms (Goldberg 2013). Cirripeds (barnacles) are also abundant and they are sessile suspension-feeders that collect particles with the use of thoracic appendages called cirri. However, the most well-investigated reef crustaceans are decapods. Within the main Brazilian reef decapods are (Coelho et al. 2008; Giraldes et al. 2015; Terossi et al. 2018): (i) spiny lobsters, such as *Panulirus echinatus* (Fig. 4.1) and *P. meripurpuratus*, which inhabit crevices and venture out to feed on mollusks and other small invertebrates; (ii) caridean shrimps (mainly those within Palaemonidae, Alpheidae and Hippolytidae families), which are small omnivores that inhabit crevices, rubble and occasionally associate with cnidarians; (iii) stenopodidean shrimps such as *Stenopus hispidus*, which feed on small invertebrates; (iv) brachyuran crabs such as *Mithraculus forceps* and *Stenorhynchus seticornis*, which are also omnivores; and (v) anomurans such as hermit crabs, which are omnivores that inhabit discarded gastropod shells. The sexual pattern and reproductive strategies vary significantly within crustaceans, but nearly all of them (with the notable exception of peracarids) produce planktotrophic larvae with typical stages such as nauplius and zoea.

Crustaceans perform multiple functional roles in reef environments. They play a relevant part in benthic-pelagic coupling, as the elevated presence of detritivores promotes effective recycling of deposited organic matter (Kramer et al. 2014). Crustaceans are also a relevant trophic link because they are usually abundant and serve as prey item for numerous reef organisms, thus transferring energy to higher trophic levels (Kramer et al. 2013). Furthermore, crustaceans associate symbiotically with several distinct organisms. Brazilian hippolytid shrimps such as *Lysmata grabhami* and *L. ankeri* are cleaning organisms that remove parasites from larger fish including groupers and moray eels, in a mutualistic relationship (Quimbayo et al. 2018). Some palaemonid shrimps (mainly those within *Periclimenes* genus) associate commensally with multiple organisms such as anemones, scleractinians, sponges, crinoids and urchins (Wirtz et al. 2009; Goldberg 2013). Additional symbionts include barnacle and hermit crab species that associate with scleractinians and gorgonians (McLaughlin and Lemaitre 1993), and isopods and copepods that are parasitic of reef fish and corals (Smit et al. 2014; Christie and Raines 2016). However, some crustaceans such as barnacles may bore into the reef framework and contribute to reef erosion (Tribollet and Golubic 2011). Therefore, crustaceans are key components of reef environments.

Nearly 70,000 crustacean species have been described, with 2400 of them reported from Brazil (Serejo et al. 2006). These organisms are unevenly studied in Brazil, with more knowledge produced about decapods than for any other major crustacean taxon. In general, the Brazilian reef decapod fauna is mostly composed of tropical Atlantic species, with about 10% of endemism (Almeida and Coelho 2008).

4.6.4 Ascidians

Ascidians are invertebrate chordates that include the sessile members of the subphylum Tunicata. Their chordate characteristics are all present during their free-swimming larval stage but are mostly lost during a metamorphosis to the sessile adult form. During this metamorphosis the notochord, post-anal tail and neural tube are resorbed, while their pharynx is enlarged to enable their filter feeding habits. As adults, ascidians typically have two muscular siphons used for inhaling and exhaling seawater. Ascidians may be solitary or form colonies with variable degrees of integration, where the atrial siphon (excurrent) may be connected to canals or common cloacal openings. These animals are protected externally by a tunic, a complex tissue with a cellulose-like matrix. Their pharynx is enlarged and perforated, being internally lined by a mucous net produced by the endostyle, an organ homologous to the hypophysis. This mucous net is used to capture planktonic microalgae, their main food source (Monniot et al. 1991).

Ascidians are invariably hermaphroditic and only colonial species may reproduce asexually for colony growth. Solitary species release their gametes in the water and fertilization is external, while colonies incubate their eggs until the larvae are formed (Monniot et al. 1991). The larvae are tadpole shaped, with a trunk hosting the oozooid, and a muscular tail used for swimming. The larval duration is very short, and usually settle after one or 2 h only (Millar 1971).

There are about 3000 ascidian species described worldwide (Shenkar and Swalla 2011), and over 100 species in Brazil (Dias et al. 2013). Most of the knowledge about the group is concentrated in the southern states, where the major research groups were located (Dias et al. 2013). During the past 20 years, the tropical shallow areas have been more thoroughly studied, and many new species were described (Soares et al. 2016). Similar to many other tropical reefs, the Brazilian coral reef fauna is composed mostly of species from the families Didemnidae, Polycitoridae and Styelidae. Coral reef ascidians are usually cryptic, living in crevices and other spaces in the reef framework. A few species, however, may be conspicuous and form large epibenthic colonies, such as *Stomozoa gigantea* in the coastal reefs (da Lotufo 2002), and *Trididemnum maragogi*, at Rocas Atoll and São Pedro and São Paulo Archipelago (Paiva et al. 2015).

Ascidians may be major components of the cryptic fauna (Reis et al. 2016), including didemnids, that produce calcareous spicules. As many other efficient filter feeders, ascidians are also important elements in the bentho-pelagic coupling. Some didemnids, such as *T. maragogi*, may host symbiotic algae from the genus *Prochloron*. Other species may also present *Prochloron* living in the colony surface, giving it a bright green color (Oliveira et al. 2013). All ascidian species may be considered as holobionts, and most have a peculiar associated fauna, including copepods, amphipods, bivalves and other invertebrates (Millar 1971). Similar to most cryptofauna taxa, the functional role and its contribution to processes involving matter and energy fluxes in the reef are still poorly known and deserve attention.

4.7 Final Remaks: Perspective and Knowledge Gaps

The fauna of Brazilian reefs features a wide variety of species spread across an extensive latitudinal spectrum. While significant advancements have been made in understanding this biodiversity, its wide distribution range poses challenges. The complexity of oceanographic conditions found within the Brazilian province implies that numerous biogeographic patterns and dynamics remain unaddressed. Another critical area of focus is connectivity; our grasp on the dispersal of larvae and the connectivity of populations for numerous benthic species remains poorly understood. This limitation complicates the formulation of effective marine protected areas. Furthermore, our understanding of the ecophysiological responses of many benthic species to multiple and synergistic anthropogenic stressors remains in need of deeper investigation, to aid in predicting the broader impacts at the ecosystem level. Thus, integrated research efforts on the ecosystem roles and responses of Brazilian benthic reef fauna are warranted.

Acknowledgements We would like to thank the editorial team for kindly inviting us to take part in this edition, and the Coral Vivo Project and its sponsors, Petrobras, through the Petrobras Socioenvironmental Program, and Arraial d'Ajuda Eco Parque. We also thank Débora Pires for valuable comments and Athila Bertoncini and Ilhas do Rio Project for some of the photographs.

References

Almeida AO, Coelho PA (2008) Estuarine and marine brachyuran crabs (Crustacea: Decapoda) from Bahia, Brazil: checklist and zoogeographical considerations. Lat Am J Aquat Res 36:183–222

Amado-Filho GM, Moura RL, Bastos AC, Salgado LT, Sumida PYG, Güth AZ, Francini-Filho RB, Pereira-Filho GH, Abrantes DP, Brasileiro PS, Bahia RG, Leal RN, Kaufman L, Kleypas JA, Farina M, Thompson FL (2012) Rhodolith beds are major $CaCO_3$ bio-factories in the tropical South West Atlantic. PLoS One 7:e35171

Amaral FD, Steiner AQ, Broadhurst MK, Cairns SD (2008) An overview of the shallow-water calcified hydroids from Brazil (Hydrozoa: Cnidaria), including the description of a new species. Zootaxa 1930:56–68

Atkinson MJ (2011) Carbon fluxes of coral reefs. In: Hopley D (ed) Encyclopedia of modern coral reefs: structure, form and process. Springer, Dordrecht, pp 181–184

Babcock RC, Heyward AJ (1986) Larval development of certain gamete-spawning scleractinian corals. Coral Reefs 5:111–116

Baird AH, Guest JR, Willis BL (2009) Systematic and biogeographical patterns in the reproductive biology of scleractinian corals. Annu Rev Ecol Evol Syst 40:551–571

Bak RPM (1994) Sea urchin bioerosion on coral reefs: place in the carbonate budget and relevant variables. Coral Reefs 13:99–103

Baker AC (2003) Flexibility and specificity in coral-algal symbiosis: diversity, ecology, and biogeography of *Symbiodinium*. Annu Rev Ecol Evol Syst 34:661–689

Bandel K (1993) Caenogastropoda during Mesozoic times. Scr Geol 2:7–56

Banha TNS, Capel KCC, Kitahara MV, Francini-Filho RB, Francini CLB, Sumida PYG, Mies M (2020) Low coral mortality during the most intense bleaching event ever recorded in subtropical Southwestern Atlantic reefs. Coral Reefs 39:515–521

Bastos AC, Moura RL, Moraes FC, Vieira LS, Braga JC, Ramalho LV, Amado-Filho GM, Magdalena UR, Webster JM (2018) Bryozoans are major modern builders of South Atlantic oddly shaped reefs. Sci Rep 8:9638

Bell JJ, Barnes DK (2001) Sponge morphological diversity: a qualitative predictor of species diversity? Aquat Conserv Mar Freshwat Ecosyst 11:109–121

Bezerra LEA, Carvalho AFU, Barreira LA, Nogueira VLR, Silva JRF, Vasconcelos IM, Melo VMM (2004) The relationship between seaweed diet and purple ink production in *Aplysia dactylomela* Rang, 1828 (Gastropoda: Opisthobranchia) from Northeastern Brazil. J Shellfish Res 23:581–584

Bo M, Baker AC, Gaino E, Wirshing HH, Scoccia F, Bavestrello G (2011) First description of algal mutualistic endosymbiosis in a black coral (Anthozoa: Antipatharia). Mar Ecol Prog Ser 435:1–11

Bosellini A, Ginsburg RN (1971) Form and internal structure of recent algal nodules (rhodolites) from Bermuda. J Geol 79:669–682

Bradstock M, Gordon DP (1983) Coral-like bryozoan growths in Tasman Bay, and their protection to conserve commercial fish stocks. N Z J Mar Freshw Res 17:159–163

Brasileiro PS, Pereira-Filho GH, Bahia RG, Abrantes DP, Guimarães SM, Moura RL, Francini-Filho RB, Bastos AC, Amado-Filho GM (2016) Macroalgal composition and community structure of the largest rhodolith beds in the world. Mar Biodivers 46:407–420

Braz GB, Lacerda CH, Evangelista H, Güth AZ, Rumbelsperger AM, Capel KC, Dall'Occo PL, Mies M (2022) Unprecedented erosion of Mussismilia harttii, a major reef-building species in the Southwestern Atlantic, after the 2019 bleaching event. Coral Reefs 41:1537–1548

Breves A, Moraes FC, Pimenta AD, Junqueira AO (2016) A unique subtidal vermetid reef formation in a nuclear power plant discharge area at Ilha Grande Bay (southwestern Atlantic Ocean). Marine Biodivers Rec 9:1–4

Bruce T, Meirelles PM, Garcia G, Paranhos R, Rezende CE, de Moura RL, Coni EO, Vasconcelos AT, Amado Filho G, Hatay M, Schmieder R, Edwards R, Dinsdale E, Thompson FL (2012) Abrolhos bank reef health evaluated by means of water quality, microbial diversity, benthic cover, and fish biomass data. PLoS One 7:e36687

Brugler MR, Opresko DM, France SC (2013) The evolutionary history of the order Antipatharia (Cnidaria: Anthozoa: Hexacorallia) as inferred from mitochondrial and nuclear DNA: implications for black coral taxonomy and systematics. Zool J Linnean Soc 169:312–361

Brusca RC, Brusca GJ (2003) Invertebrates. Sinauer Associates, Sunderland

Buchsbaum R, Buchsbaum M, Pearse J, Pearse V (1987) Animals without backbones. University of Chicago Press, Chicago

Byrne M, Nakajima Y, Chee FC, Burke RD (2007) Apical organs in echinoderm larvae: insights into larval evolution in the Ambulacraria. Evol Dev 9:432–445

Cairns SD (2007) Deep-water corals: an overview with special reference to diversity and distribution of deep-water scleractinian corals. Bull Mar Sci 81:311–322

Calvo M, Templado J, Penchaszadeh PE (1998) Reproductive biology of the gregarious Mediterranean vermetid gastropod *Dendropoma petraeum*. J Mar Biol Assoc U K 78:525–549

Cardona-Gutiérrez MF, Londoño-Cruz E (2020) Boring worms (Sipuncula and Annelida: Polychaeta): their early impact on Eastern Tropical Pacific coral reefs. Mar Ecol Prog Ser 641:101–110

Carpenter RC (1997) Invertebrate predators and grazers. In: Birkeland C (ed) Life and death of coral reefs. Chapman and Hall, New York, pp 198–229

Carreiro-Silva M, McClanahan TR (2001) Echinoid bioerosion and herbivory on Kenyan coral reefs: the role of protection from fishing. J Exp Mar Biol Ecol 262:133–153

Carrier TJ, Reitzel AM, Heyland A (2018) Evolutionary ecology of marine invertebrate larvae. Oxford University Press, Oxford

Castro CB, Medeiros MS, Loiola LL (2010) Octocorallia (Cnidaria: Anthozoa) from Brazilian reefs. J Nat Hist 44:763–827

Chen W, Wang Y, Huang Y, Wang T, Yi Z, Kiessling W (2020) Reef-building red algae from an uppermost Permian reef complex as a fossil analogue of modern coralline algal ridges. Facies 66:1–4

Chiappone M, Swanson D, Miller S (2003) One-year response of Florida keys patch reef communities to translocation of long-spined sea urchins (*Diadema antillarum*). NOAA Urchin Translocation Project Report

Christie BL, Raines JA (2016) Effect of an otic milbemycin oxime formulation on *Tegastes acroporanus* infesting corals. J Aquat Anim Health 28:235–239

Clark AM (1973) Echinoderms of coral reefs. In: Jones OA, Endean R (eds) Biology and geology of coral reefs, vol 3. Academic, New York, pp 95–123

Clavico EE, Souza AT, Gama BA, Pereira RC (2007) Antipredator defense and phenotypic plasticity of sclerites from Renilla muelleri, a tropical sea pansy. Biol Bull 213:135–140

Coelho PA, Almeida AO, Bezerra LE (2008) Checklist of the marine and estuarine Brachyura (Crustacea: Decapoda) of northern and northeastern Brazil. Zootaxa 1956:1–58

Conn DB (2000) Atlas of invertebrate reproduction and development. Wiley-Liss, New York

Corazza BM, Lacerda CHF, Güth AZ, Marcançoli RKM, Bianchini A, Calderon EN, Capel KCC, Conceição E, Faria SC, Francini- Filho RB, Garrido AG, Guebert FM, Kitahara MV, Longo

GO, Lotufo TMC, Salvi KP, Segal B, Sumida PYG, Zilberberg C, Mies M (2024) No coral recovery three years after a major bleaching event in reefs in the Southwestern Atlantic refugium. Mar Biol 171:114.

Cordeiro RTS, Neves BM, Rosa-Filho JS, Pérez CD (2015) Mesophotic coral ecosystems occur offshore and north of the Amazon River. Bull Mar Sci 91:491–510

Cordeiro CAMM, Harborne AR, Ferreira CEL (2020) The biophysical controls of macroalgal growth on subtropical reefs. Front Mar Sci 7:488

Costa OS Jr, Leão ZMAN, Nimmo M, Attrill MJ (2000) Nutrification impacts on coral reefs from northern Bahia, Brazil. In: Jones MB, Azevedo JMN, Neto AI, Costa AC, Martins AMF (eds) Island, Ocean and Deep-Sea biology. Springer, Dordrecht, pp 207–315

Costa Jr OS, Attrill MJ, Nimmo M (2006) Seasonal and spatial controls on the delivery of excess nutrients to nearshore and offshore coral reefs of Brazil. J Mar Syst 60:63–74.

da Lotufo TMC (2002) Ascidiacea (Chordata: Tunicata) do Litoral Tropical Brasileiro. Universidade de São Paulo

Dal Pizzol JL, Marques JA, Fonseca JS, Costa PG, Bianchini A (2021) Metal accumulation induces oxidative stress and alters carbonic anhydrase activity in corals and symbionts from the largest reef complex in the South Atlantic Ocean. Chemosphere. https://doi.org/10.1016/j.chemosphere.2021.133216

Dang VD, Fong CL, Shiu JH, Nozawa Y (2020) Grazing effects of sea urchin *Diadema savignyi* on algal abundance and coral recruitment processes. Sci Rep 10:20346

Davy SK, Allemand D, Weis VM (2012) Cell biology of cnidarian-dinoflagellate symbiosis. Microbiol Mol Biol Rev 76:229–261

De Goeij JM, Van Oevelen D, Vermeij MJ, Osinga R, Middelburg JJ, De Goeij AF, Admiraal W (2013) Surviving in a marine desert: the sponge loop retains resources within coral reefs. Science 342:108–110

Dias GM, Rocha RM, Lotufo TMC, Kremer LP (2013) Fifty years of ascidian biodiversity research in São Sebastião, Brazil. J Mar Biol Assoc U K 93. https://doi.org/10.1017/S002531541200063X

Duarte GAS, Villela HDM, Deocleciano M, Silva D, Barno A, Cardoso PM, Vilela CLS, Rosado P, Messias CSMA, Chacon MA, Santoro EP, Olmedo DB, Szpilman M, Rocha LA, Sweet M, Peixoto RS (2020) Heat waves are a major threat to turbid coral reefs in Brazil. Front Mar Sci 7:179

Fadlallah YH (1983) Sexual reproduction, development and larval biology in scleractinian corals. Coral Reefs 2:129–150

Ferreira LC, Grillo AC, Repinaldo Filho FP, Souza FN, Longo GO (2021) Different responses of massive and branching corals to a major heatwave at the largest and richest reef complex in South Atlantic. Mar Biol 168:54

Figueiredo MAO, Horta PA, Pedrini AG, Nunes JMC (2008) Benthic marine algae of the coral reefs of Brazil: a literature review. Oecologia Brasiliensis 12:258–269

Fishelson L (1974) Ecology of the northern Red Sea crinoids and their epi-and endozoic fauna. Mar Biol 26:183–192

Fonseca JS, Marangoni LFB, Marques JA, Bianchini A (2017) Effects of increasing temperature alone and combined with copper exposure on biochemical and physiological parameters in the zooxanthellate scleractinian coral *Mussismilia harttii*. Aquat Toxicol 190:121–132

Fonseca JS, Marangoni LFB, Marques JA, Bianchini A (2019) Carbonic anhydrase activity as a potential biomarker for acute exposure to copper in corals. Chemosphere 227:598e605

Fonseca JS, Mies M, Paranhos A, Taniguchi S, Güth AZ, Bícego MC, Marques JA, Marangoni LFB, Bianchini A (2021a) Isolated and combined effects of thermal stress and copper exposure on the trophic behavior and oxidative status of the reef-building coral *Mussismilia harttii*. Environ Pollut 268:115892

Fonseca JS, Zebral YD, Bianchini A (2021b) Metabolic status of the coral *Mussismilia harttii* in field conditions and the effects of copper exposure in vitro. Comp Biochem Physiol C 240:108924

Francini-Filho RB, Asp NE, Siegle E, Hocevar J, Lowyck K, D'Avila N, Vasconcelos AA, Baitelo R, Rezende CE, Omachi CY, Thompson CC, Thompson FL (2018) Perspectives on the Great Amazon Reef: extension, biodiversity, and threats. Front Mar Sci 5:142

Garrido AG, Leite DCA, Machado LF, Peixoto RS, Zilberberg C (2022) Highly diverse and geographically differentiated Symbiodiniaceae communities associated with the hydrocoral *Millepora alcicornis* in the Atlantic Ocean. Coral Reefs 41:937–950

Gasparini JL, Floeter SR, Ferreira CEL, Sazima I (2005) Marine ornamental trade in Brazil. Biodivers Conserv 14:2883–2899

Gates RD, Edmunds PJ (1999) The physiological mechanisms of acclimatization in tropical reef corals. Am Zool 39:30–43

Gherardi DF, Bosence DW (2001) Composition and community structure of the coralline algal reefs from Atol das Rocas, South Atlantic, Brazil. Coral Reefs 19:205–219

Giglio VJ, Ternes ML, Barbosa MC, Cordeiro CAMM, Floeter SR, Ferreira CEL (2018) Reef fish associations with sea urchins in an Atlantic oceanic Island. Mar Biodivers 48:1833–1839

Giraldes BW, Coelho Filho PA, Smyth DM (2015) Decapod assemblages in subtidal and intertidal zones—importance of scuba diving as a survey technique in tropical reefs, Brazil. Global Ecol Conserv 3:163–175

Glynn PW (1993) Coral reef bleaching: ecological perspectives. Coral Reefs 12:1–17

Goldberg WM (2013) The biology of reefs and reef organisms. University of Chicago Press, Chicago

Gondim AI, Moura RB, Christoffersen ML, Dias TL (2018) Taxonomic guide and historical review of echinoids (Echinodermata: Echinoidea) from northeastern Brazil. Zootaxa 4529:1–72

Graham NA, Nash KL (2013) The importance of structural complexity in coral reef ecosystems. Coral Reefs 32:315–326

Grande H, Reis M, Carvalho NF (2020) Use of the sea urchin *Diadema antillarum* (Echinodermata, Echinoidea) as a shelter for non-cryptobenthic juvenile reef fishes. Mar Biodivers 50:1–5

Hadfield MG, Kay EA, Gillette MU, Lloyd MC (1972) The vermetidae (Mollusca: Gastropoda) of the Hawaiian islands. Mar Biol 12:81–98

Hargis WJ Jr, Haven DS (1999) Chesapeake oyster reefs, their importance, destruction and guidelines for restoring them. In: Luckenbach MW, Mann R, Wesson JA (eds) Oyster reef habitat restoration: a synopsis and synthesis of approaches. Virginia Institute of Marine Science Press, Gloucester Point, pp 329–358

Harrison PL (2011) Sexual reproduction of scleractinian corals. In: Dubinsky Z, Stambler N (eds) Coral reefs: an ecosystem in transition. Springer, Dordrecht, pp 59–85

Harrison PL, Wallace CC (1990) Reproduction, dispersal and recruitment of scleractinian corals. In: Dubinsky Z (ed) Ecosystems of the world: coral reefs. Elsevier, Amsterdam, pp 133–207

Hoegh-Guldberg O, Mumby PJ, Hooten AJ, Steneck RD, Greenfield P, Gomez ED, Harvell CD, Sale PF, Edwards AJ, Caldeira K, Knowlton N, Eakin CM, Iglesias-Prieto R, Muthiga N, Bradbury RH, Dubi A, Hatziolos ME (2007) Coral reefs under rapid climate change and ocean acidification. Science 318:1737–1742

Hooper JN, Wörheide G, Hajdu E, Erpenbeck D, De Voogd NJ, Klautau M (2021) Zootaxa 20 years: phylum Porifera. Zootaxa 4979:38–56

Hume BCC, Ziegler M, Poulain J, Pochon X, Romac S, Boissin E, de Vargas C, Planes S, Wincker P, Voolstra CR (2018) An improved primer set and amplification protocol with increased specificity and sensitivity targeting the *Symbiodinium* ITS2 region. PeerJ 6:e4816

ICMBio (2018) Livro vermelho da fauna brasileira ameaçada de extinção. Brasília, Brazil

Kerr AM, Baird AH, Hughes TP (2011) Correlated evolution of sex and reproductive mode in corals (Anthozoa: Scleractinia). Proc R Soc B Biol Sci 278:75–81

Knowlton N, Brainard RE, Fisher R, Moews M, Plaisance L, Caley MJ (2010) Coral reef biodiversity. In: McIntyre AD (ed) Life in the World's oceans: diversity, distribution, and abundance. Blackwell Publishing, Oxford, pp 65–77

Kramer MJ, Bellwood O, Bellwood DR (2013) The trophic importance of algal turfs for coral reef fishes: the crustacean link. Coral Reefs 32:575–583

Kramer MJ, Bellwood DR, Bellwood O (2014) Benthic Crustacea on coral reefs: a quantitative survey. Mar Ecol Prog Ser 511:105–116
Laborel J (1970) Les peuplements de madréporaires des côtes tropicales du Brésil. PhD dissertation at the Université d'Abidjan, Ivory Coast
LaJeunesse TC (2001) Investigating the biodivesrity, ecology, and phylogeny of endosymbiotic dinoflagellates in the genus *Symbiodinium* using the ITS region: in search of a "species" level marker. J Phycol 37:866–880
LaJeunesse TC, Parkinson JE, Gabrielson PW, Jeong HJ, Reimer JD, Voolstra CR, Santos SR (2018) Systematic revision of Symbiodiniaceae highlights the antiquity and diversity of coral endosymbionts. Curr Biol 28:2570–2580
Lawrence J (2013) Starfish: biology and ecology of the Asteroidea. Johns Hopkins University Press, Baltimore
Lawrence J (2020) Sea urchins: biology and ecology. Elsevier, Amsterdam
Leal MC, Rocha RJ, Anaya-Rojas JM, Cruz ICS, Ferrier-Pagès C (2017) Trophic and stoichiometric consequences of nutrification for the intertidal tropical zoanthid *Zoanthus sociatus*. Mar Pollut Bull 119:169–175
Leão ZMAN, Kikuchi RKP, Testa V (2003) Corals and coral reefs of Brazil. In: Cortés J (ed) Latin American coral reefs. Elsevier, Amsterdam, pp 9–52
Leão ZMAN, Kikuchi RKP, Ferreira BP, Neves EG, Sovierzoski HH, Oliveira MD, Maida M, Correia MD, Johnsson R (2016) Brazilian coral reefs in a period of global change: a synthesis. Braz J Oceanogr 64:97–116
Lesser MP (2006) Oxidative stress in marine environments: biochemistry and physiological ecology. Annu Rev Physiol 68:253–278
Lessios HA, Robertson DR, Cubit JD (1984) Spread of *Diadema* mass mortality through the Caribbean. Science 226:335–337
Lewis JB (2006) Biology and ecology of the hydrocoral *Millepora* on coral reefs. Adv Mar Biol 50:1–55
Lewis JC, Wallis EV (1991) The function of surface sclerites in gorgonians (Coelenterata, Octocorallia). Biol Bull 181:275–288
Lima MM, Cordeiro RT, Perez CD (2019) Black Corals (Anthozoa: Antipatharia) from the Southwestern Atlantic. Zootaxa 4692.
Little AF, van Oppen MJH, Willis BL (2004) Flexibility in algal endosymbioses shapes growth in reef corals. Science 304:1492–1494
Littler MM, Littler DS (2013) The nature of crustose coralline algae and their interactions on reefs. Smithson Contrib Mar Sci 39:199–212
Logan A (1988) Sediment-shifting capability in the recent solitary coral *Scolymia cubensis* (Milne-Edwards and Haime) from Bermuda. Bull Mar Sci 43:241–248
Loiola LL (2007) Black corals (Cnidaria: Antipatharia) from Brazil: an overview. Bull Mar Sci 81:253–264
Lombardi C, Taylor PD, Cocito S (2020) Bryozoans: the 'forgotten' bioconstructors. In: Rossi S, Bramanti L (eds) Perspectives on the marine animal forests of the world. Springer, Cham, pp 193–217
Love GD, Grosjean E, Stalvies C, Fike DA, Grotzinger JP, Bradley AS, Kelly AE, Bhatia M, Meredith W, Snape CE, Bowring SA, Condon DJ, Summons RE (2009) Fossil steroids record the appearance of Demospongiae during the Cryogenian period. Nature 457:718–721
Luza AL, Aued AW, Barneche DR, Dias MS, Ferreira CE, Floeter SR, Francini-Filho RB, Longo GO, Quimbayo JP, Bender MG. Functional diversity patterns of reef fish, corals and algae in the Brazilian biogeographical province. J Biogeogr 2023 Jun;50(6):1163–76
Madin JS, Hoogenboom MO, Connolly SR, Darling ES, Falster DS, Huang D et al (2016) A trait-based approach to advance coral reef science. Trends Ecol Evol 31:419–428
Marangoni LFB, Calderon EN, Marques JA, Duarte GAS, Pereira CM, Castro CB, Bianchini A (2017a) Effects of CO_2-driven acidification of seawater on the calcification process in the calcareous hydrozoan *Millepora alcicornis* (Linnaeus, 1758). Coral Reefs 36:1133–1141

Marangoni LFB, Marques JA, Duarte GAS, Pereira CM, Calderon EN, Castro CB, Bianchini A (2017b) Copper effects on biomarkers associated with photosynthesis, oxidative status and calcification in the Brazilian coral *Mussismilia harttii* (Sclaractinia, Mussidae). Mar Environ Res 130:248e257

Marangoni LFB, Dalmolin C, Marques JA, Klein RD, Abrantes DP, Pereira CM, Calderon EN, Castro CB, Bianchini A (2019a) Oxidative stress biomarkers as potential tools in reef degradation monitoring: a study case in a South Atlantic reef under influence of the 2015–2016 El Niño/Southern Oscillation (ENSO). Ecol Indic 106:105533

Marangoni LF, Mies M, Güth AZ, Banha TN, Inague A, Fonseca JD, Dalmolin C, Faria SC, Ferrier-Pagès C, Bianchini A (2019b) Peroxynitrite generation and increased heterotrophic capacity are linked to the disruption of the coral–dinoflagellate symbiosis in a scleractinian and hydrocoral species. Microorganisms 7:426

McClanahan TR, Kamukuru AT, Muthiga NA, Yebio MG, Obura D (1996) Effect of sea urchin reductions on algae, coral, and fish populations. Conserv Biol 10:136–154

McDonald JI, Hooper JN, McGuinness KA (2002) Environmentally influenced variability in the morphology of *Cinachyrella australiensis* (Carter 1886) (Porifera: Spirophorida: Tetillidae). Mar Freshw Res 53:79–84

McEdward LR, Miner BG (2001) Larval and life-cycle patterns in echinoderms. Can J Zool 79:1125–1170

McLaughlin PA, Lemaitre R (1993) A review of the hermit crab genus *Paguritta* (Decapoda: Anomura: Paguridae) with descriptions of three new species. Raffles Bull Zool 41:1–29

McManus JW, Polsenberg JF (2004) Coral–algal phase shifts on coral reefs: ecological and environmental aspects. Prog Oceanogr 60:263–279

Mies M (2019) Evolution, diversity, distribution and the endangered future of the giant clam–Symbiodiniaceae association. Coral Reefs 38:1067–1084

Mies M, Güth AZ, Tenório AA, Banha TNS, Waters LG, Polito PS, Taniguchi S, Bícego MC, Sumida PYG (2018) In situ shifts of predominance between autotrophic and heterotrophic feeding in the reef-building coral *Mussismilia hispida*: an approach using fatty acid trophic markers. Coral Reefs 37:677–689

Mies M, Francini-Filho RB, Zilberberg C, Garrido AG, Longo GO, Laurentino E, Güth AZ, Sumida PYG, Banha TNS (2020) South Atlantic coral reefs are major global warming refugia and less susceptible to bleaching. Front Mar Sci 7:514

Millar RH (1971) The biology of ascidians. Adv Mar Biol 9:1–100

Miloslavich P, Klein E, Penchaszadeh P (2010) Gametogenic cycle of the tropical vermetids *Eualetes tulipa* and *Dendropoma corrodens* (Mollusca: Caenogastropoda: Vermetidae). J Mar Biol Assoc U K 90:509–518

Monniot C, Monniot F, Laboute P (1991) Coral reef ascidians of New Caledonia. ORSTOM, Paris

Moraes FC, Cervi F, Karez CS, Salgado LT, Moura RL, Leal GA, Bastos AC, Amado-Filho GM (2019) Long-term temporal and spatial patterns in bioeroding sponge distribution at the Abrolhos Bank, Brazil, southwestern Atlantic. Mar Ecol 40:e12531

Moura RL, Amado-Filho GM, Moraes FC, Brasileiro PS, Salomon PS, Mahiques MM, Bastos AC, Almeida MG, Silva JM Jr, Araujo BF, Brito FP, Rangel TP, Oliveira BCV, Bahia RG, Paranhos RP, Dias RJS, Siegle E, Figueiredo AG Jr, Pereira RC, Leal CV, Hajdu E, Asp NE, Gregoracci GB, Neumann-Leitão S, Yager PL, Francini-Filho RB, Fróes A, Campeão M, Silva BS, Moreira APB, Oliveira L, Soares AC, Araujo L, Oliveira NL, Teixeira JB, Valle RAB, Thompson CC, Rezende CE, Thompson FL (2016) An extensive reef system at the Amazon River mouth. Sci Adv 2:e1501252

Moya A, Tambutté S, Tambutté E, Zoccola D, Caminiti N, Allemand D (2006) Study of calcification during a daily cycle of the coral *Stylophora pistillata*: implications for light-enhanced calcification. J Exp Biol 209:3413–3419

Müller WE (2006) The stem cell concept in sponges (Porifera): metazoan traits. Semin Cell Dev Biol 17:481–491

Muricy G, Lopes DA, Hajdu E, Carvalho MS, Moraes FC, Klautau M, Menegola C, Pinheiro U (2011) Catalogue of Brazilian Porifera. Museu Nacional, Rio de Janeiro

Muscatine L (1990) The role of symbiotic algae in carbon and energy flux in reef corals. Coral Reefs 25:75–87

Muscatine L, McCloskey LR, Marian RE (1981) Estimating the daily contribution of carbon from zooxanthellae to coral animal respiration. Limnol Oceanogr 26:601–611

Negri AP, Webster NS, Hill RT, Heyward AJ (2001) Metamorphosis of broadcast spawning corals in response to bacteria isolated from crustose algae. Mar Ecol Prog Ser 223:121–131

Neves E, Pires D (2002) Sexual reproduction of Brazilian coral *Mussismilia hispida* (Verrill, 1902). Coral Reefs 21:161–168

Nitschke MR, Craveiro SC, Brandão C, Fidalgo C, Serôdio J, Calado AJ, Frommlet JC (2020) Description of *Freudenthalidium* gen. nov. and *Halluxium* gen. nov. to formally recognize clades Fr3 and H as genera in the family Symbiodiniaceae (Dinophyceae). J Phycol 56:923–940

Oliveira FAS, Colares GB, Hissa DC et al (2013) Microbial epibionts of the colonial ascidians Didemnum galacteum and Cystodytes sp. Symbiosis 59:57–63. https://doi.org/10.1007/s13199-012-0210-2

Omachi CY, Asp NE, Siegle E, Couceiro MA, Francini-Filho RB, Thompson FL (2019) Light availability for reef-building organisms in a plume-influenced shelf. Cont Shelf Res 181:25–33

Padua A, Lanna E, Klautau M (2013) Macrofauna inhabiting the sponge *Paraleucilla magna* (Porifera: Calcarea) in Rio de Janeiro, Brazil. J Mar Biol Assoc U K 93:889–898

Paiva SV, Filho RRO, Lotufo TMC (2015) Ascidians from Rocas Atoll, Northeast Brazil. Front Mar Sci 2. https://doi.org/10.3389/fmars.2015.00039

Palumbi SR (1986) How body plans limit acclimation: responses of a demosponge to wave force. Ecology 67:208–214

Pawlik JR, McMurray SE (2020) The emerging ecological and biogeochemical importance of sponges on coral reefs. Annu Rev Mar Sci 12:315–337

Pereira CM, Fonseca JS, Paiva ES, Costa PG, Mies M, Silva AG, Calderon EN, Bianchini A, Castro CB (2020) Larvae of the South Atlantic coral *Favia gravida* are tolerant to salinity and nutrient concentrations associated with river discharges. Mar Environ Res 161:105118

Pereira-Filho GH, Francini-Filho RB, Pierozzi I Jr, Pinheiro HT, Bastos AC, Moura RL, Moraes FC, Matheus Z, Bahia RD, Amado-Filho GM (2015) Sponges and fish facilitate succession from rhodolith beds to reefs. Bull Mar Sci 91:45–46

Pereira PHC, Lima GV, Pontes AVF, Côrtes LGF, Gomes E, Sampaio CLS, Pinto TK, Miranda RJ, Cardoso ATC, Araújo JC, Seoane JCS (2022) Unprecedented coral mortality on Southwestern Atlantic (SWA) coral reefs following major thermal stress. Frontiers in Marine Science 9:338.

Perez CD, Neves BM, Cordeiro RTS, Williams GC, Cairns SD (2016) Diversity and distribution of Octocorallia. In: Goffredo S, Dubinsky Z (eds) The Cnidaria, past, present and future. Springer, Dordrecht, pp 109–123

Pinto TK, Benevides LJ, Sampaio CL (2016) *Cyphoma macumba* Petuch 1979 (Gastropoda: Ovulidae): a versatile predator of the Brazilian Octocorallia. Mar Biodivers 1:165–166

Pinzón AM, Orkisz M, Useche CMR, González JST, Teillaud S, Sánchez JA, Hoyos MH (2014) A semi-automatic method to extract canal pathways in 3D micro-CT images of octocorals. PLoS One 9:e85557

Pires DO, Castro CB, Ratto CC (1999) Reef coral reproduction in the Abrolhos reef complex, Brazil: the endemic genus *Mussismilia*. Mar Biol 135:463–471

Pires DO, Segal B, Caparelli AC (2011) Reproductive effort of an endemic major reef builder along an inshore-offshore gradient in South-Western Atlantic. J Mar Biol Assoc U K 91:1613–1616

Pires DO, Castro CB, Segal B, Pereira CM, Carmo EC, Silva RG, Calderon EN (2016) Reprodução de corais de águas rasas do Brasil. In: Zilberberg C, Abrantes DP, Marques JA, Machado LF, Marangoni LFB (eds) Conhecendo os Recifes Brasileiros, Série Livros n° 58. Museu Nacional, Rio de Janeiro, pp 111–128

Pochon X, LaJeunesse TC (2021) *Miliolidium* n. gen, a new Symbiodiniacean genus whose members associate with soritid foraminifera or are free-living. J Eukaryot Microbiol. https://doi.org/10.1111/jeu.12856

Pupier CA, Mies M, Fine M, Bastos Francini-Filho R, Pereira Brandini F, Zambotti-Villela L, Colepicolo P, Ferrier-Pagès C (2021) Lipid biomarkers reveal the trophic plasticity of octocorals along a depth gradient. Limnol Oceanogr 66:2078–2087

Quimbayo JP, Schlickmann OR, Floeter SR, Sazima I (2018) Cleaning interactions at the southern limit of tropical reef fishes in the Western Atlantic. Environ Biol Fish 101:1195–1204

Rädecker N, Pogoreutz C, Gegner HM, Cárdenas A, Roth F, Bougoure J, Guagliardo P, Wild C, Pernice M, Raina JB, Meibom A, Voolstra CR (2021) Heat stress destabilizes symbiotic nutrient cycling in corals. Proc Natl Acad Sci USA 118:e2022653118

Reis VM, Karez CS, Mariath R et al (2016) Carbonate production by benthic communities on shallow Coralgal reefs of Abrolhos Bank, Brazil. PLoS One 11. https://doi.org/10.1371/journal.pone.0154417

Rix L, de Goeij JM, van Oevelen D, Struck U, Al-Horani FA, Wild C, Naumann MS (2017) Differential recycling of coral and algal dissolved organic matter via the sponge loop. Funct Ecol 31:778–789

Roberts JM, Wheeler AJ, Freiwald A (2006) Reefs of the deep: the biology and geology of cold-water coral ecosystems. Science 312:543–547

Robison JD, Warner ME (2006) Differential impacts of photoacclimation and thermal stress on the photobiology of four different phylotypes of *Symbiodinium* (Pyrrhophyta). J Phycol 42:568–579

Sammarco PW (1982) Echinoid grazing as a structuring force in coral communities: whole reef manipulations. J Exp Mar Biol Ecol 61:31–55

Santana EFC, Mies M, Longo GO, Menezes R, Aued AW, Luza AL, Bender MG, Segal B, Floeter SR, Francini-Filho RB (2023) Turbidity shapes shallow Southwestern Atlantic benthic reef communities. Mar Environ Res 183:105807

Santelices B (2002) Recent advances in fertilization ecology of macroalgae. J Phycol 38:4–10

Scheuwimmer A (1979) Sperm transfer in the sessile gastropod *Serpulorbis* (Prosobranchia: Vermetidae). Mar Ecol Prog Ser 1:65–70

Schlöder C, D'Croz L (2004) Responses of massive and branching coral species to the combined effects of water temperature and nitrate enrichment. J Exp Mar Biol Ecol 313:255–268

Schönberg CHL (2008) A history of sponge erosion: from past myths and hypotheses to recent approaches. In: Wisshak M, Tapanila L (eds) Current developments in bioerosion. Springer, Berlin, pp 165–202

Segal B, Castro CB (2011) Coral community structure and sedimentation at different distances from the coast of the Abrolhos Bank, Brazil. Braz J Oceanogr 59:119–129

Serejo C, Young PS, Cardoso IA et al (2006) Capítulo 8: Filo Arthoropoda. Subfilo Crustacea. In: Lavrado HP, Ignacio BL (eds) Biodiversidade bentônica da região central da Zona Econômica Exclusiva brasileira. Museu Nacional, Rio de Janeiro, pp 299–337

Shenkar N, Swalla BJ (2011) Global diversity of Ascidiacea. PLoS One 6:1–12

Sheppard CRC, Davy SK, Pilling GM, Graham NAJ (2018) The biology of coral reefs. Oxford University Press, Oxford

Shlesinger Y, Loya Y (1985) Coral community reproductive patterns: Red Sea versus the great barrier reef. Science 228:1333–1335

Simone LR (1999) Filo Mollusca. In: Migotto AE, Thiago CG (eds) Biodiversidade do Estado de São Paulo, Brasil: síntese do conhecimento ao final do século XX: Invertebrados Marinhos, 1st edn. FAPESP, São Paulo, pp 129–136

Sissini MN, Koerich G, de Barros-Barreto MB, Coutinho LM, Gomes FP, Oliveira W, Costa IO, de Castro Nunes JM, Henriques MC, Vieira-Pinto T, Torrano-Silva BN, Oliveira MC, Le Gall L, Horta PA (2021) Diversity, distribution, and environmental drivers of coralline red algae: the major reef builders in the southwestern Atlantic. Coral Reefs. https://doi.org/10.1007/s00338-021-02171-1

Smit NJ, Bruce NL, Hadfield KA (2014) Global diversity of fish parasitic isopod crustaceans of the family Cymothoidae. Int J Parasitol Parasites Wildl 3:188–197

Smith AM, Key MM Jr, Gordon DP (2006) Skeletal mineralogy of bryozoans: taxonomic and temporal patterns. Earth Sci Rev 78:287–306
Soares MO, da Lotufo TMC, Vieira LM et al (2016) Brazilian marine animal forests: a New World to discover in the southwestern Atlantic. In: Rossi S, Bramanti L, Gori A, del Valle COS (eds) Marine animal forests. Springer International Publishing, Cham, pp 1–38
Sorokin YI (1973) On the feeding of some scleractinian corals with bacteria and dissolved organic matter. Limnol Oceanogr 18:380–386
Spotorno P, Tâmega FTS, Bemvenuti CE (2012) An overview of the recent vermetids (Gastropoda: Vermetidae) from Brazil. Strombus 19:1–8
Spotorno-Oliveira P, Coutinho R, Tâmega FTS (2018) Recent introduction of non-indigenous vermetid species (Mollusca, Vermetidae) to the Brazilian coast. Mar Biodivers 48:1931–1941
Stanley GD (2003) The evolution of modern corals and their early history. Earth Sci Rev 60:195–225
Stanley GD, Fautin DG (2001) The origins of modern corals. Science 291:1913–1914
Stat M, Morris E, Gates RD (2008) Functional diversity in coral-dinoflagellate symbiosis. Proc Natl Acad Sci USA 105:9256–9261
Steneck RS, Watling L (1982) Feeding capabilities and limitation of herbivorous molluscs: a functional group approach. Mar Biol 68:299–319
Sussman M, Loya Y, Fine M, Rosenberg E (2003) The marine fireworm *Hermodice carunculata* is a winter reservoir and spring-summer vector for the coral-bleaching pathogen *Vibrio shiloi*. Environ Microbiol 5:250–255
Swain TD, Chandler J, Backman V, Marcelino L (2017) Consensus thermotolerance ranking for 110 *Symbiodinium* phylotypes: an exemplar utilization of a novel iterative partial-rank aggregation tool with broad application potential. Funct Ecol 31:172–183
Terossi M, Almeida AO, Buranelli RC, Castilho AL, Costa RC, Zara FJ, Mantelatto FL (2018) Checklist of decapods (Crustacea) from the coast of the São Paulo state (Brazil) supported by integrative molecular and morphological data: I. Infraorder Caridea: families Hippolytidae, Lysmatidae, Ogyrididae, Processidae and Thoridae. Zootaxa 4370:76–94
Thornhill DJ, Lewis AM, Wham D, LaJeunesse TC (2014) Host specialist lineages dominate the adaptive radiation of reef coral endosymbionts. Evolution 68:352–367
Todd CD, Havenhand JN (1989) Nudibranch-bryozoan associations: the quantification of ingestion and some observations on partial predation among Doridoidea. J Molluscan Stud 55:245–259
Tribollet A, Golubic S (2011) Reef bioerosion: agents and processes. In: Dubinsky Z, Stambler N (eds) Coral reefs: an ecosystem in transition. Springer, Dordrecht, pp 435–449
Ulstrup KE, van Oppen MJH (2003) Geographic and habitat partitioning of genetically distinct zooxanthellae (*Symbiodinium*) in *Acropora* corals on the Great Barrier Reef. Mol Ecol 12:3477–3484
Uriz MJ, Turon X, Becerro MA, Agell G (2003) Siliceous spicules and skeleton frameworks in sponges: origin, diversity, ultrastructural patterns, and biological functions. Microsc Res Tech 62:279–299
Van Soest RW (2017) Sponges of the Guyana shelf. Zootaxa 4217:1–225
Van Soest RW, Boury-Esnault N, Vacelet J, Dohrmann M, Erpenbeck D, De Voogd NJ, Santodomingo N, Vanhoorne B, Kelly M, Hooper JN (2012) Global diversity of sponges (Porifera). PLoS One 7:e35105
van Woesik R (2010) Calm before the spawn: global coral spawning patterns are explained by regional wind fields. Proc R Soc B Biol Sci 277:715–722
Vermeij GJ (1977) The Mesozoic marine revolution: evidence from snails, predators and grazers. Paleobiology 3:245–258
Veron JE (2000) Corals of the World. Townsville: Australian Institute of Marine Science.
Veron JE, Stafford-Smith MG, Turak E, DeVantier LM (2016) Corals of the World. Available online at: http://www.coralsoftheworld.org/page/home/. Accessed 20 Aug 20 2021

Vieira LM, Nascimento KB, Almeida ACS (2016) Catálogo Taxonômico da Fauna do Brasil: Bryozoa [Internet]. [updated 2015 Oct 09; cited 2016 Jan 22]. Available from: http://fauna.jbrj.gov.br/fauna/faunadobrasil/157416

Wagner D, Luck DG, Toonen RJ (2012) The biology and ecology of black corals (Cnidaria: Anthozoa: Hexacorallia: Antipatharia). Adv Mar Biol 63:67–132

Webb AE, van Heuven SM, de Bakker DM, van Duyl FC, Reichart GJ, de Nooijer LJ (2017) Combined effects of experimental acidification and eutrophication on reef sponge bioerosion rates. Front Mar Sci 4:311

Weis VM (2008) Cellular mechanisms of cnidarian bleaching: stress causes the collapse of symbiosis. J Exp Biol 211:3059–3066

Wessel GM (2018) Echinodermata. In: Skinner MK (ed) Encyclopedia of reproduction. Elsevier, Amsterdam, pp 533–545

West JM (1998) The dual role of sclerites in a gorgonian coral: conflicting functions of support and defence. Evol Ecol 12:803–821

Wirtz P, Melo G, Grave S (2009) Symbioses of decapod crustaceans along the coast of Espírito Santo, Brazil. Mar Biodivers Rec 2:e162

Wood R (1999) Reef evolution. Oxford University Press, Oxford

WoRMS (2021) World register of marine species. Available online at: https://www.marinespecies.org. Accessed 20 Aug 2021

Yakovleva IM, Baird AH, Yamamoto HH, Bhagooli R, Nonaka M, Hidaka M (2009) Algal symbionts increase oxidative damage and death in coral larvae at high temperatures. Mar Ecol Prog Ser 378:105–112

Zamora S, Rahman IA, Smith AB (2012) Plated Cambrian bilaterians reveal the earliest stages of echinoderm evolution. PLoS One 7:e38296

Zebral YD, Fonseca JS, Marques JA, Bianchini A (2019) Carbonic anhydrase as a biomarker of global and local impacts: insights from calcifying animals. Int J Mol Sci 20:3092

Chapter 5
Herbivory and Competition for Space

João L. L. Feitosa, Ricardo J. Miranda, and Luisa V. M. V. Queiroz-Véras

Abstract Brazilian coral reefs are quite singular, presenting a particular set of endemic coral species supporting the construction of reefs that diverge in form and structure from any other coralline systems. These distinctive reef formations shelter a fish herbivore functional group that is low in diversity but is composed barely of endemic species. Algal assemblages in Brazil are akin to reefs worldwide, but the coexistence with such a unique assortment of endemic species provides an interesting scenario for interactions to be studied. Conversely, we are nowhere near understanding ecological relationships in Brazilian reefs compared to the Indo-Pacific and the Caribbean. Even before reaching this objective, Brazilian coral reefs have been burdened by anthropogenic impacts common to other reefs worldwide, and these species interactions are being increasingly disrupted. The coral fauna is challenged by nutrification, increased temperature, sedimentation and turbidity, herbivore overfishing and, more recently, the invasion of exotic species. Brazilian researchers have developed copious literature, most of it undertaken in the light of these impacts. This chapter compiles some of the results of coral reef studies on herbivory and competition for space, primarily performed in the past two decades. We present these findings and briefly discuss their implications to this singular reef system.

Keywords Herbivores · Endemic species · Functional ecology · Allelopathy · Competitive strategies · Invasive species · Overfishing · Phase-shifts

J. L. L. Feitosa (✉)
Departamento de Zoologia, Universidade Federal de Pernambuco, Recife / PE, Brazil

R. J. Miranda
Instituto de Ciências Biológicas e da Saúde, Universidade Federal de Alagoas, Maceió / AL, Brazil

Luisa V. M. V. Queiroz-Véras
Departamento de Zoologia, Universidade Federal de Pernambuco, Recife / PE, Brazil

Programa de Pós-Graduação em Biologia Animal - PPGBA, UFPE, Recife / PE, Brazil

R. K. P. Kikuchi et al. (eds.), *Brazilian Coral Reefs*, Brazilian Marine Biodiversity,
https://doi.org/10.1007/978-3-031-59152-5_5

5.1 Introduction

While most work on competition and herbivory was developed in Indo-Pacific and Caribbean reefs, Brazilian systems present a distinct scenario. Compared to other reef systems, Brazilian reefs retain low-diversity communities, impoverished in corals and fish species. Hermatypic corals are mostly endemic, some of those derived from a Tertiary fauna that remained part of the dominant taxa (Leão et al. 2003). These species have found a refuge of appropriate conditions to thrive in Brazil; here, the profuse turbidity and high sedimentation rates favor massive coral forms with large polyps to the detriment of ramified forms that abound in acroporid and pocilloporid corals of the Indo-Pacific and Caribbean. This relic coral fauna interacts with other sedimentation-resistant species shared with the Caribbean, such as *Millepora* fire corals, the only arborescent corals in Brazil (Maida and Ferreira 1997; Leão and Kikuchi 2005).

Given these singular conditions of Brazilian reefs, reef-associated fishes also present a high degree of endemism; about 24% of Southwestern Atlantic ichthyofauna are endemics, and most herbivorous species are exclusive to Brazilian waters (Pinheiro et al. 2018). Even though most of the algae present in Brazil are ubiquitous representatives found on reefs worldwide, the coexistence with such a distinctive reef fauna renders an interesting setting for unique interactions. Those reefs have the potential to provide very distinguishing insights on classical concepts of reef ecology, which have been developed considering the high-diversity coral reef systems of the Caribbean and Indo-Pacific realms.

Regardless of these singularities, Brazilian reefs share common adversities of reefs elsewhere, ultimately leading to the demise of coral reefs as we know. Sediment load and nutrient input are constantly increasing with poor land-use practices. Their combined effect with the chronic warming of oceans and herbivore overfishing has also led to a disruption of Brazilian reefs' ecological relationships. Despite the lack of long-term quantitative studies, there are clear indications of sharp reductions in live coral cover in some of those reefs, from approximately 25% in the 1960s (Laborel 1970; Laborel-Deguen et al. 2019) to less than 5% (Ferreira and Maida 2006). Nowadays, Brazilian reefs are profusely dominated by macroalgae, seldom surpassing 10% of total coral cover (Aued et al. 2018). On top of that, coral fauna is now defied by the presence of new competitors: exotic corals, challenging—once more—the resilience of these unique organisms to persist.

5.2 Herbivory in Brazilian Coral Reefs

5.2.1 Brazilian Algae-Eaters: A Diversity-Poor but Endemic-Rich System

As pointed out above, Brazilian herbivorous fish assemblages are unique. According to phylogenies available (Streelman et al. 2002; Smith et al. 2008; Choat et al. 2012; Siqueira et al. 2019a, b; McCord et al. 2021), Indo-Pacific and Atlantic herbivore faunas diverged during the separation of the Tethys Sea (18–12 Ma), resulting in distinctive biogeographical histories for each ocean. While the Indo-Pacific presented a continuous expansion of Tethyan lineages, the Atlantic underwent a series of extinction events (Hou and Li 2017; Floeter et al. 2008), culminating in the rise of locally-originated lineages of herbivores (Siqueira et al. 2019a, b). For instance, the labrid genera *Sparisoma*, *Nicholsina* and *Cryptotomus* shared common ancestors with parrotfish that evolved in seagrass/macroalgae habitats in the Atlantic (Streelman et al. 2002; Smith et al. 2008). Lineages originated in the east Tethys/Indo-Pacific subsequently colonized the Atlantic; some have persisted in present-date herbivore assemblages (*Acanthurus* and *Scarus* species), others are limited to fossil records (siganids and nasine surgeonfish) (Bellwood et al. 2017; Siqueira et al. 2019a, b). After the establishment of the Amazon River barrier separating the Caribbean and Brazilian faunas (ca. 11 Ma), most herbivore species diverged; allopatric pairs of herbivore species raised quite recently, during the Pleistocene (<5 Ma) (Rocha 2003; Choat et al. 2012; Siqueira et al. 2019a, b).

These historical differences between provinces resulted in unique herbivore assemblages that diverge in species richness and taxonomic composition. The Caribbean holds only a fraction of the Great Barrier Reef's herbivorous fish richness. Brazilian reefs are further less speciose (Bellwood and Wainwright 2002; Bellwood et al. 2004; Pinheiro et al. 2018; Robertson and Van Tassell 2019). For instance, Brazilian reefs present 13 tropical roving herbivorous reef fish species, contrasting with 17 in the Caribbean and more than 130 species in the Indo-Pacific (Siqueira et al. 2019a, b). Out of these 13 species, eight are endemics of Brazilian waters (Fig. 5.1), and species composition and abundance patterns are markedly divergent. In the Caribbean, *Scarus* is the most diverse genus and used to be the most abundant in several locations (Lewis and Wainwright 1985; Bruggemann et al. 1996; Posada et al. 2003; Lang 2003) but have diminished in numbers in the last decades due to overfishing (Adam et al. 2015; Burkepile et al. 2019). Differently, *Sparisoma* parrotfishes dominate both in richness and abundance over *Scarus* across all reefs in Brazil, and for the latter, only scattered populations are found in Brazilian oceanic islands (Joyeux et al. 2001; Ferreira et al. 2004a, b; Mazzei et al. 2017). Those differences are likely linked to the distinctive formations of Brazilian reefs, which ultimately led to reef types that are particular in shape and structure (see Leão et al. 2003 and Araújo et al. 2020 for descriptions).

It is feasible that *Sparisoma* species were favored in Brazilian reefs since they are supposedly habitat generalists. Their evolutionary history is marked by a

Fig. 5.1 Brazilian endemic roving herbivores: (**a**) *Sparisoma frondosum*, (**b**) *Sparisoma amplum*, (**c**) *Sparisoma tuiupiranga*, (**d**) *Sparisoma axillare*, (**e**) *Acanthurus bahianus* and (**f**) *Scarus trispinosus*. (Photos: Dráusio Véras (**a**,**b**,**d**,**e**,**f**). Luísa Véras (**c**))

transition from peripheral seagrass/macroalgal beds (still inhabited by *S. radians* and juvenile *S. axillare*, Chaves et al. 2012) to coral reefs (Bernardi et al. 2000; Streelman et al. 2002). Species of *Sparisoma* have intermediate traits within Scarini (Bellwood 1994; Bernardi et al. 2000) that allow the occupation of a broader range of marine habitats. For instance, *Sparisoma* has representatives with a preference for temperate rocky reefs: the Brazilian *S. tuiupiranga* and the Mediterranean *S. cretense* occur where other groups of herbivore fish thrive, such as chubs (*Kyphosus* spp.) and Salema porgies (*Sarpa salpa*), respectively (Gasparini et al. 2003; Ferreira et al. 2004a, b; Bariche et al. 2004; Bodilis et al. 2014). *Sparisoma axillare* is considered one the most generalist in habitat from the Brazilian tropical ichthyofauna, with remarkably high densities across Northeastern reefs (Feitosa and Ferreira 2015; Roos et al. 2019; Araújo et al. 2020); oppositely, *Scarus* seems to be typical of modern Caribbean and Indo-Pacific reefs (Streelman et al. 2002).

Atlantic acanthurids are also habitat generalists (Araújo et al. 2020). They are commonly the most abundant roving herbivores in Brazilian (frequently >15% of total fish abundance—Ferreira et al. 2004a, b) and Caribbean (Lewis 1986; Posada et al. 2003; Lang 2003) reefs, but the dominant species seems to vary at a local scale (Posada et al. 2003; Araújo et al. 2020). Historically, three species were thought to co-occur in both regions: *Acanthurus coeruleus*, *Acanthurus chirurgus* and *Acanthurus bahianus*. However, Bernal and Rocha (2011) have found that *A. bahianus* specimens from the Caribbean diverged both in coloration and genetics from Brazilian counterparts, therefore proposing the resurrection of *Acanthurus tractus* as a valid species, restricting the distribution of *A. bahianus* to Brazil.

While surgeonfishes are a common component of ichthyofaunas in both provinces, herbivore damselfish is a distinct group, and no species occur in both provinces except for *Microspathodon chrysurus*. Territorial herbivore *Stegastes* are represented by seven species in the Caribbean. In comparison, Brazilian reefs house three endemic species, yet with a caveat: *Stegastes fuscus* and *Stegastes variabilis* inhabit the coast, while *S. rocasensis* only occur in Brazilian oceanic islands. Most interestingly, the insular species seems to be phylogenetically closest to the Eastern Pacific species *S. rectifraenum* and *S. acapulcoensis* than to Brazilian coastal endemics (McCord et al. 2021). Molecular analyses revealed that *Stegastes sanctipauli*, previously considered endemic to the São Pedro and São Paulo Archipelago, is a synonym of *S. rocasensis* (Souza et al. 2017).

Habitat use by damselfishes is also quite different between Brazilian and Caribbean systems. In a diverse community of co-habiting damselfish, like in the Caribbean, species may develop a high degree of substrate specificity (Waldner and Robertson 1980; Robertson 1984; Precht et al. 2010; Chaves et al. 2012). Live coral is a major driver of damselfish abundance in many reef systems (Munday et al. 2008, Wilson et al. 2008; MacDonald et al. 2018); thus, these coral-dependent species are prone to be handicapped under the current perspective of coral demise. On the contrary, Brazilian damsels are weakly associated with any particular benthic features, and the most critical drivers of their abundance are depth and wave surge (Chaves et al. 2021). As algae are becoming the prevalent benthic feature in Brazilian reefs (Aued et al. 2018), endemic damselfishes are pretty much thriving. At some stretches of the coast, especially in reefs already dominated by algae, damselfish densely occupy reef flats, accounting for nearly 20% of total fish abundance and densities surpassing ten individuals per m^2 (Ferreira et al. 2004a, b; Attrill and Kelmo 2007).

Another essential component of herbivore functional groups in Brazilian reefs is sea urchins, whose diversity is very similar to that of the Caribbean, with over 75% of their species shared between these two provinces (Gondim et al. 2018). *Echinometra lucunter* is the prevalent species in most of the Brazilian coast in the intertidal to shallow subtidal zones, and is especially abundant in tropical areas, but is replaced by *Paracentrotus gaimardi* in the temperate rocky reefs subject to seasonal upwelling (Sánchez-Jérez et al. 2001; Cordeiro et al. 2014; Labbé-Bellas et al. 2016; Giglio et al. 2018). *Diadema antillarum* is also present in the coast and oceanic islands. Unlike the Caribbean, which has experienced mass mortality during the 1980s supposedly due to a waterborne pathogen, there are no reports of this die-off affecting Brazilian populations (Lessios 2015). Some Brazilian sites deter the densest populations of *D. antillarum*, while most of the Caribbean populations remain, to date, 12% of the densities before the die-off (Attrill and Kelmo 2007; Lessios 2015). In subtidal coastal zones of the tropics, *Lytechinus variegatus* may be relatively abundant, while in oceanic islands, there have been reports of occasional overpopulation by *Tripneustes ventricosus* (Godinho et al. 2016).

5.2.2 Herbivores Functional Roles in Brazil

The Atlantic and the Indo-Pacific share the same suite of herbivore functional groups in a broad sense. Still, the poorer richness of species in the former is traditionally believed to make functional groups more susceptible to impacts (Bellwood et al. 2004), while changes in taxonomic composition lead to differences in services provided by herbivores (Siqueira et al. 2019a, b). Notwithstanding, species richness alone does not picture herbivores' ecosystem functions or the redundancy of species within these roles. When traits are considered to identify species that deliver each service, turf and sediment removal and bioerosion are the services with less redundant species in the Atlantic, being Brazil even more impoverished (Francini-Filho et al. 2008; Lellys et al. 2019; Feitosa et al. 2023; Table 5.1, Fig. 5.2). The number of macroalgae-removing fish species in the Atlantic is less than half of those in the Indo-Pacific. However, the diversity of Brazilian herbivores delivering this service is comparable to the Caribbean. Contrary to the Indo-Pacific, Atlantic reefs lack herbivore fish specialized in crevice cleaning, a service largely dependent on echinoids and other invertebrates.

Sea urchins became the most important herbivores on many Caribbean reefs following the overfishing of herbivorous fishes (Knowlton 1992; Hughes 1994; Bellwood et al. 2004; Lessios 2015). In Northeastern Brazil, sea urchins were shown to control macroalgae overgrowth and substrate monopolization, consequently assisting in maintaining live coral cover, especially at increased temperatures (Kilpp 1999). However, due to overfishing of their predators, some sites in Brazil have become excessively populated with urchins (Dias et al. 2011; Costa 2013), leading to 'sea urchin barrens', a well-known alternative state of impacted kelp forests (Ling and Johnson 2009. Filbee-Dexter and Scheibling 2014.; Melis et al. 2019). Reefs overpopulated by sea urchins experience overgrazing and significantly increased reef erosion rates (Bellwood et al. 2004). Erosion rates for *Echinometra lucunter* are estimated to reach ~4 kg of rock removed/m^2/year when densities are high (Ogden 1977). Therefore, urchins are primary habitat modifiers in Western Atlantic reefs (Ogden et al. 1973; Hay 1984; Kilpp 1999; Eklöf et al. 2008). In other systems, damselfishes were found to buffer the formation of urchin barrens by establishing

Table 5.1 Comparison of roving herbivores assemblages between Brazilian and Caribbean provinces

Province	Family	Species	Allopatric pairs	Feeding mode	Maximum length	Turf removal	Macroalgae removal	Sediment removal	Crevice cleaning	Bioerosion	Corallivory	Spongivory
Brazil	Acanthuridae	*Acanthurus bahianus*	a	Browser	38	X		X				
	Labridae	*Scarus trispinosus*	b	Excavator	90	X		X		X	X	
		Scarus zelindae	c	Scraper	33	X		X				
		Sparisoma amplum	d	Excavator	70	X		X		X	X	
		Sparisoma axillare	e	Scraper	42	X	X	X				
		Sparisoma frondosum	f	Scraper	37	X	X	X				
		Sparisoma tuiupiranga	g	Scraper	15	X	X	X				
		Sparisoma rocha		Scraper	30	X	X	X				
Caribbean	Acanthuridae	*Acanthurus tractus*	a	Browser	38	X		X				
	Labridae	*Scarus coelestinus*		Scraper	77	X		X				
		Scarus coeruleus		Scraper	120	X		X		X	X	
		Scarus guacamaia	b	Scraper	120	X		X		X		
		Scarus iseri		Scraper	35	X		X				
		Scarus taeniopterus	c	Scraper	35	X		X				X
		Scarus vetula		Excavator	61	X		X		X	X	
		Sparisoma atomarium	g	Scraper	25	X	X	X				
		Sparisoma aurofrenatum		Scraper	28	X	X	X			X	X

(continued)

Table 5.1 (continued)

		Sparisoma chrysopterum	f	Scraper	46	X	X	X				
		Sparisoma rubripinne	e	Scraper	48	X	X	X				
		Sparisoma viride	d	Excavator	64	X		X		X	X	
Both Provinces	Acanthuridae	*Acanthurus coeruleus*		Browser	39	X						
		Acanthurus chirurgus		Browser	39	X		X				
	Labridae	*Cryptotomus roseus*		Browser	13		X					
		Nicholsina usta		Browser	30		X					
		Sparisoma radians		Browser	20		X					
Province		**Species richness**	**Mean standard length**		**Standard deviation**	**Number of species performing functional roles**						
						Turf removal	**Macroalgae removal**	**Sediment removal**	**Crevice cleaning**	**Bioerosion**	**Corallivory**	**Spongivory**
Brazil		13	38.15		21.17	10	7	9		2	2	
Caribbean		17	49.29		31.10	14	7	13		4	4	2
Indo-Pacific		136	42.35		17.91	111	19	78	18	20	5	1

Equal letters indicate allopatric pairs of sister-species based on recent Pleistocene (<5 Ma) divergences between the provinces, following bayesian-inference cladogram in Siqueira et al. (2019a, b). Functional roles provided by each species is given, adapted from data in Siqueira et al. (2019a, b), adjusting for a few equivocal omissions

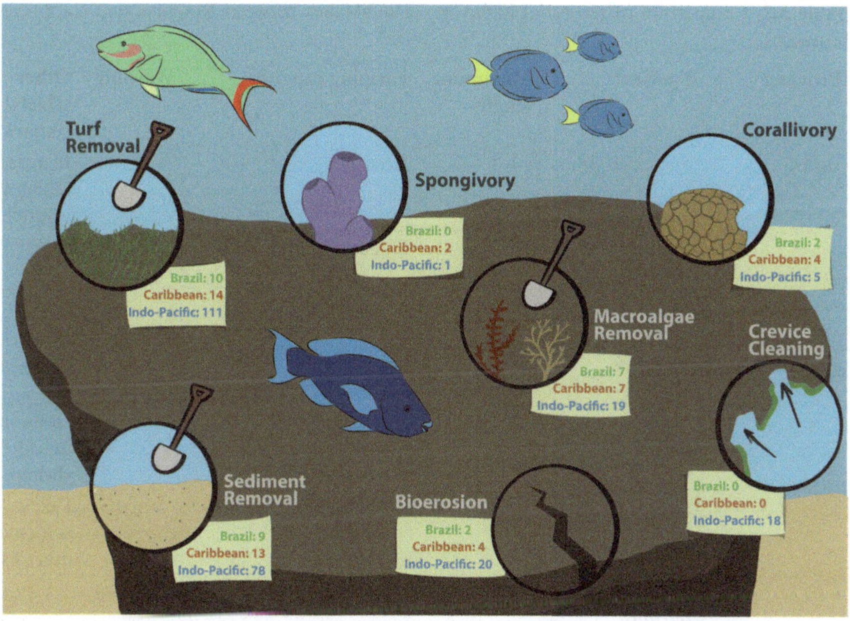

Fig. 5.2 Functional roles performed by roving herbivores (data adapted from Siqueira et al. 2019a, b). Number of species performing each role is given for biogeographic provinces (Brazil, Caribbean and Indo-Pacific)

farms that promoted erect algae growth and expelling sea urchins from their territories, thus preventing overgrazing (Irving and Witman 2009).

Damselfish are ecosystem engineers themselves, but distinctions emerge when comparing Atlantic and Indo-Pacific damselfish as habitat modifiers. While the Indo-Pacific province holds at least 27 farming damsels, the Atlantic sums 10 species, including all three Brazilian endemics. In addition, the Indo-Pacific have more species with behavioral traits that exert more remarkable changes in their habitat (Table 5.2, Fig. 5.3). The active removal of unwanted algae from territories (i.e., weeding) is observed in five Indo-Pacific species. In the Atlantic, only the Caribbean species *Stegastes planifrons* is known to weed algae out (Irvine 1980; Robertson et al. 1981; Hata and Ceccarelli 2016). Damselfish farming can be accomplished with varying intensity, from species that establish small territories dominated by one edible algae species (intensive farming) to damselfish that maintain large territories where preferred and unpalatable species are mixed (extensive farming) (Hata and Kato 2004). Most species hold either intensive or intermediate farms in the Indo-Pacific, while the Atlantic do not have species classified as intensive farmers (Hata and Ceccarelli 2016). Nevertheless, even extensive farmers can act as active habitat modifiers: the Brazilian endemic *S. fuscus* was observed to deter algal community succession at early stages, preventing the dominance by articulated calcareous algae, also increasing the diversity and abundance of algal epifauna (Ferreira

Table 5.2 Comparison of territorial herbivores assemblages between Brazilian and Caribbean provinces

Province	Species	Maximum length	Farming intensity		Weeding	Other effects/ remarks
Brazil	*Stegastes fuscus*	14	Extensive			Increases biomass of invertebrates
	Stegastes variabilis	13	Extensive			
	Stegastes rocasensis	13	Extensive			
Caribbean	*Stegastes adustus*	15	Extensive			
	Stegastes diencaeus	13	?			Domesticates mysid shrimps
	Stegastes leucostictus	10	Extensive			Decreases settlement of other fish
	Stegastes partitus	10	Extensive			
	Stegastes planifrons	13	Intermediate		X	Can kill corals
	Stegastes xanthurus	12	?			
Both Provinces	*Microspathodon chrysurus*	21	Extensive			
Province	**Species richness**	**Species performing weeding**	**Number of species by farming intensity**			
			Intensive	**Intermediate**	**Extensive**	**Unknown**
Brazil	4				4	
Caribbean	7	1		1	4	2
Indo-Pacific	27	5	6	6	8	7

Adapted from Hata and Ceccarelli (2016)

et al. 1998). Additionally, the same species was also observed to comply with late stages of algae succession inside their territories, taking advantage of edible epiphytic algae growing over articulated calcareous algae (Feitosa et al. 2012); thus, its influence observed on benthic communities seems to be variable and context-dependent.

Specific roles of individual species in delivering ecosystem services are important as well. A growing body of evidence supports the view that herbivore richness does not equal functional redundancy but complementary feeding. Experimental studies revealed the effects of herbivory performed by species combined to be responsible for a greater overall macroalgal consumption than isolated species could provide (Burkepile and Hay 2008, 2011). Once broadly categorized as functionally redundant, herbivorous fish have been demonstrated as complementary, as

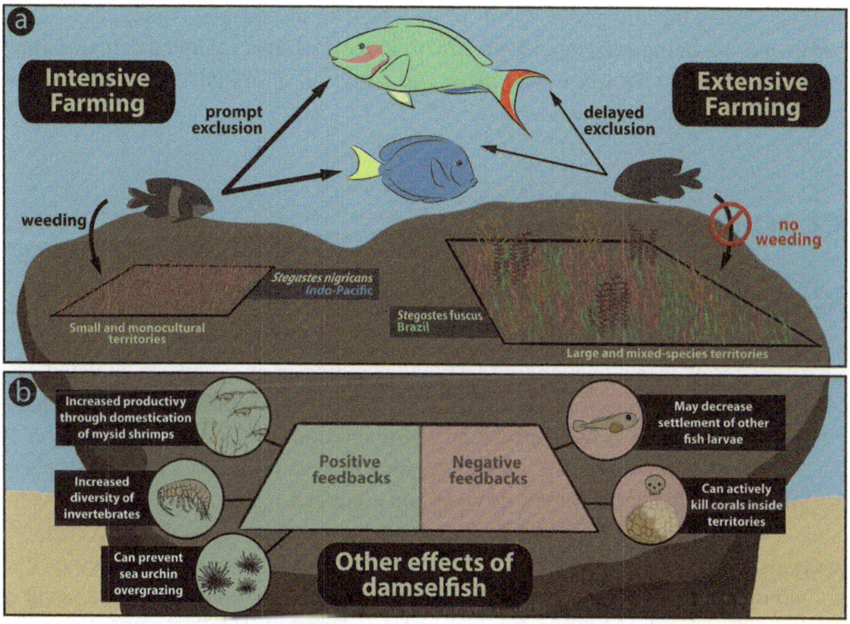

Fig. 5.3 Ecological effects of damselfish territoriality. (**a**) Differences between "intensive" and "extensive farming" (adapted from Hata and Kato 2004). (**b**) Other effects of damselfish territorial behavior

fine-scale divergences in feeding behavior unveil spatial differences in performing their functional roles (Semmler et al. 2021). Other studies have identified instances where macroalgal consumption was achieved barely by a single species (Bellwood et al. 2006; Tebbett et al. 2020), including instances with Brazilian representatives (Mendes et al. 2015, 2019). These examples revealed that unexpected actors, such as omnivores, are critical macroalgae consumers. Therefore, the commonplace of attributing richness to functional redundancy may backfire: what if the high richness of herbivore reflects the degree to which each species performs a specific, integrative, complementary and irreplaceable role in macroalgal consumption? Then, could less speciose herbivore assemblages indicate more generalized, overlapping, variable and adaptive feeding preferences for each component? There is evidence of intraspecific differences in feeding rates and preferences according to body size, ontogeny, food availability and social organization, which also modify the roles of individual herbivore species (reviewed in Bonaldo et al. 2014, but reports are increasing, e.g. Feitosa et al. 2021).

Determining how herbivores participate in reef dynamics is indeed a complex task. Still, pertinent new findings are continuously added to this narrative, each time driving us closer to better understanding the roles of fish in coral reef ecosystems. Brazilian reefs certainly can provide new insights on how herbivory works in coralline systems. One thing is for sure: herbivores are essential to coral reefs, whether in high-diversity systems or not. Moreover, specific components of herbivory have been widely overlooked; the roles of gastropods and crustaceans can be essential in distinct scales, but this piece is still to be added to the puzzle. Nonetheless, large herbivores are being extirpated from Brazilian reefs in an accelerated course. Corals are left to struggle for space with algae and a myriad of organisms under the pressure of human-based impacts.

5.3 Competition for Space

5.3.1 Brazilian Corals and Their Struggle for Space

Brazilian corals face a challenging task withstanding in an herbivore-depleted scenario, but this is not the sole menace to their persistence. Nutrification favors macroalgae and zoanthid growth, enhanced sedimentation rates smother reefs closer to the coast, and temperature rises make bleaching increasingly frequent. Nevertheless, the unique Brazilian coral fauna survives and competes against other benthic organisms. The acquisition of competitive mechanisms requires high energetic demands to corals, which would otherwise be allocated to other biological processes such as growth or reproduction (Tanner 1997). Competitive interactions frequently occur at the individual level and affect survival and development. Benthic competition can be influenced by species traits such as growth mode, reproduction strategies and source of nutrition. Hence, a solid knowledge of competitors' biology and interactive environmental factors is necessary.

Even though the literature of competition in Brazilian reefs has advanced, studies addressing competition are still very scarce. They have focused primarily on the direct physical and chemical interactions between coral species (dos Santos et al. 2013, Miranda et al. 2016a, Grillo et al. 2018) and other abundant benthic organisms, such as zoanthids (Cruz et al. 2016), cyanobacteria and macroalgae (Ramos et al. 2014; Ribeiro et al. 2018). Although ascidians, bryozoans and sponges are abundant in Brazil—sometimes encountering corals (Lorders et al. 2018)—competitive aspects are poorly known. In most cases, these competitive interactions have been investigated in a context of anthropogenic disturbance: when phase shifts have occurred, coral are facing bleaching, sedimentation rates are intense, or invasive species are widespread.

5.3.2 Who Competes for Space in Brazilian Coral Reefs?

Scleractinian corals are some of the most notorious and important reef competitors due to their abundance, morphology and size on the substrate. They are also major contributors to carbonate accretion in reef structure building (Knowlton and Jackson 2001; McField and Kramer 2007). Corals reefs of Brazil are unique due to the dominance of archaic species with massive morphology and high endemism: out of 23 scleractinian corals, 17 are zooxanthellate, five are azooxanthellate, four are endemic; from five calcareous hydrocorals, three are endemic (see also Chap. 4). Most studies investigating space competition in Brazilian reefs used scleractinians as models, such as the Brazilian endemics *Mussismilia hispida* and *Mussismilia braziliensis*, the more widely distributed *Siderastraea stellata*, *Montastraea cavernosa* and *Madracis decactis*, and the exotic invasives *Tubastraea tagusensis* and *T. coccinea*. These first three species occupy shallow and brighter reef habitats such as tops and edges, where they can be locally dominant. In shallow sites, the aspects related to light use (e.g., light incidence angle) and the relationship with symbiont zooxanthellae are fundamental to autotrophy and vitality of coral species. However, some zooxanthellate species develop a high capacity to rely on heterotrophy; *M. cavernosa* and *M. decactis* can be less dependent on light availability, occurring mainly in shaded or darker zones on reef walls and slopes or at greater depths, where they reach greater abundances (Lesser et al. 2010).

Brazilian corals are mostly massive-shaped with low growth rates (2.83–3.71 mm/year *M. hispida*, Silva et al. 2019; 8 mm/year *M. braziliensis*, Kikuchi et al. 2013; 5–7 mm/year *S. stellata*, Evangelista et al. 2018), but they can reach large colony sizes (up to 2 m in diameter for *M. braziliensis*) and high fecundity (Zilberberg et al. 2016). Reproductive modes and larval development vary between species: nearly half are brooders, and half are broadcasters. Brooder species, such as *S. stellata* and *M. decactis*, release immediately-competent planula larvae, capable of settling a few hours after spawning. These larvae have a relatively short dispersion, are less exposed to predation risk and recruits are established near the parental habitat (Lins de Barros et al. 2003). In contrast, broadcasting species, such as *M. hispida*, *M. braziliensis*, and *M. cavernosa*, have larger planktonic phase duration (at least 7 days, Pires et al. 1999; Neves and Pires 2002) and lower recruitment rates. Species with larvae settling near parental colonies can establish aggregated adult colonies, e.g., the native *S. stellata* and the invasive *Tubastraea* spp.

Tubastraea coccinea and *T. tagusensis* (popularly known as sun corals) are considered invasive corals in Brazil and have been studied since their introduction in the late 1980s (Castro and Pires 2001; Creed et al. 2017). They are native to the Indo-Pacific and supposedly arrived in Brazil via ship or oil platform fouling. Their distribution is still expanding along the Brazilian coast, including coral reefs, rocky reefs and artificial substrates (Creed et al. 2017). Sun corals have reached up to 30% cover on reef substrate at some sites in Todos os Santos Bay, Bahia state (Miranda et al. 2016a). *Tubastraea coccinea* reproduction occurs both asexually by budding and sexually by releasing gametes or fully mature larvae due to self-fertilization (Paula 2007; Luz et al. 2020). Larvae usually settle close to parental colonies forming a gregarious distribution, but they may also travel with currents to colonize other reefs (Mizrahi et al. 2014; Miranda et al. 2018a). As azooxanthellate corals, sun corals prefer shaded habitats, such as vertical or inclined substrates, where their heterotrophy is advantageous over light-dependent zooxanthellate corals. However, *Tubastraea* spp. were also reported in shallow and lighted habitats and even resisting desiccation for some hours after colonizing intertidal areas (Miranda et al. 2016b).

Zoanthids are common and abundant competitors in shallow reefs (1–7 m depth, Aued et al. 2018) frequently favored by coastal nutrient input (Cruz et al. 2016).

Palythoa cf. *variabilis* reached 46% of substrate cover at some sites and were found to double in cover after 4 years of increased nutrification (Dutra et al. 2006). *Palythoa caribaeorum*, *Zoanthus sociatus* and *Palythoa variabilis* are the most common and abundant zoanthids inhabiting reef flats in Brazil, and the first two show high growth rates (Rabelo et al. 2013). *Palythoa caribaeorum* is an aggressive competitor, growing up to 4 mm/day, frequently smothering other species (Suchanek and Green 1981). *Palythoa* species can also produce palytoxin that can inhibit the settlement of benthic species, including scleractinian corals. Moreover, *P. caribaeorum* was demonstrated to assimilate sediment (up to 6 mm/year), even at silt and clay fractions, which can be advantageous under high sedimentation rates (Haywick and Mueller 1997).

Octocorals are important structural components of the benthic community and can be used as food or provide habitat for fish, snails, algae and other marine species (McFadden et al. 2010). They can be found in both deep and shallow reefs in Brazilian waters and are often dominant space-occupiers. Although hundreds of octocorals species are registered in Brazil, only the endemic *Phyllogorgia dilatata* had its competitive mechanisms studied, which took place in areas invaded by exotic octocoral species (Lages et al. 2006). Octocorals, including *P. dilatata*, produce a diverse range of secondary metabolites, applied as a defense against predation (Van Alstyne et al. 1994; Epifânio et al. 1999), anti-fouling agents (Pereira et al. 2002) and as allelopathic chemicals when competing with scleractinian corals (Sammarco et al. 1983, Fleury et al. 2004). *Neospongodes atlantica*, an abundant azooxanthellate octocoral, is another great competitor advancing over scleractinian corals in relevant reef areas as Abrolhos, Todos os Santos Bay and other reefs on the Eastern coast of Brazil.

Invasion by octocorals has been increasingly common in Brazilian reefs. *Sarcothelia* sp., *Briareum harmum*, *Sansibia* sp., *Clavularia cf. viridis*, and *Chromonephthea braziliensis* were reported on rocky reefs (Lages et al. 2006; Mantelatto et al. 2018). *Sarcothelia* sp., for instance, forms a massive "blue carpet" near tropical coral reefs in Todos os Santos Bay (Menezes et al. 2021). These species are still understudied but have high growth rates and reproduce both sexually and asexually by budding and fission, with a great potential to spread quickly (Kahng et al. 2011). *Chromonephthea braziliensis* is an exotic species that originated in the Indo-Pacific and has been reported in subtropical rocky reefs of Brazil since the 1990s, occupying sandy substrates between 8 and 10 m depth. Their growth is fast, and the production of secondary metabolites contributes to the establishment process (Lages et al. 2006). It was most likely introduced by oil platform fouling (Ferreira et al. 2004a, b).

Although corals interact competitively with this plethora of organisms, algae are currently the dominant benthic group covering Brazilian reefs (Aued et al. 2018). Coral larvae cannot settle as the substrate becomes covered by algae (Birkeland 1977; Richmond 1997). Beyond directly competing with corals for space, algae can host pathogenic bacteria that cause indirect mortality on recruit and adult corals (Nugues and Roberts 2003; Birrell et al. 2005; Smith et al. 2006). However, encrusting calcareous algae are one of the main reef builders (Littler and Littler 1997). Encrusting algae can also facilitate the settlement of coral larvae and inhibit the recruitment of other macroalgae; therefore, they are key components for reef resilience (Antonius and Afonso-Carillo 2001; Vermeij 2005).

Although recognized as a ubiquitous component of benthic communities, cyanobacteria have been historically regarded as limited to intertidal or very shallow water habitats (Fong and Paul 2011) or as a component of the turf algae (Littler and Littler 1988). In Brazil, cyanobacteria cover significant portions of the reef substrate at greater depths on the Northern coast (~30 m, Cordeiro et al. 2021). Studies on benthic cyanobacteria, although very incipient, have shown their importance in bacterial communities associated with endemic coral species (Castro et al. 2010) and as significant constituents of turf algae formations (Walter et al. 2016). Brazilian cyanobacteria have been demonstrated to have a cryptic diversity to be studied. A new—albeit widely distributed along Brazilian coast—genus of cyanobacteria was recently described, along with six new species (Caires et al. 2018). Additionally, they are unfavorable to coral growth rates, carbonate deposition (Thacker and Paul 2001), coral recruitment (Birkeland 1977) and can be related to coral disease, for example, to the black-band disease caused by *Phormidium corallyticum* (Peters 1997).

5.3.3 Competition Between Native Species

Effective mechanisms for outcompeting rivals in the struggle for space on reef substrates are key life traits for benthic taxa, including dynamic tactics of interference competition (e.g. allelopathic interactions). Assumed the current perspective of worldwide coral downfall, increasing attention has been given to the role of other taxa as potential competitors. Macroalgae emerge as one of the major competitors, especially when herbivory rates are low due to human impacts (e.g., overfishing). Under stable conditions, corals are very effective competitors against most of their antagonists, but the multiple human-induced impacts have reduced their vitality, influencing their capabilities for competition. Nutrification and herbivore overfishing may enhance algal growth, unbalancing the competition odds in favor of macroalgae. In Brazil, studies indicate that algae have taken advantage of nutrient pollution sources to cover more bare areas, reducing space availability for coral settlement, with adverse effects on the coral cover (Costa Jr. et al. 2002, 2008). During a bleaching event, the Brazilian endemic coral *Siderastrea stellata* did not offer any resistance to filamentous algae overgrowth when grazing pressure is low (Ramos et al. 2014; Fig. 5.4). Thus, the accumulation of these impacts may be very pernicious to corals. At the same time, life traits of most algae, such as rapid growth, fast colonization and profuse reproduction methods, may be advantageous for exploitation competition under nutrification and coral bleaching events.

Several mechanisms are identified for algae to inflict direct damage to corals (which are reviewed in McCook et al. 2001; Birrell et al. 2008; Ritson-Williams et al. 2009; Chadwick and Morrow 2011). These competitive corals-macroalgae interactions occur by physical and chemical processes and can impact coral at all

Fig. 5.4 Benthic competitors in direct encounters in Brazilian coral reefs: (**a**) intraspecific competition of *Mussismilia hispida*, (**b**) *Mussismilia hispida* colony exuding mesenteric filaments (arrow) onto *P.* cf. *variabilis*, (**c**) a sweeper tentacle of *Montastraea cavernosa* (arrow) near to competing neighbour *Millepora alcicornis*, (**d**) overshading of frondose macroalgae on coral *Mussismilia braziliensis*, (**e**) turf overgrowing *Siderastarea stellata* (arrow) after bleaching episode, (**f**) cyanobacteria overgrowing coral *Mussismilia harttii*. (Photos: Ricardo Miranda (**a**, **e**, **f**), Igor Cruz (**b**, **c**), Andrei Cardoso (**d**))

stages of their life cycle but seem to be quite variable across the different species of macroalgae and corals (Chadwick and Morrow 2011; Fong and Paul 2011). Long-term monitoring has identified the contact with cyanobacteria to have the most detrimental effects on the endangered endemic *Mussismilia braziliensis*, compared with contact with other abundant organisms, such as other corals, crustose coralline and foliose macroalgae (Ribeiro et al. 2018). These negative consequences have been attributed to allelopathy (Ribeiro et al. 2018 and also observed by Rasher et al. 2011). Yet, cyanobacteria can also compete against corals by overgrowth (Fig. 5.4; Bender et al. 2012), shading and abrasion (McCook et al. 2001).

Besides the increase in coral-macroalgae competitive interactions associated with potential phase-shift events, other alternative states dominated by corallimorphs, soft corals, sponges and sea urchins have been described as well for coral reefs worldwide (Norström et al. 2009). Recently, zoanthid-dominated reefs were identified in Brazil as another possible stable state (Cruz et al. 2011). Studies investigating the effect of contact with over-abundant zoanthids on local coral fauna have shown a substantial increase in colony mortality of coral species. *Mussismilia hispida* and *Montastraea cavernosa* presented aggressive competition strategies when directly contacting *Palythoa variabilis*. Sweeper tentacles, with more nematocysts, were observed in *M. cavernosa*, and *M. hispida* applied mesenteric filaments (extrusions of digestive tubes) during competitive encounters (Cruz et al. 2016). Even though these strategies were employed, none of these species—nor *Siderastrea stellata*, which did not apply such energy-costing defenses—resisted the harmful effects of competition with *Palythoa variabilis* and contacts caused necrosis in 78% of the cases in just about 4 months (Cruz et al. 2016). Zoanthid dominance also affects reef fish, which showed a decrease in species richness and an increase in mobile invertivores densities (Cruz et al. 2015). The drivers behind zoanthid dominance are yet to be defined. However, *Palythoa* colonies are expected to exhibit allelopathic strategies (Bastidas and Bone 1996); they may also grow rapidly, gaining an advantage over coral on nutrient-enriched sites of nearshore reefs (Costa Jr. et al. 2002, 2008).

5.3.4 The Interaction with Invasive Species

With the human-mediated species invasions, a new set of competitors, including sun and soft corals, have come into play in Brazilian competition networks. Field experiments and *in situ* observations have recorded *T. tagusensis* developing elongated polyps to compete against the native species *S. stellata*, *M. cavernosa*, *M. decactis* and *M. hispida* (Miranda et al. 2016b). These invaders can also produce allelochemicals to attack competitors (dos Santos et al. 2013). Native species responses to invader attacks are variable: *S. stellata* increased tissue necrosis, whereas *M. cavernosa* not only resisted invader attacks but responded, damaging *T. tagusensis* colonies (Fig. 5.5). *Montastrea cavernosa* and *M. decactis* also are capable of developing sweeper tentacles and mesenterial filaments in response to contact with

Fig. 5.5 Competition between exotic invasive and native corals species: (**a**) natural encounter between *Montastraea cavernosa* and *Tubastraea tagusensis*, (**b**) sweeper tentacles of *M. cavernosa* (arrow) projected on *T. tagusensis*, (**c**) sweeper tentacles of *Madracis decactis* (arrows) near to contact zone with *T. tagusensis*, (**d**) experimental contact between *T. tagusensis* and *Siderastrea stellat*a that suffered tecidual mortality and (**e**) *M. cavernosa* that caused mortality on invader colony (arrow), (**f**) *T. tagusensis* causing tissue mortality with a new recruit settled on *M. hispida*, (**g**) octocoral *Sansibia* sp. overgrowing coral *M. hispida*, (**h**) *Sarcothelia* sp. overgrowing *Zoanthus sociatus*. (Photos: Ricardo Miranda (**a**, **b**, **d**, **e**, **f**), Igor Cruz (**c**), Marcelo Mantelatto (**g**) and Rodrigo Maia-Nogueira (**h**))

opponent tentacles (Chornesky 1983). The extra energy demanded to produce these mechanisms may come under the cost of renouncing energetically-costly reproduction (such as broadcasting). At the same time, enhanced use of heterotrophy can occur in *M. cavernosa* (Lapid et al. 2004). *Siderastrea stellata* can invest energy in larvae incubation to compensate for losses due to competition during its adult phase and assure effective settlement and recruitment during early life phases.

The biological traits of these invasive species make them either equivalent or superior competitors and have been determinant for their establishment's success and increasing abundance. *Tubastraea* invasions have triggered changes in native community structure, which are also prone to occur in recent soft corals invasions in Brazilian reefs (dos Santos et al. 2013; Miranda et al. 2016a; Menezes et al. 2021). *Tubastraea* prefers vertical surfaces and shaded habitats, which have optimal conditions for the rapid growth of azooxanthellate corals and experience release from competition with native zooxanthellate corals compared to high-irradiance horizontal surfaces (Paula and Creed 2005; Mizrahi et al. 2014; Miranda et al. 2016a). The dominance of more extensive areas by non-indigenous species can potentially disrupt existing trophic relationships—as fish-benthos interactions—and relevant ecosystem processes, such as coral recruitment (Miranda et al. 2018 a, b). Soft corals imported by the aquarium trade, as the octocorals *Sarcothelia* sp. and *Sansibia* sp., have increased in abundance rapidly at rocky reefs. Their encounters with native corals, octocoral,s and zoanthids increase, with some overgrowth instances. Allelochemical aggression seems to be the primary competitive mechanism used by invasive octocorals, as demonstrated by *Chromonephthea braziliensis* when competing with the endemic octocoral *Phyllogorgia dilatata*, increasing tecidual necrosis after only 2 weeks (Lages et al. 2006). The colonization of nearby coral reefs by invasive soft corals is eminent, at a great chance of similarly impacting native benthic distribution as seen in rocky reefs.

5.4 Final Remarks: Impacts and Perspectives

Brazilian corals are enduring in a system subject to several impacts: increasing temperatures results in bleaching and mortality of zooxanthellate species; terrestrial runoff in enhancing nutrification processes and sedimentation rates, burying corals faster than they can cope with. Contact with chemicals, such as fertilizers, pesticides and oil spills are becoming more frequent and cumbersome. In addition to these impacts, Brazilian herbivores have been extensively exploited for human consumption in the past few decades, and have increased as exports commodities, mainly consumed in the USA (Freire and Pauly 2010; Cunha et al. 2012; Carvalho et al. 2013; Roos et al. 2020a). The increase in their consumption is associated with the depletion of carnivore fish stocks, such as groupers and snappers (Floeter et al. 2006; Freire and Pauly 2010). Moreover, the overfishing of these higher trophic levels may also trigger cascade effects to herbivore populations (McCauley et al. 2010), which are still to be acknowledged in Brazil. Surgeonfishes, parrotfishes and damselfishes have also been collected and exported by the marine ornamental fish trade (Monteiro-Neto et al. 2003).

The potential impacts of exotic species, such as the sun coral (*Tubastraea* spp.) (Miranda et al. 2018a) and the predator lionfish (Pterois spp.) (Luiz Jr. et al. 2021), to Brazilian reefs must also be acknowledged. Increased *Tubastrea* cover is related to drastic reductions in parrotfish and surgeonfish feeding rates (Miranda et al. 2018a). Lionfish have established populations in Brazilian waters (Luiz Jr. et al. 2021; Soares et al. 2022; Maggioni et al. 2023), and if lionfish biomass reaches extremely high numbers as observed in some regions of the Caribbean (~10 t/km^2) (Côté and Maljkovic 2010), the chances are that herbivore populations face further reductions (Bumbeer et al. 2017).

Habitat degradation and overfishing have led to sharp decreases in parrotfish size and abundance. Consequently, five species now appear in the Brazilian Red List of Threatened Species (IBAMA Decree 445/2014): *S. zelindae*, *S. axillare*, *S. frondosum* and *S. rocha* are listed as Vulnerable; *S. trispinosus* is categorized as Endangered, both by the Brazilian Red List and by the IUCN Red List (Ferreira et al. 2012), that being the worst category reached by any parrotfish species (Comeros-Raynal et al. 2012). This Decree also established a national fishing ban for all Brazilian threatened species, including five endemic parrotfish; nonetheless, the fishing ban lasted for less than a year (MMA Decree 98/2015). Due to conflicts among fishing industries and environmental agencies, Decree 445/2014 underwent several modifications; currently, the fishing of each threatened species have specific regulations, for parrotfish, fishing is permitted, with the exception of *S. trispinosus*, whose capture is currently prohibited due to its endangered status (Ordinance MMA 354/2023), but should be regulated by local stock management plans (Freitas et al. 2019; Queiroz-Véras et al. 2023; Roos et al. 2020b). For most of the coast, there are no management plans implemented yet. Due to the lack of control, fishing is still irregularly and profusely occurring, including exports of parrotfish, which are prohibited. Overfishing of large herbivores is a reality for most reefs along the Brazilian coast; however, no studies to identify the implications of removing those large herbivores have been published so far. With the lack of data before herbivore depletion, the actual consequences of such removal may never be fully accessed.

Marine Protected Areas (MPAs) are the main conservation strategy for the protection of large herbivores and corals but have not been able to increase coral cover (Bruno et al. 2019) or to promote the recovery of herbivore populations in Brazil (Roos et al. 2020b). Inefficiency in fisheries management strategies and the chronic impacts of ocean warming are some of the reasons why MPAs fail to restore the resilience of Brazilian coral reefs (Roos et al. 2020b; Brandão et al. 2021). Reducing global warming and nutrient inputs is essential to MPA effectiveness, particularly in light of the recent dismantling of environmental laws and the weakening of environmental protection agencies in Brazil (Abessa et al. 2019; Araújo 2020). Drastic reductions in funds destined for both research and environmental law enforcement have been constant (Araújo 2020), and budget cuts for these areas are still increasing. If environmental policies are not restored and enforced, serious actions to reduce climate change, control invasive species and regulate agricultural and human waste are not taken, Brazilian coral reefs, as we know, are sentenced to perish. Alternative states, such as sea urchin-barrens or algae/zoanthid-dominated reefs, are very likely to become prevalent under the growing influence of these impacts and synergistic effects of depleting herbivore assemblages.

Aknowledgements We would like to thank researchers Dráusio Véras, Igor Cruz, Andrei Cardoso, Marcelo Mantelatto and Rodrigo Maia-Nogueira, who kindly contributed with photographs to this chapter. We are grateful to Petrobras for the financial support of Brazilian parrotfish research by funding the *Projeto Budiões* and to the Rufford Foundation for funding sun coral research through the *Projeto Guarda-Sol*.

References

Abessa D, Famá A, Buruaem L (2019) The systematic dismantling of Brazilian environmental laws risks losses on all fronts. Nat Ecol Evol 3:510–511

Adam TC, Kelley M, Ruttenberg BI, Burkepile DE (2015) Resource partitioning along multiple niche axes drives functional diversity in parrotfishes on Caribbean coral reefs. Oecologia 179:1173–1185

Antonius A, Afonso-Carillo J (2001) Pheophyllum conicum killing reef-Croals in Mauritius: a new indo-Pacific syndrome? Bull Mar Sci:613–618

Araújo SMVG (2020) Environmental policy in the Bolsonaro government: the response of environmentalists in the legislative arena. Bras Political Sci Rev 14:1–20

Araújo ME, Mattos FMG, Melo FPL, Chaves LCT, Feitosa CV, Lippi DL, Hackradt FCF, Hackradt CW, Nunes JLS, Leão ZM de AN, RKP K, Ferreira Junior AV, PHC P, CHR M, CLS S, JLL F (2020) Diversity patterns of reef fish along the Brazilian tropical coast. Mar Environ Res 160:105038. https://doi.org/10.1016/j.marenvres.2020.105038

Attrill MJ, Kelmo F (2007) Opportunistic responses of Diadema antillarum (Echinodermata: Echinoidea) populations following the 1997-98 El Niño event in Bahia, Brazil. Estuar Coast Shelf Sci 73:243–248

Aued AW, Smith F, Quimbayo JP, Cândido DV, Longo GO, Ferreira CEL, Witman JD, Floeter SR, Segal B (2018) Large-scale patterns of benthic marine communities in the Brazilian province. PLoS One 13:1–15

Bariche M, Letourneura Y, Harmelin-Viviena M (2004) Temporal fluctuations and settlement patterns of native and Lessepsian herbivorous fishes on the Lebanese coast (eastern Mediterranean). Environ Biol Fishes 70:81–90

Bastidas C, Bone D (1996) Competitive strategies between *Palythoa caribaeorum* and *Zoanthus sociatus* (Cnidaria: Anthozoa) at a reef flat environmental in Venezuela. Bull Mar Sci 59:543–555

Bellwood DR (1994) A phylogenetic study of the parrotfish family Scaridae (Pisces: Labroidea), with a revision of genera. Records of the Australian Museum, Supplement, 20(7):1–86

Bellwood DR, Wainwright PC (2002) The history and biogeography of fishes on coral reefs. In: Sale PS (ed) Coral reef fishes: dynamic and diversity in a complex ecosystem. Elsevier, San Diego, pp 5–32

Bellwood DR, Hughes TP, Folke C, Nyström M (2004) Confronting the coral reef crisis. Nature 429:827–833

Bellwood DR, Hughes TP, Hoey AS (2006) Sleeping functional group drives coral-reef recovery. Curr Biol 16:2434–2439

Bellwood DR, Goatley CHR, Bellwood O (2017) The evolution of fishes and corals on reefs: form, function and interdependence. Biol Rev 92:878–901

Bender D, Diaz-Pulido G, Dove S (2012) Effects of macroalgae on corals recovering from disturbance. J Exp Mar Bio Ecol 429:15–19. https://doi.org/10.1016/j.jembe.2012.06.014

Bernal MA, Rocha LA (2011) *Acanthurus tractus* Poey, 1860, a valid western Atlantic species of surgeonfish (Teleostei, Acanthuridae), distinct from *Acanthurus bahianus* Castelnau, 1855. Zootaxa:63–68

Bernardi G, Robertson DR, Clifton KE, Azzurro E (2000) Molecular systematics, zoogeography, and evolutionary ecology of the Atlantic parrotfish genus *Sparisoma*. Mol Phylogenet Evol 15:292–300

Birkeland CE (1977) The importance of rate of biomass accumulation in early successional stages of benthic communities to the survival of coral recruits. In: Proceeding of 3rd international coral reef symposium, pp 15–21

Birrell CL, McCook LJ, Willis BL (2005) Effects of algal turfs and sediment on coral settlement. Mar Pollut Bull:408–414

Birrell CL, McCook LJ, Willis BL, Harrington L (2008) Allelochemical effects of macroalgae on larval settlement of the coral Acropora millepora. Mar Ecol Prog Ser:129–137

Bodilis P, Louisy P, Draman M, Arceo HO, Francour P (2014) Can citizen science survey non-indigenous fish species in the eastern Mediterranean Sea? Environ Manag 53:172–180

Bonaldo R, Hoey A, Bellwood D (2014) The ecosystem roles of parrotfishes on tropical reefs. Oceanogr Mar Biol An Annu Rev 52:81–132

Brandão CS, da Rocha W, Zapelini C, Teixeira JL, Schiavetti A (2021) Does the respondent's role affect the final value of management effectiveness? The case of Brazilian marine protected areas. Ethnobiol Conserv 10:1–25

Bruggemann JH, Van Kessel AM, Van Rooij JM, Breeman AM (1996) Bioerosion and sediment ingestion by the Caribbean parrotfish *Scarus vetula* and *Sparisoma viride*: implications of fish size, feeding mode and habitat use. Mar Ecol Prog Ser 134:59–71

Bruno JF, Côté IM, Toth LT (2019) Climate change, coral loss, and the curious case of the parrotfish paradigm: why Don't marine protected areas improve reef resilience? Annu Rev Mar Sci 11:307–334

Bumbeer J, da Rocha RM, Bornatowski H, Robert MC, Ainsworth C (2017) Predicting impacts of lionfish (*Pterois volitans*) invasion in a coastal ecosystem of southern Brazil. Biol Invasions 20:1257–1274

Burkepile DE, Hay ME (2008) Herbivore species richness and feeding complementarity affect community structure and function on a coral reef. Proc Natl Acad Sci 105:16201–16206

Burkepile DE, Hay ME (2011) Feeding complementarity versus redundancy among herbivorous fishes on a Caribbean reef. Coral Reefs 30:351–362

Burkepile DE, Adam TC, Roycroft M, Ladd MC, Munsterman KS, Ruttenberg BI (2019) Species-specific patterns in corallivory and spongivory among Caribbean parrotfishes. Coral Reefs 38:417–423

Caires TA, de Mattos LG, Hentschke GS, de Gusmão PA, Sant'Anna CL, de Castro Nunes JM (2018) *Neolyngbya* gen. nov. (Cyanobacteria, Oscillatoriaceae): a new filamentous benthic marine taxon widely distributed along the Brazilian coast. Mol Phylogenet Evol 120:196–211. https://doi.org/10.1016/j.ympev.2017.12.009

Carvalho RAA, Cunha FEA, Montezuma AMN, Araújo ME (2013) Captura e processamento de peixes recifais no estado do Rio Grande do Norte, Brasil. Acta Fish Aquat Res 1:91–103

Castro CB, Pires DO (2001) Brazilian coral reefs: what we already know and what is still missing. Bull Mar Sci 69:357–371

Castro AP, Araújo SD, Reis AMM, Moura RL, Francini-Filho RB, Pappas G, Rodrigues TB, Thompson FL, Krüger RH (2010) Bacterial community associated with healthy and diseased reef coral *Mussismilia hispida* from Eastern Brazil. Microb Ecol 59:658–667. https://doi.org/10.1007/s00248-010-9646-1

Chadwick E, Morrow KM (2011) Competition among sessile organisms on coral reefs. In: Dubinsky Z, Stambler N (eds) Coral reefs: an ecosystem in transition. Springer, pp 347–371

Chaves LCT, Ormond CGA, Mcginty ES, Ferreira BP (2012) Space partitioning among damselfishes in the Caribbean coast of Panama: the role of habitat preferences. Neotrop Ichthyol 10:633–642. https://doi.org/10.1590/S1679-62252012000300017

Chaves LCT, Feitosa JLL, Xavier TF, Ferreira BP, Ferreira CEL (2021) Drivers of damselfishes distribution patterns in the southwestern Atlantic: tropical and subtropical reefs compared. Neotrop Ichthyol 19(4):1–22. https://doi.org/10.1590/1982-0224-2021-0010

Choat JH, Klanten OS, Herwerden LV, Robertson DR, Clements KD (2012) Patterns and processes in the evolutionary history of parrotfishes (family Labridae). Biol J Linn Soc 107:529–557

Chornesky EA (1983) Induced development of sweeper tentacles on the reef coral *Agaricia agarites*: a response to direct competition. Biol Bull 165:569–581. https://doi.org/10.2307/1541466

Comeros-Raynal MT, Choat JH, Polidoro BA, Clements KD, Abesamis R, Craig MT, Lazuardi ME, McIlwain J, Muljadi A, Myers RF, Nañola CL, Pardede S, Rocha LA, Russell B, Sanciangco JC, Stockwell B, Harwell H, Carpenter KE (2012) The likelihood of extinction of iconic and dominant herbivores and detritivores of coral reefs: the parrotfishes and surgeonfishes. PLoS One 7:e39825

Cordeiro CAMM, Harborne AR, Ferreira CEL (2014) Patterns of distribution and composition of sea urchin assemblages on Brazilian subtropical rocky reefs. Mar Biol 161:2221–2232

Cordeiro CAMM, Quimbayo JP, Nunes JA, Nunes LT, Sissini MN, Sampaio CL, Morais RA, Horta PA, Aued AW, Carraro JL, Hajdu E (2021) Conservation status of the southernmost reef of the Amazon reef system: the Parcel de Manuel Luis. Coral Reefs 40:165–185. https://doi.org/10.1007/s00338-020-02026-1

Costa AKR (2013) O efeito da exclusão da pesca em populações macrobentônicas de ambientes recifais com ênfase em ouriços *Echinometra lucunter* na baia de Tamandaré. Universidade Federal do Pernambuco, Pernambuco

Costa OS Jr, Attrill MJ, Pedrini AG, De-Paula JC (2002) Spatial and seasonal distribution of seaweeds on coral reefs from southern Bahia, Brazil. Bot Mar 45:346–355. https://doi.org/10.1515/BOT.2002.035

Costa OS Jr, Nimmo M, Attrill MJ (2008) Coastal nutrification in Brazil: a review of the role of nutrient excess on coral reef demise. J S Am Earth Sci 2:257–270

Côté IM, Maljkovic A (2010) Predation rates of Indo-Pacific lionfish on Bahamian coral reefs. Mar Ecol Prog Ser 404:219–225

Creed JC, Fenner D, Sammarco P, Cairns S, Capel K, Junqueira AOR, Cruz I, Miranda RJ, Carlos-Junior L, Mantelatto MC, Oigman-Pszczol S (2017) The invasion of the azooxanthellate coral *Tubastraea* (Scleractinia: Dendrophylliidae) throughout the world: history, pathways and vectors. Biol Invasions 19:283–305. https://doi.org/10.1007/s10530-016-1279-y

Cruz ICS, De KRKP, Longo LL, Creed JC (2011) Evidence of a phase shift to *Epizoanthus gabrieli* Carlgreen, 1951 (Order Zoanthidea) and loss of coral cover on reefs in the Southwest Atlantic. Mar Ecol 1951:1–8. https://doi.org/10.1111/maec.12141

Cruz ICS, Loiola M, Albuquerque T, Reis R, De Nunes JACC, Reimer JD, Mizuyama M, Kikuchi RKP, Creed JC (2015) Effect of phase shift from corals to zoantharia on reef fish assemblages. PLoS One 10:1–16. https://doi.org/10.1371/journal.pone.0116944

Cruz ICS, Henschen V, Kikuchi RKP, Creed JC (2016) The role of competition in the phase shift to dominance of the zoanthid *Palythoa* cf. *variabilis* on coral reefs. Mar Environ Res 115:28–35. https://doi.org/10.1016/j.marenvres.2016.01.008

Cunha FEA, Carvalho RAA, Araújo ME (2012) Exportation of reef fish for human consumption: long-term analysis using data from Rio Grande do Norte, Brazil. Bol Inst Pesca 38:369–378

Dias TLP, Neto NAL, Alves RRN (2011) Molluscs in the marine curio and souvenir trade in NE Brazil: species composition and implications for their conservation and management. Biodivers Conserv 20:2393–2405

dos Santos LAH, Ribeiro FV, Creed JC (2013) Antagonism between invasive pest corals *Tubastraea* spp. and the native reef-builder *Mussismilia hispida* in the Southwest Atlantic. J Exp Mar Bio Ecol 449:69–76

Dutra LXC, Kikuchi RKP, Leão ZMAN (2006) Todos os Santos Bay coral reefs, Eastern Brazil, revisited after 40 years. In: Proceedings of 10th international coral reef symposium, pp 1090–1095

Eklöf JS, de la Torre-Castro M, Gullström M, Uku J, Muthiga N, Lyimo T, Bandeira SO (2008) Sea urchin overgrazing of seagrasses: a review of current knowledge on causes, consequences, and management. Estuar Coast Shelf Sci 79:569–580

Epifanio RDA, Martins DL, Villaça R, Gabriel R (1999) Chemical defenses against fish predation in three Brazilian octocorals: 11β,12β-epoxypukalide as a feeding deterrent in *Phyllogorgia dilatata*. J Chem Ecol 25:2255–2265. https://doi.org/10.1023/A:1020817722885

Evangelista H, Sifeddine A, Corrège T, Servain J, Dassié EP, Logato R, Cordeiro RC, Shen CC, Le Cornec F, Nogueira J, Segal B, Castagna A, Turcq B (2018) Climatic constraints on growth rate and geochemistry (Sr/Ca and U/Ca) of the coral *Siderastrea stellata* in the southwest equatorial Atlantic (Rocas Atoll, Brazil). Geochem Geophys Geosyst 19:772–786

Feitosa JLL, Ferreira BP (2015) Distribution and feeding patterns of juvenile parrotfish on algal-dominated coral reefs. Mar Ecol 36:462–474. https://doi.org/10.1111/maec.12154

Feitosa JLL, Concentino AM, Teixeira SF, Ferreira BP (2012) Food resource use by two territorial damselfish (Pomacentridae: Stegastes) on South-Western Atlantic algal-dominated reefs. J Sea Res 70:42–49. https://doi.org/10.1016/j.seares.2012.03.006

Feitosa JLL, Chaves LCT, Queiroz-Véras LVMV, Miranda RJ, Ormond CGA, Ferreira BP (2021) Effects of social organization on the feeding of the striped parrotfish, *Scarus iseri*. Coral Reefs 40:951–957. https://doi.org/10.1007/s00338-021-02080-3

Feitosa JLL, Queiroz-véras LVM V, Maida M, Ferreira BP (2023) Going further on herbivore fishing: the removal of smaller fishes from algal-dominated reefs. Mar Ecol Prog Ser 713:117–132

Ferreira BP, Maida M (2006) Monitoring Brazilian coral reefs – status and perspectives. Minist do Meio Ambient – MMA, 250p

Ferreira CEL, Peret AC, Coutinho R (1998) Seasonal grazing rates and food processing by tropical herbivorous fishes. J Fish Biol 53:222–235

Ferreira ACEL, Floeter SR, Gasparini JL, Ferreira BP, Joyeux JC, Journal S, Jul N (2004a) Trophic structure patterns of Brazilian reef fishes: a latitudinal comparison. J Biogeogr 31:1093–1106

Ferreira CEL, Gonçalves JEA, Coutinho R (2004b) Cascos de navios e plataformas como vetores na introdução de espécies exóticas. In: Silva JSV, Souza RCCL (eds) Ballast water and bioinvasion. Interciência, Rio de Janeiro, pp 143–155

Ferreira BP, Floeter S, Rocha LA, Ferreira CE, Francini-Filho R, Moura R, Gaspar AL, Feitosa C (2012) Scarus trispinosus. The IUCN Red List of Threatened Species

Filbee-Dexter K, Scheibling RE (2014) Sea urchin barrens as alternative stable states of collapsed kelp ecosystems. Mar Ecol Prog Ser 495:1–25

Fleury BG, Coll JC, Sammarco PW, Tentori E, Duquesne S (2004) Complementary (secondary) metabolites in an octocoral competing with a scleractinian coral: effects of varying nutrient regimes. J Exp Mar Bio Ecol 303:115–131

Floeter SR, Halpern BS, Ferreira CEL (2006) Effects of fishing and protection on Brazilian reef fishes. Biol Conserv 128:391–402

Floeter SR, Rocha LA, Robertson DR, Joyeux JC, Smith-Vaniz WF, Wirtz P, Edwards AJ, Barreiros JP, Ferreira CEL, Gasparini JL, Brito A, Falcón JM, Bowen BW, Bernardi G (2008) Atlantic reef fish biogeography and evolution. J Biogeogr 35:22–47

Fong P, Paul JV (2011) Coral reef algae: the good, the bad, and the ugly. In: Dubinsky Z, Stambler N (eds) Coral reefs: an ecosystem in transition. Springer

Francini-Filho RB, Moura RL, Ferreira CM, Coni EOC (2008) Live coral predation by parrotfishes (Perciformes: Scaridae) in the Abrolhos Bank, eastern Brazil, with comments on the classification of species into functional groups. Neotrop Ichthyol 6:191–200

Freire KMF, Pauly D (2010) Fishing down Brazilian marine food webs, with emphasis on the East Brazil large marine ecosystem. Fish Res 105:57–62

Freitas MO, Previero M, Leite JR, Francini-Filho RB, Minte-Vera CV, Moura RL (2019) Age, growth, reproduction and management of Southwestern Atlantic's largest and endangered herbivorous reef fish, *Scarus trispinosus* Valenciennes, 1840. PeerJ 7:1–26

Gasparini JL, Joyeux GCCJ, Floeter SR (2003) *Sparisoma tuiupiranga*, a new species of parrotfish (Perciformes: Labroidei: Scaridae) from Brazil, with comments on the evolution of the genus. Zootaxa 384:1

Giglio VJ, Ternes MLF, Barbosa MC, Cordeiro CAMM, Floeter SR, Ferreira CEL (2018) Reef fish associations with sea urchins in an Atlantic oceanic Island. Mar Biodivers 48:1833–1839

Godinho WO, Maggioni R, Lacerda AL, Lotufo TMC (2016) Haplotype distribution and connectivity of the white sea urchin *Tripneustes ventricosus* across the Brazilian biogeographic province. PeerJ Preprints 4:e2529v1. https://doi.org/10.7287/peerj.preprints.2529v1

Gondim AI, de Moura RB, Christoffersen ML, Dias TLP (2018) Taxonomic guide and historical review of echinoids (Echinodermata: Echinoidea) from northeastern Brazil. Zootaxa 4529:1–72

Grillo AC, Bonaldo RM, Segal B (2018) Physical contact interactions with scleractinian corals in hard substrate communities. Mar Ecol 39:1–12

Hata H, Ceccarelli DM (2016) Farming behaviour of territorial damselfishes. In: Frederich B, Parmentier E (eds) Biology of damselfishes. CRC Press, Boca Raton, pp 122–152

Hata H, Kato M (2004) Monoculture and mixed-species algal farms on a coral reef are maintained through intensive and extensive management by damselfishes. J Exp Mar Bio Ecol 313:285–296

Hay ME (1984) Patterns of fish and urchin grazing on Caribbean coral reefs: are previous results typical? Ecology 65:446–454

Haywick DW, Mueller EM (1997) Sediment retention in encrusting *Palythoa* spp. – a biological twist to a geological process. Coral Reefs 16:39–46

Hou Z, Li S (2017) Tethyan changes shaped aquatic diversification. Biol Rev 93:874–896

Hughes TP (1994) Catastrophes, phase shifts, and large-scale degradation of a Caribbean coral reef. Science 265:1547–1551

Irvine GV (1980) Fish as farmers, an experimental study of herbivory in a territorial coral reef damselfish, *Eupomacentrus planifrons*. Am Zool 20:822

Irving AD, Witman JD (2009) Positive effects of damselfish override negative effects of urchins to prevent an algal habitat switch. J Ecol 97:337–347

Joyeux JC, Floeter SR, Ferreira CEL, Gasparini JL (2001) Biogeography of tropical reef fishes: the South Atlantic puzzle. J Biogeogr 28:831–841

Kahng SE, Benayahu Y, Lasker HR (2011) Sexual reproduction in octocorals. Mar Ecol Prog Ser 443:265–283

Kikuchi RKP, Oliveira MDM, Leão ZMAN (2013) Density banding pattern of the south western Atlantic coral *Mussismilia braziliensis*. J Exp Mar Bio Ecol 449:207–214

Kilpp AM (1999) Efeito da população do ouriço *Echinometra lucunter* sobre a comunidade bentônica em um recife de Tamandaré-PE. Universidade Federal de Pernambuco

Knowlton N (1992) Thresholds and multiple stable states in coral reef community dynamics. Integr Comp Biol 32:674–682

Knowlton N, Jackson JBC (2001) The ecology of coral reefs. In: Bertness MD, Gaines SD, Hay ME (eds) Marine community ecology, Sinauer As. pp. 395–422

Labbé-Bellas R, Cordeiro CAMM, Floeter SR, Segal B (2016) Sea urchin abundance and habitat relationships in different Brazilian reef types. Reg Stud Mar Sci 8:33–40

Laborel J (1970) Les peuplements de madréporaires des côtes tropicales du Brésil. Ann l'Université d'Abidjan 2:260

Laborel-Deguen F, Castro CB, Nunes FD, Pires DO (2019) Recifes brasileiros: o legado de Laborel. Museu Nacional, Rio de Janeiro

Lages BG, Fleury BG, Ferreira CEL, Pereira RC (2006) Chemical defense of an exotic coral as invasion strategy. J Exp Mar Bio Ecol 328:127–135

Lang JC (2003) Status of coral reefs in the Western Atlantic: results of initial surveys, Atlantic and Gulf Rapid Reef Assessment (AGRRA) Program. Smithsonian Institution, Washington, DC

Lapid ED, Wielgus J, Chadwick-Furman NE (2004) Sweeper tentacles of the brain coral *Platygyra daedalea*: induced development and effects on competitors. Mar Ecol Prog Ser 282:161–171

Leão ZMAN, Kikuchi RKP (2005) A relic coral fauna threatened by global changes and human activities, Eastern Brazil. Mar Pollut Bull 51:599–611

Leão ZMAN, Kikuchi RKP, Testa V (2003) Corals and coral reefs of Brazil. Latin American coral reefs. Elsevier B.V., pp 9–52

Lellys NT, de Moura RL, Bonaldo RM, Francini-Filho RB, Gibran FZ (2019) Parrotfish functional morphology and bioerosion on SW Atlantic reefs. Mar Ecol Prog Ser 629:149–163

Lesser MP, Marc S, Michael S, Michiko O, Gates RD, Andrea G (2010) Photoacclimatization by the coral *Montastraea cavernosa* in the mesophotic zone: light, food, and genetics. Ecology 91:990–1003

Lessios HA (2015) The great *Diadema antillarum* die-off: 30 years later. Annu Rev Mar Sci 8:267–283

Lewis SM (1986) The role of herbivorous fishes in the Organization of a Caribbean Reef Community. Ecol Monogr 56:183–200

Lewis SM, Wainwright PC (1985) Herbivore abundance and grazing intensity on a Caribbean coral reef. J Exp Mar Bio Ecol 87:215–228

Ling SD, Johnson CR (2009) Population dynamics of an ecologically important range-extender: kelp beds versus sea urchin barrens. Mar Ecol Prog Ser 374:113–125

Lins de Barros MM, Pires DO, Castro CB (2003) Sexual reproduction of the Brazilian reef coral *Siderastrea stellata* Verrill 1868 (Anthozoa, Scleractinia). Bull Mar Sci 73:713–724

Littler MM, Littler DS (1988) Structure and role of algae in tropical reef communities. In: Lembi CA, Waaland JR (eds) Algae and human affairs. Cambridge, pp 29–56

Littler MM, Littler DS (1997) Disease-induced mass mortality of crustose coralline algae on coral reefs provides rationale for the conservation of herbivorous fish stocks. In: Proceeding of 8th international coral reef symposium, pp 719–724

Lorders FL, Miranda RJ, Nunes JAC, Barros F (2018) Spongivory by fishes on Southwestern Atlantic coral reefs: no evidence of top-down control on sponge assemblages. Front Mar Sci 5:1–9

Luiz O Jr, dos Santos WCR, Marceniuk AP, Rocha LA, Floeter SR, Buck CE, Clautau AGCM, Ferreira CEL (2021) Multiple lionfish (*Pterois* spp.) new occurrences along the Brazilian coast confirm the invasion pathway into the Southwestern Atlantic. Biol Invasions 23:3013–3019

Luz BLP, Di Domenico M, Migotto AE, Kitahara MV (2020) Life-history traits of *Tubastraea coccinea*: reproduction, development, and larval competence. Ecol Evol 10:6223–6238

MacDonald C, Tauati MI, Jones GP (2018) Depth patterns in microhabitat versatility and selectivity in coral reef damselfishes. Mar Biol 165:1–14

Maida M, Ferreira BP (1997) Coral reefs of Brazil: overview and field guide. In: Proceedings of the 8th international coral reef symposium, vol 1, pp 263–274

Maggioni R, Rocha RS, Viana JT, Giarrizzo T, Rabelo EF, Ferreira CE, ... & Soares MO (2023) Genetic diversity patterns of lionfish in the Southwestern Atlantic Ocean reveal a rapidly expanding stepping-stone bioinvasion process. Sci Rep 13(1):13469

Mantelatto MC, da Silva AG, Louzada T dos S, CS MF, Creed JC (2018) Invasion of aquarium origin soft corals on a tropical rocky reef in the Southwest Atlantic, Brazil. Mar Pollut Bull 130:84–94

Mazzei EF, Pinheiro HT, Morais RA, Floeter SR, Veras DP, Queiroz LV, Joyeux JC, Ferreira CEL (2017) Parrotfishes of the genus *Scarus* in southwestern Atlantic oceanic reef environments: occasional pulse or initial colonization? Mar Biodivers 49:555–561

McCauley DJ, Micheli F, Young HS, Tittensor DP, Brumbaugh DR, Madin EM, Holmes KE, Smith JE, Lotze HK, DeSalles PA, Arnold SN (2010) Acute effects of removing large fish from a near-pristine coral reef. Mar Biol 157:2739–2750

McCook L, Jompa J, Diaz-Pulido G (2001) Competition between corals and algae on coral reefs: a review of evidence and mechanisms. Coral Reefs 4:5562

McCord CL, Cooper JW, Nash CM, Westneat MW (2021) Phylogeny of the damselfishes (Pomacentridae) and patterns of asymmetrical diversification in body size and feeding ecology. PLoS One:1–30

McFadden CS, Sánchez JA, France SC (2010) Molecular phylogenetic insights into the evolution of Octocorallia: a review. Integr Comp Biol 50:389–410

McField M, Kramer P (2007) Healthy reefs for healthy people: a guide to indicators of reef health and social Well-being in the Mesoamerican reef region. Heal Reefs Heal People Initiat:1–208

Melis R, Ceccherelli G, Piazzi L, Rustici M (2019) Macroalgal forests and sea urchin barrens: structural complexity loss, fisheries exploitation and catastrophic regime shifts. Ecol Complex 37:32–37

Mendes TC, Cordeiro CAMM, Ferreira CEL (2015) An experimental evaluation of macroalgal consumption and selectivity by nominally herbivorous fishes on subtropical rocky reefs. J Exp Mar Bio Ecol 471:146–152

Mendes TC, Quimbayo JP, Bouth HF, Silva LPS, Ferreira CEL (2019) The omnivorous triggerfish *Melichthys niger* is a functional herbivore on an isolated Atlantic oceanic island. J Fish Biol 95:812–819

Menezes NM, McFadden CS, Miranda RJ, Nunes JACC, Lolis L, Barros F, Sampaio CLS, Pinto TK (2021) New non-native ornamental octocorals threatening a South-west Atlantic reef. Journal of the Marine Biological Association of the United Kingdom, 101(6), 911–917. https://doi.org/10.1017/S0025315421000849

Miranda RJ, Cruz ICS, Barros F (2016a) Effects of the alien coral *Tubastraea tagusensis* on native coral assemblages in a southwestern Atlantic coral reef. Mar Biol 163:1–12

Miranda RJ, Costa Y, Lorders FL, De Anchieta CC, Barros F (2016b) New records of the alien cup-corals (*Tubastraea* spp.) within estuarine and reef systems in Todos os Santos Bay, Southwestern Atlantic. Mar Biodivers Rec 9:4–9

Miranda RJ, Tagliafico A, Kelaher BP, Mariano-Neto E, Barros F, Nunes J de ACC, Mariano-Neto E, Sippo JZ, Barros F (2018a) Do invasive corals alter coral reef processes? An empirical approach evaluating reef fish trophic interactions. Mar Ecol Prog Ser 605:125–133

Miranda RJ, Tagliafico A, Kelaher BP, Mariano-neto E, Barros F (2018b) Impact of invasive corals *Tubastraea* spp. on native coral recruitment. Mar Ecol Prog Ser 605:125–133

Mizrahi D, Navarrete SA, Flores AAV (2014) Groups travel further: pelagic metamorphosis and polyp clustering allow higher dispersal potential in sun coral propagules. Coral Reefs 33:443–448

Monteiro-Neto C, De Andrade Cunha FE, Carvalho Nottingham M, Araújo ME, Lucena Rosa I, Leite Barros GM (2003) Analysis of the marine ornamental fish trade at Ceará State, Northeast Brazil. Biodivers Conserv 12:1287–1295

Munday PL, Jones GP, Pratchett MS, Williams AJ (2008) Climate change and the future for coral reef fishes. Fish Fish 9:261–285

Neves EG, Pires DO (2002) Sexual reproduction of Brazilian coral *Mussismilia hispida* (Verrill, 1902). Coral Reefs 21:161–168

Norström AV, Nyström M, Lokrantz J, Folke C (2009) Alternative states on coral reefs: beyond coral-macroalgal phase shifts. Mar Ecol Prog Ser 376:293–306

Nugues MM, Roberts CM (2003) Coral mortality and interaction with algae in relation to sedimentation. Coral Reefs:507–516

Ogden JC (1977) Carbonate-sediment production by parrot fish and sea urchins on Caribbean reefs. Reef Biota:281–288

Ogden JC, Brown RA, Salesky N (1973) Grazing by the echinoid *Diadema antillarum* Philippi: formation of halos around West Indian patch reefs. Science 182:715–716

Paula AF (2007) Biologia reprodutiva, crescimento e competição dos corais invasores *Tubastraea coccinea* e *Tubastraea tagusensis* (Scleractinia: Dendrophylliidae) com espécies nativas. Universidade Federal do Rio de Janeiro

Paula AF, Creed JC (2005) Spatial distribution and abundance of nonindigenous coral genus *Tubastraea* (Cnidaria, Scleractinia) around Ilha Grande, Brazil. Braz J Biol 65:661–673

Pereira RC, Carvalho AGV, Gama BAP, Coutinho R (2002) Field experimental evaluation of secondary metabolites from marine invertebrates as antifoulants. Brazilian J Biol 62:311–320

Peters EC (1997) Diseases of coral-reef organisms. In: Birkeland C (ed) Life and death of coral reefs. Chapman & Hall, New York, pp 114–139

Pinheiro HT, Rocha LA, Macieira RM, Carvalho-Filho A, Anderson AB, Bender MG, Di Dario F, Ferreira CEL, Figueiredo-Filho J, Francini-Filho R, Gasparini JL, Joyeux JC, Luiz OJ, Mincarone MM, Moura RL, Nunes J de ACC, Quimbayo JP, Rosa RS, CLS S, Sazima I, Simon T, Vila-Nova DA, Floeter SR (2018) South-Western Atlantic reef fishes: zoogeographical patterns and ecological drivers reveal a secondary biodiversity centre in the Atlantic Ocean. Divers Distrib 24:951–965

Pires DO, Castro CB, Ratto CC (1999) Reef coral reproduction in the Abrolhos Reef Complex, Brazil: the endemic genus Mussismilia. Mar Biol 135:463–471

Posada JM, Villamizar E, Alvarado D (2003) Venezuela. Rapid assessment of coral reefs in the Archipelago de Los Roques National Park, Venezuela (part 2: fishes). Atoll Res Bull 496:530–543

Precht WF, Aronson RB, Moody RM, Kaufman L (2010) Changing patterns of microhabitat utilization by the threespot damselfish, *Stegastes planifrons*, on Caribbean reefs. PLoS One 5:e10835

Queiroz-Véras LVMV, Ferreira BP, Freitas M, Feitosa JLL (2023) A critical review and knowledge gaps to assess and manage threatened parrotfishes' stocks in Brazil. J Aquat Sci 85(2):44

Rabelo EF, Soares M de O, Matthews-Cascon H (2013) Competitive interactions among zoanthids (cnidaria: Zoanthidae) in an intertidal zone of northeastern Brazil. Braz J Oceanogr 61:35–42

Ramos CAC, Kikuchi RKP, Amaral FD, Fauth JE (2014) A test of herbivory-mediated coral-algae interaction on a Brazilian reef during a bleaching event. J Exp Mar Bio Ecol 456:1–7

Rasher D, Stout E, Engel S, Kubanek J, Hay M (2011) Macroalgal terpenes function as allelopathic agents against reef corals. Proc Natl Acad Sci USA 43:1772617731

Ribeiro FV, Sá JA, Fistarol GO, Salomon PS, Pereira RC, Souza MLAM, Neves LM, Amado-Filho GM, Francini-Filho RB, Salgado LT, Bastos AC, Pereira-Filho GH, Moraes FC, Moura RL (2018) Long-term effects of competition and environmental drivers on the growth of the endangered coral *Mussismilia braziliensis* (Verril, 1867). PeerJ 2018:1–20

Richmond R (1997) Reproduction and recruitment in corals: critical links in the persistence of reefs. In: Birkeland CE (ed) Life and death of coral reefs. Chapman & Hall Publishers, New York, pp 175–197

Ritson-Williams R, Arnold SN, Fogarty ND, Steneck RS, Vermeij MJA, Paul VJ (2009) New perspectives on ecological mechanisms affecting coral recruitment on reefs. Smithson Contrib Mar Sci:437–457

Robertson DR (1984) Cohabitation of competing territorial damselfishes on a Caribbean coral reef. Ecology 65:1121–1135

Robertson DR, Van Tassell J (2019) Shorefishes of the Greater Caribbean: online information system

Robertson DR, Hoffman SG, Sheldon JM (1981) Availibility of space for the territorial Caribbean damselfish *Eupomacentrus planifrons*. Ecology 62:1162–1169

Rocha LA (2003) Patterns of distribution and processes of speciation in Brazilian reef fishes. J Biogeogr 30:1161–1171

Roos NC, Pennino MG, Carvalho AR, Longo GO (2019) Drivers of abundance and biomass of Brazilian parrotfishes. Mar Ecol Prog Ser 623:117–130

Roos NC, Taylor BM, Carvalho AR, Longo GO (2020a) Demography of the largest and most endangered Brazilian parrotfish, *Scarus trispinosus*, reveals overfishing. Endanger Species Res 41:319–327

Roos NC, Longo GO, Pennino MG, Francini-Filho RB, Carvalho AR (2020b) Protecting nursery areas without fisheries management is not enough to conserve the most endangered parrotfish of the Atlantic Ocean. Sci Rep 10:1–10

Sammarco PW, Coll JC, La Barre S, Willis B (1983) Competitive strategies of soft corals (Coelenterata: Octocorallia): Allelopathic effects on selected Scleractinian corals. Coral Reefs 1:173–178

Sánchez-Jérez P, Cesar A, Cortez FS, Pereira CDS, Silva SLR (2001) Distribución espacial de las poblaciones de erizos más abundantes de la costa sudeste del litoral de São Paulo (Brasil). Ciencias Mar 27:139–153

Semmler RF, Brandl SJ, Keith SA, Bellwood DR (2021) Fine-scale foraging behavior reveals differences in the functional roles of herbivorous reef fishes. Ecol Evol:1–11

Silva ICBS, Liparini A, Pereira NS, Braga BLSS, Sial AN, Liu SC, Shen CC, Kikuchi RKP (2019) Assessing the growth rate of the South Atlantic coral species *Mussismilia hispida* (Verrill, 1902) using carbon and oxygen stable isotopes. J S Am Earth Sci 96:102346

Siqueira AC, Bellwood DR, Cowman PF (2019a) Historical biogeography of herbivorous coral reef fishes: the formation of an Atlantic fauna. J Biogeogr 46:1611–1624

Siqueira AC, Bellwood DR, Cowman PF (2019b) The evolution of traits and functions in herbivorous coral reef fishes through space and time. Proc R Soc B Biol Sci 286:20182672

Smith JE, Shaw M, Edwards RA, Obura D, Pantos O, Sala E, Sandin SA, Smriga S, Hatay M, Rohwe FL (2006) Indirect effects of algae on coral: algae-mediated, microbe-induced coral mortality. Ecol Lett:635–645

Smith LL, Fessler JL, Alfaro ME, Streelman JT, Westneat MW (2008) Phylogenetic relationships and the evolution of regulatory gene sequences in the parrotfishes. Mol Phylogenet Evol 49:136–152

Soares MO, Feitosa CV, Garcia TM, Cottens KF, Vinicius B, Paiva SV, Duarte OdS, Gurjão LM, Silva GDV, Maia RC, Previatto DM, Carneiro PBM, Cunha F, Amâncio AC, Sampaio CLS, Ferreira CEL, Pereira PHC, Rocha LA, Tavares TCL and Giarrizzo T (2022) Lionfish on the loose: Pterois invade shallow habitats in the tropical southwestern Atlantic. Front Mar Sci 9:956848. https://doi.org/10.3389/fmars.2022.956848

Souza AS, Rosa RS, Soares RX, Lima-Filho PA, Oliveira C, Shibatta OA, Molina WF (2017) A reappraisal of *Stegastes* species occurring in the South Atlantic using morphological and molecular data. Helgol Mar Res 70:20. https://doi.org/10.1186/s10152-016-0471-x

Streelman JT, Alfaro M, Westneat MW, Bellwood DR, Karl SA (2002) Evolutionary history of the parrotfishes: biogeography, ecomorphology, and comparative diversity. Evolution (NY) 56:961–971

Suchanek TH, Green DJ (1981) Interspecific competition between *Palythoa caribaeorum* and other sessile invertebrates on St. Croix reefs, U.S. Virgin Islands. Proceedings of the 4th international coral reef symposium 2:679–684

Tanner JE (1997) Interspecific competition reduces fitness in Scleractinian corals. J Exp Mar Biol Ecol:19–34

Tebbett SB, Hoey AS, Depczynski M, Wismer S, Bellwood DR (2020) Macroalgae removal on coral reefs: realised ecosystem functions transcend biogeographic locations. Coral Reefs 39:203–214

Thacker RW, Paul VJ (2001) Are benthic cyanobacteria indicators of nutrient enrichment? Relationships between cyanobacteria abundance and environmental factors on the reef flats of Guam. Bull Mar Sci:497–508

Van Alstyne KL, Wylie CR, Paul VJ (1994) Antipredator defenses in tropical Pacific soft corals (Coelenterata: Alcyonacea) II. The relative importance of chemical and structural defenses in three species of *Sinularia*. J Exp Mar Bio Ecol 178:17–34

Vermeij MJA (2005) Substrate composition and adult distribution determine recruitment patterns in a Caribbean brooding coral. Mar Ecol Prog Ser 295:123–133

Waldner R, Robertson D (1980) Patterns of habitat partitioning by eight species of territorial Caribbean damselfishes (Pisces: Pomacentridae). Bull Mar Sci 30:171–186

Walter JM, Tschoeke DA, Meirelles PM, De Oliveira L, Leomil L, Tenório M, Valle R, Salomon PS, Thompson CC, Thompson FL (2016) Taxonomic and functional metagenomic signature of turfs in the Abrolhos reef system (Brazil). PLoS One 11:1–20

Wilson SK, Burgess SC, Cheal AJ, Emslie M, Fisher R, Miller I, Polunin NVC, Sweatman HPA (2008) Habitat utilization by coral reef fish: implications for specialists vs. generalists in a changing environment. J Anim Ecol 77:220–228

Zilberberg C, Abrantes DP, Marques JA, Machado LF, Marangoni LF de B (2016) Conhecendo os Recifes Brasileiros: Rede de Pesquisas Coral Vivo

Chapter 6
The Big Dwellers and Their Interactions

Caroline V. Feitosa, Maria Danise O. Alves, Alice F. Feitosa, Ana Lídia B. Gaspar, and Jones Santander-Neto

Abstract Large vertebrates are fundamental in maintaining the structure and dynamics of the trophic webs in the reefs. In addition, these top predators, and large herbivores (*e.g.* sharks, turtles and marine mammals) are umbrella species. In Brazil, about 44 species of Chondrichthyes, either demersal or pelagic, are reef-associated. Some species of sharks tend to play the same trophic role as large actinopterygian fish, such as serranids and lutjanids, which inhabit caves and recesses in the reefs where they are mesopredators. Other fish assist in the survival of corals, such as schools of greenbeak parrotfish, playing an important ecological function as it is herbivore and bioeroder. Reefs and associated algae beds are considered abundant feeding sites for many organisms, including female sea turtles that spawn on islands or on beaches on the mainland. It is also in these environments where there are several algae and plants identified in the manatee diet, especially species of rhodophytes and marine angiosperms. The prey (*e.g.* actinopterygians, crustaceans and cephalopods) of the odontocetes cetaceans can be found, especially, in the reefs, demonstrating the importance of this ecosystem for the feeding of these species, mainly in oceanic islands. For large residents, reefs provide food, a nursery area, a refuge from predation, shelter against currents and removal of parasites. Thus, the

C. V. Feitosa (✉)
Instituto de Ciências do Mar, Universidade Federal do Ceará (UFC), Fortaleza, Ceará, Brazil
e-mail: carol_feitosa@ufc.br

M. D. O. Alves
Faculdade Frassinetti do Recife (FAFIRE), Recife, Pernambuco, Brazil

A. F. Feitosa
Programa de Pós-Graduação em Ciências Marinhas Tropicais, Instituto VerdeLuz, Universidade Federal do Ceará (UFC), Fortaleza, Ceará, Brazil

A. L. B. Gaspar
Instituto Recifes Costeiros, Tamandaré, Brazil

J. Santander-Neto
Instituto Federal de Educação, Ciência e Tecnologia do Espírito Santo (Ifes), Vitória, Espírito Santo, Brazil

R. K. P. Kikuchi et al. (eds.), *Brazilian Coral Reefs*, Brazilian Marine Biodiversity, https://doi.org/10.1007/978-3-031-59152-5_6

decline of reefs, caused by various stressors, such as climate change, fishing, pollution, as well as habitat conditions are of concern to the species that depend on them.

Keywords Umbrella species · Predator · Herbivore bioeroder · Threat status · Reef trophic web · Connectivity

6.1 Introduction

Top predators have a strong influence on marine communities since they are fundamental in maintaining the structure and dynamics of the trophic webs in these systems (Dulvy et al. 2004; Hutchings and Baum 2005; Baum and Worm 2009) (Fig. 6.1). Coral reefs are highly biodiverse and the effects of predators on the functioning of these ecosystems are hard to determine due to the trophic complexity (Borer et al. 2005). This complexity allows compensatory processes that resist changes in the community structure, adjusting according to the interactions between trophic levels (Connell and Ghedini 2015). The strength of predation and the potential consequences of the loss of top predators caused by fishing and habitat degradation vary between ecosystems, particularly in terms of species diversity and functional redundancy (Pace et al. 1999; Fagan 1997). The reduction in biomass and abundance of top predators can trigger irreversible cascade effects that destabilize food webs in the marine environment. The increase in invertebrate predators and mesopredators, and in small pelagic fishes and macroinvertebrates, the modification

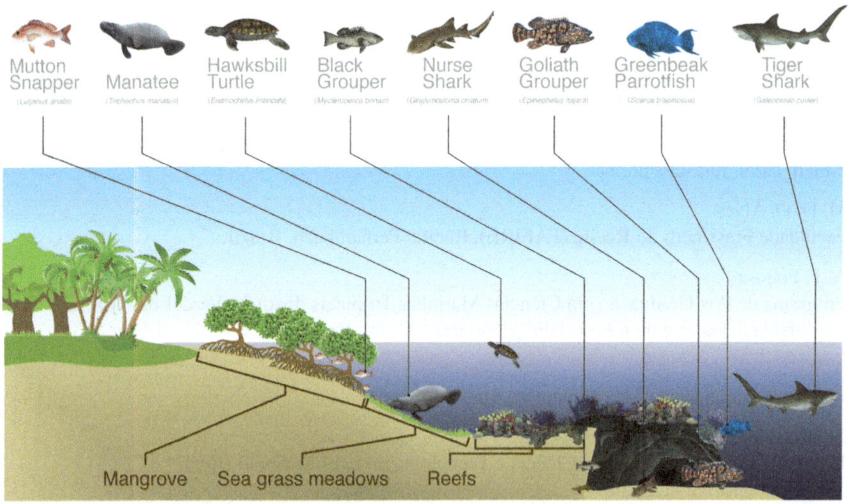

Fig. 6.1 Reefs, their big dwellers, and associated habitats. (©Charles Silva)

in productivity, species assemblage and coastal erosion, decline of reef-building corals and coralline algae are examples that cause those effects (Dulvy et al. 2004; Heithaus et al. 2008; Baum and Worm 2009).

This chapter will consider the large species that play a functional role in reefs and depend directly or indirectly on this ecosystem. In reef areas, alpha predators, and large herbivores, such as sharks, sea turtles and marine mammals, are referred to as umbrella species since they require extensive preserved areas to live, and other species that occupy the same habitat also benefit in an indirect manner (Fig. 6.1) (Benevides et al. 2017). In addition to ecological issues, the charismatic potential that megafauna awakens in society stimulates the focus on financing projects and other tools for the conservation of biodiversity (Benevides et al. 2017). Therefore, umbrella species of sharks and rays, actinopterygians (fish of the families Lutjanidae Serranidae, and Scaridae), sea turtles and manatees will be discussed regarding diversity, functional role, importance of the reefs, stressors, and threats.

6.2 Sharks and Rays

Around 1139 species of elasmobranchs are known worldwide (Weigmann 2016). In Brazil, the most recent assessment of the population status of the Chondrichthyes fauna considered 169 species (ICMBio 2016a), of which at least 44 species (Table 6.1) have some type of interaction with Brazilian reefs (Rocha and Rosa 2001; Costa 2011; ICMBio 2018a, b, c, d; Pinheiro et al. 2018). These species are generally large and several of them are the largest inhabitants of the reefs. Some are residents, others demersal or occasionally pelagic (Fig. 6.2a, a1) when they carry out reproductive migrations or aggregations for food. Among the residents, it can be highlighted the blacknose shark (*Carcharhinus acronotus*), Galapagos shark (*C. galapagensis*), Caribbean reef shark (*C. perezi*), lemon shark (*Negaprion brevirostris*) (Fig. 6.2b), nurse shark (*Ginglymostoma cirratum*) (Fig. 6.2c–c2) and the southern stingray (*Hypanus berthalutzae*) (Fig. 6.2d–d1). These species reach relatively large sizes (≥1 m) when compared to the other reef species. In addition, other non-resident species and large-sized elasmobranchs (≥3 m) have recurrent interaction with reefs, such as tiger sharks (*Galeocerdo cuvier*), hammerhead sharks (*Sphyrna* spp.), whale sharks (*Rhincodon typus*), longnose stingray (*Hypanus guttatus*) (Fig. 6.2e) and manta rays (*Mobula* spp.).

The functional role that elasmobranchs play in reefs is related to their trophic position as top predators of marine food webs. Therefore, the health condition of coastal ecosystems may depend on the integrity of shark populations and some rays. The decline in populations of these animals can cause changes in the environmental interactions and in the abundance of other species at lower trophic levels (Friedlander and DeMartini 2002; Heithaus et al. 2008, 2010; Baum and Worm 2009; Bornatowski et al. 2014). This effect is called trophic cascade or “top-down”.

The presence of elasmobranchs on reef ecosystem impacts the dynamics through predator-prey interaction, being influenced by the abundance, size and trophic level

Table 6.1 Brazilian reef-associated elasmobranch species, threat status, type of reef association and maximum length

Species	Common name	Threat status[a]	Rel. reef	ML (cm)
Carcharhinus acronotus (Poey, 1860)	Blacknose shark	VU	res	200
Carcharhinus falciformis (Bibron, 1839)	Silky shark	CR	oc-pe	330
Carcharhinus galapagensis (Snodgrass & Heller, 1905)	Galapagos shark	CR	res	370
Carcharhinus leucas (Müller & Henle, 1839)	Bull shark	VU	oc-de	340
Carcharhinus limbatus (Valenciennes, 1839)	Blacktip shark	NT	oc-de	255
Carcharhinus longimanus (Poey, 1861)	Oceanic whitetip shark	VU	oc-pe	350–395
Carcharhinus obscurus (Lesueur, 1818)	Dusky shark	EN	oc-pe	360–400
Carcharhinus perezi (Poey, 1876)	Caribbean reef shark	VU	res	295
Carcharhinus plumbeus (Nardo, 1827)	Sandbar shark	CR	oc-pe	240
Carcharhinus signatus (Poey, 1868)	Night shark	EN	oc-pe	280
Galeocerdo cuvier (Péron & Lesueur, 1822)	Tiger shark	NT	oc-de	550–740
Negaprion brevirostris (Poey, 1868)	Lemon shark	EN	res	340
Prionace glauca (Linnaeus, 1758)	Blue shark	NT	oc-pe	380
Rhizoprionodon lalandii (Müller & Henle, 1839)	Brazilian sharpnose shark	NT	oc-de	77
Rhizoprionodon porosus (Poey, 1861)	Caribbean sharpnose shark	DD	oc-de	110
Sphyrna lewini (Griffith & Smith, 1834)	Scalloped hammerhead	CR	oc-pe	370–420
Sphyrna mokarran (Rüppell, 1837)	Great hammerhead	CR	oc-pe	550–610
Sphyrna tiburo (Linnaeus, 1758)	Bonnethead	CR	oc-pe	150
Sphyrna tudes (Valenciennes, 1822)	Smalleye hammerhead	CR	oc-pe	122–150
Alopias superciliosus Lowe, 1841	Bigeye thresher	EN	oc-pe	460
Alopias vulpinus (Bonnaterre, 1788)	Thresher	CR	oc-pe	610
Cetorhinus maximus (Gunnerus, 1765)	Basking shark	CR	oc-pe	1000
Isurus oxyrinchus Rafinesque, 1810	Shortfin mako	NT	oc-pe	400
Carcharias taurus Rafinesque, 1810	Sand tiger shark	CR	oc-pe	430
Odontaspis ferox (Risso, 1810)	Smalltooth sand tiger	DD	oc-pe	410
Ginglymostoma cirratum (Bonnaterre, 1788)	Nurse shark	VU	res	300
Rhincodon typus Smith, 1828	Whale shark	VU	oc-pe	1700–2100
Hypanus berthalutzae Petean, Naylor & Lima, 2020	Southern stingray	VU	res	150
Bathytoshia centroura (Mitchill, 1815)	Roughtail stingray	CR	oc-de	220

(continued)

Table 6.1 (continued)

Species	Common name	Threat status[a]	Rel. reef	ML (cm)
Hypanus guttatus (Bloch & Schneider, 1801)	Longnose stingray	LC	oc-de	180
Hypanus marianae (Gomes, Rosa & Gadig, 2000)	Brazilian large-eyed stingray	VU	oc-de	40
Gymnura altavela (Linnaeus, 1758)	Spiny butterfly ray	CR	oc-de	260
Gymnura micrura (Bloch & Schneider, 1801)	Smooth butterfly ray	NT	oc-de	80
Mobula birostris (Walbaum, 1792)	Giant manta	VU	oc-pe	700
Mobula cf. hypostoma (Bancroft, 1831)	Lesser devil ray	VU	oc-pe	133
Mobula mobular (Müller & Henle, 1841)	Devil fish	VU	oc-pe	520
Mobula tarapacana (Philippi, 1893)	Chilean devil ray	VU	oc-pe	370
Mobula thurstoni (Lloyd, 1908)	Smoothtail mobula	VU	oc-pe	189
Aetobatus narinari (Euphrasen, 1790)	Whitespotted eagle ray	DD	oc-pe	230
Narcine bancroftii (Griffith & Smith, 1834)	Lesser electric ray	DD	oc-de	65
Narcine brasiliensis (Olfers, 1831)	Brazilian electric ray	DD	oc-de	45
Pseudobatos percellens (Walbaum, 1792)	Chola guitarfish	VU	oc-de	100
Rhinoptera bonasus (Mitchill, 1815)	Cownose ray	DD	oc-pe	110
Rhinoptera brasiliensis Müller, 1836	Brazilian cownose ray	CR	oc-pe	104

Legend: *Rel. reef* relationship with the reef, *ML* maximum length, *DD* data deficient, *LC* least concern, *NT* near threatened, *VU* vulnerable, *EN* Endangered, *CR* critically endangered; [a](ICMBio 2016a, b; Ordinances from Ministry of Environment (MMA) – MMA 148/2022 (Brasil 2022) and 354/2023 (Brasil 2023). Relationship of the species to the environment (adapted from Pinheiro et al., 2018); *res* resident, *oc-pe* occasional pelagic, *oc-de* occasional demersal. Maximum length information taken from Compagno et al. (2005) and Last et al. (2016)

of the predator, diet specialization as well as its connection with the (Wootton and Emmerson 2005; Heupel et al. 2014; Roff et al. 2016). Large sharks (e.g., tiger shark *Galeocerdo cuvier*) (Heupel et al. 2014), travel distances of over 100 km between coastal (e.g. mangroves, reefs) and pelagic environments (Meyer et al. 2010; Hazin et al. 2013; Afonso and Hazin 2015), where their diet varies according to the availability of prey in these habitats (Lowe et al. 1996; Ferreira et al. 2017). The variety of prey also at higher trophic levels (e.g. sharks, turtles and marine mammals) is a result of the ecological plasticity of these large species (Lowe et al. 1996; Heithaus et al. 2007). In the absence of these species, mesopredator sharks (e.g. medium-sized—1.5 and 3 m) (Ritchie and Johnson 2009; Heupel et al. 2014), occupy the higher trophic levels and are found less distant from the reefs (e.g. around 50 km). While the smaller elasmobranchs, including some stingray species (e.g. *Hypanus marianae*—Fig. 6.2f) are at lower trophic levels and more dependent on reef environments.

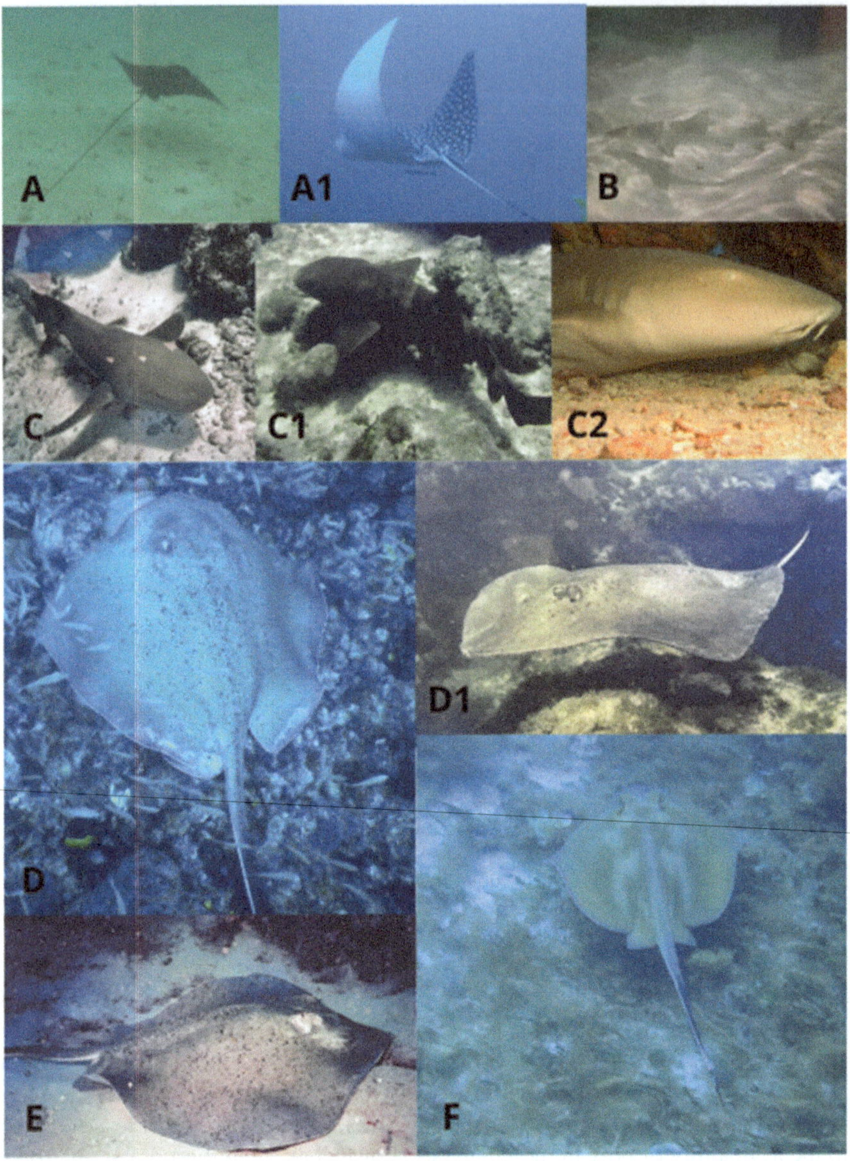

Fig. 6.2 The big Chondrichthyes dwellers: (**a**) *Aetobatus narinari,* (**b**) lemon shark *Negaprion brevirostris,* (**c**) nurse shark *Ginglymostoma cirratum,* (**d**) Southwestern stingray *Hypanus berthalutzae,* (**e**) Longnose stingray *Hypanus guttatus,* (**f**) Brazilian large-eyed stingray *Hypanus marianae.* (© **a**, **d**, **f**; © **b**, **c**, **f**—Leo Bruno)

In most studies that assessed the trophic role of elasmobranchs, it was found that several models were created with insufficient information about the diet of the studied species (Roff et al. 2016). This created a difficulty in categorizing and, consequently, added in a single top category both small and large sharks, which can feed on small crustaceans and mammals, respectively (Heupel and Bennet 1998; Ceccarelli and Ayling 2010). In addition to these problems, this trophic category also included the large predatory actinopterygian fish (e.g. Serranidae and Lutjanidae) (Roff et al. 2016).

Although the importance of maintaining elasmobranch populations for the integrity of ecosystems is recognized, recent studies suggest that the functional role of sharks and rays and the consequences of reducing their populations are more ambiguous than previously believed (Grubbs et al. 2016; Roff et al. 2016; Bornatowski et al. 2018; Rupp and Bornatowski 2021). For example, some species of sharks, whether at the top of the chain or mesopredator, may fulfill the same trophic role as some actinopterygian fish (e.g., *Caranx, Mycteroperca, Sphyraena, Lutjanus*) (Frisch et al. 2016 Roff et al. 2016; Bond et al. 2018; Barley et al. 2020). However, it is worth noting that the lack of evidence of trophic cascades caused by the absence of elasmobranchs does not indicate their inexistence, since they play some ecological functions in reef environments (Box 6.1), but rather the difficulty of assessing marine food webs.

Box 6.1: Ecological Functions Played by Elasmobranchs in Reef Environments

Nutrient cycling: due to the migration of large sharks that can increase the flow of nutrients between ocean and coastal environments. Mesopredator sharks, with lesser migration capacity, increase cycling between reef and pelagic environments (Meyer et al. 2010, Schmitz et al. 2010, McCauley et al. 2012, Roff et al. 2016).

Scavenging: top sharks and/or mesopredators remove weak and sick individuals from the environment through predation, as well as the carcasses of dead specimens (Lowe et al. 1996, Dudley et al. 2000, Pongsiri et al. 2009, Bornatowski et al. 2012).

Habitat disturbance: when looking for prey, sharks can damage the three-dimensional structure of the reef. However, these ecosystem processes are poorly quantified (Jiménez-Centeno 1997).

Removal of invasive species: as they are large predators, elasmobranchs can consider exotic species as their prey (Albins and Hixon 2008).

The high diversity of prey in reefs most certainly influenced the diversification, distribution and behavior of sharks that have become specialized to live in these environments (Sorenson et al. 2014). Only a few studies revealed the ecological benefits of reefs for elasmobranchs. However, Roff et al. (2016) compiled reports that indicated some functional roles of reefs for sharks such as food source, nursery, refuge from predation, shelter against currents and removal of parasites.

The population decline of elasmobranchs has been documented through observations of decreases in their abundance or reductions in the size of individuals captured in Brazil (Vooren and Klippel 2005; Santander-Neto et al. 2011; Barreto et al. 2016; ICMBio 2018a, b, c, d). Reflections of this condition are seen both in the increase in the number of species classified under threat status over time, and in relation to the increase in the level of threat, ascertained through the management instruments (Brasil 2004, 2014) and report of the evaluation process of elasmobranch species (ICMBio 2016a).

In Brazil, of the 169 species of Chondrichthyes evaluated by ICMBio, 57 are in some threat criteria: vulnerable (VU), threatened (EN) or critically endangered (CR). This number is equivalent to 34% of this fauna (ICMBio 2016a), being like global estimates of elasmobranchs in threat categories (Dulvy et al., 2014, 2021). When considering only the 44 species with some association with Brazilian reefs, the situation is worse, with about 34% being classified as VU, 9% EN and 29.5% CR, representing 72.7% of species in some threat criteria. The reductions in abundance and average length of individuals may have changed the ecological role of species in the reef environment (Roff et al. 2016). Fishing stands out among the various threats that these species suffer as they do not support high levels of exploitation due to their biological characteristics as k-strategists. Fishing is highlighted in the justifications presented for the classification of elasmobranch species in the red list of endangered species (ICMBio 2018a, b, c, d).

Many Brazilian reefs are included in Marine Protected Areas (MPA) and some of them, classified as sustainable use, allow the use and exploitation of their resources. This is the case of Parrachos de Maracajaú (State Environmental Protection Area of Coral Reefs) and many others. Some reefs are in MPAs furthermore restrictives, where activities such as fishing, and tourist visitation are prohibited (ex: Atol das Rocas Biological Reserve). Hence, there is a relationship between the complexity of the reef and the diversity of elasmobranchs because a greater availability of habitat is associated with a greater diversity and quantity of fish that serve as food for top predators (Hixon and Beets 1993; Rogers et al. 2014; Roff et al. 2016).

Studies indicate that the abundance of sharks in reefs in legally unprotected areas is at least 75% less than in protected areas (Nadon et al. 2012). Therefore, this protection of reefs has the potential to improve the conservation of elasmobranchs in Brazil, since they support several species and prevent a reduction in their biomass. In the Brazilian National Action Plan for the conservation of sharks and marine rays threatened with extinction (ICMBio 2016b) the Abrolhos National Marine Park (BA), Coral Reef Environmental Protection Area (RN), the archipelagos of Fernando de Noronha (PE) and Atol das Rocas (RN), Coral Coast Environmental Protection Area (PE/AL) and Parcel Manuel Luiz State Marine Park(MA) were considered as strategic areas for the conservation of this group for providing high diversity, nursery areas, migration, growth and feeding of several species of elasmobranchs.

6.3 Actinopterygian Fish

The "large-bodied residents" of reef ecosystem were sorted by the species size/weight ratio and the functional role they play as mesopredators (species of Serranidae and Lutjanidae) (Fig. 6.3) or in controlling algae and as bioeroders (species of Scaridae), according to Nelson et al. (2016) (Fig. 6.3g) (Table 6.2).

Groupers and snappers are widely distributed in marine environments in tropical and subtropical regions, including all Brazilian coral reefs. These fish have a typical demersal habit and usually inhabit caves and crevices (Claro and Lindeman 2008).

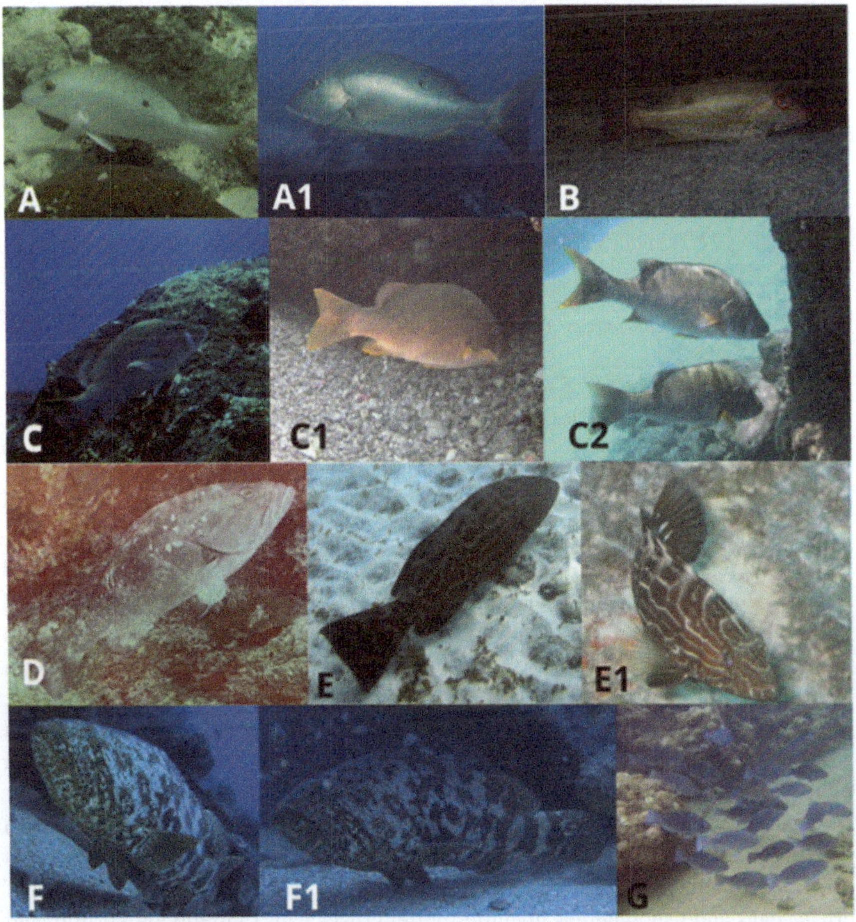

Fig. 6.3 The big Actinopterygii dwellers: (**a, a1**) mutton snapper *Lutjanus analis*, (**b**) lane snapper *Lutjanus synagris*, (**c, c1, c2**) dog snapper *Lutjanus jocu*, (**d**) red grouper *Epinephelus morio*, (**e, e1**) black grouper *Mycteroperca bonaci*, (**f, f1**) goliath grouper *Epinephelus itajara*, (**g**) greenbeak parrotfish *Scarus trispinosus*. (© **a, a1, c1, e1, f, f1**—Leo Bruno, © **b**—Yara Tibiriçá, © **c, c2, d**—Caroline Feitosa, © **g**—Liana Mendes)

Table 6.2 List of large-bodied reef fish species sorted alphabetically by family, with their respective common names, global-scale threat status and maximum length

Family	Species	Common name	Threat status	ML (cm)
Scaridae	*Scarus trispinosus* Valenciennes, 1840	Greenback parrotfish	EN	90
Serranidae	*Dermatolepis inermis* (Valenciennes, 1833)	Marbled grouper	DD	91
	Epinephelus guttatus (Linnaeus, 1758)	Red hind	LC	76
	Epinephelus itajara[1] (Lichtenstein, 1822)	Goliath grouper	VU	250
	Epinephelus morio (Valenciennes, 1828)	Red grouper	VU	125
	Mycteroperca bonaci[2] (Poey, 1860)	Black grouper	NT	177
	Mycteroperca interstitialis (Poey, 1860)	Yellowmouth grouper	VU	84
	Mycteroperca venenosa[3] (Linnaeus, 1758)	Yellowfin grouper	NT	100
Lutjanidae	*Etelis oculatus* (Valenciennes, 1828)	Queen snapper	DD	100
	Lutjanus analis (Cuvier, 1828)	Mutton snapper	NT	94
	Lutjanus buccanella[4] (Cuvier, 1828)	Blackfin snapper	DD	75
	Lutjanus cyanopterus (Cuvier, 1828)	Cubera snapper	VU	160
	Lutjanus jocu[5] (Bloch & Schneider, 1801)	Dog snapper	DD	130
	Lutjanus purpureus[2] Poey, 1876	Southern red snapper		100
	Lutjanus vivanus[5] (Cuvier, 1828)	Silk snapper	LC	83

Legend: *DD* data deficient, *LC* least concern, *NT* near threatened, *VU* vulnerable, *EN* Endangered. 1—CR critically endangered, 2—VU in Brazil; 3—VU in Brazil; 4—LC in Brazil; 5—NT in Brazil; Maximum length information was obtained from Froese and Pauly (2024)

They are associated with hard bottoms on the continental shelves and oceanic islands and can occur at depths of around 200 m (Martins et al. 2005; Frédou and Beatrice 2005; Parenti and Randall 2020). The deeper reefs (>50 m) are inhabited by adults (Heemstra and Randall 1993), while younger specimens occur both in ecosystems associated with these reefs (e.g., mangroves and submerged angiosperm beds) (Aschenbrenner et al. 2016) as well as in tide pools (Azevedo et al. 1995; Machado et al. 2003) and shallow coastal reefs (Bussotti and Guidetti 2009). In addition, in the reefs, their juveniles can also be seen using the protection provided by the branches of fire coral (*Millepora* spp.) (Leal et al. 2013).

Characterized as mesopredators (Polovina 1984), species of groupers (Serranidae) (Fig. 6.3d, e, e1, f, f1) find their prey in reefs, feed on fish and a variety of invertebrates. Most adult snappers (Lutjanidae) (Fig. 6.3a, a1, b, c, c1, c2) are predators of crustaceans and fish, whereas juveniles prefer crustaceans (amphipods, shrimp, and crabs) (Franks and VanderKooy 2000; Freitas et al. 2011) and are zooplanktivores during the larval phase (Claro and Lindeman 2008; Froese and Pauly 2024). The species of these families have a strong influence on the reef food webs, since it controls the abundance of species in lower trophic levels (Parrish 1987; Sluka et al. 2001).

Groupers are an easy target for artisanal, industrial and sport fishing because they remain holed up in the reefs during the day (Olavo et al. 2005; Craig et al. 2011). They are susceptible to overfishing, and reduced populations of groupers may elicit changes in the functioning of the reef ecosystem, with significant commercial (*e.g.* collapsed fisheries with highly exploited species) and ecological consequences (*e.g.* alteration in abundance of small fish and mobile invertebrate of the reef community (Parrish 1987; Brulé et al. 2005). The vulnerability to overfishing can be explained by their long-life cycle (*e.g,* low growth rates, sex change), and this is accentuated in species that display spawning aggregations (Bueno et al. 2016; Ellis et al. 2023; Sadovy and Colin 2012; Teixeira et al. 2004). Lutjanids are also considered important fishing resources. Commercial exploitation in Brazil began between the 1950s and 1960s and was dominated by the snapper *Lutjanus purpureus* (Rezende et al. 2003). However, official statistics recorded landings of lutjanids in multi-specific categories, with the "snapper" category comprising five species: *Etelis oculatus*, *L. bucanella*, *L. purpureus*, *L. vivanus* and *Rhomboplites aurorubens* (Rezende et al. 2003). At the beginning of the 1980s, this fishery already showed signs of overfishing and other species started to be exploited by the snapper rig fleet in the Northeast, such as: *L. analis* (Fig. 6.3a, a1), *L. synagris* (Fig. 6.3b), *L. jocu* (Fig. 6.3c, c1, c2) and *Ocyurus chrysurus* (Rezende et al. 2003).

Currently, many species of large fish living on the reefs have shown drastic reductions in their populations and are listed at some level of threat. In view of the number of species in this situation, the global threat scenario for the serranids is significant (Sadovy de Mitcheson et al. 2013, Sadovy de Mitcheson et al. 2020). About 12% of all species of groupers are categorized as "Critically Endangered" (CR), "Endangered" (EN) or "Vulnerable" (VU), and another 13% categorized as "Near Threatened" (NT) (Freitas et al. 2018). However, the advice of the scientists has been ignored (Sadovy de Mitcheson et al. 2013). After a decade, the status of grouper populations has been reassessed. Of the 167 groupers assessed, 19 species have been ranked as threatened in the following categories: Vulnerable (VU), Endangered (EN) or Critically Endangered (CR). To worsen this scenario, 15% of the species lack the necessary data for population assessments (Data deficient—DD). Therefore, the global estimate is 13% of threatened grouper species (ranging from 11% to 26%, based on whether DD species are included), and the threats are not declining (Sadovy de Mitcheson et al. 2020).

In the regional assessment carried out in Brazil by the Ministry of the Environment in 2011, some species such as the red grouper *Epinephelus morio* (Fig. 6.3d) and black grouper *Mycteroperca bonaci* (Fig. 6.3e, e1) were categorized as "Vulnerable" (VU), while others such as the goliath grouper *Epinephelus itajara* (Fig. 6.3f, f1) and warsaw grouper *Hyporthodus nigritus* were considered endangered. For lutjanids, of 13 species assessed for risk of extinction, six are included in the "Nearly Threatened" (NT) category (*L. analis, L. jocu, L. synagris, L. vivanus, Ocyurus chrysurus, R. aurorubens*) and two other species *L. cyanopterus* and *L. purpureus* are considered "Vulnerable" (VU), which means a higher risk of extinction than the others (ICMBIO 2018a, b, c, d).

In Brazil, a ban on goliath grouper fishing was established in 2002, which prohibits its capture and sale and is in force until 2023. However, there are still goliath grouper landings in some locations, as it was observed in Bahia and Sergipe states. In addition, there was an increase in its capture in the state of Maranhão, where no landings had been reported before the ban was established (Giglio et al. 2014). In the case of red and black groupers, there are practically no fishing regulations. The condition of their stocks is unknown due to the scarcity of data on biology, landings, catches per unit effort (CPUE) and composition by size of landings (Freitas et al. 2018). Management and recovery plans and actions for species like these are urgently needed for implementation in Brazil before a total ban on fishing is required (MMA 2014).

In the Northeast of Brazil spawning aggregations of lutjanids (*Lutjanus cyanopterus, L. analis and L. jocu*) were identified and confirmed (Ferreira et al. 2018; França et al. 2021). These aggregations are intensely targeted by industrial fishing and, therefore, the lack of regulation of these fisheries is one of the greatest threats to stocks. Among the lutjanids, only *L. cyanopterus* has a population recovery plan, published through Federal Ordinance No. 292, July 18, 2018 (Brazil 2018a), but not yet in effect.

Unlike the mesopredators, the greenbeak parrotfish *Scarus trispinosus* (Scaridae) (Fig. 6.3g) may be considered the only large sized herbivore in Brazil. This species can reach 90 cm and is the largest herbivore in the southwest Atlantic, being endemic to Brazil and occurring throughout the coast, most commonly in the Northeast (Moura et al. 2001; Freitas et al. 2019) in warmer waters (Francini-Filho et al. 2008; Roos et al. 2016). These fish can be found solitary or in large schools (Fig. 6.3g), living on offshore reefs as adults, with sizes over 40 cm, while juveniles are more common on coastal reefs (Roos et al. 2019). The species plays an important ecological role in reefs because they are grazers or excavators (Bonaldo et al. 2014; Lellys et al. 2019), feeding mainly on crustose coralline algae, turf and fleshy macroalgae and thus, helping corals to survive (Francini-Filho et al. 2010; Tâmega et al. 2016). In addition, they occasionally graze on live corals, removing thin layers of mucus and cyanobacteria that remain on the coral tissue (Francini-Filho et al. 2008).

With the decline of mesopredators, herbivores, especially this parrotfish, became the targets of fishing. The greenbeak parrotfish has been intensively exploited by large fishing operations in the states of Rio Grande do Norte (Cunha et al. 2012) and Bahia (Freitas et al. 2019), where this species is abundant, being responsible for a large portion of the biomass in these reefs (Francini-Filho and Moura 2008; Roos et al. 2016). However, due to its high longevity (22 years) and slow growth, this species is highly vulnerable to overexploitation (Freitas et al. 2019), where in the last decades a population decline of 50% has been recorded in some areas, placing *Scarus trispinosus* as "Endangered" (ICMBIO 2018a, b, c, d). The National Plan for the Recovery of Endangered Species, which includes the greenbeak parrotfish, was also published in 2018 (Brazil 2018b), but has not yet been effectively implemented.

Fisheries management measures need to be taken as soon as possible to prevent a total ban on fishing as the only way to recover their populations (Roos et al. 2020).

The decline of reefs, as well as the habitat conditions, are worrisome for species that depend on this ecosystem for food and shelter (Padovani-Ferreira et al. 2012), as presented with the actinopterygians listed here. Studies show that the protection of pristine habitats promotes the persistence of adult populations, especially in species that use different habitats depending on life stage (Maida and Ferreira 1997; Francini-Filho et al. 2008). This scenario is particularly worrying for some coral-eating excavators that play important roles in the dynamics of reefs and sedimentation (Comeros-Raynal et al. 2012). Among the main threats faced by fishes of the Serranidae and Lutjanidae families, overfishing, climate change and habitat loss (Coleman and Koenig 2013), especially those important in the early stages of the life cycle, are among the main threats that reduce their populations and have direct negative consequences for coastal communities and ecosystems (Coleman and Koenig 2013).

6.4 Sea Turtles

Sea turtles have a wide geographical distribution and move between different habitats to complete their long and complex life cycle, including coral reefs (Sforza et al. 2017). Of the seven existing species, five use the Brazilian coast to develop. They are members of two families: Cheloniidae (*Caretta caretta* (Linnaeus 1758); *Chelonia mydas* (Linnaeus 1758) (Fig. 6.4a); *Eretmochelys imbricata* (Linnaeus 1766); (Fig. 6.4b); *Lepidochelys olivacea* (Eschscholtz 1829) (Fig. 6.4c); and Dermochelyidae (*Dermochelys* coriacea (Vandelli 1761)) (Sforza et al. 2017). In Brazil, the spawning areas of the five existing species extend across the entire coastline (Fig. 6.4d) and the oceanic islands (e.g., *C. mydas* in Atol das Rocas—Fig. 6.4e) and *Car. caretta* in Abrolhos), with priority spawning sites depending on the species (Santos et al. 2011). In the northeast, the main nesting areas are in Sergipe and northern Bahia (ICMBio 2018a, b, c, d).

Sea turtles are skilled predators as adults and their predation in this life stage is unlikely since the adult size of the smallest species, the olive ridley (*Lepidochelys olivacea*), is at least 55 cm in curvilinear carapace length (NMFS and USFWS 1991). This factor makes them top animals in food chains. Younger animals have a smaller and more malleable shell, allowing predation by medium to large fish, ocean birds and sharks (Stancyk 1995; Heithaus 2013).

These animals have various interactions with other species and function as a habitat for hundreds of epibionts. The extensive migrations of turtles provide dispersion for organisms such as barnacles, algae, tunicates and mollusks, in addition to benefiting, and benefiting from, "cleaning" organisms (Heithaus 2013), such as some species of fish and shrimp. The group also functions as a controller of over-populations, since the various turtle species feed on sponges, algae, small fish and on the rocky structure of the reef, where the corals that form part of their diet are

Fig. 6.4 The dwellers turtles: (**a**) green turtle *Chelonia mydas*, (**b**) hawksbill *Eretmochelys imbricata*, (**c**) olive ridley *Lepidochelys olivacea*, (**d**) hawksbill nesting and hatchling, (**e**) green turtle hatchling(© **a**, **b**, **e**—Caroline Feitosa, **c**, **d**—Instituto Verdeluz)

attached. They are considered as bioturbators, affecting the structure and functioning of habitats due to foraging, contributing to the recycling of nutrients (Heithaus 2013).

These animals use different marine habitats for food, rest and reproduction (Musick and Limpus 1997). The behavior of philopatry, designated by the memorization of the chemical composition of the water and the magnetic field of the land where they were born, allows the occurrence of females in shallow coastal waters, with potential reefs (Santos et al. 2011; Bond and James 2017). In the northeast of Brazil, a strong correlation was found between species of the Cheloniidae family with coastal reef ecosystems, especially in the Coral Reef Environmental Protection Area on the coast of Rio Grande do Norte (Alves et al. 2013).

Habitats with little or no anthropogenic influence are essential for the continuity of reproduction of the species. In this context, reefs are a key part of the conservation of this group, since they are used, mainly by the species *Chelonia mydas*

(Fig. 6.4b), for rest and food purposes. This species is herbivorous in adulthood and searches the algae beds associated with the reef for a safe place for foraging (Bjorndal and Jackson 2002; Santos et al. 2011). These ecosystems are considered abundant feeding places, either for females that spawn on islands or on beaches on the continent, as with *C. mydas* in Atol das Rocas and *Eretmochelys imbricata* in Pipa beach in Rio Grande do Norte, respectively (Santos et al. 2011). Therefore, it can be indirectly considered the importance of this habitat for the reproduction of this taxon.

The species *Caretta caretta* and *Lepidochelys olivacea* are usually seen on the reefs in search of shelled invertebrates, such as bivalves and other mollusks, in addition to small fish (Bjorndal and Jackson 2002). Especially for the hawksbill turtle *Eretmochelys imbricata* (Fig. 6.4c), marine sponges, abundant in reefs, are an indispensable part of its survival, a fact justified by the horny beak specialized for digging and feeding on poriferans (Bjorndal and Jackson 2002).

With regard to impacts and threats, the decline in sea turtles is mainly due to human activities that influence at all stages of their life cycle. Among the impacts include the loss of spawning and foraging areas, emphasizing the sensitivity of reefs to anthropogenic impacts, up to high mortality rates through interaction with fishing and ingestion of marine and polluting debris (Lutçavage et al. 1997). According to the Marine Turtle Specialist Group (MTSG), the main threats to sea turtles are currently coastal development, incidental capture by fishing, direct use for human consumption, climate change, pollution and pathogens (Santos et al. 2011).

Today, turtle populations have been drastically reduced when compared to the records of past centuries, in such a way that all species that occur in Brazil fall under some level of threat (Lutcavage et al. 1997; ICMBIO 2018a, b, c, d). The Red Book of Endangered Brazilian Fauna contains the species of turtles that are classified under some level of threat (Table 6.3).

Table 6.3 List of sea turtle species, in evolutionary order, that occur in the Northeast, with their respective maximum lengths and national (Brazil) and international threat status

Species	CN	Threat status (Brazil)	Threat status (Global)	ML (cm)
Chelonia mydas (Linnaeus 1758)	Green	NT (Brasil 2022)	EN (Seminoff 2004, 2023)	150
Caretta caretta (Linnaeus 1758)	Loggerhead	VU (Brasil 2022)	VU (Casale and Tucker 2017)	130
Eretmochelys imbricata (Linnaeus 1766)	Hawksbill	EN (Brasil 2022)	CR (Mortimer and Donnelly 2008)	125
Lepidochelys olivacea (Eschscholtz 1829)	Olive ridley	VU (Brasil 2022)	VU (Abreu-Grobois and Plotkin 2008)	100
Dermochelys coriacea (Vandelli 1761)	Leatherback	CR (Brasil 2022)	VU (Wallace et al. 2013)	245

Legend: *CN* common name, *NT* near threatened, *ML* maximum length, *CR* critically endangered, *EN* endangered, *VU* vulnerable

In Brazil, more effective research in favor of the conservation of sea turtles began in the 1980s with the creation of the TAMAR project (Santos et al. 2011) and continues until today (see Chap. 12 of this book). Another milestone was the establishment of the National Action Plan (PAN) to reduce the incidental capture of sea turtles by fishing activity (Santos et al. 2011). The publications coming from TAMAR, as well as the PAN Sea Turtles, bring together several ways of mitigating the anthropogenic impacts suffered by these species, including that the preservation of sea turtles is intrinsically linked to the conservation of reefs.

TAMAR's efforts, coupled with more recent projects, are known to have resulted in reduced threats (*e.g.* slaughter of females, capture and trade of eggs), and increase in the number of spawning and recovery of populations of all five species along the Brazilian coast (ICMBio 2017). The Federal Government supported TAMAR to act against the alarming mortality of young and adult individuals, since they are animals with a long-life cycle and late sexual maturation (ICMBio 2017). However, it is believed that the number of spawnings observed so far will not be maintained in the future, due to anthropogenic impacts on the populations of juveniles and adults that have interfered in recruitment (ICMBio 2017).

6.5 Marine Mammals

The marine mastofauna on the coast of northeastern Brazil and its oceanic islands consists of members of the orders Sirenia (manatees), Cetartiodactyla (whales and dolphins) and Carnivora (pinnipeds) (Fig. 6.5a). These orders occur in estuarine (Fig. 6.5b) and pelagic ecosystems, and their movement can be considered frequent or sporadic, depending on their life stage, adaptations and migratory or erratic movements (Roman and Estes 2017). The records of this biodiversity are carried out through strandings (Parente et al. 2004; Meirelles et al. 2009), aerial surveys (Alves et al. 2013; Andriolo et al. 2010), and beach monitoring (Paludo and Langguth 2002; Pinedo et al. 2000). Pinnipeds have a strong interaction with reef ecosystems for the capture of benthic food (e.g., demersal fish and cephalopods) (Costa et al. 2004; Kirkwood et al. 2008). However, ecological information on the use of these environments in northeastern Brazil is scarce due to accidental occurrences of debilitated animals (ICMBio 2011). Therefore, this group will not be considered in this chapter.

The order Sirenia is represented in the region by the species *Trichechus manatus* (West Indian manatee) (Fig. 6.5b1), which is a generalist herbivore that occurs in warm, shallow coastal waters (Hartman 1979) of up to 15 m deep (Alves et al. 2013). Its distribution extends from the coast of Amapá to Alagoas, with gaps in Ceará, Rio Grande do Norte, Pernambuco and Alagoas (Alves et al. 2013, 2015; Lima et al. 2011). The local extinction of its populations is documented on the coast of Espírito Santo, Bahia and Sergipe (Albuquerque and Marcovaldi 1982; Borobia and Lodi 1992; Lima et al. 1992).

Fig. 6.5 The marine mammals: (**a**) South American sea lions, (**b**, **b1**) West Indian manatee *Trichechus manatus,* Guiana dolphin *Sotalia guianensis*, (**c**, **c1**, **c2**) West Indian manatee *Trichechus manatus* (© **a** Caroline Feitosa, **b**, **b1**, **c**, **c1**, **c2** Cinthya Leite/Acervo Aquasis)

Among the cetaceans, the most common coastal species are *Sotalia guianensis* (Guiana dolphin) (Fig. 6.5c), *Steno bredanensis* (rough-toothed dolphin) and *Tursiops truncatus* (bottlenose dolphin). The species *Stenella longirostris* (spinner dolphin), *Megaptera novaeangliae* (humpback whale) and *T. truncatus*, are found in the oceanic islands of Fernando de Noronha (state of Pernambuco), Abrolhos (state of Bahia) and São Pedro and São Paulo (state of Rio Grande do Norte), respectively.

Marine mammals have a high influence on oceanic trophodynamics, acting as large primary consumers (e.g., manatees) to alpha predators (e.g., *Orcinus orca*) and exerting a strong effect on the structure and functioning of marine communities. The high potential for locomotion and migration associated with the high consumption of biomass and subsequent excretion and distribution of nutrients and energy throughout marine ecosystems enable the balance of biogeochemical cycles (Doughty et al. 2016; Estes et al. 2016). Therefore, its representatives can be considered key species, causing great impacts when there is a reduction or extinction of the population in a certain area (Chatwin 2007). In addition, they are sensitive to changes in their habitats and the distribution and abundance of their prey, being called sentinels or bioindicators of environmental quality (Moore 2008). Coastal ecosystems, especially estuaries and reefs, directly influence the reproductive, feeding, and resting behavior of many neritic species. This interaction highlights the importance and ecological interdependence between animals and their habitats (Roman and Estes 2017).

Manatees play an important role in nutrient cycling and in cleaning shallow estuarine and coastal waters by consuming aquatic macrophytes, macroalgae and marine angiosperms (Moore 2008). Feces of large herbivores, such as sirenians and sea turtles, provide an increase in primary coastal productivity, as the released fecal organic matter is intensely absorbed by filtering organisms or densified in the unconsolidated substrate, fertilizing the marine soil (Aragones and Marsh 2000). In reef environments of northeastern Brazil, several vegetables identified in their diet are found, especially species of rhodophyta algae (Borges et al. 2008). The marine angiosperm *Halodule wrightii* (needle grass), considered one of the main food items of *Trichechus manatus* in Brazil (Alves et al. 2013) is often associated with shallow coastal reefs, as they need the physical protection of these substrates against wave and tidal hydrodynamism (Copertino et al. 2016).

Odontocete cetaceans are considered top predators, exerting a strong influence on populations of actinopterygians (Fig. 6.5c1), crustaceans and cephalopods (Moore 2008). These prey are found in reefs, making these ecosystems essential for food. In the northeast of Brazil, the presence of species such as *Tursiops truncatus* and *Stenella longirostris* in oceanic islands is associated with the capture of prey (Barbosa et al. 2008; Silva-Jr et al. 2007).

Marine mammals are threatened along the entire Brazilian coast by fishing activities and intense urbanization (Fig. 6.5c2), making this region one of the most vulnerable in South America (Chatwin 2007). Among the main consequences are entanglement in fishing nets, loss of habitats, chemical and solid waste pollution, and collisions with vessels. Therefore, the overlapping of coastal habitats makes them threatened with extinction (IUCN 2012).

The sirenians are considered a sentinel animal for the ecological health of its habitats because they are sensitive to ecosystem changes and pollution (Bonde et al. 2004), and all species face higher extinction risk in the world (Davidson et al. 2012). This species is one of the most endangered marine mammals in Brazil, categorized as "endangered" in the country by the Red Book of Endangered Brazilian Fauna (ICMBio 2018a) and "vulnerable" internationally (IUCN 2012). In the northeast of the country, with the exceptions of the states of Maranhão, Sergipe and Bahia, its population is estimated at just over 1000 specimens (Alves et al. 2013). Important areas of release of specimens rehabilitated in captivity in the states of Alagoas and Paraíba allow for a population increase in northeastern Brazil (ICMBio 2018b). These areas are included in Marine Protected Areas: "APA da Barra do Rio Mamanguape" (mangroves and linear reef) and "Costa dos Corais", Brazil's largest MPA (coral and sandstone reefs, estuarine mangroves), with high density of manatees (Alves et al. 2013). Although manatees reintroduced in the states of Alagoas have been observed in Sergipe (Borges et al. 2019) and Bahia (Attademo et al. 2015), there are no stablished populations in these areas, where the species is currently considered extinct (ICMBio 2018a).

Among coastal cetaceans, the international conservation status of *Sotalia guianensis* according to IUCN is "almost threatened" (Secchi et al. 2018), and *Megaptera novaeangliae* (Cooke 2018), *Tursiops truncatus* (Wells et al. 2019), *Stenella longirostris* (Braulik and Reeves 2018) and *Steno bredanensis* (Kiszka et al. 2019) as "least concern". Intense hunting for the acquisition of skin and fat has been the main threat to tropical pinnipeds in the past (Alava 2017). After the implementation of conservation actions and the ban on intentional capture, some populations have recovered. In Brazil, this population increase may justify the occurrence of these animals in areas that were not previously registered. Therefore, marine mammals on the northeastern coast of the country are sporadic frequenters of reef ecosystems, with the presence of manatees being more common. The diet of these species determines this relationship, intensifying the negative impacts suffered due to the overlap of habitats between them and humans.

6.6 Final Remarks

Reefs provide abundant food sources, act as a nursery area, as a refuge from predation, shelter against currents and provide removal of parasites for top predators, mesopredators and herbivorous species of actinopterygian fish, from the greenbeak parrotfish to manatees. The species mentioned in this chapter promote the maintenance of the reef trophic web, connectivity and nutrient cycling between different coastal and oceanic ecosystems and can be considered as sentinels of the quality of these ecosystems. The ecological interconnection and the influence of anthropogenic actions therefore affect the health of reef environments and their large marine inhabitants.

Both the reef ecosystem and these umbrella species are subject to the major environmental stressors of industrial fishing, degradation and loss of habitats. Seeking to mitigate the impacts of these stressors, Brazil has environmental management tools that include National Action Plans for Conservation directed, for example, to coral reef environments, sharks and marine rays threatened with extinction, sea turtles, manatees, Recovery of Aquatic Species (e.g. parrotfish), fisheries management instruments (e.g. fishing ban, closed season and minimum catch size), normative rulings and the creation of Marine Protected Areas.

Although the large reef species are mostly classified in some level of threat, they used to be considering as umbrella species in management tools and conservationist actions. It is worth highlighting the important role of conservation programs carried out by some agencies and civil society aimed at research and environmental education actions for the protection of species. Among them are the National Center for Research and Conservation of Aquatic Mammals (CMA/ICMBio), National Center for Research and Conservation of Marine Biodiversity in the Northeast (CEPENE/ICMBio), Non-Governmental Organizations such as the TAMAR Project, the Association for Research and Preservation of Aquatic Ecosystems (AQUASIS), and the Aquatic Mammals Foundation (FMA).

In theory, these management actions are efficient and applied worldwide, but in Brazil they lack enforcement and inspection, therefore, in many cases are not efficient. For many species there is a lack of robustness and constancy in data sampling that would allow assessment of their stocks for subsequent inference about population declines. When certain species have data and are placed in some risk category in species conservation status assessments, the publication of the red list of species takes years. For example, the second-to-last elasmobranch assessment was conducted between the years 2010/2011 and the list of threatened species was published in 2014. The last assessment was done in 2016/2017 and, up to date the list has not been published. Meanwhile, these species continue to be fished.

For cetaceans and manatees, efforts are needed for robust and periodic population estimates that guide research and conservation actions, and assist in categorizing the conservation status of species in a more reliable way. For example, for *T. manatus*, the status that was once "critically endangered" dropped to "endangered" without evidence of justifiable population calculations, raising questions by many experts (Balensiefer et al. 2017). In the case of turtles and dolphins, the public policies focused on solid waste management are necessary, because ingestion by these species –among others—is common and can lead the animal to death.

In addition to the above, it is necessary to increase the number of effective marine protected areas that include the coral reef ecosystem, since it is a key ecosystem for these species. Besides this, more efficient enforcement and participatory management efforts are needed with the general public and local communities to minimize damage to reefs and their inhabitants.

References

Afonso AS, Hazin FH (2015) Vertical movement patterns and ontogenetic niche expansion in the tiger shark, Galeocerdo cuvier. PLoS One 10(1):e0116720

Abreu-Grobois A, Plotkin P (IUCN SSC Marine Turtle Specialist Group) (2008) Lepidochelys olivacea. The IUCN Red List of threatened species 2008. https://doi.org/10.2305/IUCN.UK.2008.RLTS.T11534A3292503.en. Accessed 6 Sept 2020

Alava JJ (2017) Tropical pinnipeds: bio-ecology, threats, and conservation. CRC Press, Boca Raton, 325 pp. ISBN 9781498741392

Albins MA, Hixon MA (2008) Invasive Indo-Pacific lionfish Pterois volitans reduce recruitment of Atlantic coral-reef fishes. Mar Ecol Progr Ser 367:233–238

Albuquerque C, Marcovaldi GM (1982) Ocorrência e Distribuição do Peixe-boi Marinho no Litoral Brasileiro (Sirenia, Trichechidade, Trichechus manatus, Linnaeus 1758). In: Simpósio Internacional sobre a Utilização de Ecossistemas Costeiros: Planejamento, Poluição e Produtividade. FURG/Duke University, Rio Grande, p 27

Alves MDO, Schwamborn R, Borges JCG, Marmontel M, Costa AF, Schettini CAF, Araújo ME (2013) Aerial survey of manatees, dolphins and sea turtles off northeastern Brazil: correlations with coastal features and human activities. Biol Conserv 161:91–100

Alves MDO, Kinas PG, Marmontel M, Borges JCG, Costa AF, Schiel N, Araújo ME (2015) First abundance estimate of the Antillean manatee (*Trichechus manatus manatus*) in Brazil by aerial survey. J Mar Biol Assoc U K 96(4):955–966

Andriolo A, Kinas PG, Engel MH, Martins CCA, Rufino AM (2010) Humpback whales within the Brazilian breeding ground: distribution and population size estimate. Endang Species Res 11:233–243

Aragones LV, Marsh HD (2000) Impact of dugong grazing and turtle cropping on tropical seagrass communities. Pac Conserv Biol 5:277–288

Aschenbrenner A, Ferreira BP, Rooker JR (2016) Spatial and temporal variability in the otolith chemistry of the Brazilian snapper *Lutjanus alexandrei* from estuarine and coastal environments. J Fish Biol 89:753–769

Attademo FLN, Balensiefer DC, da Bôaviagem Freire A C, de Sousa GP, da Cunha FAGC, Luna FO (2015) Debris ingestion by the Antillean Manatee (Trichechus manatus manatus). Mar Pol Bull 101(1):284–287. https://doi.org/10.1016/j.marpolbul.2015.09.040

Azevedo JMN, Rodríguez JB, Mendizábal M, Arruda LM (1995) Study of a sample of dusky groupers, Epinephelus marginatus (Lowe, 1834), caught in the tide pool at Lajes do Pico, Azores. Bol Mus Mun Funchal 4:55–64

Balensiefer DC, Attademo FLN, Souza GP, Freire ACB, Cunha FAGC, Alencar AEB, Silva FJL, Luna FO (2017) Three decades of Antillean manatee (Trichechus manatus manatus) stranding along the Brazilian coast. Trop Cons Sci 10:1–9

Barbosa MMC, Cruz FS, Lodi L (2008) Comportamento e organização de grupo do golfinho-flíper, *Tursiops truncatus* (Cetacea, Delphinidea) no arquipélago das Cagarras, Rio de Janeiro. Rev Bras Zoociências 10:213–220

Barley SC, Clark TD, Meeuwig JJ (2020) Ecological redundancy between coral reef sharks and predatory teleosts. Rev Fish Biol Fish 30(1):153–172

Barreto R, Ferretti F, Flemming JM, Amorim A, Andrade H, Worm B, Lessa R (2016) Trends in the exploitation of South Atlantic shark populations. Conserv Biol 30(4):792–804

Baum JK, Worm B (2009) Cascading top-down effects of changing oceanic predator abundances. J Anim Ecol 78:699–714

Benevides FCM, Franco JLA, Braz VS (2017) História dos projetos de conservação de espécies da fauna no Brasil. Hist R 22(2):83–106

Bjorndal KA, Jackson JB (2002) 10 roles of sea turtles in marine ecosystems: reconstructing the past. Biol Sea Turt 2:259

Bonaldo RM, Hoey AS, Bellwood DR (2014) The ecosystem roles of parrotfishes on tropical reefs. Oceanogr Mar Biol Annu Rev 52:81–132

Bond EP, James MC (2017) Pre-nesting movements of Leatherback Sea turtles, Dermochelys coriacea, in the Western Atlantic. Front Mar Sci 4;223

Bond ME, Valentın-Albanese J, Babcock EA, Hussey NE, Heithaus MR, Chapman DD (2018) The trophic ecology of Caribbean reef sharks (Carcharhinus perezi) relative to other large teleost predators on an isolated coral atoll. Mar Biol 165(4):67

Bonde RK, Aguirre AA, Powell J (2004) Manatees as sentinels of marine ecosystem health: are they the 2000-pound canaries? EcoHealth 3:255–262

Borer ET et al (2005) What determines the strength of a trophic cascade? Ecology 86:528–537

Borges JCG, Araújo PG, Anzolin DG, Miranda GEC (2008) Identificação de itens alimentares constituintes da dieta dos peixes-boi marinhos (Trichechus manatus) na região Nordeste do Brasil. Biotemas 21:77–81

Borges JCG, Moreira AL, Batista RLG, Melo BJA, Reis EC, Vergara- Parente JE (2019) Mamíferos marinhos da Bacia Sergipe-Alagoas. In: Reis EC, Carneiro MER (Eds.), Quelônios, aves e mamíferos marinhos da Bacia de Sergipe-Alagoas. Editora UFS, São Cristóvão, pp. 116–149

Bornatowski H, Wedekin LL, Heithaus MR, Marcondes MCC, Rossi-Santos MR (2012) Shark scavenging and predation on cetaceans at Abrolhos Bank, eastern Brazil. J Mar Biol Assoc UK 92(8):1767

Bornatowski H, Navia AF, Braga RR, Abilhoa V, Corrêa MFM (2014) Ecological importance of sharks and rays in a structural foodweb analysis in southern Brazil. ICES J Mar Sci. https://doi.org/10.1093/icesjms/fsu025

Bornatowski H, Angelini R, Coll M, Barreto RR, Amorim AF (2018) Ecological role and historical trends of large pelagic predators in a subtropical marine ecosystem of the South Atlantic. Rev Fish Biol Fish 28:241–259. https://doi.org/10.1007/s11160-017-9492-z

Borobia M, Lodi L (1992). Recent observations and records of the West Indian manatee Trichechus manatus in northeastern Brazil. Biol Conserv 59:37–43

Brasil (2004) Instrução Normativa N°05 de 21 de maio de 2004

Brasil (2014) Portaria N°445 de 17 de dezembro de 2014

Brasil (2018a) Portaria interministerial no 59-C, de 9 de novembro de 2018. Define regras para o uso sustentável e recuperação dos estoques das espécies Mycteroperca interstitialis, conhecido como Badejo-Amarelo; Mycteroperca bonaci, conhecido como Sirigado; Epinephelus morio, conhecido como Garoupa-de-São-Tomé e Lutjanus cyanopterus, conhecido como Caranha. Available at http://www.in.gov.

br/web/guest/materia/-/asset_publisher/Kujrw0TZC2Mb/content/id/50539596/dole-2018-11-16-portaria-interministerial-n-59-c-de-9-de-novembro-de-2018-50539462

Brasil (2018b) Portaria interministerial no 59-B, de 9 de novembro de 2018. Define regras para o uso sustentável e recuperação dos estoques da espécie Scarus trispinosus (budião-azul). Available at http://www.in.gov.br/web/guest/materia/-/asset_publisher/Kujrw0TZC2Mb/content/id/50539608/dole-2018-11-16-portaria-interministerial-n-59-b-de-9-de-novembro-de-2018-50539540

Brasil (2022) Portaria MMA N°148 de 7 de junho de 2022

Brasil (2023) Portaria MMA N°354 de 27 de janeiro de 2023

Braulik G, Reeves R (2018) Stenella longirostris (errata version published in 2019). The IUCN Red List of Threatened Species 2018: e.T20733A156927622. https://doi.org/10.2305/IUCN.UK.2018-2.RLTS.T20733A156927622.en. Downloaded on 07 November 2020. Braz. j. oceanogr. [online]. 2016, vol.64, n.spe2, pp 53–80. ISSN 1982-436X. https://doi.org/10.1590/S1679-875920161036064sp2

Brulé T, Puerto-Novelo E, Perez-Dıaz E, Renan Galindo X (2005) Diet composition of juvenile black grouper (Mycteroperca bonaci) from coastal nursery areas of the Yucatan peninsula, Mexico. Bull Mar Sci 77:441–452

Bueno LS, Bertoncini AA, Koenig CC, Coleman FC, Freitas MO, Leite JR, De Souza TF, Hostim-Silva M (2016) Evidence for spawning aggregations of the endangered Atlantic goliath grouper Epinephelus itajara in southern Brazil. J Fish Biol 89:876–889. https://doi.org/10.1111/jfb.13028

Bussotti S, Guidetti P (2009) Do Mediterranean fish assemblages associated with marine caves and rocky reefs differ? Estuar Coast Shelf S 81:65–73

Casale P, Tucker AD (2017) Caretta caretta (amended version of 2015 assessment). The IUCN Red List of Threatened Species 2017. https://doi.org/10.2305/IUCN.UK.2017-2.RLTS.T3897A119333622.en. Accessed 6 Sept 2020

Ceccarelli D, Ayling T (2010) Role, importance and Vulnera-bility of top predators on the great barrier reef—a review. Great Barrier Reef Marine Park Authority

Chatwin A (2007) The nature conservancy's marine ecoregional assessments methodology in South American. In: Chatwin A (ed) Priorities for coastal and marine conservation in South America. The Nature Conservancy, Virginia, EUA, 63p

Claro R, Lindeman KC (2008) Biología y Manejo de los Pargos (Lutjanidae) en el Atlántico Occidental. Instituto de Oceanología, La Habana, 472p

Coleman FC, Koenig CC (2013) The effects of fishing, climate change, and other anthropogenic disturbances on red grouper and other reef fishes in the Gulf of Mexico. Integr Comp Biol 50(2):201–212

Comeros-Raynal MT, Choat JH, Polidoro BA, Clements KD, Abesamis R, Craig MT, Lazuardi ME, McIlwain J, Muljadi A, Myers RF, Nañola Jr CL, Pardede S, Rocha LA, Russell B, Sanciangco JC, Stockwell B, Harwell H, Carpenter KE (2012). The likelihood of extinction of iconic and dominant components of coral reefs: the parrotfishes and surgeonfishes. PLoS One. http://dx.plos.org/10.1371/journal.pone.0039825

Compagno LJV, Dando M, Fowler S (2005) A field guide to the sharks of the world. Collins, London, 496p

Connell SD, Ghedini G (2015) Resisting regime-shifts: the stabilising effect of compensatory processes. Trends Ecol Evol 30:513–515

Cooke JG (2018) Megaptera novaeangliae. The IUCN red list of threatened species 2018: e.T13006A50362794. https://doi.org/10.2305/IUCN.UK.2018-2.RLTS.T13006A50362794.en. Downloaded on 07 November 2020

Copertino MS, Creed JC, Lanari MO, Magalhães K, Barros K, Lana PC, Sordo L, Horta PA (2016) Seagrass and submerged aquatic vegetation (VAS) habitats off the coast of Brazil: state of knowledge, conservation and main threats. Braz J Oceanogr 64:53–80

Costa TLDA (2011) Projeto raia de fogo: aspectos populacionais da raia de fogo (Dasyatis marianae Gomes, Rosa and Gadig, 2000) e pesca de elasmobrânquios no complexo recifal do Parracho de Maracajaú. Dissertation Universidade Federal do Rio Grande do Norte)

Costa DP, Kuhn CE, Weise MJ, Shaffer SA, Arnould JPY (2004) When does physiology limit the foraging behaviour of freely diving mammals. Int Congr Series 1275:359–366

Craig MT, Sadovy de Mitcheson YJ, Heemstra PC (2011) Groupers of the world: a field and market guide. Boca Raton, Florida

Cunha FEA, Carvalho RAA, Araújo ME (2012) Exportation of reef fish for human consumption: long term analysis using data from Rio Grande do Norte, Brazil. Bol Inst Pesca 38:369–378

Davidson AD, Boyer AG, Kim H, Pompa-Mansilla S and others (2012) Drivers and hotspots of extinction risk in marine mammals. Proc Natl Acad Sci USA 109: 3395–3400

Doughty CE, Roman J, Faurby S, Wolf A, Haque A, Bakker ES, Malhi Y, Dunning JJB, Svenning JC (2016) Global nutrient transport in a world of giants. PNAS 113(4):868–873. https://doi.org/10.1073/pnas.1502549112

Dudley SF, Anderson-Reade MD, Thompson GS, McMullen PB (2000) Concurrent scavenging off a whale carcass by great white sharks, Carcharodon carcharias, and tiger sharks, Galeocerdo cuvier. Fish Bull 98(3):646–649

Dulvy NK, Freckleton RP, Polunin NVC (2004) Coral reef cascades and the indirect effects of predator removal by exploitation. Ecol Lett 7:410–416

Dulvy NK, Fowler SL, Musick JA et al (2014) Extinction risk and conservation of the world's sharks and rays. eLife 3:e00590. 34p. https://doi.org/10.7554/eLife.00590

Dulvy NK, Pacoureau N, Rigby CL, Pollom RA, Jabado RW, Ebert DA et al (2021) Overfishing drives over one-third of all sharks and rays toward a global extinction crisis. Curr Bio 31:4773–4787

Ellis RD, Koenig CC, Locascio JV, Malinowski CR, Coleman FC (2023) Spawning migrations of the Atlantic goliath grouper along the Florida Atlantic Coast. Aust Fish 8:398. https://doi.org/10.3390/fishes8080398

Estes JA, Heithaus M, McCauley DJ, Rasher DB, Worm B (2016) Megafaunal impacts on structure and function of ocean ecosystems. Annu Rev Environ Resour 41:83–116

Fagan WF (1997) Omnivory as a stabilizing feature of natural communities. Am Nat 150:554–567

Ferreira LC, Thums M, Heithaus MR, Barnett A et al (2017) The trophic role of a large marine predator, the tiger shark Galeocerdo cuvier. Sci Rep 7(1):1–14

Ferreira BP, Maida M, Olavo G, Strenzel GR, França AR, Malafaia PN (2018) Marine geomorphology of cubbera snapper spawning aggregation sites in the SW Atlantic. In: Schroeder DM, Dunkel CA, Cochrane GR (eds) Marine geological and biological habitat mapping conference—GEOHAB abstracts, Santa Barbara, USA

França AR, Olavo G, Rezende SM, Ferreira BP (2021) Spatio-temporal distribution of mutton and dog snappers spawning aggregations in the Southwest Atlantic. Aquatic Conserv Mar Freshw Ecosyst. https://doi.org/10.1002/aqc.3536

Francini-Filho RB, Moura RL (2008) Dynamics of fish assemblages on coral reefs subjected to different management regimes in the Abrolhos Bank, eastern Brazil. Aquat Conserv Mar Freshwat Ecosyst 18:1166–1179

Francini-Filho RB, Moura RL, Ferreira CM, Coni EOC (2008) Live coral predation by parrotfishes (Perciformes: Scaridae) in the Abrolhos Bank, eastern Brazil, with comments on the classification of species into functional groups. Neotrop Ichthyol 6:191–200

Francini-Filho RB, Ferreira CM, Coni EOC, Moura RL, Kaufman L (2010) Foraging activity of roving herbivorous reef fish (Acanthuridae and Scaridae) in eastern Brazil: influence of resource availability and interference competition. J Mar 90:481–492

Franks J, VanderKooy K (2000) Feeding Habits of Juvenile Lane Snapper Lutjanus synagris from Mississippi Coastal Waters, with Comments on the Diet of Gray Snapper Lutjanus griseus. Gulf and Carib Res 12. https://doi.org/10.18785/gcr.1201.02

Frédou T, Beatrice PF (2005) Bathymetric trends of Northeastern Brazilian snappers (Pisces, Lutjanidae): implications for the reef fishery dynamic. Braz Arch Biol Technol 48(5):787–800

Freitas MO, Albihoa V, Silva GHC (2011) Feeding ecology of Lutjanus analis (Teleostei: Lutjanidae) from Abrolhos Bank, Eastern Brazil. Neotrop Ichthyol 9(2):411–418

Freitas MO, Previero M, Minte-Vera CV, Spack HL, Francini-Filho RB, Moura RL (2018) Reproductive biology and management of two commercially important groupers in the SW Atlantic. Environ Biol Fish 101:79–94

Freitas MO, Previero M, Leite JR, Francini-Filho RB, Minte-Vera CV, Moura RL (2019) Age, growth, reproduction and management of Southwestern Atlantic's largest and endangered herbivorous reef fish, Scarus trispinosus Valenciennes, 1840. PeerJ 7:e7459

Friedlander A, Demartini EE (2002) Contrasts in density, size, and biomass of reef fishes between the northwestern and the main Hawaiian Islands: the effects of fishing down apex predators. Mar Ecol Prog Ser 230:253–264

Frisch AJ, Ireland M, Rizzari JR, Lönnstedt OM, Magnenat KA, Mirbach CE, Hobbs JPA (2016) Reassessing the trophic role of reef sharks as apex predators on coral reefs. Coral Reefs 35(2):459–472

Froese R, Pauly D (eds) (2024) FishBase. World Wide Web Electronic Publication. www.fishbase.org

Giglio VJ, Bertoncini AA, Ferreira BP, Hostim-Silva M, Freitas MO (2014) Landings of goliath grouper, Epinephelus itajara, in Brazil: despite prohibited over ten years, fishing continues. Nat Conserv 12:118–123

Grubbs RD, Carlson JK, Romine JG, Curtis TH, McElroy WD, McCandless CT, Cotton CF, Musick JA (2016) Critical assessment and ramifications of a purported marine trophic cascade. Sci Rep 6:20970. https://doi.org/10.1038/srep20970

Hartman DS (1979) Ecology and behavior of the manatee (Trichechus manatus) in Florida. Am Soc Mammalog Special Publ 5:153

Hazin FH, Afonso AS, Castilho PC, Ferreira LC, Rocha BC (2013) Regional movements of the tiger shark, Galeocerdo cuvier, off northeastern Brazil: inferences regarding shark attack hazard. An Acad Bras Cienc 85(3):1053–1062

Heemstra PC, Randall JE (1993) FAO species catalogue. Groupers of the world (Family Serranidae, Subfamily Epinephelinae): an annotated and illustrated catalogue of the grouper, rockcod, hind, coral grouper lyretail species known to date. FAO fisheries synopsis no. 125, vol 16. FAO, Rome

Heithaus MR (2013) Predators, prey and the ecological roles of sea turtles. In: Lutz PL, Musick JA (eds) The biology of sea turtle, Marine science series, vol 3. CRC Press, pp 249–284

Heithaus MR, Frid A, Wirsing AJ et al (2007) State-dependent risk-taking by green sea turtles mediates top-down effects of tiger shark intimidation in a marine ecosystem. J Anim Ecol 76(5):837–844

Heithaus MR, Frid A, Wirsing AJ, Worm B (2008) Predicting ecological consequences of marine top predator declines. Trends Ecol Evol 23:202–210

Heithaus MR, Frid A, Vaudo JJ, Worm B, Wirsing AJ (2010) Unraveling the ecological importance of elasmobranchs. In: Carrier JC, Musick JA, Heithaus MR (eds) Sharks and their relatives II: Biodiversity, adaptive physiology, and conservation. CRC Press, Boca Raton, FL, p 607–633

Heupel MR, Knip DM, Simpfendorfer CA, Dulvy NK (2014) Sizing up the ecological role of sharks as predators. Mar Ecol Progr Ser 495:291–298

Hixon MA, Beets JP (1993) Predation, prey refuges, and the structure of coral reef fish assemblages. Ecol Monogr 63:77–101

Hutchings JA, Baum JK (2005) Measuring marine fish biodiversity: temporal changes in abundance, life history and demography. Philos Trans R Soc B Biol Sc 360(1454):315–338

ICMBio (2011) Sumário executivo do Plano de Ação Nacional para a conservação dos sirênios. Instituto Chico Mendes. Available on: https://www.gov.br/icmbio/pt-br/assuntos/biodiversidade/pan/pan-sirenios/1-ciclo/pan-sirenios-sumario.pdf

ICMBio (2016a) Avaliação do risco de extinção dos elasmobrânquios e quimeras no Brasil: 2010–2012. http://www.ICMBio.gov.br/cepsul/images/stories/biblioteca/download/trabalhos_tecnicos/pub_2016_avaliacao_elasmo_2010_2012.pdf

ICMBio (2016b) Sumário executivo do Plano de Ação Nacional para a conservação dos Tubarões e Raias Marinhos Ameaçados de Extinção. Instituto Chico Mendes. Available on: https://www.ICMBio.gov.br/portal/images/stories/docs-pan/pan-tubaroes/1-ciclo/pan-tubaroes-sumario.pdf
ICMBio (2017) Sumário executivo do Plano de Ação Nacional para Conservação das Tartarugas Marinhas. Instituto Chico Mendes. Available on: https://www.gov.br/icmbio/pt-br/assuntos/biodiversidade/pan/pan/pan-tartarugas-marinhas
ICMBio (2018a) Livro Vermelho da Fauna Brasileira Ameaçada de Extinção. Instituto Chico Mendes de Conservação da Biodiversidade/MMA–Ministério do Meio Ambiente. ICMBio, Brasília, 252p
ICMBio (2018b) Livro Vermelho da Fauna Brasileira Ameaçada de Extinção: Volume I / 1. ed. ICMBio/MMA, Brasília, 492 p
ICMBio (2018c) Livro Vermelho da Fauna Brasileira Ameaçada de Extinção. Instituto Chico Mendes de Conservação da Biodiversidade/ Ministério do Meio Ambiente, Brasília
ICMBio (2018d) Plano de ação nacional para a conservação do peixe-boi marinho. Portaria N° 249, de 4 de abril de 2018
IUCN (2012) The IUCN red list of threatened species. Version 2012.2. http://www.iucnredlist.org
Jiménez-Centeno CE (1997) Coral colony fragmentation by whitetip reef sharks at Coiba Island National Park, Panama. Fragmentación de la colonia de corales en el Parque Nacional Isla Coiba, Panamá, por parte de tiburones de arrecife punta blanca. Rev Biol Trop 45(1B):698–700
Kirkwood R, Hume F, Hindell M (2008) Sea temperature variations mediate annual changes in the diet of Australian fur seals in Bass Strait. Mar Ecol Prog Ser 369:297–309
Kiszka J, Baird R, Braulik G (2019) Steno bredanensis. The IUCN red list of threatened species 2019: e.T20738A50376703. https://doi.org/10.2305/IUCN.UK.2019-2.RLTS.T20738A50376703.en. Downloaded on 7 November 2020
Last P, Carvalho M, Corrigan S, Gavin N, Seret B, Yang L (2016) The Rays of the World project - an explanation. In: Last PR, Yearsley GK (Eds) Rays of the World, Supplementary Information, Chapter: 1. CSIRO Australian National Fish Collection, Australia.
Leal I, Pereira P, Araújo M (2013) Coral reef fish association and behaviour on the fire coral Millepora spp. in north-east Brazil. J Mar Biolog 93(6):1703–1711
Lellys NT, Moura RL, Bonaldo RM, Francini-Filho RB, Gibran FZ (2019) Parrotfish functional morphology and bioerosion on SW Atlantic reefs. Mar Ecol Prog Ser 629:149–163. https://doi.org/10.3354/meps13102
Lima RP, Paludo D, Soavinski RJ, Silva KG, Oliveira EMA (2011) Levantamento da distribuição, ocorrência e status de conservação do Peixe-Boi Marinho (Trichechus manatus, Linnaeus, 1758) no litoral nordeste do Brasil. Nat Resour Aquidabã 1:41–57
Lima RP, Paludo D, Soavinski RJ, Silva KG, Oliveira EMA (1992) Levantamento da distribuição, ocorrência e status de conservação do peixe-boi marinho Trichechus manatus (Linnaeus 1758) no litoral Nordeste do Brasil. Period. Peixe-Boi, 1:47–72
Lowe CG, Wetherbee BM, Crow GL, Tester AL (1996) Ontogenetic dietary shifts and feeding behavior of the tiger shark, Galeocerdo cuvier. Hawaiian Waters Environ Biol Fish 47(2):203–211
Lutcavage ME, Plotkin P, Witherington B, Lutz PL (1997) Human impacts on sea turtle survival. In: Lutz PL, Musick JA (eds) The biology of sea turtles, 1:387–409. CRC Press
Machado L, Bertoncini A, Hostim-Silva M, Barreiros JP (2003) Habitat use by the juvenile dusky grouper Epinephelus marginatus and its relative abundance, in Santa Catarina, Brazil. J Ichthyol Aquat Biol 6:133–138
Maida M, Ferreira BP (1997) Coral reefs of Brazil: an overwiew. Proc 8th Int Coral Reef Symp 1:263–274
Martins AS, Olavo G, Costa PAS (2005) Recursos demersais capturados com espinhel de fundo no talude superior da região entre Salvador (BA) e o Cabo de São Tomé (RJ). In: PAS C, Martins AS, Olavo G (eds) Pesca e potenciais de exploração de recursos vivos na região central da

Zona Econômica Exclusiva brasileira. Museu Nacional, Série Livros n. 13, Rio de Janeiro, pp 109–128
McCauley DJ, Young HS, Dunbar RB, Estes JA, Semmens BX, Micheli F (2012) Assessing the effects of large mobile predators on ecosystem connectivity. Ecol Appl 22(6):1711–1717
Meirelles ACO, Monteiro-Neto C, Martins AMA, Costa AF, Barros HMDR, Alves MDO (2009) Cetacean strandings on the coast of Ceará, northeastern Brazil (1992-2005). J Mar Biol Assoc UK 89:1083–1090
Meyer CG, Papastamatiou YP, Holland KN (2010) A multiple instrument approach to quantifying the movement patterns and habitat use of tiger (Galeocerdo cuvier) and Galapagos sharks (Carcharhinus galapagensis) at French Frigate Shoals. Hawaii Mar Biol 157(8):1857–1868
Ministério do Meio Ambiente (2014) Lista Nacional Oficial de Espécies da Fauna Ameaçadas de Extinção – Peixes e Invertebrados Aquáticos. Portaria N° 445 de 17 de dezembro de 2014
Moore SE (2008) Marine mammals as ecosystem sentinels. J Mammal 89:534–540
Mortimer JA, Donnelly M (IUCN SSC Marine Turtle Specialist Group) (2008) Eretmochelys imbricata. The IUCN Red List of Threatened Species 200. https://doi.org/10.2305/IUCN.UK.2008.RLTS.T8005A12881238.en. Accessed 6 Sept 2020
Moura RL, Figueiredo JL, Sazima I (2001) A new parrotfish (Scaridae) from Brazil, and revalidation of Sparisoma amplum (Ranzani, 1842), Sparisoma frondosum (Agassiz, 1831), Sparisoma axillare (Steindachner, 1878) and Scarus trispinosus Valenciennes, 1840. Bull Mar Sci 68(3):505–524
Musick JA, Limpus CJ (1997) Habitat utilization and migration in juvenile sea turtles. In: Lutz PL, Musick JA (eds) The biology of sea turtles, 1. CRC Press, Boca Raton, pp 137–164
Nadon MO, Baum JK, Williams ID, Mcpherson JM et al (2012) Re-creating missing population baselines for Pacific reef sharks. Conserv Biol 26(3):493–503
Nelson JS, Grande TC, Wilson MV (2016) Fishes of the world. Wiley, Hoboken, p 752
NMFS and USFWS (U.S. Fish and Wildlife Service) (1991) Recovery plan for U.S. population of Atlantic green turtle Chelonia mydas. Washington, DC
Olavo G, Costa PA, Martins AS (2005) Caracterização da pesca de linha e dinâmica das frotas linheiras da Bahia. In: Costa PAS, Martins AS, Olavo G (eds) Pesca e potenciais de exploração de recursos vivos na região central da Zona Econômica Exclusiva brasileira: 13–34. Museu Nacional. Serie livros n.13, Rio de Janeiro
Pace ML, Cole JJ, Carpenter SR, Kitchell JF (1999) Trophic cascades revealed in diverse ecosystems. Trends Ecol Evol 14:483–488
Padovani-Ferreira B, Floeter S, Rocha LA, Ferreira CE, Francini-Filho R, Moura R, Gaspar AL, Feitosa CV. 2012. Scarus trispinosus. The IUCN Red List of ThreatenedSpecies2012:e.T190748A17786694.https://doi.org/10.2305/IUCN.UK.2012.RLTS.T190748A17786694.en. Downloaded on 07 April 2021
Paludo D, Langguth A (2002) Use of space and temporal distribution of Trichechus manatus manatus Linnaeus in the region of Sagi, Rio Grande do Norte State, Brazil (Sirenia, Trichechidae). Rev Bras Zool 19:205–215
Parente CL, Vergara-Parente JE, Lima RP (2004) Strandings of Antillean Manatees, Trichechus manatus manatus, in Northeastern Brazil. Lat Am J Aquat Mamm 3:69–75
Parenti P, Randall JE (2020) An annotated checklist of the fishes of the family Serranidae of the world with description of two new related families of fishes. FishTaxa 15:1–170
Parrish JD (1987) The trophic biology of snappers and groupers. In: Polovina JJ, Ralston S (eds) Tropical snappers and groupers: biology and fisheries management. Westview Press, Colorado
Pinedo MC, Rosas FW, Marmontel M (2000) Cetáceos e Pinípedes do Brasil. UNEP/FUA, Manaus, 213p
Pinheiro HT, Rocha LA, Macieira RM et al (2018) South-Western Atlantic reef fishes: zoogeographical patterns and ecological drivers reveal a secondary biodiversity Centre in the Atlantic Ocean. Divers Distrib 24(7):951–965
Polovina JJ (1984) Model of a coral reef ecosystem I. The ecopath model and its application to French Frigate Shoals. Coral Reefs 3:1–11

Pongsiri MJ, Roman J, Ezenwa VO et al (2009) Biodiversity loss affects global disease ecology. Bioscience 59(11):945–954

Rezende SM, Ferreira BP, Fredou T (2003) A pesca de lutjanídeos no nordeste do Brasil: histórico das pescarias, características das espécies e relevância para o manejo. Bol Tec Cient Cepene 11:1–17

Ritchie EG, Johnson CN (2009) Predator interactions, mesopredator release and biodiversity conservation. Ecol Lett 12:982–998

Rocha LA, Rosa IL (2001) Baseline assessment of reef fish assemblages of Parcel Manuel Luiz Marine State Park, Maranhão, Northeast Brazil. J Fish Biol 58(4):985–998

Roff G, Doropoulos C, Rogers A et al (2016) The ecological role of sharks on coral reefs. Trends Ecol Evol 31(5):395–407

Rogers A, Blanchard JL, Mumby PJ (2014) Vulnerability of coral reef fisheries to a loss of structural complexity. Curr Biol 24(9):1000–1005

Roman J, Estes JE (2017) Ecology of marine mammals. In: Bernd WJGM, Thewissen KMK (eds) Encyclopedia of marine mammals, 3rd edn. Academic

Roos NC, Pennino MG, Lopes PFM, Carvalho AR (2016) Multiple management strategies to control selectivity on parrotfishes harvesting. Ocean Coastal Manage 134:20–29

Roos NC, Pennino MG, Carvalho AR, Longo GO (2019) Drivers of abundance and biomass of Brazilian parrotfishes. Mar Ecol Prog Ser 623:117–130

Roos NC, Longo GO, Pennino MG, Francini-Filho RB, Carvalho AR (2020) Protecting nursery areas without fisheries management is not enough to conserve the most endangered parrotfish of the Atlantic Ocean. Sci Rep 10:19143. https://doi.org/10.1038/s41598-020-76207-x

Rupp A, Bornatowski H (2021) Food web model to assess the fishing impacts and ecological role of elasmobranchs in a coastal ecosystem of southern Brazil. Environ Biol Fish 104:905–921

Sadovy de Mitcheson Y, Colin PL (eds) (2012) Reef fish spawning aggregations: Biology, research and management. Fish and Fisheries Series 35. Springer Science+Business Media B.V

Sadovy de Mitcheson YJ, Craig MT, Bertoncini AA, Carpenter K, Cheung W, Choat J, Cornish A, Fennessy S, Ferreira BP, Heemstra P, Liu M, Myers R, Pollard D, Rhodes K, Rocha L, Russel B, Samoilys M, Sanciangco J (2013) Fishing groupers towards extinction: a global assessment of threats and extinction risks in a billion dollar fishery. Fish Fish 14:119–136

Sadovy de Mitcheson YJ, Linardich C, Barreiros JP, Ralph GM, Aguilar-Perera A, Afonso P, Erisman BE, Pollard DA, Fennessy ST, Bertoncini AA, Nair RJ, Rhodes KL, Francour P, Brulé T, Samoilys MA, Ferreira BP, Craig MT (2020) Valuable but vulnerable: over-fishing and under-management continue to threaten groupers so what now? Mar Pol 116:103909. https://doi.org/10.1016/j.marpol.2020.103909

Santander-Neto J, Shinozaki-Mendes RA, Silveira LM, Jucá-Queiroz B, Furtado-Neto MA, Faria VV (2011) Population structure of nurse sharks, Ginglymostoma cirratum (Orectolobiformes), caught off Ceará State, Brazil, south-western Equatorial Atlantic. J Mar Biol Assoc UK 91(6):1193

Santos AS, Almeida AP, Santos AJB, Gallo B, Giffoni B, Baptistotte C, Coelho CA, Lima EHSM, Sales G, Lo Pez GG, Stahelin G, Becker H, Castilhos JC, Thomé JCSA, Wanderlinde J, Marcovaldi MAAGDE, Mendilaharsu MMP, Damasceno MT, Barata PCR, Sforza R (2011) Plano de ação nacional para a conservação das Tartarugas Marinhas. Org. Marcovaldi MAAG, Santos AS. ICMBIO, Brasília, 120p

Schmitz OJ, Hawlena D, Trussell GC (2010) Predator control of ecosystem nutrient dynamics. Ecol Lett 13(10):1199–1209

Secchi E, Santos MC de O, Reeves R (2018) Sotalia guianensis (errata version published in 2019). The IUCN Red List of Threatened Species 2018: e.T181359A144232542. https://doi.org/10.2305/IUCN.UK.2018-2.RLTS.T181359A144232542.en. Downloaded on 07 November 2020

Seminoff JA (Southwest Fisheries Science Center, U.S.) (2004) Chelonia mydas. The IUCN Red List of Threatened Species 2004. https://doi.org/10.2305/IUCN.UK.2004.RLTS.T4615A11037468.en. Accessed 6 Sept 2020

Seminoff JA (2023) Chelonia mydas (amended version of 2004 assessment). The IUCN Red List of Threatened Species 2023: e.T4615A247654386. https://dx.doi.org/10.2305/IUCN.UK.2023-1.RLTS.T4615A247654386.en. Accessed on 29 July 2024.

Sforza R, Marcondes ACJ, Pizetta GT (org) (2017) Guia de Licenciamento Tartarugas Marinhas: Diretrizes para Avaliação e Mitigação de Impactos de Empreendimentos Costeiros e Marinhos. Instituto Chico Mendes de Conservação da Biodiversidade (ICMBio), Brasília. 130 p

Silva-Jr JM, Silva FJL, Sazima C, Sazima I (2007) Trophic relationships of the spinner dolphin at Fernando de Noronha Archipelago, SW Atlantic. Sci Mar 71(3):505–511

Sluka RD, Chiappone M, Sullivan Sealey KM (2001) Influence of habitat on grouper abundance in the FloridaKeys, USA. J Fish Biol 58:682–700

Sorenson L, Santini F, Alfaro ME (2014) The effect of habitat on modern shark diversification. J Evol Biol 27(8):1536–1548

Stancyk SE (1995) Non-human predators of sea turtles and their control. In: Bjorndal KA (ed) Biology and conservation of sea turtles. Smithsonian Institution Press, Washington, DC, pp 139–151

Tâmega F, Figueiredo M, Ferreira C, Bonaldo R (2016) Seaweed survival after consumption by the greenbeak parrotfish, Scarus trispinosus. Coral Reefs 35:329–334

Teixeira SF, Ferreira BP, Padovan IP (2004) Aspects of fishing and reproduction of the black grouper Mycteroperca bonaci (Poey 1860) (Serranidae: Epinephelinae) in the northeastern Brazil. Neot Ichth 2:19–30

Vooren CM, Klippel S (2005) Ações para a conservação de tubarões e raias no sul do Brasil. Sandro Klippel

Wallace BP, Kot CY, DiMatteo AD, Lee T, Crowder LB, Lewison RL (2013) Impacts of fisheries bycatch on marine turtle populations worldwide: toward conservation and research priorities. Ecosphere 4:1–19. https://doi.org/10.1890/ES12-00388.1

Weigmann S (2016) Annotated checklist of the living sharks, batoids and chimaeras (Chondrichthyes) of the world, with a focus on biogeographical diversity. J Fish Biol 88(3):837–1037

Wells RS, Natoli A, Braulik G (2019) Tursiops truncatus (errata version published in 2019). The IUCN red list of threatened species 2019: e.T22563A156932432. https://doi.org/10.2305/IUCN.UK.2019-1.RLTS.T22563A156932432.en. Downloaded on 07 November 2020

Wootton JT, Emmerson M (2005) Measurement of interaction strength in nature. Annu Rev Ecol Evol Syst 36:419–444

Chapter 7
Genetic and Demographic Connectivity in Brazilian Reef Environments

Douglas F. M. Gherardi, Katia C. C. Capel, Cesar A. M. M. Cordeiro, Linda Eggertsen, Clarissa A. K. Endo, Luciana S. Lima, Tito M. C. Lotufo, and Carla Zilberberg

D. F. M. Gherardi (✉)
Laboratory of Ocean and Atmosphere Studies (LOA), Earth Observation and Geoinformatics Division, National Institute for Space Research (INPE), São José dos Campos, São Paulo, Brazil
e-mail: douglas.gherardi@inpe.br

K. C. C. Capel
Institute of Biodiversity and Sustaintability (NUPEM), Universidade Federal do Rio de Janeiro, Rio de Janeiro, Brazil

Departamento de Genética e Evolução, Universidade Federal de São Carlos, São Carlos, Brazil

Centro de Biologia Marinha, Universidade de São Paulo, São Sebastião, Brazil

Cesar A. M. M. Cordeiro
Laboratory of Environmental Sciences, Universidade Estadual do Norte Fluminense, Campos dos Goytacazes, RJ, Brazil

L. Eggertsen
Hawaiian Institute of Marine Biology, University of Hawaii at Manoa, Hawaii, Kaneohe, USA

Reef Systems and Conservation Lab, Universidade Federal de Fluminense, Niteroi, Brazil

C. A. K. Endo
National Institute for Space Research (INPE), São José dos Campos, São Paulo, Brazil

Åkerblå AS, Bergen, Norway

L. S. Lima
National Institute for Space Research (INPE), São José dos Campos, São Paulo, Brazil

T. M. C. Lotufo
Instituto Oceanográfico, Universidade de São Paulo, São Paulo, Brazil

Instituto Coral Vivo, Arraial d'Ajuda, Brazil

C. Zilberberg
Institute of Biodiversity and Sustaintability (NUPEM), Universidade Federal do Rio de Janeiro, Rio de Janeiro, Brazil

R. K. P. Kikuchi et al. (eds.), *Brazilian Coral Reefs*, Brazilian Marine Biodiversity, https://doi.org/10.1007/978-3-031-59152-5_7

Abstract Biological entities present multiple forms of connections between them in space and time, and at various scales. These links are established through the exchange of matter, energy, or information and produce different forms of connectivity. In this chapter we focus on three of these forms that were studied on Brazilian coral reefs. A first section deals with the exchange of information through genetic connectivity within an evolutionary timescale. We discuss the relevance of biogeographical barriers and marine currents to genetically structure populations of coral and fish species along the Brazilian coast. The second section concerns demographic connectivity within the biophysical context, on an ecological timescale. Current biophysical modeling tools were used to infer connectivity for fish and lobster species shedding light on the most relevant involved processes. In the third section, the focus is shifted to the ecological connectivity between coral reefs and adjacent habitats, involving mainly the movement of individuals. These migrations may be related to ontogenetic processes or foraging behavior. For Brazilian reef fish species, mangroves and estuaries are particularly important, and vertical and shore to shelf migrations are also discussed. Finally, we highlight that all these forms of connectivity must be considered within a sound conservation strategy. We hope this chapter not only summarizes the research on connectivity along Brazilian coral reefs, but also pinpoint questions that must be urgently addressed to produce a more complete picture of the functioning of these systems.

Keywords Demographic connectivity · Sea anemones · Biophysical · Microsatellites

7.1 Introduction

All living things and all ecosystems have multiple forms of connections that may involve the exchange of energy, matter or genetic information. In the marine realm, connectivity is a more complicated and less studied subject, in comparison with terrestrial ecosystems (Virtanen et al. 2020). Additionally, connectivity has multiple biological meanings, and all these concepts involve some sort of linkage between different elements across space and/or time. Connectivity patterns and processes interplay with many other aspects of an ecosystems' structure and function. However, the diversity of meanings and uses of "connectivity" carries a potential for confusion and misconceptions (Selkoe et al. 2016). Hence, we need to clearly define the core concepts used throughout this chapter.

Connectivity is at the core of the population definition and delimitation (Harwood 2009). In terms of population biology and conservation, it is essential to understand how dispersal affects population size in addition to its genetic or phenotypic variability and determine the *population connectivity*. Conversely, *Demographic connectivity* is the actual exchange of individuals between populations through dispersal, and as such is one of the most relevant types of connectivity. Demographic connectivity contributes to the demographic structure, and its determination is more complex than assessing a single process, involving the knowledge of many population parameters, such as death, birth, and migration (Lowe and Allendorf 2010).

Determining the demographic connectivity of a species is always challenging, as it demands information not easily obtainable. Hence, alternative approaches that shed light on population connectivity are more commonly used to describe patterns or test population connectivity hypotheses. The most direct and frequently used method worldwide in the marine realm involves the use of population genetics to infer patterns of gene flow between populations and may be defined as "the degree to which gene flow affects evolutionary processes within populations" (Lowe and Allendorf 2010), which can be estimated based on metrics, such as fixation indexes or from Bayesian structure analyses. As the name suggests, *genetic connectivity* is the main mechanism behind the genetic differentiation of populations of species living in different areas. Organism's dispersal is treated within this approach over evolutionary time scales (centuries to millennia) and is generally associated with the formation of broad biogeographic patterns that could, eventually, lead to populations developing unique genetic characteristics.

Demographic connectivity, on the other hand, focuses on processes within ecological timescales (days to decades), involving eggs and larval dispersal. However, juveniles and adults should also be considered in those processes, assuming that recruits will reproduce and, therefore, contribute to the gene pool at the newly settled location.

The exchange of individuals among marine populations is mainly governed by their interaction with the flow field and its different scales of motion. These are the fundamental drivers of larval dispersal, which is the reason why this process is better described in a biophysical context (Cowen and Sponaugle 2009). Connectivity integrates biological characteristics of organisms and physical mechanisms acting along their dispersal trajectory that need to be considered and are associated with reproduction. This may include reproductive aggregations, lifetime offspring production, growth, survival of lethal temperatures and larval behavior, such as active swimming, diel vertical movement (DVM) and pelagic larval duration (PLD).

Brooding corals, like *Favia gravida*, tend to recruit close to their parent colony (Gleason and Hofmann 2011), while the lecithotrophic larvae of the broadcast spawners *Mussismilia hispida* may have a longer PLD of up to 10 days (Pires et al. 2016). Spending more time in the plankton has the potential to increase the dispersion distance of eggs and larvae and, consequently, the chance to colonize new areas. All these processes are influenced by horizontal and vertical water velocities (advection and diffusion), temperature, salinity, and their variability at different spatial and temporal scales. Broadly speaking, water flow controls larval dispersal according to the balance between its mean and fluctuating components. The mean flow component is typically of large spatial scales, whereas the turbulent mesoscale (10–100 km) is related to the time of fluctuating flow. The former influences particles by advection and the latter by turbulent diffusive transport (Bauer et al. 1998).

The adaptive processes of marine organisms tend to respond to the mean ocean flow, but recruitment variability may reflect the stochastic component of the environment such as transient atmospheric systems, eddy formation and internal waves

(Sundby and Kristiansen 2015). This is generally the case for near-shore populations, which are submitted to stochastic variability with a characteristic timescale of 2–5 days, consequently, larval transport in coastal areas tend to be a spatially heterogeneous process (Siegel et al. 2008).

There is another form of connectivity involving the transfer or movement of individuals between different habitats. Processes like migration at the individual scale, and not only at the populational level, are of major interest for conservation. This type of connectivity highlights the importance of different habitats in the maintenance of the biodiversity and is especially useful for conservation planning, as policies often focus on a single habitat.

Reef fish metapopulation dynamics (see Box 7.1) are being studied using a diverse suite of dispersion models and continued refinements of genetic tools. Also important is the multidisciplinary integration of behavioral studies and tagging (natural and artificial) methods for pelagic larvae (Mora and Sale 2002; Sale and Kritzer 2003). At the end of the day, the dispersion of individuals tends to strike a balance between local extinction and the colonization of empty areas, which can be occupied or reoccupied over time.

The real challenge lies in determining whether metapopulation structure exists among individuals and, if so, what form it takes. Not surprisingly, these are the least understood aspects of coral reef fish ecology. Behavioral reef fish studies indicate that larvae are not inanimate particles, but rather have quite developed behavioral and sensory capabilities (Leis 2007; Paris et al. 2007; Putman 2016). These capabilities include active swimming of larvae, which can influence both the competence to recruit back to the original population and/or to another population (Figueiredo et al. 2013).

For Brazilian coral reefs, connectivity is particularly relevant because of its discontinuous distribution along the shelf (see Chap. 3). Hence, we intend to focus on these three different forms of connectivity, and how they uncover patterns and processes relevant for Brazilian coral communities. We also want to show how these concepts may in fact be related to each other and contribute to a better assessment of coral reefs, highlighting that this knowledge is essential in planning effective conservation strategies.

Box 7.1: Metapopulation

Metapopulation represents the assemblage of discrete local populations in the same region that may become connected by different processes (e.g. migration, colonization, and extinction), forming a regional group of connected populations of a species (Levins 1970). It results from complex processes at local and regional scales. Local populations may increase or decrease as environmental conditions fluctuate or when submitted to a stochastic event. These may affect their growth and establishment as a spatial arrangement of habitat patches at a regional scale. In fact, population persistence to local or regional extinction may depend on spatial-temporal changes in connectivity. Therefore, levels of connectivity among local populations defines the vulnerability to local extinction (Hanski and Gilpin 1991).

The metapopulation structure is determined by the spatial arrangement of local populations', considering the species traits (Hanski 1998; van Nouhuys 2016). Distinct local populations must be identifiable, and organisms generally need to remain within those subpopulations, but there must be some mechanism for interpopulation dispersal. Additionally, in a metapopulation, the dynamics of local populations are determined by local demography and self-recruitment. It is difficult to evaluate the local population dynamics by itself, since the dynamics can be modified by replenishment from external sources (nearby populations) (Fig. 7.1).

The importance of the metapopulation concept to conservation is that it identifies a group of structured populations that can persist over time, isolated by inadequate areas to settle, in addition to the recognition of areas that are source or sink of migrants.

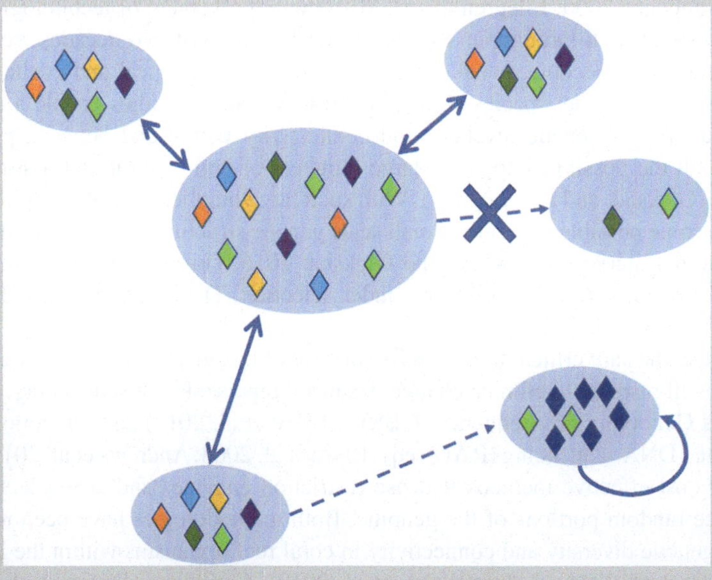

Fig. 7.1 Level of connection among populations in a system of habitat patches: connectivity may allow recovery after local extinction, while absence of connection results in increase of vulnerability and decrease of species diversity (island biogeography). Colored diamonds represent different species; arrows indicate typical mainland-island dispersal direction between populations (blue circles); dashed lines indicate weak connections directed to sink populations; cross indicate hard barrier

7.2 Gene Flow Between Reefs—Connectivity in the Evolutionary Timescale

In the last 50 years, the development of genetic tools has impacted many fields of the biological sciences, from basic unicellular functioning to ecosystem ecology. Following this development, an increasing number of studies aimed at unveiling aspects of organismic evolution using molecular tools, including marine organisms.

The first studies of population genetics using coral reef organisms in Brazil were conducted in the 1990s, using allozymes to evaluate the connectivity of sea anemones and sponges, rather than coral species, while working generally on a more regional scale (Russo et al. 1994; Klautau et al. 1999). With the turn of the century, and the availability of a plethora of molecular markers, the number of studies increased consistently, shifting the focus in the last decades to reef fish and scleractinian corals.

New studies have switched then from allozymes to the use of nuclear and mitochondrial loci (Santos et al. 2006; Neves et al. 2008; Nunes et al. 2011; de Souza et al. 2017), in addition to microsatellites (Peluso et al. 2018; Padua et al. 2018), providing more accuracy to unravel population genetic patterns (Bruford and Wayne 1993; Sunnucks 2000; Ellegren 2014). With the improvement of technologies and lower costs of high-throughput sequencing during the past two decades, genomic approaches have become a viable possibility to study ecological and evolutionary changes in non-model species (Ellegren 2014). Genomic studies enable a deeper understanding of genetic diversity and connectivity within and between populations, with the possibility to study large numbers of both neutral and non-neutral regions (Oleksiak and Rajora 2019). With such large numbers of polymorphic loci, it has become possible to identify small scale genetic structure, often not found with conventional markers (Crawford and Oleksiak 2016; Drury 2016), in addition to finding specific genes that might be under selection (Devlin-Durante and Baums 2017; Nielsen et al. 2020).

Such studies are critical to our understanding of the adaptation process and how species will respond to climate change. Reduced representation sequencing, which includes Genotyping-by-sequence (GBS) (Elshire et al. 2011) and restriction site-associated DNA sequencing (RAD-seq) (Baird et al. 2008; Andrews et al. 2016), are popular cost-effective methods that use restriction enzymes and size selection to sequence random portions of the genome. Both methodologies have been used to access genetic diversity and connectivity in coral reef organisms within the Pacific (Shinzato et al. 2016), Indian (Salas et al. 2019), Caribbean (Drury et al. 2016; Hammerman et al. 2018; Sturm et al. 2020), and Atlantic Oceans (Volk et al. 2021).

Within the Brazilian coast, the most comprehensive study focusing on scleractinian corals connectivity have used microsatellite markers, which are highly polymorphic markers, to evaluate genetic diversity and connectivity of the endemic species *Mussismilia hispida*, an important reef builder coral along the coast (Peluso et al. 2018). The study found five genetically differentiated populations along 3500 km of coastline, with gene flow being restricted mainly by ocean currents, and also raising evidence for the hypothesis that oceanic islands might have acted as refugia for corals during low sea-level periods in the Last Glacial Maximum (Leão et al. 2003; Pinheiro et al. 2017; Peluso et al. 2018). While similar patterns of gene flow breaks have been observed for other scleractinian species (e.g. *Favia gravida* and *Siderastrea radians*), this pattern is not a general rule. The scleractinians *Montastraea cavernosa* and *S. siderea*, and the hydrocoral *Millepora alcicornis* showed high genetic connectivity throughout their geographic range (Nunes et al. 2011; Souza et al. 2017), although these differences might actually be due to the mitochondrial and nuclear genetic markers with different levels of variability.

Discordances on connectivity patterns have also been observed for coral reef fishes, with species such as the yellowtail snapper *Ocyurus chrysurus* and the butterflyfish *Chaetodon striatus* displaying strong gene flow among a wide geographic range (da Silva et al. 2015; Liedke et al. 2020), while small cryptic fishes, such as the bridled goby *Coryphopterus glaucofraenum* and the rockpool blenny *Entomacrodus vomerinus* seem to be more affected by geographic and environmental barriers (Neves et al. 2016; Volk et al. 2021).

The most recurrent genetic pattern within the Brazilian coast is the differentiation between the Oceanic Islands (e.g., Fernando de Noronha and Rocas Atoll) and coastal regions, observed for both reef fishes and corals, and driven mainly by ocean currents and geographic distance (Nunes et al. 2011; Cunha et al. 2014; Neves et al. 2016; Peluso et al. 2018; Volk et al. 2021) (Fig. 7.2). Likewise, southern Brazil frequently harbors genetically distinct populations, which has been associated with the presence of a strong upwelling in the Cabo Frio region (22° S) that separates warm tropical reefs in the north from colder subtropical rocky reefs in the south, representing a region of ecological transition (Santos et al. 2006; Peluso et al. 2018; Volk et al. 2021). Additionally, ocean currents have been shown to prevent gene flow between the central and southern regions of Brazil (Volk et al. 2021). A third recurrent pattern is the isolation of the reef fish fauna from Saint Peter and Saint Paul Archipelago (0° N) a remote area in northern Brazil (Cunha et al. 2014; Peluso et al. 2018; Volk et al. 2021). Within the central region, different patterns of connectivity have been observed among corals and fishes (Neves et al. 2016; Peluso et al. 2018; Volk et al. 2021), which suggests the presence of semi-permeable barriers, such as the São Francisco River outflow. For instance, this outflow is the distribution break between the two species of hydrocorals (*M. brazilensis* and *M. nitida* (Souza et al. 2017)), while is not a barrier for larval dispersal of the coral *M. hispida* (Peluso et al. 2018).

Patterns of connectivity in Brazilian corals seem to be regulated by the following factors: geographical isolation, oceanic currents directing or restricting gene flow, the occurrence of upwelling likely limiting dispersal and settlement, freshwater discharges, fluctuations on the sea level, and the availability of substrate acting as a stepping-stone (Nunes et al. 2011; Neves et al. 2016; Peluso et al. 2018; Pinheiro et al. 2018). The main biogeographic barriers are shown in Fig. 7.2. However, a set of additional factors can influence dispersion and genetic connectivity of corals, fishes, and other invertebrates, including: differences in reproductive mode, despite not being a good predictor of dispersion (Guzmán et al. 2011; Peluso et al. 2018),

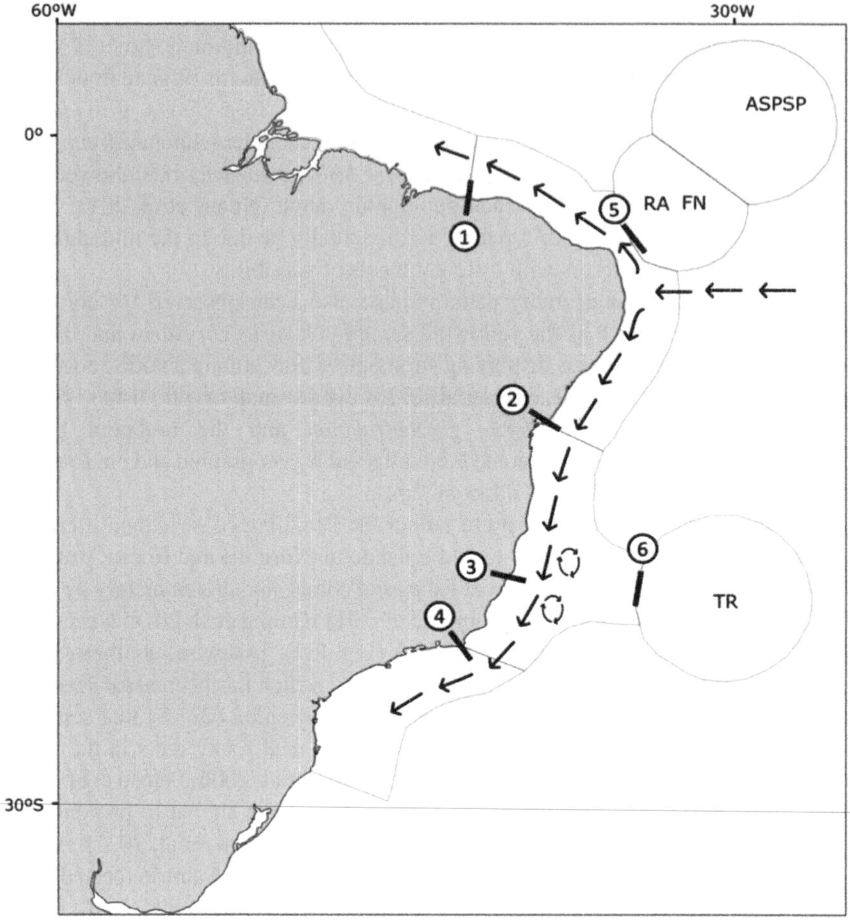

Fig. 7.2 Outline of the seven marine ecogeographic regions in Brazil according to Spalding et al. 2007 (see also Chap. 1 of this book). Arrows indicate the mean surface ocean current directions. Black bars indicate possible biogeographic barriers: (1) Amazon River; (2) São Francisco River; (3) Doce River; (4) strong upwelling; (5–6) long distances from available hard substrates

transport by rafting and artificial structures (Capel et al. 2019; Mantelatto et al. 2020), and adaptability to the new environment (Sexton et al. 2014).

Nevertheless, there is still scarce data on genetic connectivity of coral reef organisms along the Brazilian coast, and most studies have used only a few nuclear or mitochondrial loci. New approaches are now targeting a more comprehensive understanding of the ecological forces contributing to spatial patterns on population connectivity, combining genetic and environmental data in the recent field of seascape genomics (Selkoe et al. 2016). Future studies should focus on more sensitive markers, such as microsatellites or SNPs, and increase the number of target species to provide a fine scale analysis of the main patterns of connectivity within the Brazilian coast.

7.3 Biophysical Context of Demographic Connectivity in the Tropical South Atlantic

As previously mentioned, demographic connectivity involves the exchange of individuals between populations through the interaction of propagules (eggs and larvae) with the flow field. In most marine organisms, this process may involve a pelagic dispersal phase and ontogenetic migration of the juvenile phase to an adult habitat (Brown et al. 2016). As such, it tends to be highly variable both in space and time, making it very difficult to assess empirically. This is particularly true in the tropical South Atlantic where oceanic islands are connected with the north and northeast coast of Brazil (Peluso et al. 2018; Endo et al. 2019). Physical and biological models with different levels of complexity, from individual-based to end-to-end models, can be used not only to overcome practical limitations of empirical studies but also to test ecological hypotheses through *in silico* simulations (Box 7.2).

The use of models is justified by the fact that connectivity plays an important role in the maintenance and persistence of populations and their recovery from disturbance (Hughes et al. 2003; Green et al. 2015). In fact, model results suggest that persistence can be directly related to a certain threshold for the total amount of replacement through gain or loss of larvae over many generations, considering all possible paths (Botsford et al. 2009). There is also a more subtle, but not less important, consequence of connectivity that results from the influence of environmental conditions of the metapopulation on the mean phenotype of a subpopulation (Marshall and Morgan 2011). This implies that larvae moving from one site to another will likely have their post settlement fitness and survival reduced, compared to local recruits, especially along the nearshore-offshore gradient (Hamilton et al. 2008), as it is the case for the tropical South Atlantic (Endo et al. 2019).

Demographic connectivity in the tropical South Atlantic has been estimated using different methodological approaches, from single to multi-species, and from simple advective-diffusive to conservation planning models (e.g., Rudorff et al. 2009; Magris et al. 2016). A thorough review of the literature is beyond the scope of this chapter. Instead, we refer to the published research to synthesize what is known about biophysical modelling to estimate demographic connectivity in the region and point to some issues we consider deserve more attention in the future.

From a biophysical standpoint, the problem of demographic connectivity has been constrained by the survival of eggs and larvae, the complex nature of surface ocean circulation and the geographic distribution of reefs and other hard bottoms (e.g., rhodolith beds) habitats. In Brazil, these may form geographically isolated ecosystems in the oceanic islands, such as the São Pedro and São Paulo Archipelago (SPSP), Atol das Rocas (AR), Fernando de Noronha Archipelago (FN) and Trindade and Martim Vaz islands (TR). All of these are today important MPAs that ideally should form a network with other protected coastal reef habitats, including Parcel do Manuel Luis (ML), Recife dos Corais (RC), Costa dos Corais (CC), Abrolhos (AB), Arraial do Cabo and Cabo Frio (CF) (see Fig. 7.3). However, coastal MPAs are only some of the possible destinations of larvae from the oceanic islands, as they are part of, and behave as, an environmental continuum.

The ecological concept behind this idea has been summarized for large marine ecosystems (Sherman 1991), defined as large (>200,000 km^2) geographical regions sharing common submarine topography, productivity and trophically linked populations that are submitted to the same hydrographic regime. The idea of targeting large recruitment areas (LRAs) outside MPAs for conservation, as proposed by Endo et al. (2019) for Brazil, is supported by evidences suggesting that MPAs network designs should include other relevant recruitment areas and migration corridors (Green et al. 2015).

Results from biophysical simulations of dispersal using the grazing reef fish *Sparisoma* as a model species (Endo et al. 2019) indicate that austral summertime mortality rates are significantly higher than winter, but interannual variabilities are not significant. In this study. Mortality by lethal temperature varied for different sites, with hypothermia being more prevalent especially where upwelling of cooler waters occurs, such as in CF and SPSP. The southern branch of the South Equatorial Current (sSEC) is in its northernmost position (~10°S) during the austral spring/summer coincident with the southernmost position of the Intertropical Convergence Zone (ITCZ) and a positive wind stress curl (Rodrigues et al. 2007). This configuration induces the sSEC bifurcation to be positioned at lower latitudes and increases the southward water transport of BC carrying larvae from CC to AB (Fig. 7.3). During the winter this connection seems to be lost, as the sSEC bifurcation moves southwards and the BC transport around CC is reduced.

The low demographic connectivity among MPAs in the tropical South Atlantic, estimated by biophysical modelling, raises concern about the network fragility (Fig. 7.4). Particularly, the low level of local retention (CC and TR, only) suggests

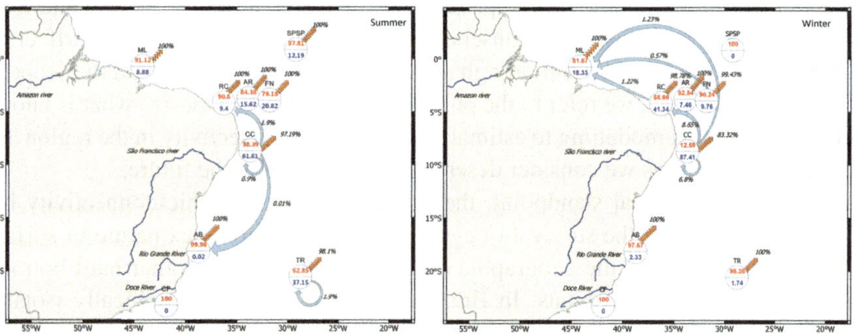

Fig. 7.3 Summary maps of larval mortality (red) and survival (green) per site for summer and winter. The orange arrows indicate the percentage of larvae that is exported from each area and does not recruit in any MPA. Blue arrows indicate the origin and destination sites of those larvae that actually recruit. Figures near blue arrows are total percentages of recruitment and local retention. *ML* Parcel do Manuel Luis, *RC* Recife dos Corais, *CC* Costa dos Corais, *SPSP* São Pedro and São Paulo Archipelago (SPSP), *AR* Atol das Rocas, *FN* Fernando de Noronha Archipelago, *TR* Trindade and Martim Vaz islands, *AB* Abrolhos, *CF* Cabo Frio

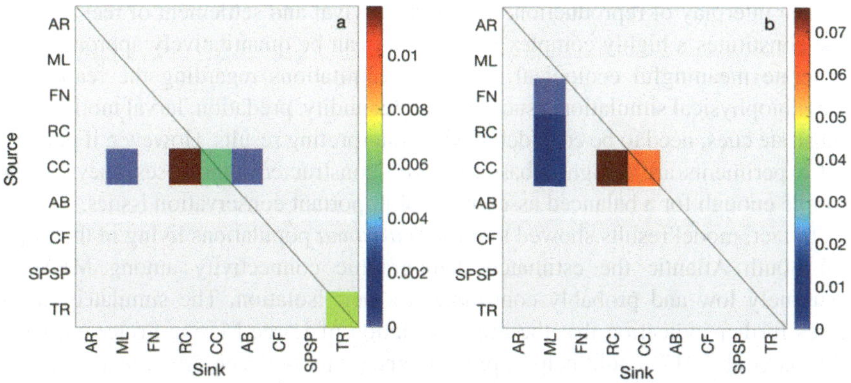

Fig. 7.4 Transition probability matrix for the summer (**a**) and winter (**b**) considering the five coastal areas and the four oceanic islands. The five coastal MPAs: Parcel do Manuel Luis (ML), Recife dos Corais (RC), Costa dos Corais (CC), Abrolhos (AB) and Arraial do Cabo and Cabo Frio (CF); and the four oceanic islands: São Pedro and São Paulo Archipelago (SPSP), Atol das Rocas (AR), Fernando de Noronha Archipelago (FN) and Trindade and Martim Vaz islands (TR). Note the different color scales for summer and winter (from Endo et al. 2019)

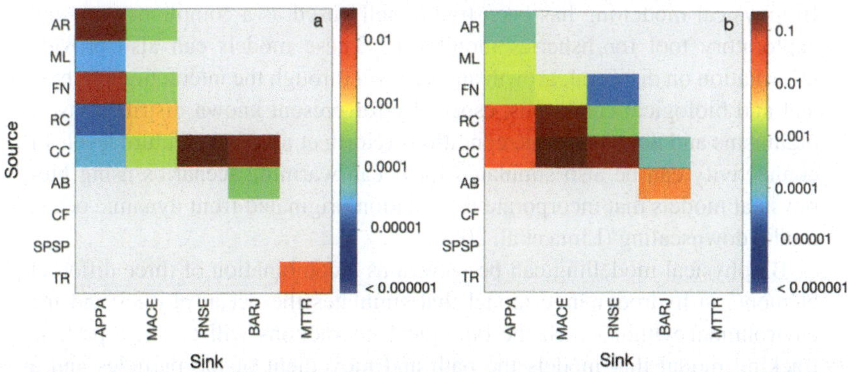

Fig. 7.5 Transition probability matrix for the summer (**a**) and winter (**b**) computed from simulations including non-protected coastal environments called Large Recruitment Areas (LRAs). Spawning (source) sites are the same as in Fig. 7.3. The LRAs are APPA, Amapá to Pará; MACE, Maranhão to Ceará; RNSE, Rio Grande do Norte to Sergipe; BARJ, Bahia to Rio de Janeiro and MTTR, Vitória-Trindade chain. Note the different color scales for summer and winter (from Endo et al. 2019)

that this important self-seeding process can be a limiting factor for long-term resilience of some MPAs. Despite this scenario of fragility, overall connectivity in the tropical South Atlantic increases if simulations consider Large Recruitment Areas (LRAs), that include non-protected coastal environments (Fig. 7.5). Under such conditions, it becomes evident that both CC and RC form an important hub connecting northern and southern MPAs. Also, CC and AB (further south) appear as important sources of recruitment for both subtropical and temperate environments.

The interplay of reproduction, transport, survival and settlement of reef fish larvae constitutes a highly complex process that can be quantitatively approached to generate meaningful ecological scenarios. Limitations regarding the realism of these biophysical simulations, such as adult fecundity, predation, larval motility and substrate cues, need to be considered when interpreting results. However, if numerical experiments are designed based on well-constructed hypotheses, they can be simple enough for a balanced assessment of important conservation issues.

In fact, model results showed that for *Sparisoma* populations living in the tropical South Atlantic the estimated demographic connectivity among MPAs is extremely low and probably contributes to their isolation. The simulated LRAs gives further indication that the implementation of Large Marine Protected Areas (Lewis et al. 2017) could help improve connectivity between offshore and coastal environments and strengthen the north-south connection along the tropical Brazilian coast. It is important to understand whether these patterns of connectivity hold true for simulations with other organisms, to determine if this is a general process.

Box 7.2: Biophysical Modelling

Biophysical modelling has been traditionally used as a complementary and exploratory tool for fisheries monitoring. These models can also provide information on dispersal, growth and survival through the interaction of physical and biological conditions, especially for present known distributions of organisms and oceanographic conditions (North et al. 2009). Future levels of connectivity can be also simulated for ocean warming scenarios using biophysical models that incorporate information originated from dynamic ocean model downscaling (Lima et al. 2021).

Biophysical modelling can be viewed as a combination of three different elements: a hydrodynamic model that simulates the ocean physics and the environment within which the biological interactions will occur; a particle tracking model that models the path and movement of the particles and a coupled individual based model (IBM) that incorporates as many biological traits, behavior and processes of the targeted species as possible.

The hydrodynamic model consists of a set of discrete equations that resolve variables as 3D velocities, turbulence, salinity, temperature, density among other variables in different pre-set spatio-temporal scales. The particle tracking model estimates the trajectories of the released particles, which requires that the hydrodynamical model is previously validated with observational measurements.

The IBM or the biological model may include a multitude of biological factors, from larval pelagic duration to growth rates and predation. The IBM is also set to consider processes over a certain spatio-temporal variation. From the biological perspective it can include processes evolving the reproduction process, such as spawning activity and egg production. It is also possible to include egg buoyancy and mortality. As the egg hatches and starts its development as a larva it is possible to include growth, dispersal, pelagic larval duration, feeding, horizontal and vertical swimming behavior, mortality, predation and many others.

Despite all complexity that can be included in biophysical models, the larvae can also be considered as passive particles that are only transported by the ocean currents (Treml et al. 2008). By only including the coupling between the hydrodynamical and the particle tracking model, it is still possible to emulate a few biological characteristics, such as spawning time, initial density and pelagic larval duration. However, the particles do not have any behavior and are transported passively by the main flow from one point to another. For example, the connectivity of spiny lobster in the Tropical Atlantic was studied using a simple advective-diffusive model using Lagrangian dispersion (Rudorff et al. 2009). By using an IBM more biological characteristics are considered, increasing the biological realism of the model and providing more details of the factors mediating mortality in the early life stages (Peck and Hufnagl 2012).

From spawning to larval dispersal and settlement in distant ecosystems there are a lot of smaller processes involved. Although the use of models helps to shed some light into these complex and unknown processes, the uncertainties associated with the different modeling approaches must be considered when using the results for management settings.

7.4 Reefs and Their Neighbors—Connectivity at the Scale of Individuals

7.4.1 Ecological Connectivity

The tropical coastal seascape comprises a mosaic of habitats such as mangroves, seagrass meadows, macroalgal beds, mudflats, rocky and biogenic reefs. Many species migrate between those habitats for various reasons. It is common especially among fish and crustaceans to utilize multiple habitats during their life, on varying temporal and spatial scales. This section will focus mainly on fish species because of their intrinsic higher mobility and because they are among the most studied group in Brazilian seascapes.

Migrations include foraging, tidal and ontogenetic migrations, where fish utilize separate habitats as juveniles and adults (Nagelkerken 2009). These movements happen on different temporal and spatial scales, such as short-term foraging migrations during a few hours, up to ontogenetic migrations, which may take years to complete (Pina-Amargós and González-Sansón 2009). Several species that spend their adult life stages on reefs utilize mangroves and seagrass meadows as foraging or nursery grounds (Nagelkerken et al. 2000). These movements result in transfer of energy, nutrients and carbon between habitats and is commonly referred to as *ecological connectivity* (Nagelkerken 2009). Multi-habitat seascapes allow species to take advantage of resources in several habitats and usually support higher fish abundance and diversity compared to less diverse seascapes (Olds et al. 2012). This has implications for management of coastal areas, since alterations or degradations of one habitat also affect adjacent habitats through these mobile links (Goodridge Gaines et al. 2020). Additionally, more than 40% of the reef fish species that use multiple habitats (e.g., mangroves, seagrass meadows, estuaries) are considered threatened in Brazil (Vila-Nova et al. 2011). Several species from both commercially and ecologically important families utilize non-reef habitats during their life. In Brazilian waters, species from the families Lutjanidae, Epinephelidae, Carangidae and Haemulidae are recorded to move between estuaries and mangroves to nearby reefs (Xavier et al. 2012; Chaves et al. 2013).

7.4.2 Main Coastal Habitats

Most research on connectivity has focused on ontogenetic migrations. In the Caribbean, mangroves and seagrasses serve as nursery and foraging habitats for a variety of species including different trophic groups and several keystone species, such as parrotfish (Ogden and Quinn 1984). In the Pacific, seagrass meadows seem to be primarily used as nurseries, while mangroves contain a less diverse juvenile fish assemblage (Igulu et al. 2014). Mangroves in Brazil are in general subjected to low salinity and their roots are usually colonized by oysters and brown algae (*Bostrichia* spp., Cordeiro-Marino et al. 1992) which provides little structural complexity when compared to sponges, for instance, and are not very palatable for herbivorous fish (Pereira et al. 2017).

Brazilian tropical reefs are patchily distributed along the tropical coast and formed by cemented terraces, fringing reefs, patch reefs and pinnacle reefs (Chap. 1). Bank and fringing reefs are the dominant features among Brazilian reefs and most of them have a tidal regime that affects the access to the leeward side of reefs (Chap. 1). Globally, the reef flat is a common feature of fringing reefs, where soft sediment and macroalgae beds flourish in waters sheltered by the reef formations. These calm habitats are used by several reef species as feeding grounds, functioning as large tidepools that may shelter many juvenile individuals, or even larger predators during high tide (Chaves et al. 2013; Harborne 2013).

Seagrass meadows are commonly cited in literature as important nursery and foraging areas for fish and crustaceans in the Pacific and the Caribbean (Boström et al. 2006; Huijbers et al. 2013) but they may play a less important role for fish along the Brazilian seascapes (Eggertsen et al. 2017). The seagrasses occurring in Brazil are small species (of the genera *Halodule*, *Halophila* and *Ruppia*) that provide limited structural complexity (Copertino et al. 2016). Large seagrass meadows are uncommon in Brazil and also underrepresented in coastal protected areas which makes them vulnerable to human pressure.

7.4.3 Types of Migrations in Brazilian Seascapes

7.4.3.1 Within-Habitat Movements

Movements within habitats, either ontogenetic or daily foraging migrations, have only been inferred for Brazilian reef systems based on indirect evidence, supported by field observations of behavior (feeding, mating, and/or agonistic). Few studies have used methods such as tagging techniques to directly quantify these migrations. Most of these techniques are expensive and time consuming (Box 7.3; Whoriskey and Hindell 2016) and, therefore, have rarely been applied in studies along Brazilian reef environments. However, passive acoustic telemetry has recently been conducted to study the movements of parrotfishes of the *Sparisoma* genus on reefs in Tamandaré (Pernambuco) (Giancalone et al. 2016) and Scarid parrotfishes and groupers (*Mycteroperca bonaci*) on reefs in Abrolhos (Bahia) (Hackradt et al. in prep.).

Ontogenetic migrations are the most cited movements inferred by size distribution of species. Cordeiro (2009) observed an increase in the average size of fishes from reef flats towards the fore reef zone in shallow coastal reefs of Paraíba state, especially for Acanthurids, *Chaetodon striatus* (Chaetodontidae), *Anisotremus virginicus* (Haemulidae), *Abudefduf saxatilis* (Pomacentridae) and *Haemulon parra* (Haemulidae) (Fig. 7.6). The same pattern was observed by Querino (2011) on other reefs in Paraíba where the mean size of Acanthurids, *Abudefduf saxatilis*, *Halichoeres poeyi* (Labride) and *L. jocu* (Lutjanidae) was larger on reefs than in macroalgal beds closer to the shore. Intertidal pools are also cited as temporary habitats for juveniles of several reef fish species who migrate to other areas of the reefs after growing to larger sizes (Dumaresq 2019), reinforcing the existence of ontogenetic changes in spatial distribution of reef fishes. However, all authors agree that it is not possible to discard the effects of different predation rates, fishing pressure and other anthropogenic stressors that usually are higher close to shore and may influence the observed

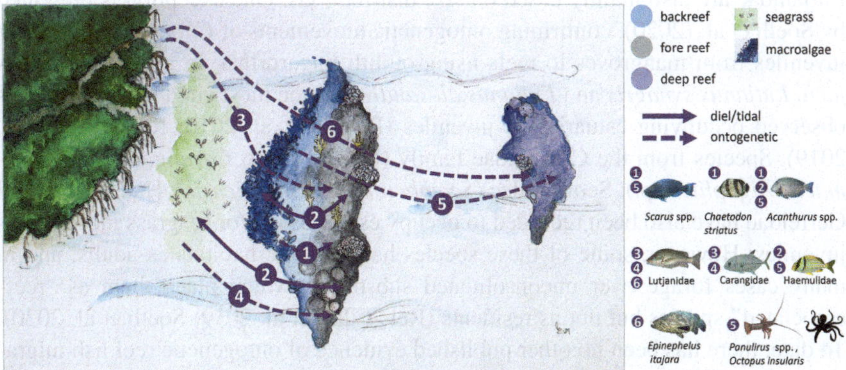

Fig. 7.6 Ecological connectivity between mangroves, seagrass meadows and reefs through migrations of some reef species

size distributions. Still, as the same pattern has been consistently observed on multiple reefs in various locations and studies, it is likely that juveniles from a number of species settle in more sheltered areas such as macroalgae and/or rhodolith beds, in the reef flat zone, later moving to deeper areas when they outgrow the benefits of their nursery (Costa et al. 2020).

7.4.3.2 Between-Habitat Movements: Fishes and Other Animals

Most of the tropical reefs along the Brazilian coast are influenced by coastal rivers with seasonal regimes (Chap. 1). Because of varying freshwater influxes, most estuaries are dominated by tides, with the higher salinities during high tide facilitating migration of transient marine species also to upper zones of estuaries. Such movements vary in time, either as ontogenetic migrations or daily foraging movements (Fig. 7.6). The first is better reported based on inferences of uneven spatial distribution of different size categories. The second, daily foraging migrations, has been observed and reported by fishermen but is poorly described in literature. Similar to what is reported for the Caribbean and the Pacific, grunts (Haemulidae), barracudas (Sphyraenidae) and snappers (Lutjanidae), as well as the goliath grouper (*Epinephelus itajara*) have been observed in estuary mouths close to reefs in Brazil (Ferreira et al. 2014), usually as juveniles. Contrastingly, in Brazil, juvenile parrotfish have been little recorded in non-reef habitats.

At the same time, there are differences in habitat use for the same species between biogeographic provinces. The yellow-tailed snapper (*Ocyurus chrysurus*), which in the Caribbean almost exclusively utilize *Thalassia* seagrass meadows as a nursery habitat, has not been recorded in shallow seagrass (*Halodule*) meadows in Brazil (Eggertsen et al. 2017). Differences in seascape configuration and habitat complexity may have shaped these patterns. There is still limited evidence for obligatory dependency of Brazilian reef fish species on seagrass meadows as a nursery habitat.

Inferences about ontogenetic migration of Carangidae, Sphyraenids and Lutjanidae are also mainly based on size distributions. One exception is presented by Soeth et al. (2020), confirming ontogenetic movements of *Chaetodipterus faber* juveniles from mangroves to reefs using otolith fingerprints. In general, *Lujtanus jocu*, *Lutjanus synagris* and *Lutjanus alexandrei* are the most cited reef fish species observed occupying estuaries as juveniles (Pereira et al. 2010; Reis-Filho et al. 2019). Species from the Carangidae family (*Caranx latus*, *Carangoides bartholomaei*, *Oligoplites* sp.), Scombridae (*Scomberomorus brasiliensis*), Haemulidae and Gerreidae have also been recorded to occupy estuaries and/or seagrass meadows as juveniles. However, some of these species have pelagic habitats as adults, and in many cases forage over unconsolidated substrates, which place them as "reef-associated" species but not as residents (Reis-Filho et al. 2019; Soeth et al. 2020). To date, there has been no other published evidence of ontogenetic reef fish migration using tracking or tagging methods.

Ontogenetic migrations to deeper reefs have been indicated by authors who recorded larger individuals of certain species exclusively at greater depths. Larger specimens of five Lutjanidae species (*L. analis*, *L. chrysurus*, *L. jocu*, *L. synagris* and *L. vivanus*) have consistently been detected in deeper reefs in the states of Ceará, Rio Grande do Norte, Pernambuco, Alagoas and Bahia (Frédou and Ferreira 2005), adding to the evidence of ontogenetic migrations occurring in this family.

Octopuses (*Octopus insularis*, Batista and Leite 2016) and lobsters (*Panulirus* genus, Cruz et al. 2020) have also been observed to attain larger sizes on deeper reefs while their juvenile stages are associated with shallow areas within reef seascapes. In general, large-bodied species possess higher mobility, increasing their odds to reach deep reefs further from the coast. However, it is almost impossible to distinguish the importance of anthropogenic pressure on these patterns, selecting for smaller specimens on shallower reefs and thus influencing size distributions (Harborne 2013).

Although evidence of ontogenetic movements exists, the mechanism underpinning these migrations are not completely explained. Fitness maximization, through pre-settlement preferences (Brown et al. 2016), predation pressure, growth maximization and reproductive behavior (Haywood and Kenyon 2009) are among the most accepted processes involved in ontogenetic changes in habitats.

Crustaceans such as lobsters, crabs and shrimps perform ontogenetic, reproduction and diel migrations either from reef to estuaries, or from estuaries to reefs in Caribbean and Pacific ecosystems (Krumme 2009). Megafauna such as manatees (*Trichechus manatus*, Paludo and Langguth 2002), sea turtles, sharks and rays also use multiple habitats including estuaries and reefs, due to reproduction, ontogenetic and foraging migrations. None of these species are obligatory associated with reefs, but their home range encompass a large seascape including reefs.

In general, we know little about ecological effects of ecological connectivity in Brazil, as well as implications for management. This is a vast field yet to explore in Brazilian tropical reefs.

Box 7.3: Tracking Movement

Most tagging techniques consist of attaching an identifiable mark to individuals of a target species and then recapture or observations the tagged individuals to analyze their movements in space and time. The most popular techniques involve acoustic tags that emit signals detectable either passively, by stationary hydrophones, or actively, by a human operated hydrophone (Hussey et al. 2015), or with a hydrophone mounted on an underwater autonomous vehicle (White et al. 2016). Global Positioning System (GPS) tags use a similar principle, but utilize orbital satellites as receivers that transmit the captured information to a land-based center. The choice of tagging method will depend on the main objectives of the study (e.g., diel activity, home range, long distance movements) and the available budget.

Acoustic tagging has been widely applied to reef fishes (Pittman et al. 2014), but also to sea turtles (Chevis et al. 2017), marine mammals (Lydersen et al. 2002) and invertebrates (e.g., jellyfish—Mooney et al. 2015). Additional information can also be obtained from sensors included in some tags that register and transmit environmental information (e.g., depth, salinity, temperature, dissolved oxygen), the animal's 3D acceleration or physiological state (e.g., heart rate, temperature) (Cooke et al. 2004; Nassar et al. 2018). The movement information is then analyzed to extrapolate the path or area used by the studied animals based on the frequency of detections and their coordinates (Calenge et al. 2009). Information such as site fidelity, energy expenditure, feeding activity and reproduction among others can also be obtained from the combination of detections and complementary data from associated sensors and visual observations, when present.

7.5 Connectivity in the Conservation Context: The Missing Parts in the Puzzle

At this point, it should be clear that all types of connectivity discussed in this chapter are relevant to coral reef conservation. However, the conservation strategy continuously evolves, resulting in a temporal mosaic of initiatives conceived under different sets of paradigms in a timeline. For instance, the first reef areas included in MPAs were created in 1979 for Rocas Atoll biological reserve. By then, the singularity of the target areas was the most important reason to propose the creation of a conservation unit. As a more complete picture of the biology of most marine species emerged with the use of a wide array of technological advancements, stemming from satellite imagery to high throughput DNA sequencing, paradigms have shifted to include connectivity as one of the key processes to be considered (D'Aloia et al. 2017; Virtanen et al. 2020).

Even though connectivity is considered an important aspect for marine spatial planning in conservation, it still focuses on a few target species, lacking a true integrative or multi-specific approach (Magris et al. 2016, 2018). Additionally, conservation demands from multilateral agreements have also influenced not only the location of new MPAs, but also their size and level of legal protection (Giglio et al. 2018). Hence, despite the technical and academic discussions and proposed guidelines, the outcomes of conservation actions is the result of complex negotiations, frequently asymmetrical in terms of influence.

Within the scope of Brazilian Coral Reefs, there is a consistent body of knowledge that has accumulated during the past decade. There is a scientific consensus about the importance of connectivity processes for the design of marine protected areas (MPA) networks, especially when the replenishment of local populations become dependent on nearby habitats (Planes et al. 2009). It has been advocated that MPAs should ensure the persistence of metapopulations by conserving the mechanisms that drive adult replacement and the survival rate at very low levels of recruitment in local subpopulations (Sale et al. 2010; Burgess et al. 2014; Mcleod et al. 2019; Roberts et al. 2021). This usually involves taking decisions about the size, spacing and location of individual MPAs to maximize connectivity and conservation benefits (Green et al. 2015). Factors, such as larval dispersal and self-recruitment (as the proportion of locally produced offspring that remain in the same population), species home range, ontogenetic habitat shifts, and resource availability should influence the choice of MPA's size.

The effective protection of habitats and management of fishery reserves, based on MPAs networks, can be particularly sensitive to their spatial distribution and sizes. For reef fish larvae that settle close to their parents and or have high rates of self-recruitment, their connectivity via the exchange of individuals may be limited to short distances, demanding a closely spaced network. Obviously, optimum MPA spacing depends on the availability of suitable habitats for recruitment, especially for species that undergo ontogenetic habitat shifts or spawning migrations. The general principle, however, that guides the definition of MPAs to sustain fisheries yields states that harvested fish species should be able to persist inside MPAs and sustain yield outside through spillover of larvae and adults (Burgess et al. 2014). Despite the mounting scientific evidence highlighting the role of larval dispersal to connectivity and, consequently, to MPA effectiveness, it remains a challenge to measure it in the field and it is a major source of uncertainty (for a review, see Burgess et al. 2014).

Within the Brazilian context, Ferreira et al. (2022) showed that connectivity was indeed a key factor for MPAs effectiveness, regardless of their fully or partially protected status. It seems now obvious or even intuitive to think of connectivity as an integral part of conservation spatial planning, however, even in recent years, this was not at the core of the Brazilian strategy. For instance, the Environment Ministry has implemented and regularly updated the designation of areas prioritized for conservation actions. The methods followed explicitly the systematic conservation planning strategy, proposed by Margules and Pressey (2000). Connectivity was considered as a process that may be targeted for conservation however, it was not a core property of the areas to be prioritized. Hence, the maps of areas and proposed actions rarely, if at all, take connectivity into account (Environment Ministry Ordinance 463, December 18th 2018).

Resulting from these combined processes, the country has a mosaic of MPAs and conservation actions that, combined with the lack of resources, has produced poor results. The creation of two large MPAs around sensitive oceanic islands that exclude their most critical habitats clearly illustrates this problem (Giglio et al. 2018). Brazilian coral reefs are biodiversity treasures that are deeply connected to virtually all marine and coastal biomes. Therefore, connectivity must be further studied and fully incorporated into conservation strategies and methods.

7.6 Final Remarks: The Way Forward

Connectivity is a cornerstone for coral reef biology and conservation (Magris et al. 2016) because local population persistence can be strongly dependent on external subsidies via larval supply (network persistence). From the basic biological knowledge to conservation and resources management, ignoring these processes certainly produce an incomplete picture, many times resulting in ineffective decisions.

Considering what is known for Brazilian reef organisms, there are still many blanks to fill. Accurate data and reliable biophysical models, genetic population structure and assessments of migration and individual movements of key species are essential to advance the knowledge on Brazilian coral reefs. Beyond the need for basic information about the evolutionary history and biology of reef organisms, the maintenance of essential ecosystems services depends on this knowledge. So, we can identify some relevant questions that need to be addressed to turn the conservation of reef ecosystems in the South Atlantic Basin more effective:

(1) Do current biophysical models explain adequately the genetic structure of keystone species in reef ecosystems?
(2) How connectivity regulates/modulates the populational structure of keystone species?
(3) Is the functional structure of reef ecosystems regulated or modulated by connectivity processes?
(4) How connectivity processes are affected by climate changes and what are the projected consequences for the resilience and conservation of reef ecosystems?
(5) Will climate change impacts on connectivity compromise the effectiveness of current MPAs?

It is beyond the scope of this chapter to tackle on all aspects of conservation strategies, but one must now and again claim for more effective plans, considering not only connectivity issues but all the aspects indicated by the large body of work continuously produced.

References

Andrews KR, Good JM, Miller MR et al (2016) Harnessing the power of RADseq for ecological and evolutionary genomics. Nat Rev Genet 17:81–92. https://doi.org/10.1038/nrg.2015.28

Baird NA, Etter PD, Atwood TS et al (2008) Rapid SNP discovery and genetic mapping using sequenced RAD markers. PLoS One 3:e3376. https://doi.org/10.1371/journal.pone.0003376

Batista AT, Leite TS (2016) Octopus insularis (Cephalopoda: Octopodidae) on the tropical coast of Brazil: where it lives and what it eats. Braz J Oceanogr 64:353–364. https://doi.org/10.1590/s1679-87592016123406404

Bauer S, Swenson MS, Griffa A et al (1998) Eddy-mean flow decomposition and eddy-diffusivity estimates in the tropical Pacific Ocean: 1. Methodology 103:30855–30871. https://doi.org/10.1029/1998JC900009

Boström C, Jackson EL, Simenstad CA (2006) Seagrass landscapes and their effects on associated fauna: a review. Estuar Coast Shelf Sci 68:383–403. https://doi.org/10.1016/j.ecss.2006.01.026

Botsford LW, White JW, Coffroth M-A et al (2009) Connectivity and resilience of coral reef metapopulations in marine protected areas: matching empirical efforts to predictive needs. Coral Reefs 28:327–337. https://doi.org/10.1007/s00338-009-0466-z

Brown CJ, Harborne AR, Paris CB, Mumby PJ (2016) Uniting paradigms of connectivity in marine ecology. Ecology 97:2447–2457. https://doi.org/10.1002/ecy.1463

Bruford MW, Wayne RK (1993) Microsatellites and their application to population genetic studies. Curr Opin Genet Dev 3:939–943. https://doi.org/10.1016/0959-437X(93)90017-J

Burgess SC, Nickols KJ, Griesemer CD et al (2014) Beyond connectivity: how empirical methods can quantify population persistence to improve marine protected-area design. Ecol Appl 24:257–270. https://doi.org/10.1890/13-0710.1

Calenge C, Dray S, Royer-Carenzi M (2009) The concept of animals' trajectories from a data analysis perspective. Eco Inform 4:34–41. https://doi.org/10.1016/J.ECOINF.2008.10.002

Capel KCC, Creed J, Kitahara MV et al (2019) Multiple introductions and secondary dispersion of Tubastraea spp. in the Southwestern Atlantic. Sci Rep 9:13978. https://doi.org/10.1038/s41598-019-50442-3

Chaves LTC, Pereira PHC, Feitosa JLL (2013) Coral reef fish association with macroalgal beds on a tropical reef system in North-Eastern Brazil. Mar Freshw Res 64:1101. https://doi.org/10.1071/MF13054

Chevis M, Godley B, Lewis J et al (2017) Movement patterns of juvenile hawksbill turtles Eretmochelys imbricata at a Caribbean coral atoll: long-term tracking using passive acoustic telemetry. Endanger Species Res 32:309–319. https://doi.org/10.3354/esr00812

Cooke SJ, Hinch SG, Wikelski M et al (2004) Biotelemetry: a mechanistic approach to ecology. Trends Ecol Evol 19:334–343. https://doi.org/10.1016/J.TREE.2004.04.003

Copertino MS, Creed JC, Lanari MO et al (2016) Seagrass and submerged aquatic vegetation (VAS) habitats off the coast of Brazil: state of knowledge, conservation and main threats. Braz J Oceanogr 64:53–80. https://doi.org/10.1590/S1679-875920161036064sp2

Cordeiro CAMM (2009) Estrutura da comunidade de peixes recifais do litoral sul da Paraíba

Cordeiro-Marino M, Braga MRA, Eston VR et al (1992) Mangrove macroalgal communities of Latin America: the state of art and perspectives. In: Seelinger U (ed) Coastal plant communities of Latin America. Elsevier, pp 51–64

Costa ACP, Garcia TM, Paiva BP et al (2020) Seagrass and rhodolith beds are important seascapes for the development of fish eggs and larvae in tropical coastal areas. Mar Environ Res 161:105064. https://doi.org/10.1016/j.marenvres.2020.105064

Cowen RK, Sponaugle S (2009) Larval dispersal and marine population connectivity. Annu Rev Mar Sci 1:443–466. https://doi.org/10.1146/annurev.marine.010908.163757

Crawford DL, Oleksiak MF (2016) Ecological population genomics in the marine environment. Brief Funct Genomics 15:342–351. https://doi.org/10.1093/bfgp/elw008

Cruz R, Santana JVM, Barreto CG et al (2020) Towards the rebuilding of spiny lobster stocks in Brazil: a review. Crustaceana 93:957–983. https://doi.org/10.1163/15685403-bja10073

Cunha IMC da, Souza AS de, Dias EA, et al (2014) Genetic multipartitions based on D-loop sequences and chromosomal patterns in Brown Chromis, *Chromis multilineata* (Pomacentridae), in the Western Atlantic. Biomed Res Int 2014:1–11. https://doi.org/10.1155/2014/254698

D'Aloia CC, Daigle RM, Côté IM et al (2017) A multiple-species framework for integrating movement processes across life stages into the design of marine protected areas. Biol Conserv 216:93–100. https://doi.org/10.1016/j.biocon.2017.10.012

da Silva R, Veneza I, Sampaio I et al (2015) High levels of genetic connectivity among populations of yellowtail snapper, Ocyurus chrysurus (Lutjanidae—Perciformes), in the Western South Atlantic revealed through multilocus analysis. PLoS One 10:e0122173. https://doi.org/10.1371/journal.pone.0122173

de Souza JN, Nunes FLD, Zilberberg C et al (2017) Contrasting patterns of connectivity among endemic and widespread fire coral species (Millepora spp.) in the tropical Southwestern Atlantic. Coral Reefs 36:701–716. https://doi.org/10.1007/s00338-017-1562-0

Devlin-Durante MK, Baums IB (2017) Genome-wide survey of single-nucleotide polymorphisms reveals fine-scale population structure and signs of selection in the threatened Caribbean Elkhorn coral, *Acropora palmata*. PeerJ 5:e4077. https://doi.org/10.7717/peerj.4077

Drury C, Dale KE, Panlilio JM et al (2016) Genomic variation among populations of threatened coral: Acropora cervicornis. BMC Genomics 17:286. https://doi.org/10.1186/s12864-016-2583-8

Dumaresq GM (2019) Assembleias de peixes recifais em poças de maré tropicais: elementos estruturadores e mudanças em sua composição (1999 e 2018)

Eggertsen L, Ferreira CEL, Fontoura L et al (2017) Seaweed beds support more juvenile reef fish than seagrass beds in a South-Western Atlantic tropical seascape. Estuar Coast Shelf Sci 196:97–108. https://doi.org/10.1016/j.ecss.2017.06.041

Ellegren H (2014) Genome sequencing and population genomics in non-model organisms. Trends Ecol Evol 29:51–63. https://doi.org/10.1016/J.TREE.2013.09.008

Elshire RJ, Glaubitz JC, Sun Q et al (2011) A robust, simple genotyping-by-sequencing (GBS) approach for high diversity species. PLoS One 6:e19379. https://doi.org/10.1371/journal.pone.0019379

Endo CAK, Gherardi DFM, Pezzi LP, Lima LN (2019) Low connectivity compromises the conservation of reef fishes by marine protected areas in the tropical South Atlantic. Sci Rep 9. https://doi.org/10.1038/s41598-019-45042-0

Ferreira HM, Reuss-Strenzel GM, Alves JA, Schiavetti A (2014) Local ecological knowledge of the artisanal fishers on Epinephelus itajara (Lichtenstein, 1822) (Teleostei: Epinephelidae) on Ilhéus coast—Bahia State, Brazil. J Ethnobiol Ethnomed 10:51. https://doi.org/10.1186/1746-4269-10-51

Ferreira HM, Magris RA, Floeter SR, Ferreira CEL (2022) Drivers of ecological effectiveness of marine protected areas: a meta-analytic approach from the Southwestern Atlantic Ocean (Brazil). J Environ Manag 301:113889. https://doi.org/10.1016/j.jenvman.2021.113889

Figueiredo J, Baird AH, Connolly SR (2013) Synthesizing larval competence dynamics and reef-scale retention reveals a high potential for self-recruitment in corals. Ecology 94:650–659. https://doi.org/10.1890/12-0767.1

Frédou T, Ferreira BP (2005) Bathymetric trends of northeastern Brazilian snappers (Pisces, Lutjanidae): implications for the reef fishery dynamic. Braz Arch Biol Technol 48:787–800. https://doi.org/10.1590/S1516-89132005000600015

Giancalone VM, Simon T, Ferreira BP, et al (2016) A pilot study on movement patterns of Brazilian reef fish using acoustic telemetry

Giglio VJ, Pinheiro HT, Bender MG et al (2018) Large and remote marine protected areas in the South Atlantic Ocean are flawed and raise concerns: comments on Soares and Lucas (2018). Mar Policy 96:13–17. https://doi.org/10.1016/j.marpol.2018.07.017

Gleason DF, Hofmann DK (2011) Coral larvae: from gametes to recruits. J Exp Mar Biol Ecol 408:42–57. https://doi.org/10.1016/J.JEMBE.2011.07.025

Goodridge Gaines LA, Olds AD, Henderson CJ et al (2020) Linking ecosystem condition and landscape context in the conservation of ecosystem multifunctionality. Biol Conserv 243:108479. https://doi.org/10.1016/j.biocon.2020.108479

Green AL, Maypa AP, Almany GR et al (2015) Larval dispersal and movement patterns of coral reef fishes, and implications for marine reserve network design. Biol Rev 90:1215–1247. https://doi.org/10.1111/brv.12155

Guzmán BE, Nuñes JJ, Vejar A et al (2011) Genetic diversity and population structure of two South American marine gastropods, Crepipatella dilatata and C. fecunda (Gastropoda : Calyptraeidae): distinct patterns based on developmental mode. Ita J Zoo 78:444–454

Hamilton SL, Regetz J, Warner RR (2008) Postsettlement survival linked to larval life in a marine fish. Proc Natl Acad Sci 105:1561–1566. https://doi.org/10.1073/pnas.0707676105

Hammerman NM, Rivera-Vicens RE, Galaska MP et al (2018) Population connectivity of the plating coral Agaricia lamarcki from Southwest Puerto Rico. Coral Reefs 37:183–191. https://doi.org/10.1007/s00338-017-1646-x

Hanski I (1998) Metapopulation dynamics. Nature 396:41–49. https://doi.org/10.1038/23876

Hanski I, Gilpin M (1991) Metapopulation dynamics: brief history and conceptual domain. Biol J Linn Soc 42:3–16. https://doi.org/10.1111/j.1095-8312.1991.tb00548.x

Harborne AR (2013) The ecology, behaviour and physiology of fishes on coral reef flats, and the potential impacts of climate change. J Fish Biol 83:417–447. https://doi.org/10.1111/jfb.12203

Harwood TD (2009) The circular definition of populations and its implications for biological sampling. Mol Ecol 18:765–768. https://doi.org/10.1111/j.1365-294X.2008.04066.x

Haywood MDE, Kenyon RA (2009) Habitat shifts by decapods—an example of connectivity across tropical coastal ecosystems. In: Nagelkerken I (ed) Ecological connectivity among tropical coastal ecosystems. Springer, Dordrecht, pp 229–269

Hughes TP, Baird AH, Bellwood DR et al (2003) Climate change, human impacts, and the resilience of coral reefs. Science 301:929–934. https://doi.org/10.1126/science.1085046

Huijbers CM, Nagelkerken I, Debrot AO, Jongejans E (2013) Geographic coupling of juvenile and adult habitat shapes spatial population dynamics of a coral reef fish. Ecology 94:1859–1870. https://doi.org/10.1890/11-1759.1

Hussey NE, Kessel ST, Aarestrup K et al (2015) Aquatic animal telemetry: a panoramic window into the underwater world. Science (New York, NY) 348:1255642. https://doi.org/10.1126/science.1255642

Igulu MM, Nagelkerken I, Dorenbosch M et al (2014) Mangrove habitat use by juvenile reef fish: meta-analysis reveals that tidal regime matters more than biogeographic region. PLoS One 9:e114715. https://doi.org/10.1371/journal.pone.0114715

Klautau M, Russo CAM, Lazoski C et al (1999) Does cosmopolitanism result from Overconservative systematics? A case study using the marine sponge Chondrilla nucula. Evolution 53:1414–1422

Krumme U (2009) Diel and tidal movements by fish and decapods linking tropical coastal ecosystems. In: Nagelkerken I (ed) Ecological connectivity among tropical coastal ecosystems. Springer, Dordrecht, pp 271–324

Leão ZMAN, Kikuchi RKP, Testa V (2003) Corals and coral reefs of Brazil. Latin American Coral Reefs 9–52. https://doi.org/10.1016/B978-044451388-5/50003-5

Leis J (2007) Behaviour as input for modelling dispersal of fish larvae: behaviour, biogeography, hydrodynamics, ontogeny, physiology and phylogeny meet hydrography. Mar Ecol Prog Ser 347:185–193. https://doi.org/10.3354/meps06977

Levins R (1970) Extinction. In: Gerstenhaber M (ed) Some mathematical problems in biology. American Mathematical Society, Providence, RI, pp 77–107

Lewis N, Day J, Aulani W et al (2017) Large-scale marine protected areas: guidelines for design and management. IUCN, International Union for Conservation of Nature

Liedke AMR, Pinheiro HT, Floeter SR, Bernardi G (2020) Phylogeography of the banded butterflyfish, Chaetodon striatus, indicates high connectivity between biogeographic provinces and ecosystems in the western Atlantic. Neotrop Ichthyol 18. https://doi.org/10.1590/1982-0224-2019-0054

Lima LS, Gherardi DFM, Pezzi LP et al (2021) Potential changes in the connectivity of marine protected areas driven by extreme ocean warming. Sci Rep 11. https://doi.org/10.1038/s41598-021-89192-6

Lowe WH, Allendorf FW (2010) What can genetics tell us about population connectivity? Mol Ecol 19:3038–3051. https://doi.org/10.1111/j.1365-294X.2010.04688.x

Lydersen C, Nøst OA, Lovell P et al (2002) Salinity and temperature structure of a freezing Arctic fjord-monitored by white whales (*Delphinapterus leucas*). Geophys Res Lett 29:34-1–34-4. https://doi.org/10.1029/2002GL015462

Magris RA, Treml EA, Pressey RL, Weeks R (2016) Integrating multiple species connectivity and habitat quality into conservation planning for coral reefs. Ecography 39:649–664. https://doi.org/10.1111/ecog.01507

Magris RA, Andrello M, Pressey RL et al (2018) Biologically representative and well-connected marine reserves enhance biodiversity persistence in conservation planning. Conserv Lett 11

Mantelatto MC, Póvoa AA, Skinner LF et al (2020) Marine litter and wood debris as habitat and vector for the range expansion of invasive corals (Tubastraea spp.). Mar Pollut Bull 160:111659. https://doi.org/10.1016/J.MARPOLBUL.2020.111659

Margules CR, Pressey RL (2000) Systematic conservation planning. Nature 405:243–253. https://doi.org/10.1038/35012251

Marshall DJ, Morgan SG (2011) Ecological and evolutionary consequences of linked life-history stages in the sea. Curr Biol 21

Mcleod E, Anthony KRN, Mumby PJ et al (2019) The future of resilience-based management in coral reef ecosystems. J Environ Manag 233:291–301. https://doi.org/10.1016/J.JENVMAN.2018.11.034

Mooney TA, Katija K, Shorter KA et al (2015) ITAG: an eco-sensor for fine-scale behavioral measurements of soft-bodied marine invertebrates. Anim Biotelem 3:31. https://doi.org/10.1186/s40317-015-0076-1

Mora C, Sale PF (2002) Are populations of coral reef fish open or closed? Trends Ecol Evol 17:422–428. https://doi.org/10.1016/S0169-5347(02)02584-3

Nagelkerken I (ed) (2009) Ecological connectivity among tropical coastal ecosystems. Springer, Dordrecht

Nagelkerken I, Dorenbosch M, Verberk W et al (2000) Importance of shallow-water biotopes of a Caribbean bay for juvenile coral reef fishes: patterns in biotope association, community structure and spatial distribution. Mar Ecol Prog Ser 202:175–192. https://doi.org/10.3354/meps202175

Nassar JM, Khan SM, Velling SJ et al (2018) Compliant lightweight non-invasive standalone "marine skin" tagging system. npj flexible. Electronics 2:13. https://doi.org/10.1038/s41528-018-0025-1

Neves EG, Andrade SCS, da Silveira FL, Solferini VN (2008) Genetic variation and population structuring in two brooding coral species (Siderastrea stellata and Siderastrea radians) from Brazil. Genetica 132:243–254. https://doi.org/10.1007/s10709-007-9168-z

Neves JMM, Lima SMQ, Mendes LF et al (2016) Population structure of the Rockpool Blenny Entomacrodus vomerinus shows source-sink dynamics among ecoregions in the tropical Southwestern Atlantic. PLoS One 11:e0157472. https://doi.org/10.1371/journal.pone.0157472

Nielsen ES, Henriques R, Beger M et al (2020) Multi-model seascape genomics identifies distinct environmental drivers of selection among sympatric marine species. BMC Evol Biol 20:121. https://doi.org/10.1186/s12862-020-01679-4

North EW, Gallego A, Petitgas P, et al (2009) ICES Cooperative Research Report Rapport des Recherches Collectives Manual of recommended practices for modelling physical-biological interactions during fish early life International Council for the Exploration of the Sea Conseil International pour l'Exploration de la Mer

Nunes FLD, Norris RD, Knowlton N (2011) Long distance dispersal and connectivity in Amphi-Atlantic corals at regional and basin scales. PLoS One 6. https://doi.org/10.1371/journal.pone.0022298

Ogden JC, Quinn TP (1984) Migration in coral reef fishes: ecological significance and orientation mechanisms. In: McCleave JD, Arnold JP, Dodson JJ, Neill WH (eds) Mechanisms of migration in fishes. Springer, Boston, pp 293–308

Olds A, Connolly R, Pitt K, Maxwell P (2012) Primacy of seascape connectivity effects in structuring coral reef fish assemblages. Mar Ecol Prog Ser 462:191–203. https://doi.org/10.3354/meps09849

Oleksiak MF, Rajora OP (2019) Marine population genomics: challenges and opportunities. Springer, Cham, pp 3–35

Padua A, Cunha HA, Klautau M (2018) Gene flow and differentiation in a native calcareous sponge (Porifera) with unknown dispersal phase. Mar Biodivers 48:2125–2135. https://doi.org/10.1007/s12526-017-0742-z

Paludo D, Langguth A (2002) Use of space and temporal distribution of Trichechus manatus manatus Linnaeus in the region of Sagi, Rio Grande do Norte State, Brazil (Sirenia, Trichechidae). Revista Brasileira de Zoologia 19:205–215. https://doi.org/10.1590/S0101-81752002000100019

Paris C, Chérubin L, Cowen R (2007) Surfing, spinning, or diving from reef to reef: effects on population connectivity. Mar Ecol Prog Ser 347:285–300. https://doi.org/10.3354/meps06985

Peck MA, Hufnagl M (2012) Can IBMs tell us why most larvae die in the sea? Model sensitivities and scenarios reveal research needs. J Mar Syst 93:77–93. https://doi.org/10.1016/J.JMARSYS.2011.08.005

Peluso L, Tascheri V, Nunes FLD et al (2018) Contemporary and historical oceanographic processes explain genetic connectivity in a Southwestern Atlantic coral. Sci Rep 8:1–12. https://doi.org/10.1038/s41598-018-21010-y

Pereira PHC, Ferreira BP, Rezende SM (2010) Community structure of the ichthyofauna associated with seagrass beds (Halodule wrightii) in Formoso River estuary—Pernambuco, Brazil. An Acad Bras Cienc 82:617–628. https://doi.org/10.1590/S0001-37652010000300009

Pereira PHC, dos Santos MVB, Lippi DL et al (2017) Difference in the trophic structure of fish communities between artificial and natural habitats in a tropical estuary. Mar Freshw Res 68:473. https://doi.org/10.1071/MF15326

Pina-Amargós F, González-Sansón G (2009) Movement patterns of goliath grouper Epinephelus itajara around Southeast Cuba: implications for conservation. Endanger Species Res 7:243–247. https://doi.org/10.3354/esr00192

Pinheiro HT, Bernardi G, Simon T et al (2017) Island biogeography of marine organisms. Nature 549:82–85. https://doi.org/10.1038/nature23680

Pinheiro HT, Rocha LA, Macieira RM et al (2018) South-Western Atlantic reef fishes: zoogeographical patterns and ecological drivers reveal a secondary biodiversity Centre in the Atlantic Ocean. Divers Distrib 24:951–965. https://doi.org/10.1111/ddi.12729

Pires D d O, Castro CB, Segal B et al (2016) Reprodução de corais de águas rasas do Brasil. In: Zilberberg C, Abrantes DP, Marques JA et al (eds) Conhecendo os Recifes Brasileiros, 1st edn. Museu Nacional UFRJ, Rio de Janeiro, pp 111–128

Pittman SJ, Monaco ME, Friedlander AM et al (2014) Fish with chips: tracking reef fish movements to evaluate size and connectivity of Caribbean marine protected areas. PLoS One 9:e96028. https://doi.org/10.1371/journal.pone.0096028

Planes S, Jones GP, Thorrold SR (2009) Larval dispersal connects fish populations in a network of marine protected areas. Proc Natl Acad Sci USA 106:5693–5697. https://doi.org/10.1073/pnas.0808007106

Putman NF (2016) An ecological perspective on the migrations of marine fishes. Environ Biol Fish 99:801–804. https://doi.org/10.1007/s10641-016-0512-y

Querino LAC (2011) Composição e estrutura da comunidade de peixes recifais do Parque Estadual Marinho Areia Vermelha. Cabedelo, PB

Reis-Filho JA, Schmid K, Harvey ES, Giarrizzo T (2019) Coastal fish assemblages reflect marine habitat connectivity and ontogenetic shifts in an estuary-bay-continental shelf gradient. Mar Environ Res 148:57–66. https://doi.org/10.1016/J.MARENVRES.2019.05.004

Rodrigues RR, Rothstein LM, Wimbush M (2007) Seasonal variability of the South Equatorial Current bifurcation in the Atlantic ocean: a numerical study. J Phys Oceanogr 37:16–30

Roberts KE, Cook CN, Beher J, Treml EA (2021) Assessing the current state of ecological connectivity in a large marine protected area system. Conserv Biol 35:699–710. https://doi.org/10.1111/cobi.13580

Rudorff CAG, Lorenzzetti JA, Gherardi DFM, Lins-Oliveira JE (2009) Modeling spiny lobster larval dispersion in the tropical Atlantic. Fish Res 96:206–215. https://doi.org/10.1016/J.FISHRES.2008.11.005

Russo CAM, Solé-Cava AM, Thorpe JP (1994) Population structure and genetic variation in two tropical sea anemones (Cnidaria, Actinidae) with different reproductive strategies. Mar Biol 119:267–276. https://doi.org/10.1007/BF00349566

Salas EM, Bernardi G, Berumen ML et al (2019) RADseq analyses reveal concordant Indian Ocean biogeographic and phylogeographic boundaries in the reef fish *Dascyllus trimaculatus*. R Soc Open Sci 6:172413. https://doi.org/10.1098/rsos.172413

Sale PF, Kritzer JP (2003) Determining the extent and spatial scale of population connectivity: decapods and coral reef fishes compared. Fish Res 65:153–172. https://doi.org/10.1016/J.FISHRES.2003.09.013

Sale PF, van Lavieren H, Ablan Lagman MC et al (2010) Preserving reef connectivity: a handbook for marine protected area managers. Connectivity Working Group, Coral Reef Targeted Research & Capacity Building for Management Program, UNU-INWEH, Melbourne

Santos S, Hrbeck T, Faris IP et al (2006) Population genetic structuring of the king weakfish, Macrodon ancylodon (Sciaenidae), in Atlantic coastal waters of South America: deep

genetic divergence without morphological change. Mol Ecol 15:4361–4373. https://doi.org/10.1111/j.1365-294X.2006.03108.x
Selkoe K, D'Aloia C, Crandall E et al (2016) A decade of seascape genetics: contributions to basic and applied marine connectivity. Mar Ecol Prog Ser 554:1–19. https://doi.org/10.3354/meps11792
Sexton JP, Hangartner SB, Hoffmann AA (2014) Genetic isolation by environment or distance: which pattern of gene flow is most common? Evolution 68:1–15. https://doi.org/10.1111/evo.12258
Sherman K (1991) The large marine ecosystem concept: research and management strategy for living marine resources. Ecol Appl 1:349–360. https://doi.org/10.2307/1941896
Shinzato C, Mungpakdee S, Arakaki N, Satoh N (2016) Genome-wide SNP analysis explains coral diversity and recovery in the Ryukyu archipelago. Sci Rep 5:18211. https://doi.org/10.1038/srep18211
Siegel DA, Mitarai S, Costello CJ et al (2008) The stochastic nature of larval connectivity among nearshore marine populations. Proc Natl Acad Sci USA 105:8974–8979. https://doi.org/10.1073/pnas.0802544105
Soeth M, Spach HL, Daros FA et al (2020) Use of otolith elemental signatures to unravel life-time movement patterns of Atlantic spadefish, Chaetodipterus faber, in the Southwest Atlantic Ocean. J Sea Res 158:101873. https://doi.org/10.1016/J.SEARES.2020.101873
Spalding MD, Fox HE, Allen GR, et al. (2007) Marine ecoregions of the world: a bioregionalization of coastal and shelf areas. BioScience 57(7):573–583. https://doi.org/10.1641/B570707
Sturm AB, Eckert RJ, Méndez JG et al (2020) Population genetic structure of the great star coral, Montastraea cavernosa, across the Cuban archipelago with comparisons between microsatellite and SNP markers. Sci Rep 10:15432. https://doi.org/10.1038/s41598-020-72112-5
Sundby S, Kristiansen T (2015) The principles of buoyancy in marine fish eggs and their vertical distributions across the world oceans. PLoS One 10:e0138821. https://doi.org/10.1371/journal.pone.0138821
Sunnucks P (2000) Efficient genetic markers for population biology. Trends Ecol Evol 15:199–203. https://doi.org/10.1016/S0169-5347(00)01825-5
Treml EA, Halpin PN, Urban DL, Pratson LF (2008) Modeling population connectivity by ocean currents, a graph-theoretic approach for marine conservation. Landsc Ecol 23:19–36. https://doi.org/10.1007/s10980-007-9138-y
van Nouhuys S (2016) Metapopulation ecology. In: eLS. John Wiley & Sons, Ltd, Chichester, pp 1–9
Vila-Nova DA, Bender MG, Carvalho-Filho A et al (2011) The use of non-reef habitats by Brazilian Reef Fish species: considerations for the design of marine protected areas. Natureza & Conservação 9:79–86. https://doi.org/10.4322/natcon.2011.010
Virtanen EA, Moilanen A, Viitasalo M (2020) Marine connectivity in spatial conservation planning: analogues from the terrestrial realm. Landsc Ecol 35:1021–1034. https://doi.org/10.1007/s10980-020-00997-8
Volk DR, Konvalina JD, Floeter SR et al (2021) Going against the flow: barriers to gene flow impact patterns of connectivity in cryptic coral reef gobies throughout the western Atlantic. J Biogeogr 48:427–439. https://doi.org/10.1111/jbi.14010
White CF, Lin Y, Clark CM, Lowe CG (2016) Human vs robot: comparing the viability and utility of autonomous underwater vehicles for the acoustic telemetry tracking of marine organisms. J Exp Mar Biol Ecol 485:112–118. https://doi.org/10.1016/J.JEMBE.2016.08.010
Whoriskey F, Hindell M (2016) Developments in tagging technology and their contributions to the protection of marine species at risk. Ocean Dev Int Law 47:221–232. https://doi.org/10.1080/00908320.2016.1194090
Xavier JH d A, Cordeiro CAMM, Tenório GD et al (2012) Fish assemblage of the Mamanguape environmental protection area, NE Brazil: abundance, composition and microhabitat availability along the mangrove-reef gradient. Neotrop Ichthyol 10:109–122. https://doi.org/10.1590/S1679-62252012000100011

Chapter 8
Food for Us

Maria Elisabeth de Araújo, Sérgio M. G. Mattos, Bárbara P. M. M. Lippi, Paulo W. Melo, and Nemer E. Narchi

Abstract This chapter deals with artisanal fishers and fish as food for humans in a book about Brazilian reefs. In doing so, we reflect on the urgent need to realize that healthy eating is the right of all of us and that these fishers are responsible for bringing the fish to our tables. The conservation of coastal ecosystems depends on public policies that affect these traditional populations and sometimes cause socioeconomic and environmental imbalances. Regarding the reefs, such problems arise concerning fishing marine fish with great commercial value, such as lobsters and octopuses. However, the drastic reduction of fisheries resources and environmental pollution provoke a change of habit and risk to food security, seriously damaging human health. Contaminated by heavy metals and recently by oil, many organisms that form the trophic webs of the fish species become seriously ill or die, affecting human beings. The impoverishment of nature affects small-scale fishers and other peoples in the primary sector that directly depend on nature to thrive. With poor public management and disorderly growth, coastal peoples lose their fishing territories and fish, as they end up selling the catch to tourists and distant markets, with disadvantageous profits or no profit at all. Their families start consuming industrialized foods with high content of sodium, lipids, and carbohydrates. This diet may cause chronic diseases and nutritional deficiencies and replace a diet rich in proteins, unsaturated fatty acids, and vitamins found in fish. Food quality is directly linked to fisheries policy, marked by territorial conflicts and disrespect for

M. E. de Araújo (✉)
Departamento de Oceanografia, Universidade Federal de Pernambuco (UFPE),
Grupo de Ictiologia Marinha Tropical (IMAT), Pernambuco, Brazil

S. M. G. Mattos
Grupo de Ictiologia Marinha Tropical (IMAT), Independent Researcher, Pernambuco, Brazil

B. P. M. M. Lippi
Independent Researcher, Pernambuco, Brazil

P. W. Melo
Grupo de Ictiologia Marinha Tropical (IMAT), Independent Researcher, Recife, Brazil

N. E. Narchi
El Colegio de Michoacán, Zamora, Mexico

R. K. P. Kikuchi et al. (eds.), *Brazilian Coral Reefs*, Brazilian Marine Biodiversity,
https://doi.org/10.1007/978-3-031-59152-5_8

environmental justice, and needs to be discussed in a multidisciplinary forum. We have found this book to be such a venue.

Keywords Coastal reefs · Small-scale fishers · Food security and nutrition · Fisheries management

8.1 Introduction

For several reasons, we find it essential to address the challenges of food studies and issues related to "food for all of us" within the context of Brazilian coral reefs. First, reef environments are essential for high biodiversity, where many species are relevant as fishing resources. Second, food is a right for all, an everyday act, in its most straightforward way of exercising citizenship and commitment to the population's basic needs. Third, it also allows us to understand the precepts on food and nutritional security, in the necessary reasonableness, a human right to healthy food.

As a worldwide food source in the most diverse aquatic ecosystems, fish products are essential against hunger and misery their importance and accessibility subject aquatic resources to intense anthropogenic pressures. Therefore, aquatic resource extraction requires a well-managed fishery system, although these pressures are not exclusively for fishing activity. Among the most diverse ecosystems, coastal reefs stand out as areas that need rules and norms for multi and interdisciplinary approaches under the eyes of the social, economic, and environmental sciences in the search for sustainability.

In tropical seas, reefs, mangroves, marshes, and marine grasslands form a mosaic of interconnected coastal ecosystems providing services for socioeconomic well-being, such as fishing (Moberg and Folke 1999). However, several authorities impose temporary or permanent closure of these areas to fishing, hoping to prevent fish stocks from running out and maintain or even increase production in adjacent areas. Yet, the imposition of no-take areas on coral reefs for conservation purposes does not lead to positive benefits to fishing (Cochrane 2002; Hall 2002).

The preservation of small-scale fisheries and their associated traditions, along with maintaining a social, economic, and environmental balance, presents itself as a fundamental pathway for food security and poverty eradication, visibly manifested spatiotemporally (Diegues 1998; Loring et al. 2019). Despite their access to sometimes limited fishing resources, small-scale fishers often find themselves in situations of vulnerability and food insecurity. In places like Colombia, Chile, and Peru, there are records of the annual fish consumption of families directly dependent on small-scale fisheries. However, self-consumption dimensions still need to be addressed; the economic and nutritional value of all fishing resources that go straight to fishers'families' dining tables is unknown (Espinoza-Tenorio et al. 2021). Stimulating food traditions among small-scale fishing populations to the use of fishing resources is a necessary legal and institutional concern for the state and society to ensure that food, including fish, reaches the Brazilian people.

In light of food availability to the population, this chapter discusses the importance of fishing resources exploited in coastal reefs, which are fundamental environments for the life cycle of many marine species. Due to the intense pressure and the need to mitigate anthropogenic impacts on these reef environments, it is argued to favor multi and interdisciplinary management measures, committing to socio-environmental actions, and attending to and respecting vulnerable populations. Therefore, we seek to answer to what extent the exploitation of reef fish is contributing to maintaining (1) the diversity of life in these environments at healthy levels, (2) the population's food and nutritional security, and (3) the social well-being of fishers.

8.2 Reefs and Connected Ecosystems: Most Caught Species' Life Cycles

Reefs are ecosystems that house a large number of organisms during certain stages of their life cycle, including fish, crustaceans, and mollusks, the main fish resources. The fish distribution patterns depend on rare species, represented by 3/4 of the diversity and specialists in the Brazilian five morphotypes' reef habitats (Araújo et al. 2020a). These biogeographic particularities and ecological connectivity result from the complex physical, hydrological, and biological characteristics (Nagelkerken 2009), favoring the occurrence of different conditions at distinct stages of life (Parrish 1989; Roberts et al. 2002).

The degree of dependence of a species on a given ecosystem, where its individuals spend most of their life, can characterize them as "reef fishes". This identification, therefore, does not refer to a taxonomic grouping but to an ecological terminology that alludes to the main environment used as residence (Araújo et al. 2018). Many reef fish are typically marine migrants, moving between different ecosystems throughout their life cycle. As a general pattern, it usually starts spawning in the open sea (Chittaro et al. 2005), followed by a dispersal phase of eggs and larvae towards nursery grounds (roots of mangroves in estuaries, seaweed meadows, and phanerogams) where they set. Back to reef environments, juveniles reach the reproductive period and then migrate to the ocean, where the cycle is repeated (Boehlert and Mundy 1988).

Marine fish species of commercial importance commonly follow this described pattern, including those of the Lutjanidae, Carangidae, Haemirhamphidae, Serranidae, Lobotidae, Haemulidae, Sparidae, Scianidae, and Scombridae families (Vasconcelos-Filho and Oliveira 2000). Silva-Falcão (2012) confirmed this interconnection between ecosystems for marine migratory fish but found another pattern for an estuarine zone in Pernambuco, where adults reproduce inshore, and the larvae are dispersed in the mangrove region, meadows, or even reefs, depending on the species (Silva-Falcão et al. 2013).

Crustaceans (lobsters, shrimp, and crabs) are important human foods and follow the ecological pattern of connectivity of marine migrants, in which larvae undergo several metamorphoses in the plankton and share the same benthic nurseries as fish species (Leis and McCormick 2002). Juvenile lobster and shrimp thrive in coastal reefs and bays, while crabs remain closer to the estuarine region (Souza-Filho et al. 2015). Studies conducted by Giraldes et al. (2015) show that over 70 species of decapods live on the reef bench in Porto de Galinhas, Pernambuco, a major tourist hub. Many species are exploited by artisanal fishers, such as the crabs of the Carpiliidae, Ocypodidae, Menippidae, and Panopeidae families, of the stone-crab (aratu-de-pedra) (family Plagusiidae), besides the common lobsters—families Panulinuridae and Scyllaridae.

Cephalopod mollusks have a migration pattern restricted to marine environments because they are stenohaline (Boletzky and Hanlon 1983). Juvenile squids are found in reefs, growing in deeper coastal waters. Octopuses, however, depend on reefs to feed, camouflage or mimic, escape predators, and reproduce, with species of *Octopus* being commonly explored in Brazilian waters (Vaske-Júnior 2011). The eggs form clusters in the reef structures, and as soon as they hatch, the paralarvae—miniatures of adults—swim for about a month and disperse in the sea, returning later to the benthic habitats (Iglesias et al. 2000).

For most bivalves of commercial interest, fertilization occurs in the open sea, and spawning is annual and continuous. In addition to their meat, the shell of these gastropods has a high sale value to tourists and collectors; they are used for decoration, tools, folk medicine, and magical or religious rituals (Dias et al. 2011; Costa-Neto 2006; Leo-Neto et al. 2012). The king helmet, a large sea snail belonging to the Cassidae family, also serves as subsistence food for fishing families (Mota et al. 2020).

8.2.1 Fisheries in Brazilian Coastal Reefs

Since humankind had access to coastal environments, fish has become an essential food source. These "sea workers" who carry out artisanal fishing have their peculiar way of life as an art of fishing, mainly based on multispecific techniques. The bibliographic records on the commercialization of reef animals extracted directly for human consumption are contemporary and result from overfishing (Ribeiro 2004; Cunha et al. 2012; Carvalho et al. 2013).

There are cultural, social, health, and environmental issues to be understood in this scenario. On the one hand, there are living organisms and their relationships with each other and the environment. On the other hand, humans have different knowledge, practices, and beliefs (Ferreira-Júnior et al. 2015). Such interactions between the environment and traditional human populations show they live not only "in" nature but "from" nature (Marques 2001), with a vital dependency on the social and ecological systems. The strong interaction with nature allows fishers to develop sensitive space-temporal knowledge of an aquatic environment and to manage fishing costs. Due to the three-dimensional structure of the reefs, fishing directly on these rocky environments becomes more selective (Fig. 8.1). It demands the use of techniques such as lines, traps, or harpoons in freediving, in addition to small fishing cast nets (*puçás*) to catch small fishes to ornamental trade (Nottingham et al. 2005).

Hand-gathering is carried out to catch crabs, and they can be attracted with a beam of light during the night (Freitas and Santos 2007), aided by a small net with a stick and bait or a skewer to chase away the animal from its hole to a nearby place where collection is easier (Fig. 8.1). In the case of lobsters, because of their prickly shells, fishers may wear gloves for manual captures, and place them in a bag-net, or using a small net to cover the reef used by the lobsters as shelters and an artisanal fishing coat hook (*bicheiro*) to chase them away from their loci, and the lobsters get stuck in the net. Regarding preferred resource extraction ranking, the octopuses are behind lobsters, shrimp, and groupers. These cephalopods are usually caught using the artisanal *bicheiro* (Fig. 8.1): (1) in areas free from reefs through diving,

Fig. 8.1 Artisanal fishing techniques and practices mainly observed in reef environments in Brazil. (1) Hook and line targeting fishes and squids; (2) Lobster fishery through freediving using a small net to embrace the reefs where lobsters shelter; (3) Lobsters, crabs, and fishes caught by traps; (4) Spearfishing targeting fishes and (5) Octopus captures using an artisanal harpoon with coat hook

(2) walking on the reef plateau in search of their locations, or (3) caught outside when trying to flee when chlorine (Jesus et al. 2015) or a large amount of sea salt is thrown on their shelters.

In addition to direct extraction, shrimp fishing by trawling near the reefs has been causing incidental capture (bycatch) and overexploitation of shells (Mota et al. 2020) and box crab (*guajá*), also caught in lobster traps (*manzuás* or *covos*) (Fig. 8.1) (Fausto-Filho 1966, 1974; Costa 1968). Many squids (Loliginidae) are bycatch of shrimp trawling but are also fished with rods in reef environments (Vaske-Júnior 2011). Some fishing gear types can negatively affect coastal habitats, which is critical for juvenile target species that aggregate in these areas for protection from predators (Hall 2002). An example is the use of "marambaias", an artificial shelter previously manufactured with mangroves and today using metal carcasses of drums. Its use has become widespread, possibly degrading the surrounding reefs and often attracting animals to potentially toxic environments. However, it should be noted that paragraph 6.8 of the Code of Conduct for Responsible Fishing (FAO 1995) references explicitly the protection of these habitats as a guiding principle for sustainability.

8.3 Contemporary Humans' Impacts[1]: Food security and Consumption

According to fishers, in addition to overfishing, other activities that strongly threaten environmental health and fishing resources are recreational nautical traffic, dumping of pollutants of continental origin, improper disposal of solid waste, lack of sanitation, and mangrove destruction (Araújo et al. 2014; Collier et al. 2015; Correia 2016; Melo 2018; Sulaiman et al. 2018). There are also silent anthropogenic risks, such as introducing exotic species (see Chap. 10). This is the case of the lionfish (*Pterois volitans*), which potentially changes reef fish assemblages' composition and structure throughout the American tropical Atlantic (Cobián Rojas et al. 2018).

The inappropriate use of reef environments for recreation ranges from trampling and artificial feeding of fish, which negatively interfere with the species' behavior and ecology (see Chap. 11 of this book), to significant urban impacts. The impacts caused by tourism are due to disorderly development, encouraged by public policies implemented since the 1970s (Melo et al. 2020) for resort expansions throughout the Brazilian coast. As the fishing communities' villages become under pressure from the socioeconomic influence of ventures destined for vacationers and tourist activities (Alcântara et al. 2004), the fish trade changes its dynamics, often leading to marginalizing fishers (Diegues 1998). Large business complexes are intermediaries, usually known to fishers, to pressure for the exclusiveness of fish production

[1] The remaining chapters of Part 3 of this book provide detailed studies on the different anthropogenic impacts on reef environments.

acquisition. Still, without interaction with these traditional communities, an impasse for life conservation in coastal environments is reached.

In August 2019, the coast of northeastern Brazil (about 3300 km long) suffered the largest environmental disaster in its history caused by an oil spill. So far, there is no confirmation of its origin nor a scientific study that points to a solution (Araújo et al. 2020b). It is known, however, that the main transport routes for large ships carrying more than 2×10^6 tons of oil pass around the reefs (Santos et al. 2016). In contact with the sea, part of the mixture of molecules that form the crude oil is volatilized, another is solubilized in the water column and/or also emulsified, adhering directly to rocky structures, or forming pockets below the sandy substrate and, thus, polluting the reef environments and causing death and disease throughout the biota (Araújo et al. 2020b).

In the aftermath, at the end of 2019, there was a collapse in the artisanal fisheries economy, paralyzing the value chain, including those communities not directly affected by oil (Ramalho 2020; Araújo et al. 2020b). In 2020, there were signs of fishing activity recovery, although fishing communities faced another crisis due to the COVID-19 pandemic that again affected the fish trade. All aggravated by the lack of direct small-scale public fishing policies (Ramalho 2020). The stored fish production became the main food item for their families, most of whom defaulted on payments of necessary home expenses (Araújo et al. 2020b).

The toxins accumulated in fish tissues by the oil spill in 2019 disseminated throughout the trophic web, causing health issues for human consumption (Araújo et al. 2020b). Research is ongoing, and some publications show damaging effects on Brazilian reefs. However, the presence of oil in marine species' digestive and respiratory tracts (G1 BA 2019) and turtles' feces (Da Silva et al. 2024) has been registered. The consumption of contaminated animals represents a calamity for the environment and public health since most petroleum compounds are carcinogenic and can cause numerous other diseases, even death (Santos et al. 2022). The ingestion of fish contaminated with HPA is carcinogenic due to the inability to biotransform polycyclic aromatic hydrocarbons accumulated in fish tissues (Meador et al. 1995), demanding caution about its consumption in situations of risk of contamination.

However, plastic pollutants also represent a hazardous menace to the reef builders, its inhabitants, and its users. Severe risks to human health arise from many other exponentially increasing anthropogenic activities, such as the release of contaminating microplastics (MPs) in natural environments (Cox et al. 2019). Unfortunately, we still know very little about the effects that the accumulation of microplastics has on humans (Barboza et al. 2020). Since 2014, the United Nations Environment Assembly (UNEA) has recognized that plastic material constitutes a global ocean threat. In its forums, its members consider global governance ineffective and believe that, with the strengthening of UNEA coordination and scientific and technological knowledge (UNEA, 2016), it will be possible to establish "a zero long-term vision" when no plastic waste should be thrown into the ocean. Meanwhile, very little is known about the effects of microplastic accumulation on humans (Barboza et al. 2020).

Pollution by plastics and their contaminants is one of the most significant challenges of the Anthropocene, requiring a classification to study new plastic formations (De-la-Torre et al. 2021). Of the four new categories, biota is directly affected by "plastic rust" due to the possibility of contamination by ingesting plastics, which can harm the food chain. Practically all marine animals are subject to the ingestion of microplastics, some nanoscopic, which are becoming more widespread and in greater quantity every day. Barboza et al. (2020) calculated that marine fish can become contaminated at a rate of 842 MP items/year (fibers, fragments, or pellets) and suffer from neurotoxic diseases or lipid oxidative damage. In the fishes' bodies, MPs were found mainly in the gastrointestinal tract but also gills, dorsal muscle, and other organs.

In addition to fishes, crustaceans and mollusks (Barboza et al. 2020) also suffer diseases and mutations caused by the accumulation of these microplastics, in particular evident in bivalve mollusks (Smith et al. 2018). Corals can also be affected and even though these animals can select their food, leaving aside inedible objects (Hankins et al. 2022)—except when these objects are biofouled (Corona et al. 2020)—it is also possible that reefs' health is compromised by the stress caused by microplastics exposure (Hall et al. 2015; Tang et al. 2018), especially when considering it will be synergistic with other stressors such as coral bleaching.

8.3.1 (Un)Sustainable Fishing

We live in a period of declining availability of fishing resources. What is the conservation status of reef fish? Paragraph 6.3 of the Code of Conduct for Responsible Fishing (FAO 1995) articulates the general principle that "states should prevent overfishing and excess fishing capacity and should implement management measures to ensure that fishing effort is commensurate with the productive capacity of the fishery resources and their sustainable utilization."

Natural fishing stock reduction imposes a vigorous and widespread struggle against waste, aiming at changes in production and consumption patterns. Pauly et al. (1998) argue that the declining populations of some targeted top-predator fish species have directed fishing efforts toward other species at lower trophic levels. The fast pace of anthropic-driven environmental changes brings an additional challenge to the fisher's way of life and relationship with nature. These combined impacts cause the decline of populations, individual size, spawning, fish stocks, and even the extinction of coastal species (McGill et al. 2015). Animals with a large theoretical maximum size and low growth rate (K-strategists) tend to be at a high trophic level. They are subject to small natural predation but potentially vulnerable to indiscriminate fishing, worldwide "threatened" and overfished (Seyboth et al. 2011).

Increased reef fishing may be due to reasons such as top predators overfishing and considered by Ivo et al. (2010) an alternative activity to overexploited lobsters in the Northeast Region. Previously discarded or kept for fishers' consumption, reef fishes shifted from bycatch to target species (Ribeiro 2004). In addition, due to the strong international market demand for human consumption, the fishing effort sharply increased (Campos and Oliveira 2001; Cunha et al. 2012; Carvalho et al. 2013) through the stimulus given to fish processing companies to the artisanal fishing fleet on the northeastern coast (Carvalho 2009). A striking case is the Spotted goatfish (*Pseudupeneus maculatus*—Bloch, 1793), which is sharply declining in reefs. Then abundant, this species started to be exported—replacing the parrotfish (Scaridae)—and was very well accepted in restaurants due to its culinary attributes. Table 8.1 shows consensus results obtained from Cunha et al. (2012) and Mattos et al. (2017) on the composition of trap-caught fishes in Northeastern Brazil reef environments.

Redirecting catches of target species, previously with low economic value, is a trend in the world market and results from the decline in traditionally exploited stocks (Cunha et al. 2012). The excessive fishing effort in reef environments, coupled with the need for more information on species' biomasses and the loss of coastal habitats, resulted in fishing unsustainability in the short term (Carvalho et al.

Table 8.1 Fish used for human consumption, caught with traps directly from Northeastern Brazil reef environments

Species	Family	Category[a]	Common names
Acanthurus spp.	Acanthuridae	2nd	Doctorfish
Carangoides bartholomaei (Cuvier 1833)	Carangidae	2nd	Yellow jack
Cephalopholis fulva (Linnaeus 1758)	Serranidae	2nd	Coney
Epinephelus morio (Valenciennes 1828)[b]	Serranidae	2nd	Red grouper
Hyporthodus nigritus (Holbrook 1855)[b]	Serranidae	2nd	Warsaw grouper
Haemulon melanurum (Linnaeus 1758)[c]	Haemulidae	2nd	Cottonwick grunt
Haemulon plumierii (Lacepède 1801)[c]	Haemulidae	2nd	White grunt
Hyporthodus niveatus (Valenciennes 1828)	Serranidae	2nd	Snowy grouper
Lutjanus analis (Cuvier 1828)	Lutjanidae	1st	Mutton snapper
Haemulon parra (Desmarest 1823)	Haemulidae	2nd	Sailor's grunt
Lutjanus jocu (Bloch and Schneider 1801)	Lutjanidae	1st	Dog snapper
Lutjanus purpureus (Poey 1866)	Lutjanidae	1st	Southern red snapper
Lutjanus synagris (Linnaeus 1758)	Lutjanidae	1st	Lane snapper
Lutjanus vivanus (Cuvier 1828)	Lutjanidae	1st	Silk snapper
Mycteroperca bonaci (Poey 1860)	Serranidae	1st	Black grouper
Ocyurus chrysurus (Bloch 1791)	Lutjanidae	1st	Yellowtail snapper
Pseudupeneus maculatus (Bloch 1793)	Mullidae	1st	Spotted goatfish
Selene setapinnis (Mitchill 1815)—*S. vomer* (Linnaeus 1758)	Carangidae	2nd	Atlantic moonfish
Sparisoma spp	Scaridae	1st	Parrotfish

The table includes their species and families and classification for sale (category)
Legend: [a]Value in which each species is categorized for commercialization, according to local consumption demand; [b]Genus: *Epinephelus*, *E. niveatus* ou *E. adscencionis;* [c]Others species of the genus *Haemulon*

2013). The demand is not restricted to adult fish for human consumption. Still, it affects juveniles of the same species for aquarium trade (Nottingham et al. 2005), leading to overfishing with devastating effects on most reef fish species (see Chap. 11 of this book).

8.4 Reef Fishes: Social Stratification and Food Consumption

Food for all of us! As we—*Homo sapiens*—are an essential part of the socio-ecological systems, including reef environments, what is the importance of fish's nutritional aspects for our survival? Even if fishing guarantees food and a source of income, the living conditions of fishing communities can be considered vulnerable on social, economic, and political aspects due to the low average income obtained and access to essential health and educational services (Hellebrandt et al. 2014; Silva et al. 2019). In Brazil, when implementing the Organic Law on Food and Nutritional Security (LOSAN)—Law No. 11,346 of 2006, it was challenging to involve the four basic needs of food resources: access, availability, stability, and utilization. The State must ensure that food, including fish, is promoted in public health policies through initiatives to tackle food and nutritional insecurity in the Brazilian population, especially for women and children, and in school meals, respecting cultures and stimulating the local market. In addition, the State must also guarantee the active participation of fishing communities in decisions that impact the right to food (HLPE 2014).

Food consumption in fishing communities is driven by production and market conditions, obeying "economic rationality" criteria. The nutritional value of food is not the main factor in food choice for this population. The economic issues, such as the options available after marketing, are a greater influence on this choice for the fishing community (Woortmann 1978). Furthermore, current habits in fishing communities are influenced and modified by existing trends in society's eating habits, individual preferences, and cultural taboos. However, fish plays an important role in the subsistence of these families beyond direct consumption and sales as barter exchanges. For Woortmann (1978), in less favorable circumstances, part of the production destined for the family is exchanged for other items of basic needs, *e.g.*, medicines, changing their food consumption by options of lesser market value.

Fish consumption varies year-round, especially when tourists influence changing habits in these fishing communities. The commercial value of fishing resources increases considerably during the high season in northeast Brazil, prioritizing tourists' demands to the detriment of local families´ consumption. Consequently, these communities replaced fisheries consumption with processed meats to save money (Pinto et al. 2017). This changing habit is specific neither to fishing communities nor in this country. Narchi et al. (2020) reported the inclusion of industrialized sugary products as substitutes for natural sweets in the diets of the traditional peoples of northwestern Mexico, which is a risk to the community's health and culture. Industrialized foods are often high in fat, sugar, sodium, and dyes while low in fiber and vitamins (Brazil 2014). Thus, its frequent consumption, associated with reduced physical activity, is related to the early development of obesity, dyslipidemia, increased blood pressure, reduced glucose intolerance, and risk of associated diseases, such as chronic cardiovascular diseases, diabetes, and cancer (Brasil 2014, WHO 1990). There is also an increased risk of alimentary deficiencies due to reduced consumption of fresh foods (Brasil 2014).

Caution is necessary in offering industrialized foods to children. Besides worsening the population's health and well-being indicators, culinary acculturation, arising from economic and market interests, is also taking young people away from the local culture (Narchi et al. 2020). The additives of industrialized food stimulate the perception of flavor when habits are being established, potentially impacting an individual's health throughout life (WHO 1990). In addition to health risks, the frequent consumption of these foods in childhood may allow for their intermission and later the adoption as local food with lower nutritional quality (Narchi et al. 2020). On the other hand, the increase in fish consumption by adolescents in fishing communities could be associated with the adequate consumption of total lipids in diet and health (Liberalino 2011).

8.4.1 Nutritional Value of Reef Fishing Resources

Edible parts and nutritional composition vary among fishing resources. However, regardless of species, their composition indicates the following nutritional qualities: (a) availability of proteins rich in essential amino acids and bioavailability superior to vegetable sources from 5% to 15%; (b) richness in biological quality lipids, such as unsaturated fatty acids (e.g., Omega-3); and (c) essential micronutrients (vitamins A, B and D, and minerals). It is also a source of protein that is easy to chew and digest, which is especially interesting for the diet of the elderly and children (Brasil 2014; HLPE 2014).

The diet composition of a population predicts the incidence of chronic noncommunicable diseases and nutritional deficiencies (Brasil 2014). A diet rich in polyunsaturated fatty acids from the omega-3 series has been protective against coronary artery disease, preventing thrombosis, increased blood pressure, and inflammation. In contrast, saturated and trans fatty acids are related to growing such risks (Santos and Bortolozo 2008; Lottenberg 2009).

The current dietary pattern of animal sources includes fish, crustaceans, and mollusks, but also beef, chicken, and processed meats (sausages, salami, hams). In coastal fishing communities in Brazil, aquatic resources are the primary source of animal protein and an essential source of micronutrients, homogeneously among men and women in different age groups (Pinto et al. 2017). Processed meats go through industrial mechanisms aimed at increasing shelf life and food flavor, raising the content of lipids, sodium, and carbohydrates in these preparations (Brasil 2014). Natural meats hardly have carbohydrates, unlike sausages, which are often composed of corn starch, cassava starch, maltodextrin, glucose syrup, and sugar (Angelini 2011).

Sodium acts as a preservative, enhancing flavor and masking some undesirable flavors from additives involved in processed foods (Brasil 2014). Since its consumption must be controlled to avoid cardiovascular diseases, including ultra-processed foods in the diet has been challenging. The daily limit for human sodium consumption is 2000 mg, according to the World Health Organization (WHO 2012). However, processed foods can exceed this daily limit by up to 72%, even when consumed in small portions, such as low as 100 g a day. Therefore, high sodium intake in the diet is related to the higher prevalence of cardiovascular diseases (Brasil 2014).

Industrialized meats are more caloric in composition when compared to natural options, mainly fish. The sausage group is an option with lower protein and higher fat content and is often present in the population's dishes. On the other hand, fish are high protein options with reduced amounts of fat. The chicken egg is similar to processed meat options when its protein content and omega-6/omega-3 ratio are evaluated. However, it consists of a natural protein source option, without the addition of preservatives and featuring low sodium content (Table 8.2).

As for the lipid content, fish meat has approximately 12% of its caloric value from these macronutrients, while industrialized options usually have between 60–80% in their composition (IBGE 2011). This is a worrying scenario, especially when assessing that the lipidic composition of fish is rich in unsaturated fatty acids. At the same time, sausages contain primarily saturated fat, monounsaturated fatty acids, and trans-fat (Table 8.2). The caloric level of these foods and the low quantity and quality of protein do not stimulate satiety and promote high consumption, favoring body weight gain (Brasil 2014).

Although fish's sodium content is higher than that of meat and eggs, it derives from their origin in the sea. On the other hand, the amount of this micronutrient present in processed meats is at least three times higher, mostly added during processing (Table 8.2). Figure 8.2 below compares fishing products' and ultra-processing foods' main nutritional importance.

Another relevant aspect is the omega-6/omega-3 ratio. Omega 6 is a fatty acid known for its pro-inflammatory action, and its consumption is related to inflammatory and autoimmune pathologies, while omega 3 has anti-inflammatory action and is known for its immunomodulatory and cardioprotective function (Simopoulos 2002; Martin et al. 2020). Recommendations on this ratio are, on average, between 4:1 and 5:1 (Martin et al. 2020), with fish having the best ratio (maximum of 3:1) and sausages with a minimum ratio of 6:1, confirming the inflammatory power of these foods (Table 8.2).

The calorie contents of the reef fish described vary between 66.03 kcal.100 g^{-1} and 106.7 kcal.100 g-1, in the wrasse-parrot and mackerel, respectively. Fish can vary in fat content between 0.5 and 25% and can be classified into four different categories: lean (less than 2% fat), low fat (2–4% fat), semi-fat (4–8% fat), and highly fat (greater than 8% fat) (Ackman 1990). Lean fish have lighter-colored flesh and are considered to have deeper water habits (Barbosa et al. 2020). The reef fish

Table 8.2 Nutritional composition of different protein sources in eating habits of the Northeastern Brazil fishing population

Protein source (100 g)	Kcal[a]	CHO[a] (g)	PTN[a] (g)	LIP[a] total (g)	LIP sat.[a] (g)	A.G.[a] mono-ins.[a] (g).	A.G. poli-ins.[a] (g)	A.G. Trans (g)	ω6[a] (g)	ω3[a] (g)	[a]Prop. ω6: ω3	Na[a] (mg)
Meat												
Pork	289	–	25.34	20.06	7.36	9.02	1.89	0.04	1.71	0.06	28:1	59
Beef	242	–	24.22	15.42	5.74	7.00	0.61	0.68	0.42	0.16	3:1	67
Pountry	239	–	27.30	13.6	3.79	5.34	2.97	0.45	2.57	0.11	23:1	82
Processed meats												
Salami	336	2.4	21.85	25.9	9.32	11.13	2.53	0.59	2.26	0.13	17:1	1449
Sausage	321	3.61	9.72	29.51	10.86	12.6	2.98	0.8	2.55	0.29	9:1	1175
Mortadella	307	3.2	11.10	27.7	10.21	12.81	2.92	0.8	2.48	0.44	6:1	930
Ham	226	0.42	20.53	15.20	5.04	7.07	1.78	0.04	1.62	0.16	10:1	941
Fisheries												
Fish	117	–	24.16	1.53	0.36	0.28	0.65	–	0.01	0.02	0.5:1	106
Crab	102	–	20.20	1.77	0.23	0.28	0.68	–	0.03	0.02	1.5:1	279
Blue crab	102	–	20.20	1.77	0.23	0.28	0.68	–	0.03	0.02	1.5:1	279
Shrimp	99	–	20.91	1.08	0.29	0.2	0.44	0.02	0.02	0.01	2:1	224
Egg												
Chicken egg	155	–	12.58	10.61	3.27	4.08	1.41	–	1.19	0.04	30:1	124

Source: IGBE Food Composition Table, 2011

[a]*Kcal* Calories, *CHO* carbohydrate, *PTN* protein, *LIP* lipids, *LIP sat* saturated fat, *A.G* fatty acids, *Mono-ins* mono-unsaturated, *Poly-ins* poly-unsaturated, *Co.* cholesterol, *ω3* omega-3 fatty acids, Prop. ω6: ω3: omega-6 and omega-3 ratio, *Na* Sodium

Fig. 8.2 Comparison between the main nutritional sources of fishing origin (left) and industrialized (right), consumed by families of artisanal fishers. The size of the letters (in bold) represents the importance of the items

Table 8.3 Nutritional composition of the different species of reef fish as the dietary habits of the Northeastern Brazil fishing population

Fish common names (100 g)	Calories (kcal)	LIP total (%)	PTN (g)	Fe (mg)	Ref.
Lane snapper	100.00	0.43	19.69	4.5	[1]
Cottonwick grouper	89.95	1.61	19.96	4.15	[1]
Sand tilefish[a]	85.06	1.54	17.72	4.85	[1]
Southern red snapper	83.6	1.00	18.4	–	[3]
Yellowtail snapper	83.33	0.82	19.48	0.71	[4]
Coney	79.07	0.52	19.13	5.3	[1]
Write grunt	75.72	1.54	18.15	4.85	[1]
Sailor's grunt	74.08	0.11	20.51	5.35	[1]
Parrotfish	66.03	0.08	18.05	4.9	[1]
Spotted goatfish	–	0.25	19.22	–	[5]

Source: [1] Moura (2019), [2] Menezes et al. (2008), [3] Caula et al. (2008), [4] Andrade et al. (2009), [5] SÁ Júnior et al. (2020)

Legend: PTN protein, *Total LIP* total lipid. Percentage value (gram of lipid per 100 g of fish). Fe: Iron, Ref. (Bibliographic references), [a]*Malacanthus plumieri* (Bloch, 1786)

described have low-fat levels, with a maximum content of 2.5% lipids in their composition (Table 8.3).

The reef species *Sparisoma* chrysopterum (parrotfish), *Haemulon parra* (cottonwick grouper), and *Lutjanus synagris* (lane snapper) can be considered the leanest, while the pelagic species *Centropomus undecimalis* (common snook) and S*comberomorus cavalla* (king mackerel) are those with the highest fat content. The lipid composition of Common snook mainly contains saturated fatty acids (65.71%). Still, it has a cholesterol content of 187.52 mg.100 g-1 (Menezes et al. 2008), below the recommended limit of 300 mg/day for the Brazilian population diet (Santos et al. 2013). Despite the higher fat content in mackerel fish, the composition of this fat is mostly unsaturated fatty acids (56.05%), including 15.9% of this fat from polyunsaturated fatty acids and 5.66% eicosapentaenoic acid (omega 3); therefore, it can be considered a good source of these nutrients (Moura 2019) and protective against cardiovascular risk.

The nutritional composition of the described reef fish demonstrates a similar amount of protein for different species (Table 8.3). The reef fishes described have between 3.9% and 29.72% of the recommended daily intake (RDI) of iron per 100 g. According to the Brazilian Ministry of Health, foods above 15% and 30% of the RDI can be considered a source or rich food for a particular micronutrient, respectively (RDC No. 54, November 12, 2012). Thus, except for the yellowtail snapper (*Ocyurus chrysurus*), all reef fish presented can be considered sources of iron, with sailor's grunt being the richest species on this micronutrient.

8.5 Fishing Management

Food quality is directly linked to fisheries policy, which has always been marked by conflicts between fishers'ecological-traditional and the academy's technical-scientific knowledge, besides government political-ideological issues. The regulatory frameworks resulting from these discussions led to institutionally constant social, economic, and environmental imbalances. What is in question here is how much this "political" process contributed to improving fishers' quality of life precisely by ignoring them in management actions. Accompanying Narchi et al. (2018), it is necessary to criticize the intervention and management processes without community participation, which inhibit the collective creation of knowledge that offers local alternatives to sustainability, environmental justice, and human dignity.

Fisheries management measures are established to protect exploited species, and coastal reefs must be understood as fishing areas with ecological characteristics relevant to adequate advice and decision-making. Based on multidisciplinary knowledge concerning ecological and fishing aspects, these measures must recognize fishers' behavior and tantamount fishing strategies to succeed. Spatial-temporal restrictions can help fisheries managers by limiting the capture of specific life stages, protecting genetic reservoirs of fishery resources and the most critical habitats for its sustainability, and restricting the fishing effort to optimize the value of the catch, avoiding stock over-exploitation (Hall 2002). The Code of Conduct (FAO 1995) warns that management measures should ensure that "excess fishing capacity is avoided and the exploitation of stocks remains economically viable" (paragraph 7.2.2a).

Fishing management requires a historical understanding of strategies to establish fisheries control rules, enabling the convergence of objectives for sustainable development. It was not before the 1960s that debates on fishing activity began, determining its reorganization and regulation. However, with the industrialization of fisheries, there was an increase in fishers' participation in the political decision-making arena. Unfortunately, this policy chronology led to an overall uncontrolled increase in catch, leaving aside the sustainability of fishing resources in favor of the short-term economic benefit. Thus, a scarce technical foundation, combined with the turbulent political moment, was responsible for the failure of fisheries development plans elaborated during the 1960s and 1970s (Mattos et al. 2017; Wojcieschoswki et al. 2019).

This long and tortuous path has generated injustices and exclusions through a continuous intervention process that results in the lack of recognition of fishers' ways of life, on land and in water, and of the know-how about fisheries resources and the environment as a greater good. As Melo et al. (2020) warn, the constant changes in the institutional framework of the federal government to regulate fishing hinder the continuity of public policies and cause the loss of information essential to the fisheries sector management. This situation was aggravated at the end of the twentieth century when public policies highlighted environmental issues without posing the social and economic development at the same level, leaving fishers and their families practically helpless (Carvalho and Callou 2008).

New scenarios and regulatory frameworks followed the government's political determinations through guidelines for developing and promoting fisheries and aquaculture production in the country (Mattos 2011). This moment, at least in the eyes of the small-scale fishing sector, was enthusiastic given the new expectations brought by the Fishers' Letter of Commitment of August 2002, with contributions to policies related to the sustainable use of fishing resources. However, the management system did not succeed, marked by a political-ideological process for many reasons, mainly negligence and the lack of a fishery management plan and the sectoral conflicts of attributions between environmental conservation and fisheries development (Mattos et al. 2017; Melo et al. 2020).

8.5.1 Territoriality

The creation of new environmental management provisions also proved to be conflicting, as they focused on limiting fishing efforts and managing conflicts with social development for environmental conservation. Pedrosa and Lessa (2017) affirmed: "This new arrangement allowed new management instruments and the organization of the environment to emerge under the aspect of protection of natural heritage to encompass the multidimensional character exposed, such as the Marine Protected Areas (MPA)," that incorporate several coastal reefs. Fishers' rules of access and use were created based on ecological concepts, without seeing them as natives and considering the art of fishing developed in these environments for generations and the socioeconomic and cultural aspects inherent to the community. As stressed before, fair and effective fisheries management requires collaboration between disciplines.

In Brazil, fisheries management does not allow specifically exclusive access rights, except for Protected Areas of sustainable use such as *Reservas estrativistas—RESEX* (Marine Extractive Reserves) and *Áreas de Proteção Ambiental—APA* (Environmental Protected Areas), which give rights to traditional communities to maintain fishing activities and their territories (Wojciechowski et al. 2019). However, the participation of fishers' representatives in APA's Councils is generally minimal, unlike when it comes to a RESEX. The creation of RESEX improves the empowerment and active participation of traditional populations in their territory by becoming deliberative, instead of consultative, in the councils of environmental bodies (OECO 2015). This principle currently integrates other co-management approaches, such as Fisheries Agreements, Conformity Terms, and Conservation Units Management Councils.

The Brazilian territorial sea, beaches, mangroves, estuarine systems and rivers with tidal influences are commonage use areas of the Brazilian people. However, for traditional communities and others, transitional possession may be granted (Wojciechowski et al. 2019). Reef systems are fishing areas close to the shore and are on these coastal beaches where small-scale fishers'residences are established and are under the control of the State. These areas are undoubtedly considered property of the Federal State, imposed by the Constitution, and any private title is null and void. However, the total and unrestricted guarantee that MPAs could bring benefits to artisanal fishing, with the intricate existing governance relationships and the complexity of the legal and institutional framework, is a matter of reflection. Some protected areas prove to be remarkably effective, while the creation of others lacks sufficient scientific justification (see Narchi and Price 2015 and references therein).

The construction of nursery grounds, for example, could be essential marine areas but never as a guarantee of the full development of fisheries in the maintenance of fishing territories and the cultural heritage of fishers (Appolloni et al. 2018). Approaches such as ethno-conservation (Diegues 2014) and biocultural conservation (Gavin et al. 2018) often argue that ecosystem protection strategies ignore the central question of who should be involved in the conservation process. Therefore, it is necessary to integrate local/traditional knowledge and popular participation and create partnerships that provide efficient ecosystem conservation strategies, considering the local socio-ecological context (Diegues 2014; Gavin et al. 2018).

In protected territories, where extractive and exploration activities are prohibited, opportunity costs are measured as the most significant economic value of these activities if no protection system is applied (Naidoo et al. 2006). However, assessing the opportunity costs of portions of the maritime territory can be very difficult since marine systems are typically for everyday use for many activities (Beck et al. 2004; Ostrom et al. 1999). Therefore, calculating the total turnover of the area occupied by each activity can be a way to get around this problem. The overestimation resulting from the valuation of artisanal fishing activity in the use of a fishing territory is due to the monetary sum coming from ecosystem services and economic activity data (Appolloni et al. 2018). However, orders of magnitude of other economic activities (i.e., tourism) are much higher and sometimes may underestimate artisanal fisheries'values. Therefore, environmental management must move towards understanding the use of different spaces by groups of users to establish actions for marine spatial planning, which is currently conceived as a fisheries management tool (Gerhardinger et al. 2020).

8.6 Final Remarks

The fishing activity represents great social, economic, and cultural diversity, with multispecific catches showing rapid fluctuations in time and space. Because of this diversity, although it seems like a truism, it is necessary to make more social observations about the daily diet of artisanal fishers and their dependence on the health of coastal ecosystems. The reef environments function as fishing territories that extend from the bases on land, where the traditional fishing communities are located, to the areas where fish, mollusks, and crustaceans are caught. They are facilitated by the proximity to these resources for family consumption and commercialization. Although fish are an essential protein food with great nutritional value, the fishers' families are deprived of consumption to attend to tourists' demands, mainly during the holidays when the prices go up.

How much would a fisher need to spend to fill his table if he stopped fishing for months because of the oil spill or the coronavirus pandemic? Socioeconomic studies could strengthen the arguments in favor of environmental conservation. It would be interesting if we marine scientists could help managers figure out how much the sea provides to its population regarding food capital. It is essential, therefore, to integrate data on nutrition, ecology, sociology, and economics and discuss them with decision-makers to bring subsidies to fisheries management. The complexity of managing fishery resources in the production system means managing stocks at healthy levels and implementing policies that consider sustainable fishing practices, combating activities harmful to coastal ecosystems, and the autonomy of small-scale fishing communities. In an ideal world, this subsidy should be sufficient to place a greater interest in the sustainable conservation of reef resources. In this world, this chapter is just a step, a reflection on these topics, but many other steps need to be taken as quickly as possible.

Acknowledgements Many thanks to the artisanal fishers who teach us how to conserve nature and fishing resources as food necessary for the survival of the human species.

References

Ackman RG (1990) Seafood lipids and fatty acids. Food Rev Intl 6:617–646

Alcântara R, Ferreira BP, Travassos P (2004) A pesca artesanal e o turismo em Porto de Galinhas, estado Pernambuco. Boletim técnico-científico do CEPENE 12:195–207

Andrade Q, Bispo S, Druzian JI (2009) Avaliação da qualidade nutricional em espécies de pescado mais produzidas no Estado da Bahia Quality evaluate nutritional the fishes more consumed in State of Bahia—Brazil. Cienc Tecnol Aliment 29:721–726

Angelini APR (2011) Quantificação do Colágeno, da Composição Centesimal e Estudo do Balanço de Massa dos Nutrientes Declarados, na Avaliação da Qualidade de Salsichas. Dissertação de mestrado Pós-Graduação em Ciência de Alimentos da Faculdade de Farmácia da Universidade Federal de Minas Gerais. Belo Horizonte 2011

Appolloni M, Smeriglio C, Amati B et al (2018) Catalogue of the primary types of marine molluscan taxa described by Tommaso Allery Di Maria, Marquis of Monterosato. Zootaxa 4477. https://doi.org/10.11646/zootaxa.4477.1.1

Araújo ME, Alves MDO, Simões L (2014) Histórias de pescadores: Meio ambiente, recursos pesqueiros e tradição em Rio Formoso, Pernambuco. Editora UFPE (BC2014-176) 104 p, il, Recife

Araújo ME, Mattos FGM, Feitosa CV et al (2018) (2018) Histórico do conhecimento dos 'peixes ósseos'—Capítulo 6. In: ME de Araújo, CV Feitosa, SMG de Mattos (Org.). Ecologia de peixes recifais de Pernambuco. 1ed. Recife: UFPE 1:193–228

Araújo ME, Mattos FMG, Melo FPL et al (2020a) Diversity patterns of reef fish along the Brazilian tropical coast. Mar Environm Reach 160:105038. Elsevier Ltd. https://doi.org/10.1016/j.marenvres.2020.105038

Araújo ME, Ramalho CWN, Melo PW (2020b) Artisanal fishers, consumers and the environment: immediate consequences of the oil spill in Pernambuco, Northeast Brazil. Cad Saúde Pública [online] 36(1):e00230319. ISSN 1678-4464. https://doi.org/10.1590/0102-311X00230319

Barbosa SE, Silva DL, Trindade-Filho EM et al (2020) Risk of cardiovascular disease in fishermen in a community. Revista Pesquisa em Fisioterapia 10:376–384

Barboza LGA, Lopes C, Oliveira P, et al. (2020) Microplastics in wild fish from North East Atlantic Ocean and its potential for causing neurotoxic effects, lipid oxidative damage, and human health risks associated with ingestion exposure. Sci Total Environ [online], 717, 15/May/2020, 134625. https://www.sciencedirect.com/science/article/pii/S0048969719346169?via%3Dihub

Boehlert GW, Mundy BC (1988) Roles of behavioral and physical factors in larval and juvenile fish recruitment to estuarine nursery areas. Am Fish Soc Symp 3:51–67

Boletzky SV, Hanlon RT (1983) A review of the laboratory maintenance, rearing and culture of cephalopod molluscs. Mem Natl Mus Vic 44:147–187

Brasil (2014) Guia alimentar para a população brasileira/Ministério da Saúde, Secretaria de Atenção à Saúde, Departamento de Atenção Básica.—2. ed., 1. reimpr.—Brasília: Ministério da Saúde

Campos CEC, Oliveira JEL (2001) Caracterização biométrica e merística do saramunete, Pseudupeneus maculatus (Osteichthys: Mullidae), em Ponta de Pedras, Pernambuco. Bol Inst Pesca 27:185–189

Carvalho RAA (2009) Peixes recifais para consumo humano: captura, processamento e exportação no Rio Grande do Norte, Brasil [Dissertação de Mestrado]. Universidade Federal do Ceará, Fortaleza (CE)

Carvalho FEA d, Callou ABF (2008) Extensão pesqueira e desenvolvimento local: a experiência da Secretaria Especial de Aqüicultura e Pesca no Estado de Pernambuco, 2003–2006. INTERAÇÕES, Campo Grande 9:65–76

Carvalho RAA, Cunha FEA, Montezuma AMN et al (2013) Captura e processamento de peixes recifais no estado do Rio Grande do Norte, Brasil. Acta Fish Aquat Res 1:91–103. https://doi.org/10.2312/ActaFish.2013.1.1.91-103

Caula FCB, Oliveira MP, Maia EL (2008) Teor de colesterol e composição centesimal de algumas espécies de peixes do estado do Ceará. Ciênc. Tecnol. Aliment. Campinas 28:959–963

Chittaro PM, Usseglio P, Sale PF (2005) Variation in fish density, assemblage composition and relative rates of predation among mangrove, seagrass and coral reef habitats. Environ Biol Fish 72(175):187

Cobián Rojas D, Schmitter-Soto JJ, Aguilar-Perera A et al (2018) Diversidad de las comunidades de peces en dos áreas marinas protegidas del Caribe y su relación con el pez león. Rev Biol Trop. 66(1)189–203. http://dx.doi.org/10.15517/rbt.v66i1.28197

Cochrane KL (2002) A fishery manager's guidebook. Management measures and their application. In: Cochrane KL (ed) FAO fisheries technical paper, vol 424. FAO, Rome, p 231

Collier CA et al (2015) Integrated approach to the understanding of the degradation of an urban river: local perceptions, environmental parameters and geoprocessing. J Ethnobiol Ethnomed 11:1–13

Corona E, Martin C, Marasco R, Duarte CM (2020) Passive and active removal of marine microplastics by a mushroom coral (Danafungia scruposa). Front Mar Sci 7:128

Correia JRM de B et al (2016) Poluição em recifes de coral por vinhoto da cana-de-açúcar. In: Zilberberg C et al (eds) Conhecendo os recifes brasileiros. Rio de Museu Nacional, UFRJ, p 360

Costa RS (1968) Estudo preliminar sobre a biologia e a pesca do caranguejo Carpilius corallinus (Herbst), no Estado do Ceará. Arquivos de estudos biológicos da Universidade Federal do Ceará 8:211–219

Costa-Neto EM (2006) The mollusks in zootherapy: traditional medicine and clinical-pharmacological importance. Biotemas 19(3):71–78

Cox K, Covernton G, Davies, H, Dower J, Juanes F, Dudas S (2019) Human Consumption of Microplastics. Environ Sci Technol 53. https://doi.org/10.1021/acs.est.9b01517

Cunha FEA, Carvalho RAA, Araújo ME (2012) Exportation of reef fish for human consumption: long-term analysis using data from Rio Grande do Norte, Brazil. Bol Inst Pesca 38:369–378

De-la-Torre GE, Dioses-Salinas DC, Pizarro-Ortega CI, Santillán L (2021) New plastic formations in the Anthropocene. Sci Total Environ 754:14221–14226

Dias TL, LeoNeto NA, Alves RRN (2011) Molluscs in the marine curio and souvenir trade in NE Brazil: species composition and implications for their conservation and management. Biodivers Conserv. https://doi.org/10.1007/s1053101199915

Da Silva MFSD, Zanardi-Lamardo E, Valcarcel Rojas LA, de Oliveira Alves MD, Chimendes da Silva Neves V, de Araújo ME (2024) Traces of oil in sea turtle feces. Mar Pollut Bull. 200:116088. https://doi.org/10.1016/j.marpolbul.2024.116088

Diegues ACS (1998) Environmental impact assessment: the point of view of artisanal fishermen communities in Brazil. Ocean Coast Manag:119–133. Evolução da terra, São Paulo. https://br.pinterest.com/pin/267049452879125112/

Diegues ACS (2014) The role of ethnoscience in the build-up of ethnoconservation as a new approach to nature conservation in the tropics. Revue d'ethnoécologie 6:1–18

Espinoza-Tenorio A, Ehuan-Noh RG, Cuevas-Gómez GA et al (2021) Between Uncertainty and Hope: Young Leaders as Agents of Change in Sustainable Small-Scale Fisheries. Ambio 51:1–15. https://doi.org/10.1007/s13280-021-01639-2

FAO (1995) Code of conduct for responsible fisheries. Food and Agriculture Organization of the United Nations, Rome

Fausto-Filho J (1966) Primeira contribuição ao inventário dos crustáceos decápodos marinhos do Nordeste brasileiro. Arquivos de Ciências do Mar 6:31–37

Ferreira Júnior WS et al (2015) Resilience and adaptation in social-ecological systems. In: Albuquerque UP, De Medeiros PM, Casas A (eds) Evolutionary ethnobiology, New York, pp 105–120

Freitas AETS, Santos MCF (2007) Aspectos da biologia pesqueira do aratu-de-pedra Plagussia depressa (Fabricius, 1775) (Crustacea, Brachyura, Plagusiidae) capturado em Tamandaré (Pernambuco—Brasil). Boletim Técnico-Científico do CEPENE. Tamandaré (PE) 15:39–46

Gavin MC, Mccarter J, Berkes F et al (2018) Effective biodiversity conservation requires dynamic, pluralistic, partnership-based approaches. Sustainability (Switzerland) 10:1–11

Gerhardinger LC, Andrade MM, Corrêa MR, Turra A (2020) Crafting a sustainability transition experiment for the Brazilian blue economy. Mar Policy 120. https://doi.org/10.1016/j.marpol.2020.104157

Giraldes BW, Silva AZ, Corrêa FM et al (2015) Artisanal fishing of spiny lobsters with gillnets—a significant anthropic impact on tropical reef ecosystem. Glob Ecol Conser 4:72–580

Hall S (2002) The use of technical measures in responsible fisheries: area and time restrictions. In: Cochrane KL (ed) A fishery manager's guidebook. Management measures and their application. FAO Fisheries Technical Paper. No. 424. FAO, Rome, 231p

Hall NM, Berry, KLE, Rintoul, L et al. (2015) Microplastic ingestion by scleractinian corals. Mar Biol 162:725–732. https://doi.org/10.1007/s00227-015-2619-7

Hankins C, Raimondo S, Lasseigne D (2022) Microplastic ingestion by coral as a function of the interaction between calyx and microplastic size. Sci Total Env 810:152333. https://doi.org/10.1016/j.scitotenv.2021.152333

Hellebrandt D, Allison EH, Delaporte A (2014) Segurança alimentar e pesca artesanal: análise crítica de iniciativas na América Latina. Desenvolvimento e Meio Ambiente 32:7–27

HLPE (2014) Sustainable fisheries and aquaculture for food security and nutrition. A report by the High Level Panel of Experts on Food Security and Nutrition of the Committee on World Food Security, Rome

IBGE (2011) Pesquisa de Orçamentos Familiares 2008–2009: Tabela de Composição Nutricional dos Alimentos Consumidos no Brasil. [sl: sn] v. 39

Iglesias J, Sanchez FJ, Otero JJ (2000) Culture of octopus (Octopus vulgaris, Cuvier): present knowledge, problems and perspectives. Recent advances in Mediterranean aquaculture finfish species diversification. CIHEAM, Cah. Options Mediterr 47:313–322

Ivo CTC, de Vasconcelos JA, Osório FM (2010) Pesca de peixes com covos no estado do Rio Grande do Norte. Boletim Técnico Científico do CEPENE Tamandaré 18:75–85

Jesus MD, Gouveia MTJ, Zapelini C et al (2015) Pesca Artesanal e Cadeia Produtiva de Octopus Insularis: O Caso dos Ambientes Recifais do Sul da Bahia. Brasil Gaia Scientia 9:195–204

Leis J, McCormick MI (2002) The Biology, Behavior, and Ecology of the Pelagic, Larval Stage of Coral Reef Fishes. In: Coral Reef Fishes: Dynamics and Diversity in a Complex Ecosystem. Elsevier, pp 171–199

LeoNeto NA, Voeks RA, Dias TLP, Alves RRN (2012) Mollusks of Candomblé: symbolic and ritualistic importance. J Ethnobiol Ethnomed. https://doi.org/10.1186/17464269810

Liberalino LCP (2011) Consumo de alimentos entre adolescentes de comunidades pesqueiras do Rio Grande do Norte (RN). Dissertação de mestrado. In: Pós graduação em Ciências da Nutrição, Centro de Ciências da Saúde, Universidade Federal da Paraíba. João Pessoa

Loring PA, Fazzino DV, Agapito M, Chuenpagdee R et al (2019) Chapter 4: Fish and Food Security in Small-Scale Fisheries. In R. Chuenpagdee, & S. Jentoft (Eds.), Transdisciplinarity for Small-Scale Fisheries Governance: Analysis and Practice (pp. 55–73). Springer. https://doi.org/10.1007/978-3-319-94938-3_4

Lottenberg AMP (2009) Importância da gordura alimentar na prevenção e no controle de distúrbios metabólicos e da doença cardiovascular. Arq Bras Endocrinol Metab, São Paulo 53:595–607. https://doi.org/10.1590/S0004-27302009000500012

Marques JG (2001) A pesquisa/teoria e método. In: Pescando Pescadores, 2nd edn. NUPAUB, São Paulo, pp 47–52

Martin CA, Almeida VV, Ruiz MR et al (2020) Ácidos graxos poliinsaturados ômega-3 e ômega-6: importância e ocorrência em alimentos. Rev Nutr Campinas 19(6):761–770. https://doi.org/10.1590/S1415-52732006000600011

Mattos SMG (2011) Desafios à implementação de políticas públicas e à gestão participativa da pesca artisanal. In SMG Mattos, RT Moura, WM Maia Jr. (Eds.), Gestão de Pescarias Costeiras e da Maricultura. Anais da II Oficina de Trabalho de Aquicultura e Pesca do Nordeste, p. 79–96

Mattos SMG, Wojciechowski MJ, Macnaughton AE et al (2017) Implementing the small-scale fisheries guidelines: lessons from Brazilian clam fisheries. In: Jentoft S, Chuenpagdee R, Barragán-Paladines MJ, Franz N (eds) The small-scale fisheries guidelines: global implementation, vol 14. Springer, Heidelberg, pp 473–494

McGill BJ, Dornelas M, Gotelli NJ et al (2015) Fifteen forms of biodiversity trend in the Anthropocene. Trend Ecol Evol 30:104–113

Meador JP, Stein JE, Reichert WL et al (1995) Bioaccumulation of polycyclic aromatic hydrocarbons by marine organisms. Rev Environ Contam Toxicol 143:79–165

Melo PW (2018) Pesca artesanal: diversidade, ameaça e destino do pescado em Rio Formoso (PE). Recife, UFPE

Melo PW, Mattos SMG, Araújo ME (2020) Disminución de los pescadores artesanales en Rio Formoso, Pernambuco: Una contextualización histórica en Brasil. In: América Profunda: visiones y convergencias en la oceanografía social del continente. Editores: Nemer E. Narchi y Christine M. Beitl. Mexico

Menezes MES, Lira GM, Omena CMB et al (2008) Composição centesimal, colesterol e perfil de ácidos graxos dos peixes tainha (Mugil cephalus) e camurim (Centropomus undecimalis) da Lagoa Mundaú. AL/Brasil Rev Inst Adolfo Lutz 67:89–95

Moberg F, Folke C (1999) Ecological goods and services of coral reef ecosystems. Ecol Econ 29:215–233

Mota ELS, Alves RRN, Dias TLP (2020) Fishing, trade, and local ecological knowledge of the marine gastropod, Cassis tuberosa, a target species of the international shell trade. Ethnobiol Conserv 9:23. https://doi.org/10.15451/ec2020-06-9.23-1-11

Moura SMA (2019) Avaliação Nutricional e das Atividades Anticolinesterásica, Antifúngica e Antioxidante de Produtos Obtidos de Pescados Marinhos do Estado do Ceará. In: Tese de Doutorado. Curso de Pós Graduação em Biotecnologia da Rede Nordeste de Biotecnologia (RENORBIO). Fortaleza

Nagelkerken I (2009) Ecological connectivity among tropical coastal ecosystems. Springer, p 615

Naidoo R, Balmford A, Ferraro PJ, Polasky S, Ricketts TH, Rouget M (2006) Integrating economic costs into conservation planning. Trends Ecol Evol 21(12):681–687. https://doi.org/10.1016/j.tree.2006.10.003

Narchi, NE, Price LL (2015) Ethnobiology of Corals and Coral Reefs. Narchi, Nemer E. & Lisa L. Price (eds). Springer Ethnobiology Series. Springer International Publishing. ISBN 978-3-319-23763-3

Narchi NE, Cariño M, Mesa-Jurado MA et al (2018) El CoLaboratorio de Oceanografía Social: espacio plural para la conservación integral de los mares y las sociedades costeras. Sociedad y Ambiente 18:285–301

Narchi NE, Marlett CM, Hernández-Santana G (2020) What happened to seri sweets? Dietary transition through culinary acculturation. J Ethnobiol 40:348–367

Nottingham MC, Barreto LM, Araújo ME et al (2005) A explotação de peixes ornamentais marinhos no Estado do Ceará, Brasil: capturas e manutenção nas empresas e exportação. Boletim técnico-científico do CEPENE, Brasil 13:53–73

OECO (2015) O que é uma Reserva Extrativista. Retrieved November 17, 2020, from https://www.oeco.org.br/dicionario-ambiental/29258-o-que-e-uma-reserva-extrativista/

Ostrom E, Burger J, Field CB, Norgaard RB, Policansky D (1999) Revisiting the Commons: Local Lessons, Global Challenges. 284(5412):278–282. https://doi.org/10.1126/science.284.5412.278

Parrish JD (1989) Fish communities of interacting shallow-water habitats in tropical oceanic regions. Mar Ecol Prog Ser 58:143–160

Pauly D, Christensen V, Dalsgaard J et al (1998) Fishing down marine food webs. Science 279:860–863

Pedrosa BMJ, Lessa RPT (2017) O social como prioridade na pesca artesanal: Diretrizes internacionais para a pesca artesanal sustentável. Labomar Arquivos de Ciências do Mar 50:100–109

Pinto MF, Mourão JS, Alves RRN (2017) Animal source foods consumed in two fishing communities on the northeast coast of Brazil. Environ Dev Sustain 19:679–692

Ramalho CWN (2020) O colapso da pesca artesanal no litoral de Pernambuco. Revista Coletiva—Diversidade Socioambiental, Fundaj, n° 11, Recife, p. 1–10

Ribeiro FP (2004) Composição da biocenose e abundância relativa de peixes capturados com covos nos estados do Rio Grande do Norte e Pernambuco (Brasil). Boletim Técnico Científico do CEPENE 12:113–118

Roberts CM, McClean CJ, Veron JEN et al (2002) Marine biodiversity hotspots and conservation priorities for tropical reefs. Science 15:1280–1284

Sá Júnior PLS, Silva LJ, Andrade HA et al (2020) Rendimento e Composição Centesimal de Filés e Carne Mecanicamente Separada de Saramunetes (Pseudupeneus maculatus BLOCH, 1793). Arq. Ciên. Mar, Fortaleza 5:52–62

Santos RD, Gagliardi ACM, Xavier HT et al (2013) Sociedade Brasileira de Cardiologia. I Diretriz sobre o consumo de Gorduras e Saúde Cardiovascular. Arq Bras Cardiol 100:1–40

Santos HF, Carmo FL, Rosado AS, et al. (2016) Contaminação de recifes de coral por petróleo e seus derivados. In: Zilberberg C, Abrantes DP, Marques JA, Machado LF, Marangoni LFB, editores. Conhecendo os recifes brasileiros: rede de pesquisas Coral Vivo. Rio de Janeiro: Museu Nacional/Universidade Federal do Rio de Janeiro; p. 183–94

Santos M, Santos C, Alves M, et al (2022) Oil in Northeast Brazil: mapping conflicts and impacts of the largest disaster on the country's coast. Anais da Academia Brasileira de Ciências, 94. https://doi.org/10.1590/0001-3765202220220014

Seyboth E, Condini MV, Garcia AM et al (2011) Age, growth and reproductive aspects of the dusky grouper Myteroperca marginata (Actinoptserygii, Epinephelidae) in a man-made rocky habitat in southern Brazil. Neotrop Ichthyol 9:849–856

Silva MRO, Pennino MG, Lopes PFM (2019) Social-ecological trends: managing the vulnerability of coastal fishing. Ecol Soc 24:4

Silva-Falcão EC (2012) Evidências de conectividade entre habitats costeiros tropicais através do estudo de peixes em fases iniciais do ciclo de vida [Tese Doutorado]. Universidade Federal de Pernambuco, Recife (PE)

Silva-Falcão EC, Severi W, Araújo ME (2013) Spatial and temporal variation of Achirus larvae (Actinopterygii: Achiriidae) in mangrove, beach and reef habitats in northeastern Brazil. J Mar Biol Assoc U K 93:381–388

Simopoulos AP (2002) Omega 3 fatty acids in inflammation and autoimmune diseases. J Am Coll Nutr 21:495–505

Smith M, Love DC, Rochman CM et al. (2018) Microplastics in Seafood and the Implications for Human Health. Curr Envir Health Rpt 5, 375–386. https://doi.org/10.1007/s40572-018-0206-z

Souza-Filho JF, Negromonte AO, Araújo CL et al (2015) Crustáceos. In: Guia biológico do litoral de Pernambuco. Coord. Jesser Fidelis de Souza-Filho, Petrobrás, Ipojuca, Pernambuco, pp 84–101

Sulaiman SN, Carbone AS, Coutinho SMV (2018) Educando entre as águas: o convívio entre a vida natural e humana na área de proteção ambiental (APA) de Guadalupe. CPRH, Recife

Tang J, Ni X, Zhou Z et al. (2018) Acute microplastic exposure raises stress response and suppresses detoxification and immune capacities in the scleractinian coral Pocillopora damicornis. Environ Pollut, 243(A):66–74. https://doi.org/10.1016/j.envpol.2018.08.045

Vasconcelos-Filho AL, Oliveira AME (2000) Ictiologia. In: Gerenciamento participativo de estuários e manguezais. Eds: Barros, H.M.; Macedo, S.J.; Leça, E.E.; Lima, T. EdUFPE, 252p

Vaske-Júnior T (2011) Lulas e polvos da costa brasileira. Eds: Vaske-Júnior, T; Costa, FAP. Fortaleza: UFC/LABOMAR/NAVE, 184p.; il (Coleção Habitat, v. 5). Bibliografia: p. 123–132 ISBN: 978-85-7563-842-2

WHO (2012) Guideline: Sodium intake for adults and children [Internet]. Genebra: World Health Organization; 2012 [acessado em 13 dez. 2020]; 86(6): 2012. Disponível em: https://apps.who.int/iris/bitstream/handle/10665/77985/9789241504836_eng.pdf?sequence=1

Wojciechowski MJ, Ferreira BMP, Vila-Nova DA et al (2019) Defending the beach: transdisciplinary approaches in small-scale fisheries in Pernambuco, Brazil. In: Chuenpagdee R, Jentoft S (eds) Transdisciplinarity for small-scale fisheries governance. MARE Publication Series, vol 21. Springer, Heidelberg, pp 283–301. https://doi.org/10.1007/978-3-319-94938-3_15

Woortmann K (1978) Hábitos e Ideologias Alimentares em Grupos Sociais de Baixa Renda: Relatório Final. Série Antropologia 20:114

World Health Organization (1990) Diet, nutrition and prevention of chronic diseases. Geneva (WHO—Technical Report Series, 797) https://www.who.int/dietphysicalactivity/publications/trs916/download/en/

Chapter 9
Biotechnology and Reef Environment

Leticia V. Costa-Lotufo, Paula C. Jimenez, Diego V. Wilke, Angelica R. Soares, Ricardo Coutinho, and Renato C. Pereira

Abstract Coral reefs are considered the habitat exhibiting the greatest biodiversity within the marine realm. Besides their ecological functions, coral reefs shelter one quarter to one third of all marine species, encompassing microorganisms, macroalgae and invertebrates of great economic importance, since they may be a source of inspiration for marine biotechnology. But coral reefs are also vulnerable to a variety of adversities, such as eutrophication, increased sedimentation, overfishing and overexploitation of other reef-based resources. This chapter presents an overview on marine biotechnology, although not exhaustive, exploring the origin and development of the Brazilian marine biotechnology and the biotechnological potential of marine organisms from coral reefs. The main focus herein is to discuss approaches of exploiting such biotechnological potential in an ecologically friendly or sustainable manner and, furthermore, on the means by which the products of processes

---

L. V. Costa-Lotufo (✉)
Laboratório de Farmacologia Marinha, Instituto de Ciências Biomédicas, Universidade de São Paulo, São Paulo, Brazil
e-mail: costalotufo@usp.br

P. C. Jimenez
Laboratório de Bioprospecção de Organismos Marinhos, Instituto do Mar, Universidade Federal de São Paulo, São Paulo, Brazil

D. V. Wilke
Laboratório de Bioprospecção e Biotecnologia Marinha, Núcleo de Pesquisa e Desenvolvimento de Medicamentos (NPDM), Departamento de Fisiologia e Farmacologia, Faculdade de Medicina, Universidade Federal do Ceará, Fortaleza, Ceará, Brazil

A. R. Soares
Grupo de Produtos Naturais de Organismos Aquáticos (GPNOA), Instituto de Biodiversidade e Sustentabilidade (NUPEM), Universidade Federal do Rio de Janeiro, Rio de Janeiro, Brazil

R. Coutinho
Instituto de Estudos do Mar Almirante Paulo Moreira, Rio de Janeiro, Brazil

R. C. Pereira
Departamento de Biologia Marinha, Universidade Federal Fluminense e Instituto de Pesquisas Jardim Botânico do Rio de Janeiro, Rio de Janeiro, Brazil

R. K. P. Kikuchi et al. (eds.), *Brazilian Coral Reefs*, Brazilian Marine Biodiversity,
https://doi.org/10.1007/978-3-031-59152-5_9

obtained thereof can be used to improve the quality of human life and, also, in favor of the coral reefs themselves.

Keywords Blue technology · Biospropection · Marine pharmaceuticals · Antifouling · Marine natural products

9.1 Introduction

According to the Organization for Economic Co-operation and Development (OECD 2013), biotechnology is broadly defined as "the application of science and technology to living organisms, as well as parts, products and models thereof to alter living or non-living material for the production of knowledge, goods or service". While it can be thought of as a relatively recent field of study, as a practice, however, it has been around humankind for over 11,000 years. This is the approximate age of agriculture; the activity of domesticating plants is considered the inaugural biotechnological enterprise (Bhatia and Goli 2018). A more contemporary and tangible notion of the term generally involves a molecular essence to the definition, incorporating, beyond the source material, the techniques applied. As also pointed out by the OECD (2013) and well-reasoned in the 2010 report by the Marine Board of the European Science Foundation, biotechnology is defined as cutting-edge molecular or genomic biological applications where molecular or genetic material is manipulated to generate desirable products or their benefits (Qin et al. 2011). Indeed, biotechnology has always played a key role in evolution, societal organization, and well-being of the human species (Bocquet-Appel 2011). Currently, it is assessed for its application in diversified industrial segments which reach for whole or parts of living organisms to generate products or make processes feasible—among which energy and fuel supply, chemical, biologicals and materials stand out—in health and medical care and for environmental uses. In fact, there is an ongoing trend of color-coding (see Box 9.1) the different branches of biotechnology based on their purpose (Kafarski 2012).

Marine biotechnology, dubbed blue biotechnology, is the branch that looks at the biodiversity of Earth's oceans and seas as the source for innovation or as the target for biotechnological intervention. Due to access limitations and scarcity of tools to navigate the oceans' magnitude, the marine environment has been only better accessed in the last 50 years. The innovative potential is supported by organisms that have a longer evolutionary history and inhabit and interact in an environment subjected to very different conditions. In fact, the marine environment is more diversified at higher taxonomic levels, a feature regarded as taxonomic distinctness, sheltering many taxa not represented in terrestrial organisms. Such biological diversity directly reflects genetic diversity and is also the very birthplace of biotechnological innovations of nature (Appeltans et al. 2012; Snelgrove 2016).

Box 9.1: Multiple "Colors" of Biotechnology

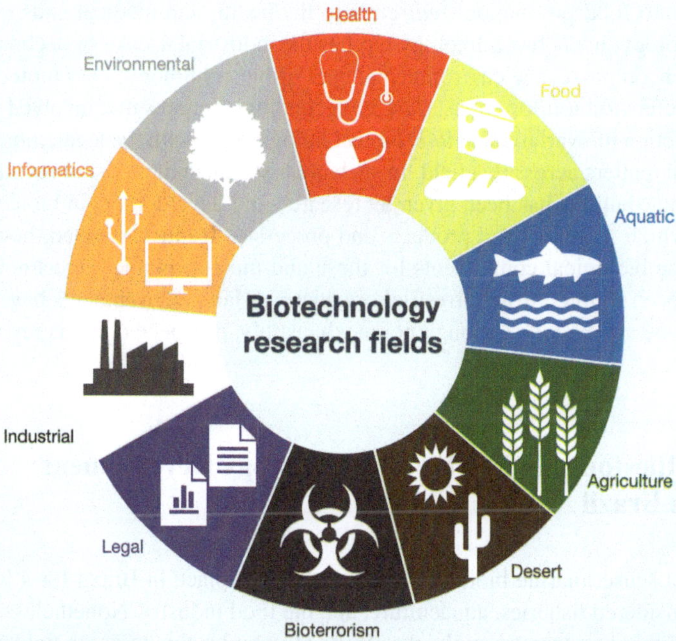

Box 1 – Multiple "Colors" of Biotechnology. Biotechnology, simply defined as *"the application of science and technology to living organism"*, aggregates many research fields and can be as broad as both science and technology mixed together. Herein, we illustrate these different areas of research using the rainbow code created by Kafarski (2012): **green** biotechnology refers to the application of biotechnology to agriculture; **blue** includes aquaculture, coastal and marine biotechnology; **brown** refers to aride zone and desert biotechnology; **yellow** denotes the biotechnology that goes into food production; **red** refers to the health and medical ends of biotechnology; **white** biotechnology typifies the industrial uses of biotechnology; **grey** encompasses the environmental uses of biotechnology; **gold** biotechnology, refers to bioinformatics; **dark** biotechnology denotes that involved in the production biowarfare; and **violet** biotechnology deals with the legal, moral and ethical matters across this field. In such contexts, an endless list of revolutionary contributions has been given by research and development in biotechnology which spans beyond products and processes. It must be noted, however, that the biological components for these and most available biotechnologies have been sourced from terrestrial organisms, as land environments have been langsyne and widely, although not inexhaustibly, prospected in this respect.

Biotechnology simple defined as "*the application of science and technology to living organism*" aggregates many research fields and can be as broad as both science and technology mixed together. Herein, we illustrate these different areas of research using the rainbow code created by Kafarski (2012): *green* biotechnology refers to the application of biotechnology to agriculture; *blue* includes aquaculture, coastal and marine biotechnology; *brown* refers to aride zone and desert biotechnology; *yellow* denotes the biotechnology that

(continued)

Box 9.1 (continued)

goes into food production; *red* refers to the health and medical ends of biotechnology; *white* biotechnology typifies the industrial uses of biotechnology; *grey* encompasses the environmental uses of biotechnology; *gold* biotechnology, refers to bioinformatics; *dark* biotechnology denotes that involved in the production biowarfare; an *violet* biotechnology deals with the legal, moral and ethical matters across this field. In such contexts, an endless list of revolutionary contributions has been given by research and development in biotechnology which spans beyond products and processes. It must be noted, however, that the biological components for these and most available biotechnologies have been sourced from terrestrial organism, as land environments have been langsyne and widely, although not inexhaustibly, prospected in this respect.

9.2 Milestones in Marine Biotechnology Development in Brazil

In a strict sense, marine biotechnology has been in place in Brazil for a long time when considered fisheries, aquaculture, and the food industry. Nonetheless, looking through a more systematic angle, the efforts to organize this research field under the marine biotechnology umbrella faced a significant growth only at the beginning of this century, with the launch of "BIOMAR" (*Programa de Levantamento e Avaliação do Potencial Biotecnológico da Biodiversidade Marinha;* Program for the Survey and Assessment of the Biotechnological Potential of Marine Biodiversity) and the creation of "REDEALGAS" (*Rede Nacional de Pesquisa em Biotecnologia de Macroalgas Marinhas*; National Network in Biotechnology of Marine Macroalgae), both in 2005. In the following years, many grant applications were opened covering mainly the bioprospection of marine pharmaceuticals and production of biodiesel. These opportunities enabled the consolidation of research groups in these and even other areas, ensuing the 2010 publication "*Caracterização do Estado da Arte em Biotecnologia Marinha no Brasil*" (State of the Art Characterization in Marine Biotechnology in Brazil), embraced by governmental organs and academic institutions, which well summarizes the proceeds up to that point. Ten years later, two Brazilian publications addressing relevant topics in these matters were launched: "*Fronteiras do Conhecimento em Ciências do Mar*" (Frontiers in Marine Sciences Knowledge), which discussed cutting-edge themes, including a large section on topics related to biotechnology, in marine sciences, and "*Biotecnologia Marinha*" (Marine Biotechnology), which presents a comprehensive overview of the field, including study designs and methods as well and products and processes developed by means of marine biotechnology (Lana and Castello 2020; Thompson and Thompson 2020). Both these 2020 publications came together through the research and organization of academics and their institutions, supported by solid grounds provided by the Ministry of Education and of Science, Technology, and Innovation as well as by the Brazilian Navy.

In this context, the Brazilian Navy also played a direct role by creating and maintaining many initiatives in the country, such as the research program "PROARQUIPELAGO" (*Programa Arquipélago de São Pedro e São Paulo*; Program for the Saint Peter and Saint Paul's Archipelago), in 1998, to support the research at the Archipelago. Later, the program "PROTRINDADE" (*Programa de Pesquisas na Ilha da Trindade*; Program for the Research at Trindade Island) has officially been advancing the knowledge of the Trindade and Martin Vaz islands and surroundings since 2007 (SECIRM 2017; Oliveira et al. 2018). Indeed, the Brazilian Navy has labeled the Brazilian EEZ as the "Blue Amazon", alluding to the Amazon rainforest, to draw attention and add value to the Brazilian marine biodiversity. Figure 9.1 shows a timeline with the main events and supports for research that catalyzed the development of marine biotechnology in Brazil.

Results emerging from the various Brazilian research groups dedicated to this field have typically been divulged within an assortment of national scientific events. However, beginning in 2005, the need for specific meetings on marine biotechnology were answered and those have been held in the country to best accommodate and, furthermore, unite and connect the field. Only to mention a few, the Redealgas Workshops have transpired through 7 editions since 2005; the Biofouling, Benthic Ecology and Marine Biotechnology Meetings were carried out in 2015, 2017 and 2019; and the I Marine Biotechnology Symposium occurred in 2018. Notably, in 2016, the XV International Symposium on Marine Natural Products (MaNaPro) was held, for the first time, in the American continent, joining over 180 scientists at Cumbuco Beach, on the Northeast coast of Brazil. Rewards collected from these early initiatives were reviewed by Thompson et al. (2018), who presented a historical overview and further discussed the state-of-art of the development of marine biotechnology in Brazil. Therein, the authors show that among the 381 research groups self-categorized under the subject "marine biotechnology" in the Brazilian Research Council (CNPq) directory, algae was the most studied group of organisms, followed by marine bacteria.

In 2018, the establishment of the National Research Network in Marine Biotechnology (Rede BIOTECMAR by the Brazilian government, through the Ministry of Science, Technology, Innovation and Communication (MCTIC) aimed at supporting Brazilian bioeconomy (www.biotecmar.sage.coppe.ufrj.br), illustrates the full-grown interest and recent national organization of this research field. Furthermore, to foster the development of human resources with technical and critical capabilities for the sustainable exploration of marine assets, 2016 witnessed the creation of the first graduate programs dedicated specifically to this subject: Marine Biotechnology, based in the coastal city Arraial do Cabo, RJ, with the participation of scientists from the Instituto de Estudos do Mar Almirante Paulo Moreira (IEAPM), Universidade Federal Fluminense (UFF) and Universidade Federal do Rio de Janeiro (UFRJ), and Bioproducts and Bioprocesses—assembled in the harbor city of Santos, SP by researchers from the Universidade Federal de São Paulo (UNIFESP).

9.3 Biotechnological Innovation from Coral Reefs

9.3.1 Marine Pharmaceuticals

Considering the broad definition of marine biotechnology and its potential impacts on human life, marine drugs are among the paramount and most coveted bioproducts. Natural products have been recognized as one of the pillars for drug development by the pharmaceutical industry, since the origins of current drugs are commonly natural products, in at least half of the substances used in therapeutics (Newman and Cragg 2020). Herein we will focus on active principles obtained from the most studied Brazilian coral reef organisms, including those exhibiting anticancer, antimicrobial, antiparasitic and anti-inflammatory properties.

In Brazil, there are no pharmaceutical products developed from a native marine source (Thompson et al. 2018). However, there are many examples of ongoing studies emphasizing the biotechnological potential of Brazilian marine biodiversity, mainly with coral reef associated organisms, including macroalgae, sponges, tunicates, cnidarians, and microorganisms (Fernandes et al. 2014; Ióca et al. 2018; Wilke et al. 2021). A recent review discussed the anticancer potential of compounds produced by invertebrates and microorganisms collected in Brazilian marine littoral. A quantitative analysis showed that 238 different marine organisms were studied, leading to the isolation of 393 compounds, from which 61 showed anticancer properties and 31 were investigated concerning the mechanism of action involved in their anticancer potential (Wilke et al. 2021). These numbers are much greater when considered studies with macroalgae, since Brazil has a privileged position in terms

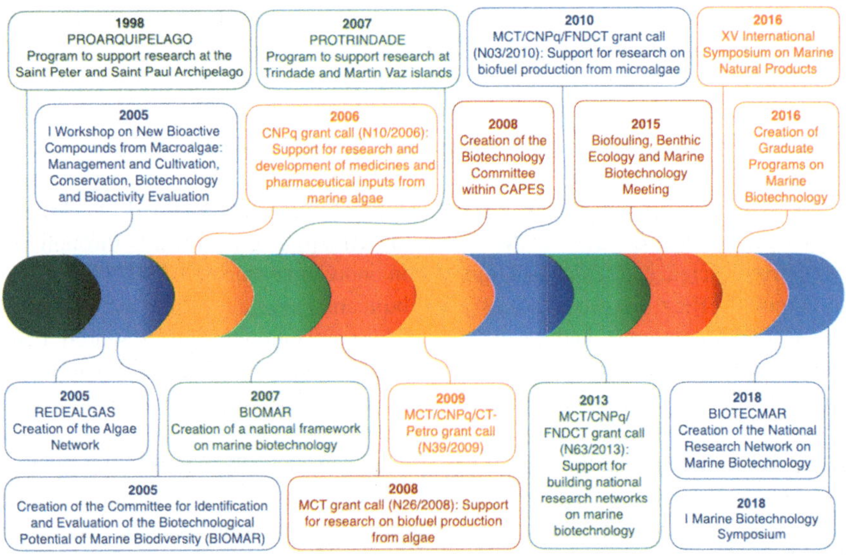

Fig. 9.1 Milestones in marine biotechnology development in Brazil

of macroalgal diversity, and the scientific community is very active in this line of research (Thompson et al. 2018).

Considering the promising anticancer compounds, sulfated glycosaminoglycans with antimetastatic activity obtained from solitary ascidians commonly found at Brazilian tropical rocky shores, *Phallusia nigra* and *Styela plicata* (Fig. 9.2), are among the most notable ones (Pavão et al. 1995, 1998; Abreu et al. 2019). Succinctly, these macromolecules showed the ability to immobilize P-selectin, an endogenous glycoprotein related to cell-cell adhesion, attenuating metastasis development in mice (Kozlowski et al. 2011; Abreu et al. 2019; Silva et al. 2020). These studies are in progress, with a particular focus on further preclinical evaluation of the compounds and cultivation of the ascidians to assure an ecologically and economically sustainable supply chain.

Zoanthids, also known as sea mat corals, are among the most common benthic marine organisms in Brazilian tropical shallow water communities, including coral reefs (Boscolo and Silveira 2005; Rabelo et al. 2015; Santos et al. 2016). The most recent survey (Santos et al. 2016) described the occurrence of at least 13 species of zoanthids in the Brazilian coast, with 4 species of *Palythoa—P. caribaeorum, P. variabilis, P. grandiflora*, and *P. aff. clavate—*, being *P. caribaeorum* the most abundant and most widespread species. There are many studies describing the biotechnological properties of *Palythoa* spp. and their associated microbiota (Table 9.1). The molecules isolated thereof belong to different chemical classes, including, but not limited to ceramides, fatty acids, mycosporins, palytoxins, peptides, sterols, zoanthoxanthins, prostaglandins, lipidic α-amino acids and pyrazines (Wilke et al. 2021). Biological activities related to the substances obtained from *Palythoa* spp. and associated bacteria are broad, ranging from those related to the known intoxicating potential of these invertebrates to a myriad of other pharmacological activities, such as antioxidant, anticancer, anti-inflammatory, antiprotozoal and neuroprotective. An interesting example can be given by the TBX2 modulatory properties displayed by the DNA-damaging agent chromomycin A5, an aureolic acid isolated from a *Streptomyces* sp. strain associated with *P. caribaeorum*. TBX2 is a transcription factor related to malignant tumor development. This recent finding is suggestive of a parallel mechanism of action between chromomycins and the anticancer marine drug trabectedin, where binding to specific sequences of DNA displace transcription factors related to cancer cell proliferation and resistance (Colmegna et al. 2014; Sahm et al. 2020; Wilke et al. 2021).

Among the macroalgae that commonly occur in Brazilian coral reefs, *Dictyota* (Phaeophyceae) and *Laurencia* (Rhodomelaceae) are the most studied genera in a biotechnological context. The *Dictyota* genus comprises 98 species (https://www.algaebase.org/search/genus/detail/?genus_id=51), of which 28 are found in Brazil (https://ala-bie.sibbr.gov.br/ala-bie/species/369934#overview), and its representatives have been long recognized as prolific producers of secondary metabolites with antiviral properties. For instance, diterpenes from *D. menstrualis* and *D. pfaffi* have been proven effective against HIV-1, inhibiting both the virus replication, due to direct inhibition of recombinant HIV-1 reverse transcriptase activity, and HIV-1 infection in mammal cells (Barbosa et al. 2004; Pereira et al. 2004; De Souza Pereira et al. 2005; Cavalcanti et al. 2011; Cirne-Santos et al. 2020) and inhibit HSV-1 replicative cycle in Vero cells as well (Abrantes et al. 2010). Furthermore, preclinical evaluation of toxicity and efficacy of the dolabelladienetriol, a diterpene isolated from *D. pfaffi*, reinforced its potential as an anti-HIV-1 drug (Garrido et al. 2017). Interestingly, macroalgal-derived compounds were also active against other pathogenic viruses, such as those that cause Zika, Chikungunya and Herpes simplex, reinforcing the biotechnological potential of this marine resource (Cirne-Santos et al. 2020).

The genus *Laurencia*, comprising 146 taxonomically accepted species, is recognized worldwide as the most profuse genus of macroalgae by extraordinary diversity of produced secondary metabolites. So far, over 1000 compounds were isolated from *Laurencia* spp., including sesquiterpenes, diterpenes, triterpenes, acetogenins, indole alkaloids, aromatic compounds, steroids, and other miscellaneous substances, and halogenation is a common feature among these compounds. Although bioactivity of the isolated compounds has not been as thoroughly investigated as their chemical diversity, there are evidence of the pharmacological potential for these metabolites, especially against cancer (cytotoxicity), microbial (bacteria, virus, and fungi) and parasitic (malaria, leishmaniasis and Chagas) diseases (revised by Harizani et al. 2016). Elatol, a halogenated chamigrane sesquiterpene first described by Sims et al. (1974) from *Laurencia elata*, was isolated from different

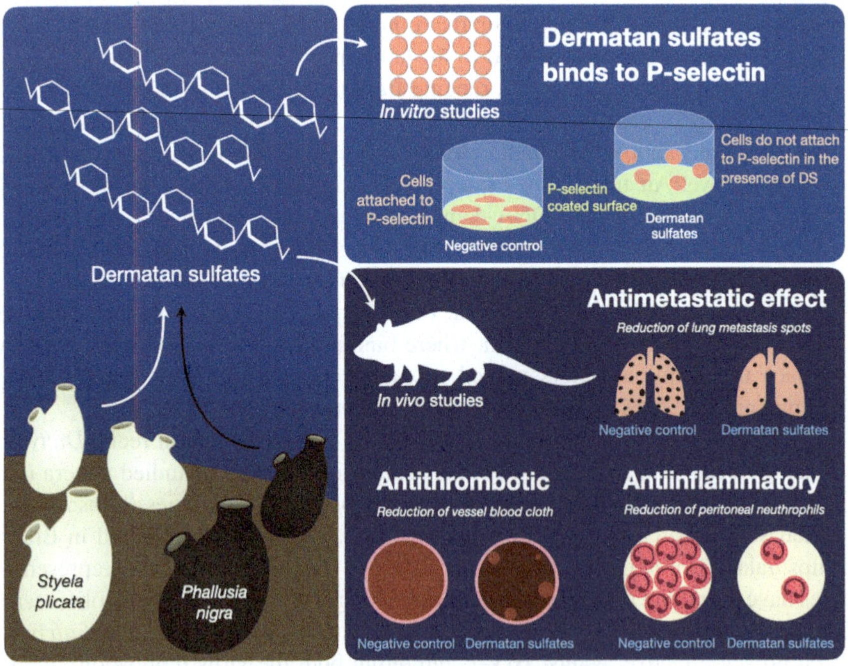

Fig. 9.2 Sulfated glycosaminoglycans obtained from solitary ascidians *Phallusia nigra* and *Styela plicata* display several pharmacological effects

Laurencia spp. collected in Brazil and one of the most studied metabolites during the last two decades. There have been reports on its anticancer, antiparasitic and antiviral activities (Dos Santos et al. 2010; Desoti et al. 2012; Cristina Desoti et al. 2014) but, mostly, this molecule has been assessed for its feeding deterrent and antifouling properties (Hay et al. 1987; Harizani et al. 2016; Sudatti et al. 2016), which will be better discussed later in this chapter.

9.3.2 Cosmetics and Personal Care

The cosmetic industry is increasingly turning to the oceans in the search for new materials or active ingredients. Marine organisms are exceptional sources of substances with different applications and of great interest for the cosmetic industry (Brunt and Burgess 2018). The large spectrum of biological properties of marine molecules such as polysaccharides, terpenes, polyketides, lipids, alkaloids, and peptides have great potential as cosmetics as well as cosmeceuticals. They exhibit photo-protective, anti-aging, antimicrobial, antioxidant, and moisturizing activities that have been shaped and naturally selected through millions of years of evolution. Beyond these properties, marine organisms are also sources of other ingredients of great importance for the cosmetic industry, such as excipients and additives (Guillerme et al. 2017).

According to the Federal Food, Drug and Cosmetic Act of the United State Food and Drug Administration and the European Commission regulation No 1223/2009 on cosmetic products, they are defined as any chemical element, its compounds in either natural or processed form and mixture of compounds intended to care for hygiene or to enhance conditions and beautify external human body parts and the oral cavity. Cosmeceuticals are cosmetic products with bioactive ingredients purported to have medical benefits.

The cosmetic properties of coral reef organisms are still poorly known but ingredients and compounds from organisms occurring in these environments have been investigated. Among them, macroalgae have stood out since the age of Hippocrates (460–370 BC) (Voultsiadou Eleni 2010; Pereira 2018). This interest is highlighted in the international market in countless products containing from raw algal material to specific macroalgal substances in their compositions (Pimentel et al. 2018). The Rodophyta *Aspararagopsis armata* is easily found in coral reef areas and they are already commercialized under the name Ysaline100® in the USA and Europe as a cosmeceutical in shampoos and skin care products due to its antimicrobial activity (Pinteus et al. 2018). This product smooths the appearance of dandruff and acne, which are caused by fungi and bacteria. It is known that halogenated organic compounds in extracts of *A. taxiformis* are responsible for the antimicrobial activity present in this alga (Pinteus et al. 2018). Invertebrates have also been used for cosmetic applications and personal care in different populations and cultures since ancient times (Voultsiadou Eleni 2010). The most successful case of cosmetics with

Table 9.1 Biotechnological potential of *Palythoa* species and its associated microbiota found in Brazil*

Biotechnological Application	Studied species	Type of chemical sample	Main findings	Sampling site	Reference
Anticancer	*P. variabilis*	Lipidic α-amino acids	Lipidic a-amino acids isolated from a natural source for the first time. Antiproliferative effect on glioblastoma (SF-295), colon cancer (HCT-8) and leukemia (HL-60) cell lines. Induction of apoptosis in HL-60 cells	Paracuru Beach, Ceara State	Wilke et al. (2010) and Wilke et al. (2009)
Anticancer	*P. variabilis*	Ergostan sterols	Induction of colon cancer (HCT 116) cell death at μM concentrations	Paracuru Beach, Ceara State	Pinto et al. (2017)
Anticancer	*Streptomyces* sp. BRA-384 recovered from *P. caribaeorum*	Chromomycins (A_5, A_6, A_7 and A_8)	Identification of three new chromomycins with antiproliferative activity at low nM concentrations against melanoma (501mel, WM293A and MM200), rhabdomyosarcoma (RD and RH30) and breast (MCF-7), colon (HCT 116) and prostate cancer (PC-3 M) cell lines. Binding of chromomycins A_5 and A_6 to TBX2, a transcription factor associated with tumor development for melanoma, sarcoma, and breast cancers	Paracuru Beach, Ceara State	Pinto et al. (2020) and Sahm et al. (2020)
Anticancer	Actinobacteria strains: BRA-479, BRA-480, BRA-481, BRA-489, BRA-490 recovered from *P. caribaeorum*	Chromomycins (A_2 and A_3) and dimethylchromomycin A_2	Ethyl acetate extracts depicted antiproliferative activity against colon cancer cells (HCT 116) at pM to low nM concentrations	Trindade Islands, Espírito Santo State	Jimenez et al. (2017)

Antioxidant	*P. caribaeorum*	Hydroalcoholic extract, dichloromethane, ethyl acetate and aqueous fractions	Samples displayed radical scavenging properties and toxicity on the *Artemia* sp. model	Paracuru Beach, Ceara State	de Alencar et al. (2015)
Antiprotozoal	*P. caribaeorum*	Ethanolic extract	Potent growth inhibition of *Leishmania amazonensis* and *Trypanossoma cruzi*	Ponta de Mangue Beach, Alagoas State	Lhullier et al. (2020)
Hemostasis and inflammation control	*P. caribaeorum*	Mucus	Crude mucus is described as a putative source of enzymes and lectins with hemostatic and anti-inflammatory properties.	Porto de Galinhas Beach, Pernambuco State.	Guarnieri et al. (2018)
Intoxication studies	*P. caribaeorum*	Palytoxin	Among the most potent marine toxins with broad biological properties: Na^+/K^+ ATPase inhibitor; potential skin toxicity associated to dinoflagellates blooms; necrosis of intestinal cells; time-dependent hemolysis; electrical changes on heart; pharmacological tool as an inducer of skin cancer	Caribbean Sea, Puerto Rico	Béress et al. (1983), Pelin et al. (2011), Sheridan et al. (2005), Tosteson et al. (1995), Valverde et al. (2008), Wattenberg (2007), and Weidmann (1977)

(continued)

Table 9.1 (continued)

Biotechnological Application	Studied species	Type of chemical sample	Main findings	Sampling site	Reference
Na^+ channel blockage	*P. caribaeorum*	Low molecular weight peptides (within a mass range of 1800 to 9000 Da)	Neurotoxins from nematocysts act as modulators of multiple voltage-gated channels on superior cervical neuron ganglion culture. The first cnidarian venom known for inhibiting calcium channels	La Gallega coral reef, Veracruz State, Mexico	Lazcano-Pérez et al. (2016)
Neuroprotective	*P. caribaeorum*	Kunitz-like peptide-3 (PcKuz-3)	PcKuz-3 is a new toxin with potent antilocomotor activity in zebrafish larvae. Additionally, PcKuz3 suppresses 6-OHDA-induced neurotoxicity in zebrafish. PcKuz-3 is also a potential voltage gated potassium-channel inhibitor according to several *in silico* studies	*In silico* data	Liao et al. (2018a, b)
Neuroprotective	*P. caribaeorum*	Polypeptide containing ShK domain, named PcShK3	PcShK3 displayed neuro- and cardio-protective effects at sub-IC_{50} concentrations and cardiac toxicity at concentrations above IC_{50} value. PcShK3 is a potential voltage gated potassium-channel inhibitor according to *in silico* studies	*In silico* data	Liao et al. (2018a)
Neuroprotective	*P. variabilis*	Pp V-shape α-helical peptide (PpVα)	PpVα has potential application of PpVα for management of epileptic seizures and for protection of dopaminergic neurons from degeneration. PpVα peptide serves as a new structural template for development of Na^+ channel blockers	*In silico* data	Liao et al. (2019)

Neurotoxic phospholipase	*P. caribaeorum*	A2-PLTX-Pcb1a	Pharmacological tool for neurotoxicity studies	La Gallega coral reef, Veracruz State, Mexico	Cuevas-Cruz et al. (2019)
Petroleum derivatives degradation	*Alcanivorax dieselolei* identified in the mucus of *P. caribaeorum*	Mucus of *P. caribaeorum*	Hydrocarbon degrading bacterium *Alcanivorax dieselolei* identified through analysis of DNA sequencing	Porto de Galinhas Beach, Pernambuco State	Campos et al. (2015)

[a]Studies with species found in the Brazilian Economic Exclusive Zone (EEZ) were included even with specimens collected abroad Brazil's EEZ or obtained by *in silico* analysis

active principles obtained from a marine invertebrate is the anti-wrinkle cream containing pseudopterosins, diterpene glycosides isolated from the Caribbean gorgonian *Pseudopterogorgia elisabethae*. In fact, Resilience®, made up from active ingredients obtained from *P. elisabethae*, represents a strong and successful line of marine cosmeceuticals developed by the American manufacturer Estée Lauder (Kijjoa and Sawangwong 2004).

In Brazil, successful initiatives for the use of algal products in cosmetic production have been observed in the last few years, revealing an enormous potential of the macroalgal industry (Marinho-Soriano 2017). Projects for the sustainable use of these organisms in hair and skin care products have been developed by small companies from traditional coastal communities in northeastern Brazil, such as "*Algas do Brasil*" and "*Mulheres de Corpo e Alga*" (www.algasdobrasil.com.br; www.brasilcidadao.org.br). *Gracilaria* spp. are the main macroalgae farmed and exploited by these communities, in addition to being one of the best studied genera among the phycological community in Brazil. Several studies associating phycologists, taxonomists, chemists, pharmacologists and professionals with other expertise has been carried out with the objective of broadening knowledge on its physiology, chemical composition, biological activities and on means for improving biomass productivity (Bezerra and Marinho-Soriano 2010; Cardozo et al. 2011; Guaratini et al. 2012; Marinho-Soriano 2017; Barros-Gomes et al. 2018). These macroalgae produce compounds such as polysaccharides, polyunsaturated fatty acids (PUFA's), carotenoids and mycosporine-like amino acids (MAAs) that are of particular interest to the cosmetic industry (Michalak et al. 2020).

Agar is one of the cell wall components in macroalgae and *Gracilaria* species are a rich source of this polysaccharide. This compound has been incorporated into creams and lotions, providing numerous cosmetic benefits, including anti-aging and anti-wrinkle properties, further enhancing dermal fibroblast proliferation, deposition of collagen and other matrix factors, in addition to inhibition of matrix metalloproteinases and serine protease-elastase (Couteau and Coiffard 2016). The use of this type of substance as cosmetics extends the perspectives for the local algal-farmers who, until the 1990s, would send all its agarophyte production to the international market (Marinho-Soriano 2017).

Photo-oxidative damage is the result of ultraviolet (UV) irradiation due to generation of reactive oxygen species (ROS). MAAs, PUFA's, phenolic compounds (*e.g.* phlorotannins in brown macroalgae), carotenoids (*e.g.* fucoxanthin and tocopherol) and also polysaccharides provide anti-oxidative and photoprotective functions against UV-induced skin damage and they are widely incorporated into anti-aging products. Numerous studies have demonstrated the abundance of these compounds in Brazilian macroalgae, as well as its potentials as antioxidants and protective action (Guaratini et al. 2012; Stein et al. 2013; Barros-Gomes et al. 2018; Torres et al. 2018). For example, Cardozo et al. (2011) showed the high amount of MAAs in three species of *Gracilaria* (*G. birdiae, G. domingensis* and *G. tenuistipitata*) and the quantitative differences between natural populations and lab-grown individuals. The level of these compounds occurring in field populations was 150 times greater than in those cultivated *in vitro*.

Photoprotective and antioxidant compounds have been isolated from marine microorganisms since they are prolific producers of substances with imminent use in cosmetics and cosmeceuticals (Corinaldesi et al. 2017). In Brazil, in a study conducted by Maciel et al. (2018) revealed that compounds from the endophytic-fungi *Annulohypoxylon stygium* isolated from the red macroalga *Bostrychia radicans* showed a strong photoprotective potential, further unveiling this fungus as an important source of chromophores to be used as UV-B-filters in sunscreen.

Nearly 20 years ago, the prospects of finding new cosmetics from the sea, especially among coral reef species and their associated fauna, was recognized as 300–400 times more likely than isolating them from a terrestrial source (Bruckner 2002) and presumably remains true today. Nonetheless, the contribution of marine ingredients to the Brazilian cosmetic industry is very limited, especially when compared to the number of plant-based products already commercialized. The popularity and marketing appeal of the natural ingredients have encouraged the industry to engage in a large-scale search for new substances exhibiting cosmetic properties of marine origin and should contribute to improve the number of marine-based products in Brazil.

9.3.3 Antifouling Compounds

After the banishment of TBT (tributyltin) in 2008 by the International Maritime Organization (IMO) following a wide set of evidence of adverse effects on non-target marine organisms—such as imposex in gastropod mollusks—maritime paint industries struggle to find a replacement strategy with low environmental toxicity and high antifouling property. Natural products are potential candidates thereof because they are already in the environment and their side toxicity should be a circumventable threat. The concept behind the use of compounds produced by marine organisms is that, theoretically, these materials would be less harmful to the environment when compared to the heavy metals currently used in antifouling paint matrices, since, in many cases, their mode of action does not implicate in direct toxicity towards the targets, but in repelling the adhering larvae (Da Gama et al. 2008). In this context, several species of macroalgae, sponges and corals that inhabit reef environments, including in Brazilian littoral, have yielded natural products with antifouling properties.

In the first studies in this area in Brazil, the bioactivity of a *Laurencia obtusa* extract (further named *L. dendroidea*), was found to be related to defensive properties against fouling (Da Gama et al. 2002) and herbivores (Pereira et al. 2003). These promising ecological and biotechnological results motivated the development of a rapid, sensitive, and inexpensive method for quantification of small amounts of the sesquiterpene elatol on the surface and within-thallus of *L. dendroidea*, which also revealed an intra-population variability in the production of this metabolite (Sudatti et al. 2006). Furthermore, differences between superficial and within-thallus contents of *L. dendroidea* raised the idea of investigating the dynamics of production and allocation of elatol in this macroalga (Sudatti et al. 2008). Storage, transport, and exudation of secondary metabolites from “corps en cerise” to the surface of *L. dendroidea* were characterized (Salgado et al. 2008), as well as the mean by which they exert their antifouling activity on the surface of this macroalga (Reis et al. 2013). Still, low concentration of secondary metabolites on the surface of *L. dendroidea* was generally insufficient to inhibit fouling by the mussel *Perna perna* and herbivory by the sea-urchin *Lytechinus variegatus* (Sudatti et al. 2008). However, diel variations of these concentrations (Sudatti et al. 2016) can be high enough to inhibit herbivory (Pereira et al. 2003) and fouling (Da Gama et al. 2002).

In a comprehensive study design, macroalgae collected in several localities along the Brazilian coast between 3 and 27 ° S and comprising 2 time periods, pre-2008 and 2008–2019 (Da Gama et al. 2008), were evaluated for antifouling (AF) activity. Throughout the pre-2008 study, 51 populations belonging to 42 species were sampled, extracted, and tested in laboratory and, in some cases, in field assays. The majority of red algal extracts (55%) exhibited strong (> 80%) AF activity, while only 14% of brown macroalgae presented a comparable effect, and 27% of the green macroalgae exhibited moderate (< 80%) AF activity. Among the 5 pure compounds tested, the halogenated sesquiterpene elatol (Salgado et al. 2008), obtained from *L. dendroidea* (Fig. 9.3), and a dolabellane diterpene, isolated from the brown macroalgae *Dictyota pfaffii* (Barbosa et al. 2007), exhibited significant AF activity, which were confirmed by further field testing.

During the second period of the study (2008–2019), a total of 60 macroalgal species were sampled and tested for AF activity of which 21 belonged to the phylum Ochrophyta (brown macroalgae) and 20 to the Rhodophyta (red macroalgae). A significant AF activity was displayed by 86% of the Ochrophyta and 70% of the Rhodophyta species. The initial observations for a higher AF activity among red algae (<2008) was not supported in the second set of data and revealed a more equitable distribution of AF activity among red and brown macroalgae.

Despite the promising results obtained in the AF tests, products obtained from non-cultivated organisms usually have limitations to their commercial availability. Elatol, for example, is produced in low yields by *L. obtusa* (Sudatti et al. 2008) and its complex structure hinders chemical synthesis. Still, the production of biocides should not be dependent on large-scale cultivation of marine organisms. Various other studies have set out to search for a solution to overcome issues related to biofouling, but in the field of marine biotechnology there is a lack of research aimed at revealing molecules analogous to those detected in marine organisms with AF activities. Nonetheless, efforts have been made to understand the biogenetic routes for the production of *Laurencia* sesquiterpenes (Batista 2015; De Oliveira et al. 2015) and obtain them through heterologous cloning (Calegario et al. 2016).

For instance, this kind of study was carried out with a synthetic PAF (platelets activator factor) analog, an ether-glycerol-phospholipid with similar chemistry to natural sponge compounds which act at the same cellular level of already tested natural products (Batista 2015). The AF activity and toxicity of this compound was tested against invertebrates’ larvae from target (*Amphibalanus amphitrite* and

Bugula neritina) and non-target species (*Lytechinus variegatus* and *Tubastraea coccinea*). The results showed low toxicity of the synthetic compound when compared with copper sulfate and a potential option for antifouling management and control of invasive species *T. coccinea*.

Revelations on microorganism communication patterns, called quorum sensing (QS), unveiled a potential strategy for biofouling control. Quorum sensing is a complex system of cellular communication, common in Gram-negative bacteria, through which the increased production of certain molecules within microbial communities signalizes for target genes to be activated or repressed. This system coordinates several processes, such as the expression of infectious factors and formation of biofilms (Dobretsov et al. 2013). Microbial biofilm increases or inhibits the fixation of larvae and spores of macro-fouling; thus, the discovery of compounds that inhibit QS bacterial communication may provide a strategy to prevent biological fouling.

On the Brazilian coast, studies have been carried out to prospect, in marine bacteria, compounds capable of inhibiting communication among *Chromobacterium violaceum* CV017, *Agrobacterium tumefaciens* NTL4 and *Pseudomonas aeruginosa* PA01 for potential application in "environmentally safe" antifouling paints (Carvalho et al. 2017; Batista et al. 2018). Assays carried out with bacteria present in macroalgae species from the Cabo Frio upwelling region, showed that over 70% of the extracts efficiently inhibited bacterial QS communication and biofilm formation at laboratory scales (Batista 2015; Carvalho et al. 2017). In addition, methanol extracts from the sponges *Darwinella* sp., *Dysidea etheria, Hymeniacidon heliophila* and *Tedania ignis* from the same region were also efficient inhibitors of QS (Batista et al. 2018). Considering that more than 40% of a sponge's biomass is made up of microsymbionts, a single individual can represent a relevant source of bacteria that produce bioactive compounds with high commercial potential. Unlike macroorganisms, microorganisms can be easily modified genetically to increase the yield of the compounds and their bioactivity, in addition to being cultivable under laboratory conditions.

9.4 Bioprospecting Using an Eco-Evolutionary Framework

More than 20,000 secondary metabolites have been identified from marine organisms, many of which with known biological activities, but only a small fraction have been described for their ecological roles. Due to the great distance between the high number of known natural products and the lack of knowledge related to their ecological roles, an ecologically centered investigative workflow (Fig. 9.4) has been proposed, prioritizing the ecological function over chemical identification (Ledoux and Antunes 2018).

According to this workflow, the definitions of ecological roles would include defense against consumers and competitors (allelopathy), interactions with microorganisms, sexual pheromones, and other ecological functions. The biosynthesis patterns may be explored in biogeographic, inter-/intra-population and

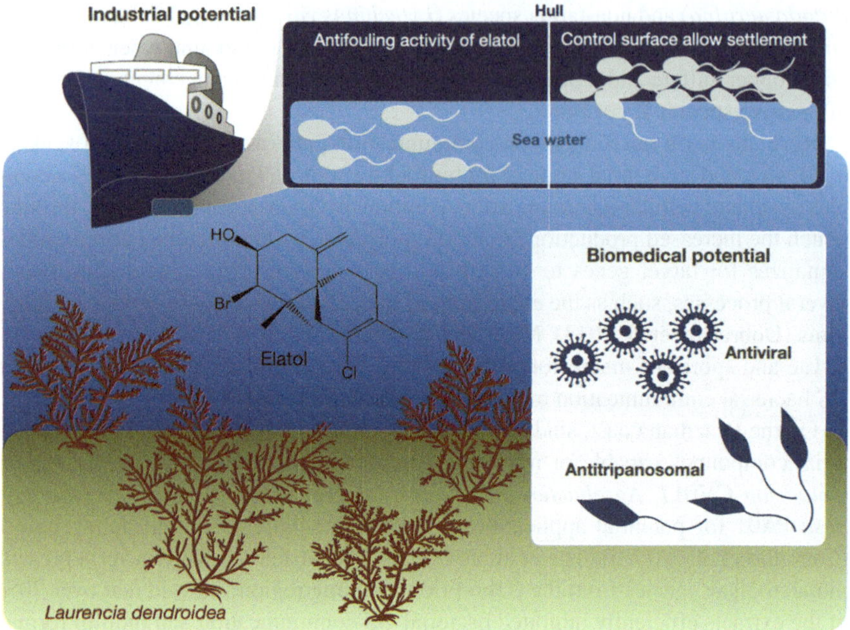

Fig. 9.3 Elatol obtained from *Laurencia dendroidea* depicts industrial and biomedical potential

intra-individual approaches, as well as temporal evaluation, such as seasonal and ontogenetic. The underlying processes within this framework would comprise the effects of biotic and abiotic ecological factors, as well as evolutionary aspects, such as local adaptation, acclimatization, and others.

Studies carried out with *Laurencia* provide unequivocal evidence that this eco-evolutionary approach can be viable and useful, since ecological studies can subsidize bioprospecting and marine biotechnology in general. An eminently ecological study evidenced chemical cues from *L. dendroidea* are detected by its grazer, the mollusk *Aplysia brasiliana*. In this context, 15 sesquiterpenes were isolated for experimental evidence that elatol is responsible for attracting the mollusk to its host (Nocchi et al. 2017) and this small library of sesquiterpenes is a valuable source for future biotechnological approaches.

Stimulated and aided by various ecological studies, the pharmacological potential of sesquiterpene elatol was evidenced. Elatol exhibited a potent antiproliferative activity against promastigote and intracellular amastigote forms of the *Leishmania amazonensis* protozoan and may be advantageous for the development of new anti-leishmanial chemotherapy treatments (Torres et al. 2018). It has also shown an effective trypanocidal activity against *Trypanosoma cruzi* and good potential for the treatment of Chagas' disease (Veiga-Santos et al. 2010; Desoti et al. 2012). Due to such high biotechnological potential of elatol, efforts have also been made to obtain this metabolite from non-wild sources (see Pereira and Costa-Lotufo 2012) by expanding the yet scarce information about genes involved in the biosynthesis of terpenoids, and specifically elatol, in *L. dendroidea*. Indeed, the biosynthesis of terpenoid precursors was shown to involve 20 genes and a further suite of different genes that catalyze chemical modifications on these precursors, thus accounting for the high diversity of terpenoids that occur in this species (de Oliveira et al. 2012; De Oliveira et al. 2015). These pioneering studies also demonstrated the role of mevalonate pathway in the biosynthesis of terpenes in *L. dendroidea*, which subsidizes a heterologous biosynthesis of its bioactive terpenes.

These examples of ecological roles connected with omics approaches clearly follow the eco-evolutionary workflow. In addition, this approach contributes to reducing the gap between the number of known natural products and knowledge regarding their ecological roles (see Ledoux and Antunes 2018). In addition, this workflow rationalizes bioprospection, since secondary metabolites are adaptive chemicals produced in response to environmental pressures and not by the action of human pathogens or diseases. By these means, ecological approaches are more likely to reveal the function of secondary metabolites, which can be then translated into a pharmacological application. Therefore, conveying an ecological-evolutionary workflow into bioprospection increases the chances of a successful outcome, while also optimizing the use of the collected biological material.

9.5 Final Remarks

Marine biotechnology is considered an area of great interest that holds a vast potential in contributing to the growth of a sustainable and efficient society (Buonocore 2012). Due to their high diversity of organisms, coral reefs represent an important source for marine biotechnology development, as they host many species of macroalgae and invertebrates, including Porifera, Cnidaria, Bryozoa and Tunicata, which represent major target groups among bioprospecting programs and have yielded several goods and services that are currently in use. Furthermore, coral reefs shelter an invisible and mostly untapped richness of microorganisms, which are revealing their capacity and versatility in affording efficacious solutions by means of a biotechnology framework.

Still, beyond their economic or even therapeutic functions, the development of marine biotechnology using resources from coral reefs must certainly unravel through a sustainable and ecologically friendly approach. It is visible and scientifically documented the unprecedented decline or loss suffered by habitat-forming hard corals in recent decades (Boström-Einarsson et al. 2020) and, unquestionably, additional impacts must not pile onto that.

In fact, the rising threats urge for measures that value coral reef conservation, improving resilience and investing in restoration efforts. The accumulation of knowledge on coral biology has made it easier to produce diagnoses and predictions based on molecular biotechnology and artificial intelligence, such as accurate models of susceptibility of corals to environmental stress (Mayfield and Chen 2019). Hence, biotechnology enables the identification of reefs that should be prioritized

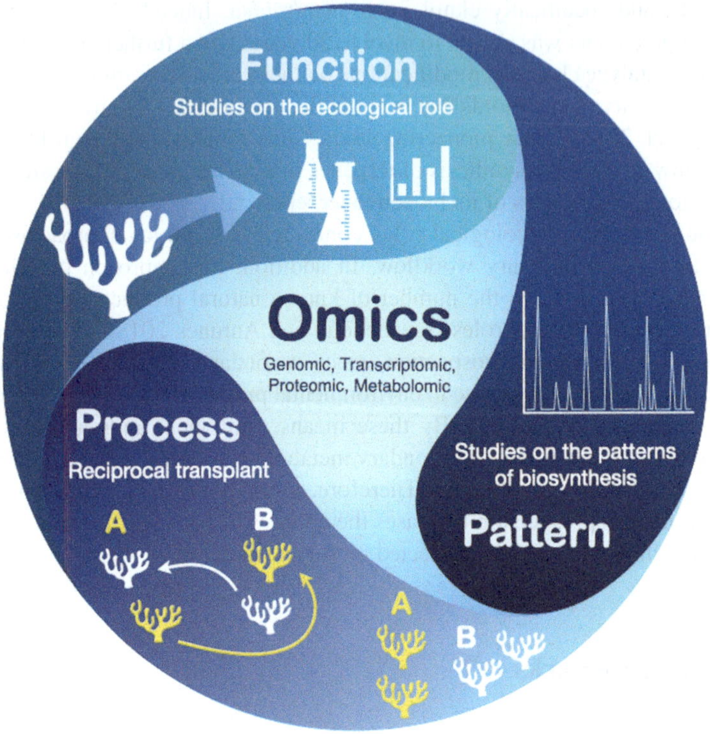

Fig. 9.4 Ecology-focused workflow, comprising three approaches: definition of ecological roles, description of biosynthesis patterns and inference of underlying processes. (Modified from Ledoux and Antunes 2018)

for conservation under the effects of ongoing global change. Understanding the genomic basis of reef coral hybridization, a known phenomenon occurring among reef-building corals in nature, and interspecific hybridization may produce resistant genotypes to ongoing global changes to be useful in coral reef restoration initiatives (Vollmer and Palumbi 2002; Chan et al. 2018; Mao 2020). In this scenario, marine biotechnology has also shown a remarkable potential to provide innovative solutions in these matters. Efforts on cryopreservation (Cirino et al. 2019), microbiota restoration (Ziegler et al. 2017; Damjanovic et al. 2017), coral gardening and propagation (Barton et al. 2015; Boström-Einarsson et al. 2020) are already proving to be efficient in coral restoration. Unfortunately, coral restoration enterprises are dominated by short-lived projects, most enduring only a few months of monitoring, and small scaled assessments, incorporating merely medium sized restored areas. Moreover, these initiatives generally use Indo-pacific fast-growing branched corals (Boström-Einarsson et al. 2020) and the vast majority do not employ state-of-art techniques. Although many Brazilian scientists have the expertise to run these kinds of experiments, they also rely on the same indo-pacific species that can be easily grown in aquariums (Peixoto et al. 2017).

It has been implied that chemically mediated interactions between benthic organisms seem to have their best expression and highest ecological repercussions upon population regulation and community structuring (Hay 2009). In this sense, greater awareness in marine chemical ecology has an extraordinary potential to enable purposeful biotechnologies to support coral reef restoration and conservation. Reciprocally, preserved ecosystems contain peculiar genetic resources and, therefore, escalate biotechnological opportunities in providing products and processes for economically and ecologically prosperous societies.

References

Abrantes JL, Barbosa J, Cavalcanti D et al (2010) The effects of the diterpenes isolated from the Brazilian brown algae Dictyota pfaffii and Dictyota menstrualis against the Herpes Simplex type-1 replicative cycle. Planta Med 76:339–344. https://doi.org/10.1055/s-0029-1186144

Abreu WS, Soares PAG, Motta JM et al (2019) Tunicate heparan sulfate enriched in 2-sulfated β-glucuronic acid: structure, anticoagulant activity, and inhibitory effect on the binding of human colon adenocarcinoma cells to immobilized P-selectin. Mar Drugs 17:351. https://doi.org/10.3390/md17060351

Appeltans W, Ahyong ST, Anderson G et al (2012) The magnitude of global marine species diversity. Curr Biol 22:2189–2202. https://doi.org/10.1016/j.cub.2012.09.036

Arrieta JM, Arnaud-Haond S, Duarte CM (2010) What lies underneath: conserving the oceans' genetic resources. Proc Natl Acad Sci 107:18318–18324. https://doi.org/10.1073/pnas.0911897107

Barbosa JP, Pereira RC, Abrantes JL et al (2004) In vitro antiviral diterpenes from the Brazilian brown alga Dictyota pfaffii. Planta Med 70:856–860. https://doi.org/10.1055/s-2004-827235

Barbosa JP, Fleury BG, Teixeira VL et al (2007) Natural products as antifoulants in a brazilian brown alga Dictyota pfaffii (Phaeophyta, Dictyotales). Biochem Syst Ecol 35:549–553

Barros-Gomes JAC, Nascimento DLA, Silveira ACR et al (2018) In vivo evaluation of the antioxidant activity and protective action of the seaweed Gracilaria birdiae. Oxidative Med Cell Longev 2018:1–12. https://doi.org/10.1155/2018/9354296

Barton JA, Willis BL, Hutson KS (2015) Coral propagation: a review of techniques for ornamental trade and reef restoration. Rev Aquac 9:238–256

Batista WR (2015) The synthesis of glycerophospholipids for use as a booster bioced in antifouling coatings. Quim Nova 38:971–923

Batista D, Costa R, Carvalho AP et al (2018) Environmental conditions affect activity and associated microorganisms of marine sponges. Mar Environ Res 142:59–68. https://doi.org/10.1016/j.marenvres.2018.09.020

Béress L, Zwick J, Kolkenbrock HJ et al (1983) A method for the isolation of the caribbean palytoxin (C-PTX) from the coelenterate (zooanthid) Palythoa caribaeorum. Toxicon 21:285–290. https://doi.org/10.1016/0041-0101(83)90013-2

Bezerra AF, Marinho-Soriano E (2010) Cultivation of the red seaweed Gracilaria birdiae (Gracilariales, Rhodophyta) in tropical waters of Northeast Brazil. Biomass Bioenergy 34:1813–1817. https://doi.org/10.1016/j.biombioe.2010.07.016

Bhatia S, Goli D (2018) Introduction to pharmaceutical biotechnology, Volume 1 Basic techniques and concepts. IOP Publishing, London

Bocquet-Appel JP (2011) When the world's population took off: the springboard of the neolithic demographic transition. Science 333:560–561. https://doi.org/10.1126/science.1208880

Boscolo HK, Silveira FL (2005) Reproductive biology of Palythoa caribaeorum and Protopalythoa valilbilis (Cnidaria, Anthozoa, Zoanthidea) from the southeastern coast of Brazil. Brazilian J Biol 65:29–41. https://doi.org/10.1590/S1519-69842005000100006

Boström-Einarsson L, Babcock RC, Bayraktarov E et al (2020) Coral restoration—a systematic review of current methods, successes, failures and future directions. PLoS One 15:e0226631. https://doi.org/10.1371/journal.pone.0226631

Bruckner AW (2002) Life-saving products from coral reefs | issues in science and technology. Issues Sci Technol 18:39–44

Brunt EG, Burgess JG (2018) The promise of marine molecules as cosmetic active ingredients. Int J Cosmet Sci 40:1–15. https://doi.org/10.1111/ics.12435

Buonocore F (2012) Marine biotechnology: developments and perspectives December. J Aquac Res Dev 4:1

Calegario G, Pollier J, Arendt P et al (2016) Cloning and functional characterization of cycloartenol synthase from the red seaweed Laurencia dendroidea. PLoS One 11:165954. https://doi.org/10.1371/journal.pone.0165954

Campos FF, Garcia JE, Luna-Finkler CL et al (2015) Alcanivorax dieselolei, uma bactéria degradadora de alcanos associada ao muco do zoantídeo Palythoa caribaeorum (Cnidaria, Anthozoa). Braz J Biol 75:431–434. https://doi.org/10.1590/1519-6984.16113

Cardozo KHM, Marques LG, Carvalho VM et al (2011) Analyses of photoprotective compounds in red algae from the Brazilian coast. Braz J Pharmacogn 21:202–208. https://doi.org/10.1590/S0102-695X2011005000047

Carvalho AP, Batista D, Dobretsov S, Coutinho R (2017) Extracts of seaweeds as potential inhibitors of quorum sensing and bacterial growth. J Appl Phycol 29:789–797. https://doi.org/10.1007/s10811-016-1014-1

Cavalcanti DN, de Oliveira MAR, de-Paula JC et al (2011) Variability of a diterpene with potential anti-HIV activity isolated from the Brazilian brown alga Dictyota menstrualis. J Appl Phycol 23:873–876. https://doi.org/10.1007/s10811-010-9601-z

Chan WY, Peplow LM, Menéndez P et al (2018) Interspecific hybridization may provide novel opportunities for coral reef restoration. Front Mar Sci 5:160. https://doi.org/10.3389/fmars.2018.00160

Cirino L, Wen ZH, Hsieh K et al (2019) First instance of settlement by cryopreserved coral larvae in symbiotic association with dinoflagellates. Sci Rep 9:1–8. https://doi.org/10.1038/s41598-019-55374-6

Cirne-Santos CC, de Souza BC, de Oliveira MC et al (2020) In vitro studies on the inhibition of replication of Zika and chikungunya viruses by Dolastane isolated from seaweed Canistrocarpus cervicornis. Sci Rep 10:1–10. https://doi.org/10.1038/s41598-020-65357-7

Colmegna B, Uboldi S, Erba E, D'Incalci M (2014) Resistance to minor groove binders. Drug Discov Today Technol 11:73–79. https://doi.org/10.1016/j.ddtec.2014.03.001

Corinaldesi C, Barone G, Marcellini F et al (2017) Marine microbial-derived molecules and their potential use in cosmeceutical and cosmetic products. Mar Drugs 15:118. https://doi.org/10.3390/md15040118

Couteau C, Coiffard L (2016) Seaweed application in cosmetics. In: Fleurence J, Kevine I (eds) Seaweed in health and disease prevention, 1st edn. Elsevier, London, pp 423–441

Cristina Desoti V, Lazarin-Bidóia D, Bueno Sudatti D et al (2014) Additional evidence of the trypanocidal action of (−)-elatol on amastigote forms through the involvement of reactive oxygen species. Mar Drugs 12:4973–4983. https://doi.org/10.3390/md12094973

Cuevas-Cruz M, Lazcano-Pérez F, Hernández-Guzmán U et al (2019) A novel phospholipase A2 isolated from palythoa caribaeorum possesses neurotoxic activity. Toxins (Basel) 11:89. https://doi.org/10.3390/toxins11020089

Da Gama BAP, Pereira RC, Carvalho AGV, Coutinho R (2002) The effects of seaweed secondary metabolites on biofouling. Biofouling 18:13–20. https://doi.org/10.1080/08927010290017680

da Gama BAP, Carvalho AGV, Weidner K et al (2008) Antifouling activity of natural products from Brazilian seaweeds. Bot Mar 51:191–201. https://doi.org/10.1515/bot.2008.027

Damjanovic K, Blackall LL, Webster NS, van Oppen MJH (2017) The contribution of microbial biotechnology to mitigating coral reef degradation. Microb Biotechnol 10:1236–1243. https://doi.org/10.1111/1751-7915.12769

de Alencar DB, Melo AA, Silva GC et al (2015) Antioxidant, hemolytic, antimicrobial, and cytotoxic activities of the tropical Atlantic marine zoanthid Palythoa caribaeorum. An Acad Bras Cienc 87:1113–1123. https://doi.org/10.1590/0001-3765201520140370

de Oliveira LS, Gregoracci GB, Silva GGZ et al (2012) Transcriptomic analysis of the red seaweed Laurencia dendroidea (Florideophyceae, Rhodophyta) and its microbiome. BMC Genomics 13:487. https://doi.org/10.1186/1471-2164-13-487

De Oliveira LS, Tschoeke DA, De Oliveira AS et al (2015) New insights on the terpenome of the red seaweed Laurencia dendroidea (florideophyceae, rhodophyta). Mar Drugs 13:879–902. https://doi.org/10.3390/md13020879

De Souza PH, Leão-Ferreira LR, Moussatché N et al (2005) Effects of diterpenes isolated from the Brazilian marine alga Dictyota menstrualis on HIV-1 reverse transcriptase. Planta Med 71:1019–1024. https://doi.org/10.1055/s-2005-873113

Desoti VC, Lazarin-Bidóia D, Sudatti DB et al (2012) Trypanocidal action of (−)-elatol involves an oxidative stress triggered by mitochondria dysfunction. Mar Drugs 10:1631–1646. https://doi.org/10.3390/md10081631

Dobretsov S, Abed R, Teplitski M (2013) Mini-review: inhibition of biofouling by marine microoganisms. Biofouling 29:423–441

Dos Santos AO, Veiga-Santos P, Ueda-Nakamura T et al (2010) Effect of elatol, isolated from red seaweed Laurencia dendroidea, on Leishmania amazonensis. Mar Drugs 8:2733–2743. https://doi.org/10.3390/md8112733

Erwin PM, Lopez-Legentil S, Schuhmann PW (2010) The pharmaceutical value of marine biodiversity for anti-cancer drug discovery. Ecol Econ 70:445–451. https://doi.org/10.1016/j.ecolecon.2010.09.030

Fernandes DRP, de Oliveira VP, Yoneshigue Valentin Y (2014) Seaweed biotechnology in Brazil: six decades of studies on natural products and their antibiotic and other biological activities. J Appl Phycol 26:1923–1937. https://doi.org/10.1007/s10811-014-0287-5

Garrido V, Barros C, Melchiades VA et al (2017) Subchronic toxicity and anti-HSV-1 activity in experimental animal of dolabelladienetriol from the seaweed, Dictyota pfaffii. Regul Toxicol Pharmacol 86:193–198. https://doi.org/10.1016/j.yrtph.2017.03.007

Guaratini T, Lopes NP, Marinho-Soriano E et al (2012) Antioxidant activity and chemical composition of the non polar fraction of gracilaria domingensis (Kützing) Sonder ex Dickie and Gracilaria birdiae (Plastino & Oliveira). Braz J Pharmacogn 22:724–729. https://doi.org/10.1590/S0102-695X2012005000063

Guarnieri MC, de Albuquerque Modesto JC, Pérez CD et al (2018) Zoanthid mucus as new source of useful biologically active proteins. Toxicon 143:96–107. https://doi.org/10.1016/j.toxicon.2018.01.012

Guillerme JB, Couteau C, Coiffard L (2017) Applications for marine resources in cosmetics. Cosmetics 4:1–15. https://doi.org/10.3390/cosmetics4030035

Harizani M, Ioannou E, Roussis V (2016) Progress in the chemistry of organic natural products. Springer, Cham

Hay ME (2009) Marine chemical ecology: chemical signals and cues structure marine populations, communities, and ecosystems. Annu Rev Mar Sci 1:193–212

Hay ME, Fenical W, Gustafson K (1987) Chemical defense against coral-reef herbivores. Ecology 68:1581–1591

Ióca LP, Nicacio KJ, Berlinck RGS (2018) Natural products from marine invertebrates and microorganisms in Brazil between 2004 and 2017: still the challenges, more rewards. J Braz Chem Soc 29:998–1031. https://doi.org/10.21577/0103-5053.20180016

Jimenez PC, Wilke DV, Ferreira KQ et al (2017) Potencial Biotecnológico de Organismos Marinhos. In: PROTRINDADE Programa de Pesquisas Cientíﬁcas na Ilha de Trindade 10 anos de pesquisas. Brasília, p 200

Kafarski P (2012) Rainbow code of biotechnology. J Chem 66:811–816
Kijjoa A, Sawangwong P (2004) Drugs and cosmetics from the sea. Mar Drugs 2:73–82
Kozlowski EO, Pavao MSG, Borsig L (2011) Ascidian dermatan sulfates attenuate metastasis, inflammation and thrombosis by inhibition of P-selectin. J Thromb Haemost 9:1807–1815. https://doi.org/10.1111/j.1538-7836.2011.04401.x
Lana P d C, Castello JP (2020) Fronteiras do conhecimento em ciências do mar. FURG, Rio Grande
Lazcano-Pérez F, Castro H, Arenas I et al (2016) Activity of Palythoa caribaeorum venom on voltage-gated ion channels in mammalian superior cervical ganglion neurons. Toxins (Basel) 8:135–145. https://doi.org/10.3390/toxins8050135
Leal MC, Puga J, Serôdio J et al (2012) Trends in the discovery of new marine natural products from invertebrates over the last two decades - where and what are we bioprospecting? PLoS One 7:e30580. https://doi.org/10.1371/journal.pone.0030580
Ledoux JB, Antunes A (2018) Beyond the beaten path: improving natural products bioprospecting using an eco-evolutionary framework—the case of the octocorals. Crit Rev Biotechnol 38:184–198. https://doi.org/10.1080/07388551.2017.1331335
Lhullier C, Moritz MIG, Tabalipa EO et al (2020) Biological activities of marine invertebrates extracts from the northeast Brazilian coast. Brazilian J Biol 80:393–404. https://doi.org/10.1590/1519-6984.213678
Liao Q, Gong G, Siu SWI et al (2018a) A novel ShK-like toxic peptide fromthe transcriptome of the cnidarian Palythoa caribaeorum displays neuroprotection and cardioprotection in zebrafish. Toxins (Basel) 10:238–253. https://doi.org/10.3390/toxins10060238
Liao Q, Li S, Siu SWI et al (2018b) Novel Kunitz-like peptides discovered in the Zoanthid Palythoa caribaeorum through transcriptome sequencing. J Proteome Res 17:891–902. https://doi.org/10.1021/acs.jproteome.7b00686
Liao Q, Li S, Siu SWI et al (2019) Novel neurotoxic peptides from Protopalythoa variabilis virtually interact with voltage-gated sodium channel and display anti-epilepsy and neuroprotective activities in zebrafish. Arch Toxicol 93:189–206. https://doi.org/10.1007/s00204-018-2334-5
Maciel OMC, Tavares RSN, Caluz DRE et al (2018) Photoprotective potential of metabolites isolated from algae-associated fungi Annulohypoxylon stygium. J Photochem Photobiol B 178:316–322
Maida M, Ferreira B (2003) Área de Proteção Ambiental Costa dos Corais. Ministério do Meio Ambiente, Brasília
Mao Y (2020) Genomic insights into hybridization of reef corals. Coral Reefs 39:61–67. https://doi.org/10.1007/s00338-019-01869-7
Marinho-Soriano E (2017) Historical context of commercial exploitation of seaweeds in Brazil. J Appl Phycol 29:665–671. https://doi.org/10.1007/s10811-016-0866-8
Mayfield AB, Chen C-S (2019) Enabling coral reef triage through molecular biotechnology and artificial intelligence. Platax 16:23–47. https://doi.org/10.29926/PLATAX.201912_2019.0002
Michalak I, Dmytryk A, Chojnacka K (2020) Algae cosmetics. In: Se-Kwon K (ed) Encyclopedia of marine biotechnology. John Wiley & Sons Ltd, pp 65–85
Moura RL, Amado-Filho GM, Moraes FC et al (2016) An extensive reef system at the Amazon River mouth. Sci Adv 2:e1501252. https://doi.org/10.1126/sciadv.1501252
Newman DJ, Cragg GM (2020) Natural products as sources of new drugs over the nearly four decades from 01/1981 to 09/2019. J Nat Prod 83:770–803. https://doi.org/10.1021/acs.jnatprod.9b01285
Nocchi N, Soares AR, Souto ML et al (2017) Detection of a chemical cue from the host seaweed Laurencia dendroidea by the associated mollusc Aplysia brasiliana. PLoS One 12:e0187126. https://doi.org/10.1371/journal.pone.0187126
OECD (2013) OECD science, technology and industry scoreboard 2013 © OECD 2013 158 4. Targeting new growth areas 4. Biotechnology R&D
Oliveira JEL, Viana D de L, Souza MAC de (2018) Arquipélago de São Pedro e São Paulo: 20 anos de pesquisa, 1st edn. Via Design Publicações, Recife

Pavão MSG, Mourão PAS, Mulloy B, Tollefsen DM (1995) A unique dermatan sulfate-like glycosaminoglycan from ascidian: its structure and the effect of its unusual sulfation pattern on anticoagulant activity. J Biol Chem 270:31027–31036. https://doi.org/10.1074/jbc.270.52.31027

Pavão MSG, Aiello KRM, Werneck CC et al (1998) Highly sulfated dermatan sulfates from ascidians. Structure versus anticoagulant activity of these glycosaminoglycans. J Biol Chem 273:27848–27857. https://doi.org/10.1074/jbc.273.43.27848

Peixoto RS, Rosado PM, Leite DC de A, et al (2017) Beneficial microorganisms for corals (BMC): proposed mechanisms for coral health and resilience. Front Microbiol 8:341. https://doi.org/10.3389/fmicb.2017.00341

Pelin M, Zanette C, De Bortoli M et al (2011) Effects of the marine toxin palytoxin on human skin keratinocytes: role of ionic imbalance. Toxicology 282:30–38. https://doi.org/10.1016/j.tox.2011.01.010

Pereira L (2018) Seaweeds as source of bioactive substances and skin care therapy-cosmeceuticals, algotheraphy, and thalassotherapy. Cosmetics 5:68–108. https://doi.org/10.3390/cosmetics5040068

Pereira RC, Costa-Lotufo LV (2012) Bioprospecting for bioactives from seaweeds: potential, obstacles and alternatives. Rev Bras 22:894–905

Pereira RC, Da Gama BA, Teixeira VL, Yoneshigue-Valentin Y (2003) Ecological roles of natural products of the Brazilian red seaweed Laurencia obtusa. Braz J Biol 63:665–672. https://doi.org/10.1590/S1519-69842003000400013

Pereira HS, Leão-Ferreira LR, Moussatché N et al (2004) Antiviral activity of diterpenes isolated from the Brazilian marine alga Dictyota menstrualis against human immunodeficiency virus type 1 (HIV-1). Antivir Res 64:69–76. https://doi.org/10.1016/j.antiviral.2004.06.006

Pimentel FB, Alves RC, Rodrigues F, Oliveira MBPP (2018) Macroalgae-derived ingredients for cosmetic industry-an update. Cosmetics 5:2–19. https://doi.org/10.3390/cosmetics5010002

Pinteus S, Lemos MFL, Alves C et al (2018) Marine invasive macroalgae: turning a real threat into a major opportunity—the biotechnological potential of Sargassum muticum and Asparagopsis armata. Algal Res 34:217–234. https://doi.org/10.1016/j.algal.2018.06.018

Pinto FCL, Almeida JGL, Silveira ER et al (2017) Steroids from the Brazilian Zoanthids Palythoa caribaeorum and Palythoa variabilis. J Braz Chem Soc 28:485–491

Pinto FCL, Silveira ER, Vasconcelos ACL et al (2020) Dextrorotatory chromomycins from the marine Streptomyces sp. associated to Palythoa caribaeorum. J Braz Chem Soc 31:143–152. https://doi.org/10.21577/0103-5053.20190144

Qin S, Müller WEG, Cooper EL (2011) Marine biotechnology: a new vision and strategy for Europe. Marine board-ESF position paper 15., vol 2011. Drukkerij De Windroos NV

Rabelo EF, Soares M d O, Bezerra LEA, Matthews-Cascon H (2015) Distribution pattern of zoanthids (Cnidaria: Zoantharia) on a tropical reef. Mar Biol Res 11:584–592. https://doi.org/10.1080/17451000.2014.962542

Reis VM, Oliveira LS, Passos RMF et al (2013) Traffic of secondary metabolites to cell surface in the red alga Laurencia dendroidea depends on a two-step transport by the cytoskeleton. PLoS One 8:e63929. https://doi.org/10.1371/journal.pone.0063929

Sahm BDB, Peres J, Rezende-Teixeira P et al (2020) Targeting the oncogenic TBX2 transcription factor with chromomycins. Front Chem 8:1–9. https://doi.org/10.3389/fchem.2020.00110

Salgado LT, Viana NB, Andrade LR et al (2008) Intra-cellular storage, transport and exocytosis of halogenated compounds in marine red alga Laurencia obtusa. J Struct Biol 162:345–355. https://doi.org/10.1016/j.jsb.2008.01.015

Santos MEA, Kitahara MV, Lindner A, Reimer JD (2016) Overview of the order Zoantharia (Cnidaria: Anthozoa) in Brazil. Mar Biodivers 46:547–559. https://doi.org/10.1007/s12526-015-0396-7

SECIRM (2017) Secretaria da Comissão Interministerial de Recursos do Mar, Marinha do Brasil. https://www.mar.mil.br/secirm/portugues/secirm.html. Accessed 1 Jan 2021

Sheridan RE, Deshpande SS, Adler M (2005) Cytotoxic actions of palytoxin on aortic smooth muscle cells in culture. J Appl Toxicol 25:365–373. https://doi.org/10.1002/jat.1080

Silva C, Motta J, Teixeira F et al (2020) Non-anticoagulant heparan sulfate from the ascidian Phallusia nigra prevents colon carcinoma metastasis in mice by disrupting platelet-tumor cell interaction. Cancers (Basel) 12:1353–1365. https://doi.org/10.3390/cancers12061353

Sims JJ, Lin GHY, Wing RM (1974) Marine natural products X elatol, a halogenated sesquiterpene alcohol from the red alga Laurencia elata. Tetrahedron Letters 15:3487–3490. https://doi.org/10.1016/S0040-4039(01)91944-6

Snelgrove PVR (2016) An ocean of discovery: biodiversity beyond the census of marine life. Planta Med 82:790–799

Stein E, Marques L, Fujii M, Colepicolo P (2013) Mycosporine-like amino acids (MAAs): qualitative analysis of photoprotective compounds in Laurencia complex seaweed species (Ceramiales, Rhodophyta) from Espírito Santo State, Brazil. Planta Med 79:PI104. https://doi.org/10.1055/s-0033-1352193

Sudatti DB, Rodrigues SV, Pereira RC (2006) Quantitative GC-ECD analysis of halogenated metabolites: determination of surface and within-thallus elatol of Laurencia obtusa. J Chem Ecol 32:835–843. https://doi.org/10.1007/s10886-006-9033-z

Sudatti DB, Rodrigues SV, Coutinho R et al (2008) Transport and defensive role of elatol at the surface of the red seaweed Laurencia obtusa (Ceramiales, Rhodophyta). J Phycol 44:584–591. https://doi.org/10.1111/j.1529-8817.2008.00507.x

Sudatti DB, Fujii MT, Rodrigues SV et al (2016) Diel variation of sesquiterpene elatol production: a chemical defense mechanism of the red seaweed Laurencia dendroidea. Biochem Syst Ecol 64:131–135. https://doi.org/10.1016/j.bse.2015.12.001

Thompson F, Thompson C (2020) Biotecnologia Marinha. FURG, Rio Grande

Thompson F, Krüger R, Thompson CC et al (2018) Marine biotechnology in Brazil: recent developments and its potential for innovation. Front Mar Sci 5:1–8. https://doi.org/10.3389/fmars.2018.00236

Torres P, Santos JP, Chow F et al (2018) Comparative analysis of in vitro antioxidant capacities of mycosporine-like amino acids (MAAs). Algal Res 34:57–67. https://doi.org/10.1016/j.algal.2018.07.007

Tosteson MT, Scriven DRL, Bharadwaj AK et al (1995) Interaction of palytoxin with red cells: structure-function studies. Toxicon 33:799–807. https://doi.org/10.1016/0041-0101(95)00010-J

Valverde I, Lago J, Vieites JM, Cabado AG (2008) In vitro approaches to evaluate palytoxin-induced toxicity and cell death in intestinal cells. J Appl Toxicol 28:294–302. https://doi.org/10.1002/jat.1278

Veiga-Santos P, Pelizzaro-Rocha KJ, Santos AO et al (2010) In vitro anti-trypanosomal activity of elatol isolated from red seaweed Laurencia dendroidea. Parasitology 137:1661–1670. https://doi.org/10.1017/S003118201000034X

Vollmer SV, Palumbi SR (2002) Hybridization and the evolution of reef coral diversity. Science 296:2023–2025. https://doi.org/10.1126/science.1069524

Voultsiadou Eleni E (2010) Therapeutic properties and uses of marine invertebrates in the ancient Greek world and early Byzantium. J Ethnopharmacol 130:237–247. https://doi.org/10.1016/j.jep.2010.04.041

Wattenberg EV (2007) Palytoxin: exploiting a novel skin tumor promoter to explore signal transduction and carcinogenesis. Am J Physiol Cell Physiol 292:C24–C32

Webb TJ, vanden Berghe E, O'Dor R (2010) Biodiversity's big wet secret: the global distribution of marine biological records reveals chronic under-exploration of the deep pelagic ocean. PLoS One 5:e10223. https://doi.org/10.1371/journal.pone.0010223

Weidmann S (1977) Effects of palytoxin on the electrical activity of dog and rabbit heart. Experientia 33:1487–1489. https://doi.org/10.1007/BF01918825

Wilke DV, Jimenez PC, Pessoa C et al (2009) Cytotoxic lipidic α-amino acids from the zoanthid Protopalythoa variabilis from the Northeastern Coast of Brazil. J Braz Chem Soc 20:1455–1459. https://doi.org/10.1590/S0103-50532009000800010

Wilke DV, Jimenez PC, Araújo RM et al (2010) Pro-apoptotic activity of lipidic alpha-amino acids isolated from Protopalythoa variabilis. Bioorganic Med Chem 18:7997–8004. https://doi.org/10.1016/j.bmc.2010.09.027

Wilke DV, Jimenez PC, Branco PC et al (2021) Anticancer potential of compounds from the Brazilian Blue Amazon. Planta Med 87:49–70

Ziegler M, Seneca FO, Yum LK et al (2017) Bacterial community dynamics are linked to patterns of coral heat tolerance. Nat Commun 8:1–8. https://doi.org/10.1038/ncomms14213

Chapter 10
Climate Change and Local Impacts Threaten Brazilian Coral Reefs

Marcelo O. Soares, Eliana Matos, Beatriz Diniz, Sandra V. Paiva, Anne L. Gurgel, Caroline C. Lucas, João E. P. Freitas, Carlos E. P. Teixeira, Renan G. P. Guerra, Eduardo L. Barros, Marcelo V. Kitahara, Kátia C. C. Capel, Luiz C. Cotovicz Jr, Igor C. S. Cruz, and Emanuelle F. Rabelo

Abstract Brazilian reefs are hypothesized as potential climate-change refugia due to the lower occurrence of coral bleaching events and mortality rates than other reefs regions worldwide. This hypothesis is called Brazilian Reef Refugia Hypothesis (BRRH). In this book chapter, we review the chronic and acute stressors that are currently affecting these unique reefs and communities, including long-term warming, marine heatwaves (MHWs), acidification, sea-level rise, mismanaged touristic and industrial activities, higher nutrient inputs, increased plastic pollution, human-caused disasters (e.g., oil spills and mining), overfishing, and invasive species such as lionfish (*Pterois* spp.) and sun corals (*Tubastraea* spp). The results show the

M.O. Soares (✉)
Instituto de Ciências do Mar (LABOMAR), Universidade Federal do Ceará (UFC), Fortaleza, Brazil

Dipartimento di Scienze e Tecnologie Biologiche e Ambientali (DISTEBA), Università del Salento, Lecce, Italy
e-mail: marcelosoares@ufc.br

E. Matos · B. Diniz · S.V. Paiva · A.L. Gurgel · C.C. Lucas · J.E.P. Freitas · C.E.P. Teixeira · R.G.P. Guerra · E.L. Barros · L.C. Cotovicz Jr
Instituto de Ciências do Mar (LABOMAR), Universidade Federal do Ceará (UFC), Fortaleza, Brazil

M.V. Kitahara
Centro de Biologia Marinha (CEBIMAR), Universidade de São Paulo (USP), São Sebastião, Brazil

K.C.C. Capel
Departamento de Genética e Evolução, Universidade Federal de São Carlos (UFSCAR), São Carlos, Brazil

I.C.S. Cruz
Instituto de Geociências, Universidade Federal da Bahia (UFBA), Salvador, Bahia, Brazil

E.F. Rabelo
Universidade Federal Rural do Semiárido (UFERSA), Mossoró, Brazil

R. K. P. Kikuchi et al. (eds.), *Brazilian Coral Reefs*, Brazilian Marine Biodiversity,
https://doi.org/10.1007/978-3-031-59152-5_10

uneven incidence and frequency of stressors, which characterizes a regional wide-variation in human pressures, vulnerability and reef resistance. These stressors are impacting these reefs as cumulative and synergistic pressures and, consequently, undermining their potential to provide long-term refugia contradicting BRRH. Overall, there was an increase of the sea surface temperature and MHWs at all reef regions in the last 39 years, including the year 2024 as the worst heat stress ever recorded. The effects of warming and MHWs, the main stressors, have been most significant in low-latitude reefs, where the sea temperature has been heated twice as much as the eastern Brazilian reefs (0.2 °C *versus* 0.1 °C per decade). Moreover, the frequency of MHWs have increased more in equatorial reefs than in higher latitude reefs (15–20 °S). Despite being hypothesized by BRRH as more resistant than the Caribbean and Indo-Pacific reefs, the Southwestern Atlantic reefs present a large number of phase-shift events and indicators of reef decline (e.g., decreasing richness, biomass, and loss of architectural complexity) which indicates significant degradation in the supply of geo-ecological functions such as reef framework production and sediment generation, the maintenance of reef habitat complexity and reef growth potential. The disproportionate distribution of rare fish species and decreased fish biomass also reveals a higher vulnerability to human impacts. Overall, all the Southwestern Atlantic reefs are threatened by a plethora of human pressures until the end of this century, have limited potential to provide climate-change refugia, and need urgent attention to reduce and mitigate impacts at local and global scale.

Keywords Shallow-water coral reef · Brazil · Phase-shift · Climate change · Heatwaves · Fisheries · Invasive species · Marine pollution · Marine debris · Ocean acidification

10.1 Introduction

The multiple impacts on tropical reefs are increasing and becoming more intense in the Anthropocene. The cumulative and/or synergistic effect from these human impacts along the Southwestern Atlantic coast (Fig. 10.1) can trigger a reduction in reef richness, biomass, loss of architectural complexity, reduced coral growth, increased mortality, disease outbreaks, nutritional and reproductive

Fig. 10.1 (continued) communities (Amazonia, Northeastern Brazil, Eastern Brazil, Fernando de Noronha and Rocas Atoll, São Pedro and São Paulo Islands, Trindade and Martin Vaz Islands, and Southeastern Brazil) according to the nomenclature of Spalding et al. (2007). Examples of some remarkable Brazilian reefs: 1 (Great Amazon Reef System—GARS, Amazon coast), 2 (Pedra da Risca do Meio Marine Park, Ceará coast). 3 (Maracajaú Reefs, Rio Grande do Norte coast), 4 (Coral Coast Environmental Protected Area, Pernambuco and Alagoas coast), 5 (Todos os Santos Bay, Salvador, Bahia coast), 6 (Abrolhos Bank, Bahia coast). See also Chap. 1 for a new proposal of marine ecoregion classification

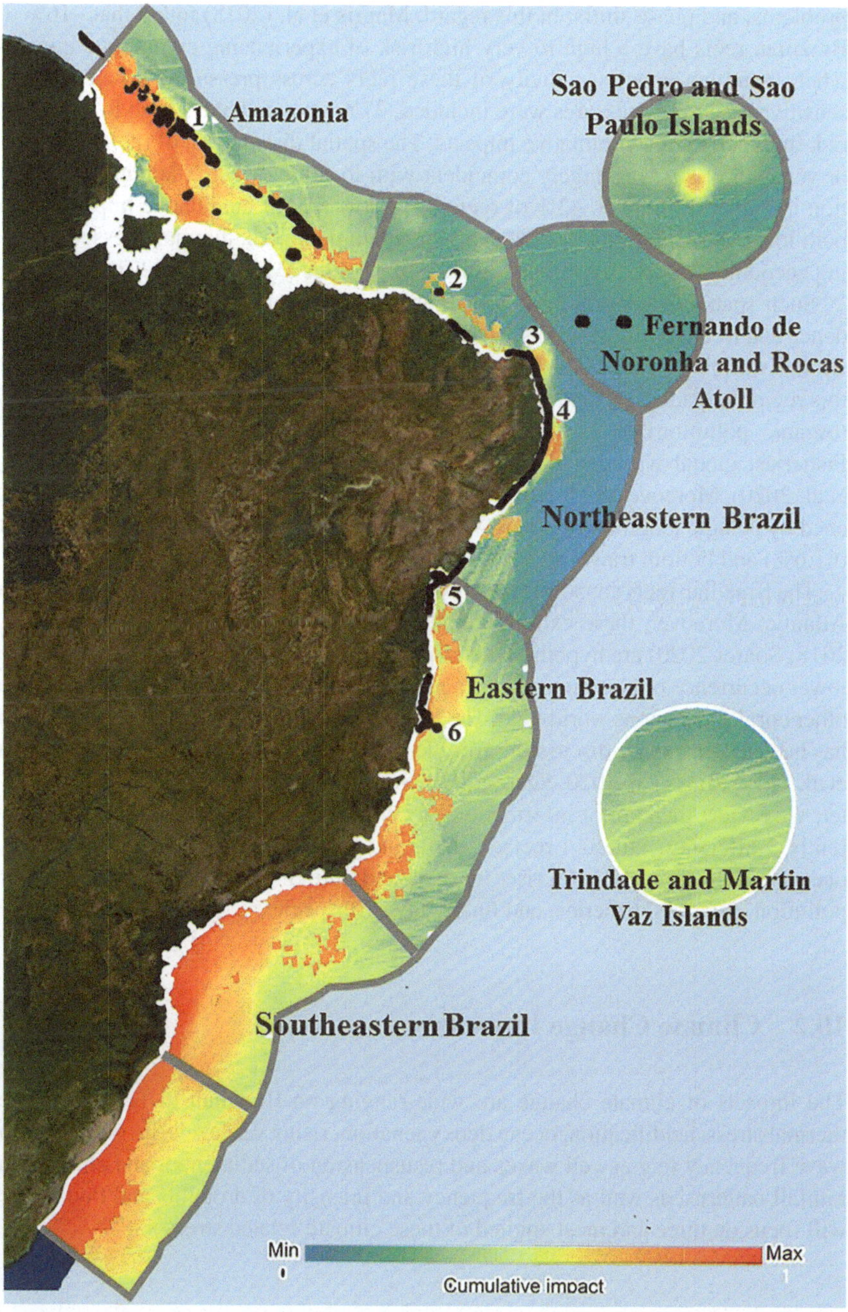

Fig. 10.1 Cumulative human impacts along the Brazilian reef regions. Black color shows the occurrence of some shallow-water reefs and mesophotic ecosystems. (Source: Spalding et al. (2007), Leão et al. (2016), Francini-Filho et al. (2018), Soares et al. (2019b), Magris et al. (2020).) Fig. 10.1 (continued) Marine ecoregions where occurs the Brazilian reefs and coral

problems, and phase shifts. In this regard, Magris et al. (2018) found that ~16% of Brazilian reefs have a high to very high risk of experiencing cumulative impacts when assuming equal sensitivity of these reefs across pressures. When variable sensitivities across pressures were included, 37% of reefs have a high to very high risk of experiencing cumulative impacts. The spatial distribution of reefs with high or very high risks was largely coincident with localities where the human population is higher. Moreover, 42% of reefs had a low risk of cumulative impacts from both local and global stressors. Among the main pressures, thermal stress and fishing accounted for the two largest values of exposures (Magris et al. 2018, 2020).

Such spatial heterogeneity is remarkable and indicates a variation on the incidence and frequency of stressors (Magris et al. 2018; Soares et al. 2021a) (Fig. 10.1). Global warming extended across over 95% of the region, while the proportion of this region impacted by land-based activities was much smaller, ranging from 7.6% (organic pollution) to 22.6% (fertilizers and pesticides) (Magris et al. 2020). Fisheries, global warming and land-based activities were severe threats (Magris et al. 2020). Moreover, 82.8% of the Brazilian economic exclusive zone was threatened by pelagic longlines, the most widespread fishery, followed by pelagic driftnet (61.6%) and bottom trawl (60.1%).

The Brazilian reefs constitute the only biogenic coralline reefs in the Southwestern Atlantic. Moreover, these extreme reefs (*sensu* Schoepf et al., 2023; Camp et al. 2018; Soares 2020) are hypothesized as potential climate-change refugia due to the lower occurrence of bleaching events and lower mortality rates in some species than other coral reef regions worldwide. The Brazilian Reef Refugia Hypothesis (BRRH) has been under intense discussion over the last few years (Leão et al. 2016; Teixeira et al. 2019; Mies et al. 2020; Soares 2020; Soares et al. 2021a, c). In this book chapter, we review the current information on chronic and acute stressors that are currently affecting these reefs, including warming, marine heatwaves, ocean acidification, sea-level rise, touristic and nutrient inputs, plastic and organic pollution, disasters, fisheries, and finally invasive species.

10.2 Climate Change Impacts

The impacts of climate change are wide-ranging on Brazilian reefs and include thermal stress, acidification, ocean deoxygenation, rising sea levels, increased storm wave frequency (e.g., swell waves and resuspension of sediments), and changes in rainfall patterns, as well as the frequency and intensity of droughts and floods. We will focus on three and most studied of these climate-related stressors.

10.2.1 Global Warming and Heatwaves

In addition to the near global long-term increases in sea surface temperatures (SST), the occurrence of Marine heatwaves (MHWs) has also increased in frequency worldwide (Oliver et al. 2018). MHWs are prolonged periods of extremely anomalously warm temperatures that also can have negative impacts (Oliver et al. 2018; Holbrook et al. 2020; Oliver et al. 2020) including coral mass bleaching (Hughes et al. 2017). MHWs will become more frequent, intense, and longer-lasting due global warming (Oliver et al. 2018, 2020; Holbrook et al. 2020). In this regard, characterization of long-term SST changes and the occurrence of MHWs in the Brazilian reefs are rare but suggest that they are one of the main threats to them (Rodrigues et al. 2019; Teixeira et al. 2019; Duarte et al. 2020) including identification of vulnerability areas to coral bleaching as a tool to inform conservation actions (Bleuel et al. 2021). Aiming to fill this information gap, unpublished data analysis is presented here.

Overall, there was an increase in the SST at all four Brazilian reef regions in the last 42 years (Fig. 10.2), including the year 2024 as the worst heat stress ever recorded (Fig. 10.4). A clear trend of ~0.2 °C increase per decade was detected in low-latitude equatorial reefs such as Pedra da Risca do Meio and Rocas Atoll (Fig. 10.2). Moreover, the results indicate 0.12 °C increase trend in Abrolhos Bank and in the Coral Coast MPA ($p < 0.01$). These values are consistent with global average trends (Oliver et al. 2018) but local physical and ecological context of reef communities have to be considered. Each site has a specific average SST and variability for which the organisms are evolutionarily adapted. Overall, the SST has increased between 9.5% and 22% of the mean seasonal SST variability of each site, with the largest trends occurring in lower latitudes where the mean SST is higher and variability is smaller such as the equatorial reefs cited above. This suggests that there is a wide-regional variation in the seawater warming exposure according to the reef region. This is expected because the spatial distribution of these reefs occurs on a continental scale (~4,000 km) spanning from equatorial to subtropical areas, and from coastal to insular areas. In this way, these results clearly indicate that the equatorial SW Atlantic reefs have been more affected by the increasing warming trend (Fig. 10.2). These results are reinforced by the temporal analysis of MHWs explained below.

The average MHWs frequency (Fig. 10.3) is close to three events per year during 1982–2023 on the four coral reef Brazilian regions, but can reach values larger than 10 events per year in Rocas Atoll and Coral Coast. The frequency of MHWs events has increased in the last 39 years in Abrolhos and Pedra da Risca do Meio($p < 0.05$). This increase is larger in low-latitude regions sucha Pedra da Risca do Meio Park (Figs. 10.3). The average duration of these MHW events is between 11.8 and 17.8 days, but it can have a maximum duration of more than 25 days in all reef sites (Fig. 10.3). The duration of the MHWs has also increased in Abrolhos and Rocas reef sites ($p < 0.05$). The average SST anomalies caused by the MHWs is between 0.7 and 0.93 °C, with maximum anomalies reaching between 2 and 2.3 °C

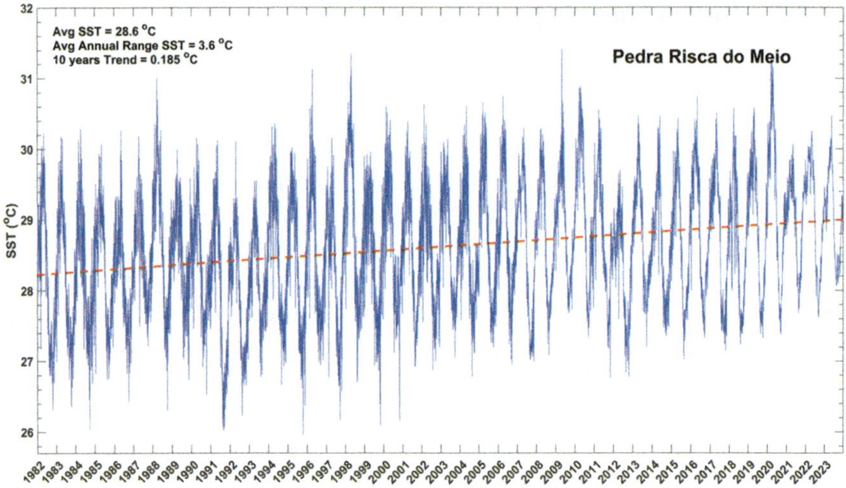

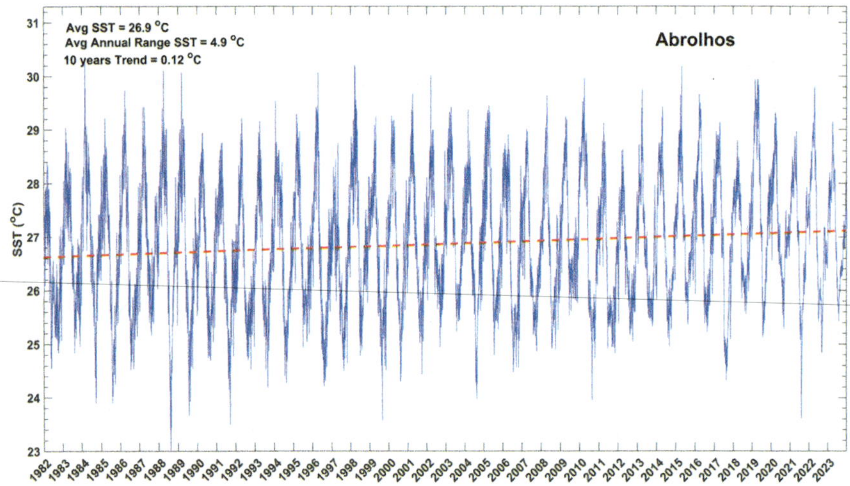

Fig. 10.2 (continued)

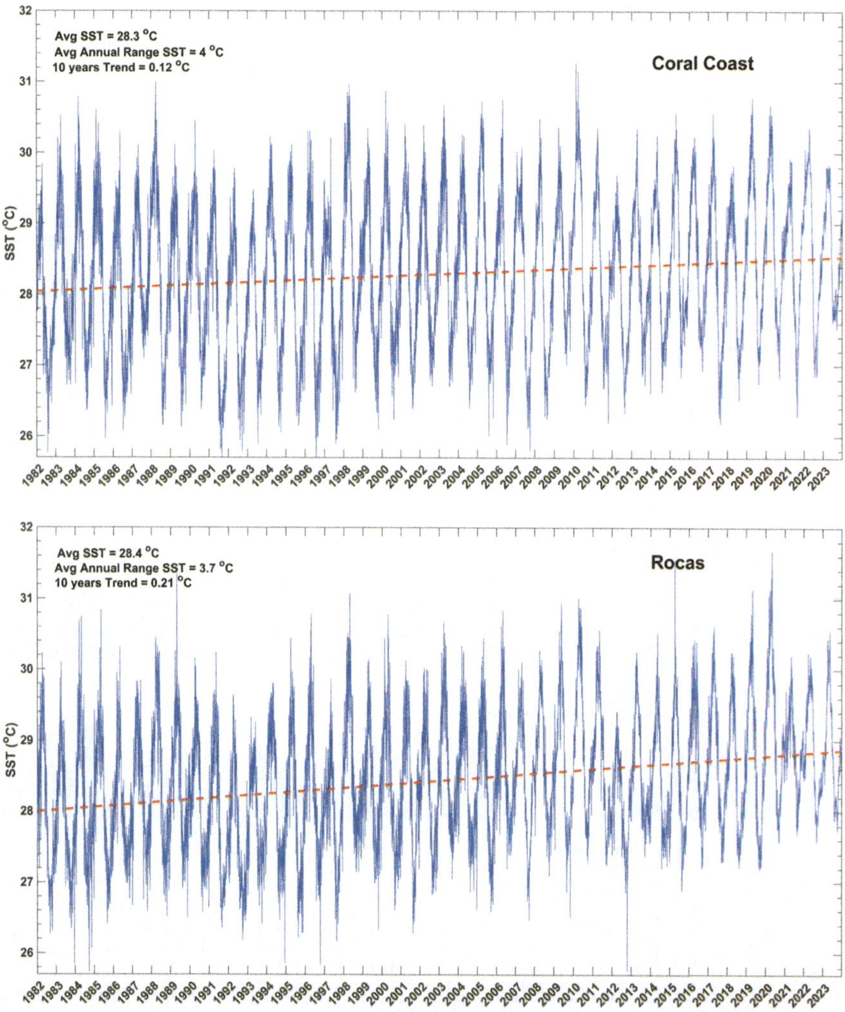

Fig. 10.2 SST time-series for the period 1982–2023 at Pedra da Risca do Meio Marine Park, Abrolhos, Coral Coast and Rocas Atoll. The locations of these sites are presented in Fig. 10.1. All coral reef areas show a SST increased trend along these years. The red line denotes the SST liner trend ($p < 0.01$). The SST averages values for the period, the mean annual range and the 10 years annual trend are shown in the figures. The SST was extracted from the OISST dataset

(Fig. 10.3). Both, average and maximum MHW intensity have not changed in the analyzed period ($p < 0.05$). The MHWs properties and its trends are consistent with the values reported in Oliver et al. (2018). These authors suggested that local trends observed in reef regions cannot be explained solely due to the rise in the mean SST. As for the SST long term changes, the effects of the MHWs have to be analyzed site by site due to its own seasonal variability.

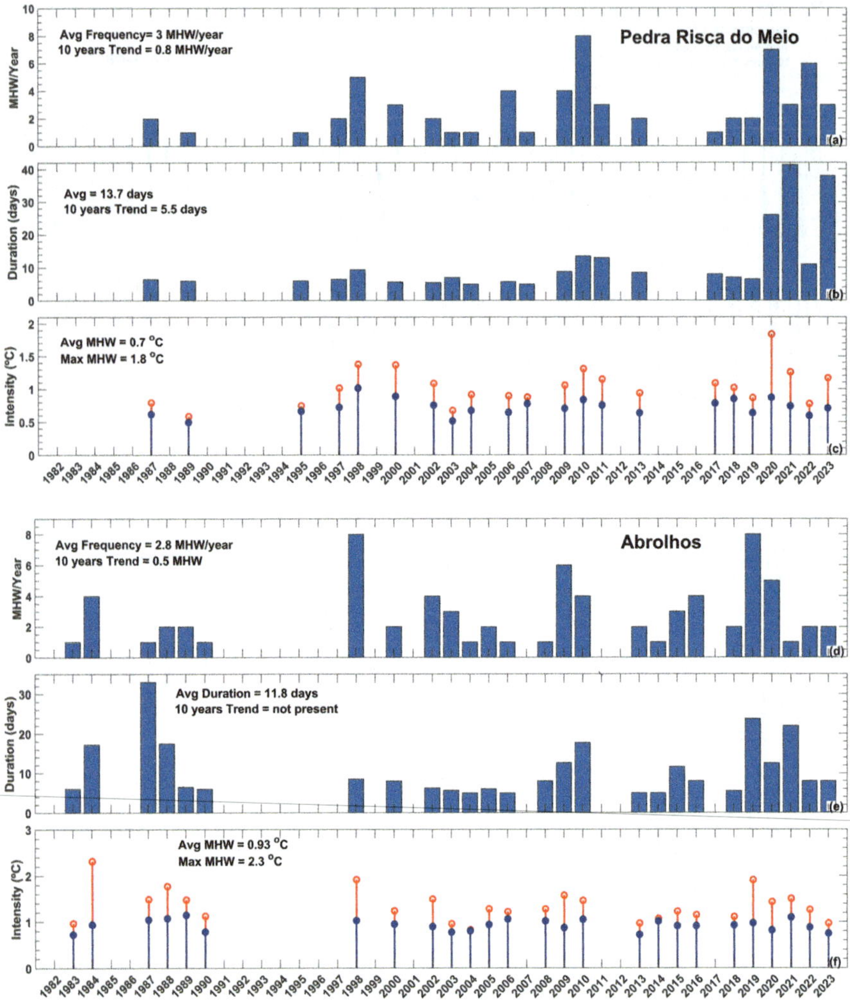

Fig. 10.3 (continued)

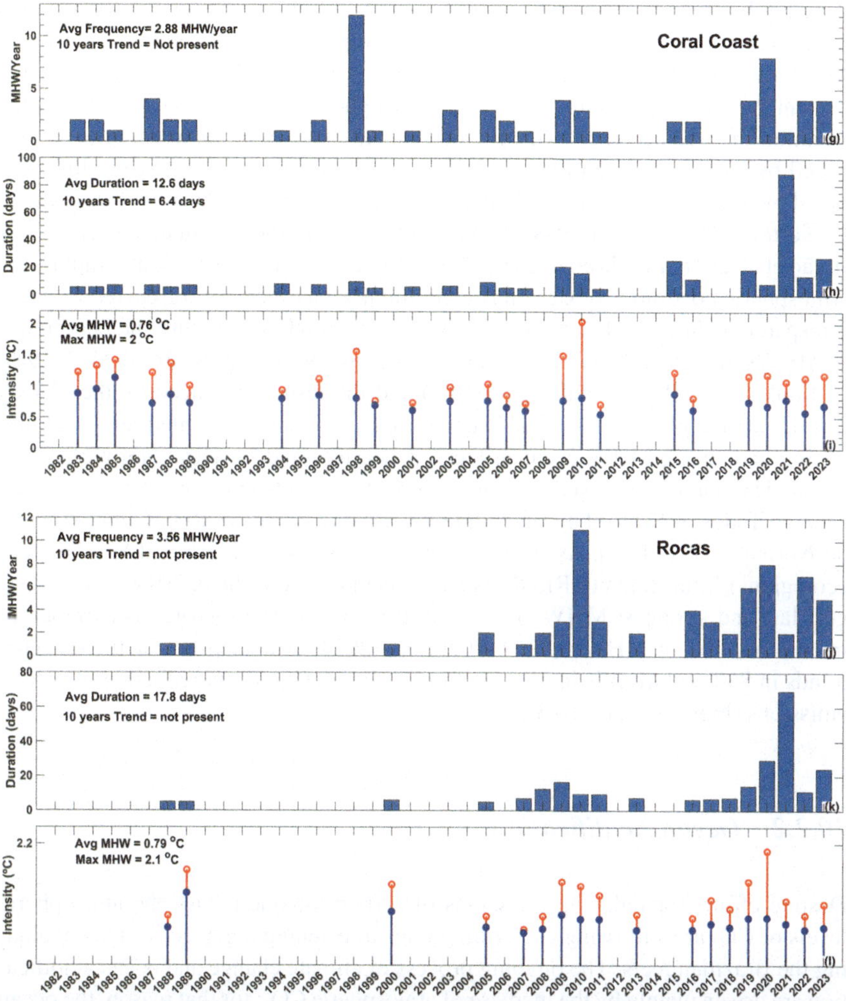

Fig. 10.3 Time series of annual averages marine heatwave properties for the period 1982–2023 at Pedra do Risca do Meio Marine park (panels **a–c**), Abrolhos Marine Park (panels **d–f**), Coral Coast marine protected area (panels **g–i**), and Rocas Atoll (panels **j–l**). (**a**), (**d**), (**g**), (**j**) shows the frequency of occurrence (MHWs/year), (**b**), (**e**), (**h**), (**k**) shows the annual average duration (days) and (**c**), (**f**), (**i**), (**l**) the MHWs annual average (blue lines) and maximum intensity (red lines) (°C). The text inside each figure shows the 1982–2023 average values for frequency, duration and average and maximum intensity. The 10 years linear trend is also shown when statistically significant at the 5% level. MHW do not present statistically significant trends. The MHWs properties are calculated and available at http://www.marineheatwaves.org/tracker.html

Although the MHWs intensity has not changed in the four analyzed regions in the last decades, it has become more frequent and longer lasting which may lead to more mass bleaching events as the ones observed in 2010 (Soares et al. 2019a), 2019 (Duarte et al. 2020), 2020 and again in 2024. Recovery of bleached corals may demand the absence of another severe bleaching event (Hughes et al. 2017), which is no longer realistic while projected global temperatures will continue to rise and MHWs will become more frequent and longer lasting in the next years under the current greenhouse gas emission scenario (Hughes et al. 2017). This is clearly the situation of Brazilian reefs which hamper their capacity to act as refugia, as proposed by BRRH, in face of increasing local and global impacts. The shading provided by moderate turbidity might buffer the effects of heat-light stress especially in turbid-tolerant corals (Van Woesik et al. 2012) such as the Brazilian massive corals (Leão et al. 2016; Loiola et al. 2019; Mies et al. 2020). However, oceanographic and atmospheric conditions (e.g., winds and rainfall) may increase (or decrease) water transparency. In case of reduction of turbidity in combination with the occurrence of MHWs, these corals will be susceptible to mass bleaching (Soares et al. 2019a).

Looking at MHWs results (Fig. 10.3) and the results produced by the NOAA Coral Watch (e.g., Fig. 10.4), 2024 appears to have an unprecedented regional long-lasting and strong MHW event in the Northeastern and Eastern coral reef ecoregions. This large-scale event spanned more than 2,000 km from the Ceará state coast to Todos os Santos bay (including the oceanic islands of Rocas and Fernando de Noronha) (Fig. 10.1). As for the other reef systems (Eastern and Southeastern ecoregions) from Bahia to Rio de Janeiro (including Abrolhos), 2019 was the longest last and strongest MHW since 1985 (Fig. 10.3). The repeated occurrence of intense and recent MHWs (2019, 2020, and 2024) is a snapshot of the detected trends in SST and MHWs in the near future without significant reduction of carbon emissions (Figs. 10.2 and 10.3).

10.2.2 Ocean Acidification

Despite efforts for reducing emissions of carbon dioxide (CO_2), the atmospheric levels of CO_2 are still rising at alarming rates threatening many ecosystems including the Brazilian reefs. The ongoing process of climate change and ocean acidification are driven mainly by the increase of atmospheric CO_2; for that reason, the ocean acidification is also referred as the "other CO_2 problem" (Doney et al. 2009), which impact reefs together with sea-level rise, global warming, MHWs (Sect. 4.1), and local/regional human impacts, as reviewed in this book chapter. Diverse types of organisms that form shells and skeletons from $CaCO_3$ are sensitive to the acidification (Gattuso et al. 2015; Doney et al. 2020), particularly that secrete aragonite (e.g., scleractinian corals) and high-magnesium calcite (e.g., coralline algae) (Kleypas and Yates 2009). Both groups (e.g., corals and coralline algae) are important reef-builders in Brazil (Leão et al. 2016). In this regard, diverse regions may experience the appearance of undersaturated surface waters with respect to the carbonate mineral aragonite ($\Omega_{ara} < 1$; tenancy of dissolution of $CaCO_3$), also called "corrosive waters".

However, the studies of carbonate chemistry and acidification are still scarce along the Brazilian shelf where we found the reefs at shallow and mesophotic

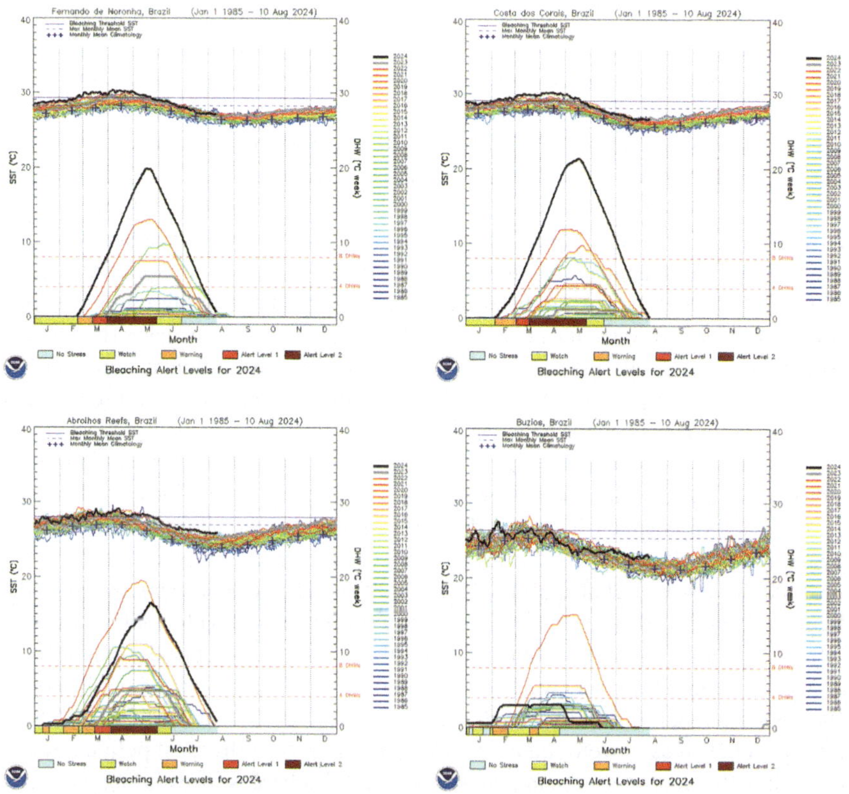

Fig. 10.4 SST and Degree Heating Week (DHW) time-series for the period 1985–2024. The figures show that 2024 was an unprecedented regional long last and strong MHW at Fernando de Noronha Island and Costa dos Corais (upper panels) low-latitude reefs while 2019 was worst at Abrolhos reefs and Buzios, Rio de Janeiro (higher latitudes sites under an upwelling effect). The DHW product shows accumulated heat stress, which can lead to coral bleaching and death. The scale ranges from 0 to 20 °C-weeks. The DHW product accumulates the instantaneous SST anomaly after the temperature exceeds the 90th percentile. Significant coral bleaching usually occurs when the DHW value reaches 4 °C-weeks and severe, widespread bleaching by the time the DHW value reaches 8 °C-weeks. (Source: https://coralreefwatch.noaa.gov—NOAA Coral Reef Watch (2018))

depths. To our best knowledge, there is only one study that has investigated the carbonate chemistry parameters in a continental shelf dominated by low-latitude reefs (Cotovicz et al. 2020a) (Pedra da Risca do Meio Marine Park, Fig. 10.1). This study showed that the waters presented lower values of pH, CO_3^{2-} and Ω_{ara}, compared to nearshore regions without the influence of coral reef waters. This pattern was attributed to the occurrence of $CaCO_3$ precipitation in coral-reef dominated waters. The $CaCO_3$ precipitation involves the consumption of total alkalinity (TA) and DIC in a ratio 2:1, with production of CO_2. This mechanism explains the lower values of pH, CO_3^{2-} and Ω_{ara} in this region and Ω_{ara} around 3.5 in these warm SW Atlantic waters (Cotovicz et al. 2020a).

These levels of Ω_{ara} indicate that the reef region presents supersaturated conditions, which favors the secretion of $CaCO_3$ by marine organisms (e.g., calcareous algae and corals). The net dissolution of $CaCO_3$ in these warm equatorial waters is predicted until the end of this century (considering the actual growth rates of atmospheric CO_2 concentrations, Cotovicz et al. 2020a). However, it is important to point out that a recent study showed that coral reefs could reach net dissolution at Ω_{ara} of around 2.3 with 100% living coral cover, and at $\Omega_{ara} > 3.5$ with 30% of living coral cover (Kline et al. 2019). For instance, a recent study has reported a mass bleaching event in the dominant coral *Siderastrea stellata* in this same equatorial region (Soares et al. 2019a). The predicted increase and expansion of bleaching events as the result of global warming and MHWs (IPCC 2019) may cause deleterious effects on the Brazilian corals and can accelerate the skeleton dissolution and erosion. The relative role of temperature versus acidification on the process of coral reef calcification is still unclear (Dove et al. 2020).

In addition to the acidification driven by the anthropogenic-derived CO_2 emissions at the global scale, some estuaries and coastal Brazilian regions are also experiencing anthropogenic CO_2-induced acidification and carbonate chemistry changes by mixing processes between freshwater and seawater, and eutrophication (Cotovicz et al. 2018; Cotovicz et al. 2020b; Cai et al. 2021). The coastal ocean responds differently from the open ocean to CO_2 inputs, and can be more vulnerable to negative effects of acidification. For example, low values of Ω_{arag} and episodic evidences of corrosive waters ($\Omega_{arag} < 1$) were found in the tropical coastal waters of Guanabara Bay (Rio de Janeiro, Brazil); a highly eutrophic estuary (Cotovicz et al. 2018) The lowest values of Ω_{arag} were found in well-mixed waters of the most polluted region of the bay that receives large amounts of effluent discharges, with strong heterotrophic activities and low buffering capacity (Cotovicz et al. 2018).

Finally, it is imperative to highlight that the detection and monitoring of acidification require long-term and high-quality assessment of carbonate chemistry parameters with interdisciplinary efforts, which is still lacking in Brazil. Furthermore, the bioassays and mesocosm experiments are important to investigate the effects of acidification and warming in reference organisms and keystone Brazilian species. It is worth mentioning the creation of the Brazilian Ocean Acidification Research Group (BrOA) and the Latin-American Ocean Acidification Network (LAOCA), aiming to study acidification and its effects on ecosystems.

10.2.3 Sea Level Rise (SLR)

Sea level variations are pivotal to the reef formation and carbonate framework, including in Brazil. SLR scenarios are predicted between 0.26 and 0.98 m by the year 2100 (IPCC 2019). Although many reefs worldwide retain carbonate accretion rates close to recent SLR trends, few reefs will have the capacity to track SLR projections under RCP4.5 climate change scenarios without sustained recovery, and under RCP8.5 scenarios most reefs are predicted to experience mean water depth increases of more than 0.5 m by 2100 (Perry et al. 2018). This change will undermine many of their geo-ecological functions (Fig. 10.5) such as reef framework production and sediment generation, the maintenance of reef habitat complexity and reef growth potential (Perry and Alvarez-Filip 2018).

In the context of Brazilian reefs, there is a heterogeneity in the susceptibility to the impacts in the face of SLR. Among the most vulnerable reef systems, we can mention the Rocas Atoll (Fig. 10.2) and the intertidal sandstone reefs in Northeastern Brazil that are hotspots and important zones for touristic activities and nursery habitats for species. These two Brazilian reef systems have a high risk of submergence due to the low height above the current sea level and proximity to the shoreline, respectively. In Brazil, studies on sea level variation have their prelude between the 1960s and 1970s with studies by Bigarella and Andrade (1964), Bigarella et al. (1965) and Bittencourt et al. (1979). The studies used dating methods, such as

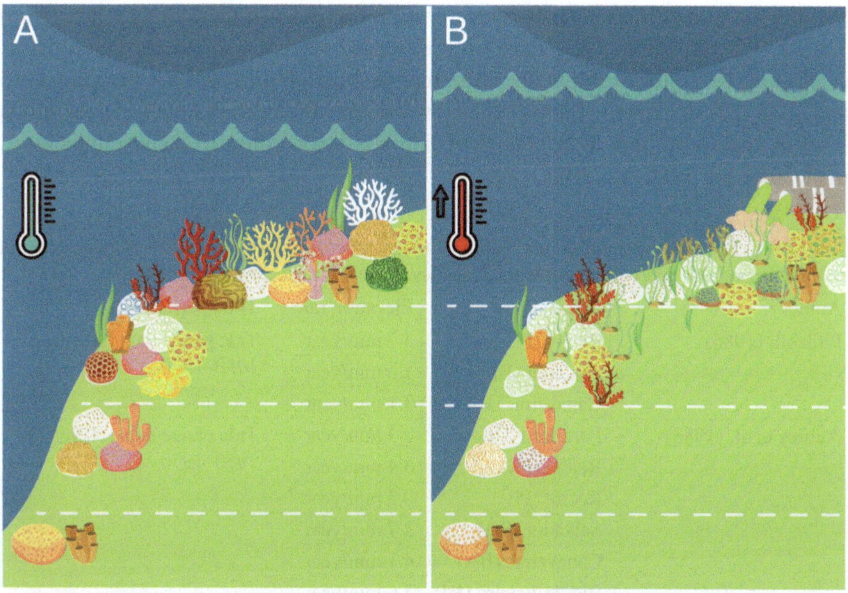

Fig. 10.5 Submergence risk in face of a sea level rise on the Brazilian coast. In the current scenario (**a**), the phenomenon of bleaching in massive and arborescent Brazilian corals already exists and there is a reef architectural complexity due to the presence of 3-D (tri dimensionality) provided by reef-building species (for example, the hydrocoral *Millepora* spp.). In the future scenario (**b**) (e.g., RCP 8.5), the rising sea level reduces the potential for bioconstruction and submergence of the reef carbonate factory. Due to global warming and other human pressures (e.g., nutrients inputs) along the Brazilian coast, the persistent phase-shift occurs more frequently with the dominance of algae and zoanthids. Moreover, severe coral bleaching (and mass mortality) becomes more frequent leading to reef decline in provision of the geo-ecological functions. In this way, with rising sea levels, more adverse conditions may reduce the protective effect (e.g., energy dissipation and heights of waves) provided by Brazilian coral reefs on their safeguarding as a "first line of defense" against beach erosion

radiocarbon, which allowed to improve the records of sea level behavior on the Brazilian coast in the last 7,000 years (Suguio 2010). In this period, studies indicate that the sea level variation reached, for the first time, its current quota (Suguio et al. 1985; Angulo et al. 2006). In general terms, there is a trend towards SLR in Brazil that will affect their nearshore and offshore reefs (e.g., Rocas Atoll) although inaccurate records occur due to the absence of historical data series and continuous and reliable indicators.

In a review conducted by Ximenes Neto (2020), distinct situations of sea level records for the Brazilian coast during the Middle Holocene are presented. Belém (2007) having as focal points the regions off Fortaleza (Ceará coast) and Salvador (Bahia coast) observed that there is a tendency of sea level rise in coral reef regions. Losada et al. (2013) evaluated the changes in sea level together with tidal data, storm waves and extreme events to reach a "total sea level" in Brazil, and thus it was found that the SLR towards the south of the country and that the events of El-Niño would be positively affecting the sea level in the coastal region (Klein and Short 2016). In the states of São Paulo and Rio de Janeiro, the southern limit of reef-building coral distribution (Leão et al. 2016;), increases in mean sea level between 1.8 and 4.2 mm respectively have been observed since the 1950s. In Pernambuco coast (Fig. 10.2), for example, there are records of a 5.6 mm increase between 1946 and 1988 (PBMC 2016). Table 10.1 below presents the rates of change in mean sea level at specific points along the Brazilian coast according to some publications made between 1986 and 2013.

Table 10.1 Sea level variation rates on the Brazilian tropical coast where reef systems or coral communities occur and will face a trend of sea level rise in the upcoming decades

Author	Site	Elevation rate	Time period
Pirazolli (1986)	Recife (PE)	3.7 mm/year^{-1}	Tide gauge data (20 years) 1950–1970
	Salvador (BA)	1.6 mm/year^{-1}	
	Canaveiras (BA)	3.1 mm/year^{-1}	
Aubrey et al. (1988)	Fortaleza (CE)	0.3 mm/year^{-1}	Tide gauge data (20 years) 1950–1970
	Belém (PA)	3.4 mm/year^{-1}	
	Recife (PE)	0.2 mm/year^{-1}	
	Salvador (BA)	2.7 mm/year^{-1}	
	Canaveiras (BA)	4.1 mm/year^{-1}	
	Rio de Janeiro (RJ)	3.6 mm/year^{-1}	
Silva (1992)	Rio de Janeiro (RJ)	12.6 mm/year^{-1}	Tide gauge data (20 years) 1965–1986
Harari e Camargo (1994)	Recife (PE)	5.6 mm/year^{-1}	Tide gauge data (42 years) 1946–1988
Franca (2000) Mesquita (2003)	Equatorial Atlantic	4.0 mm/year^{-1}	Altimetry
IHCantabria Losada et al. (2013)	Salvador (BA)	~2.0 mm/year^{-1}	Modelling data Tide gauge data (1950–2009)

Adapted from Klein and Short (2016)
In the site column, City and State (parenthesis)

While the northeastern and northern (Amazon) coasts of Brazil (Fig. 10.2) do not present a curve or harmonic pattern (Irion et al. 2012), the eastern coast of Brazil exposes a clear pattern with maximum sea level above the current one in the middle Holocene, with a subsequent fall until the present (Angulo et al. 2006). With rising sea levels, adverse conditions may reduce the protective effect (e.g., energy dissipation and heights of waves) provided by Brazilian reefs on their safeguarding as a "first line of defence" (Elliff and Silva 2017) (Fig. 10.7). This, considering that the nearshore reefs will have an increase in depth, resulting in changes in hydrodynamic conditions, erosion and damage to the urban infrastructure, including in the Brazilian coast that already suffers with the increased beach erosion process. In addition, these reefs also face a plethora of threats from other human-related stressors reviewed in this chapter (Sects. 10.3 and 10.4). These multiple human pressures contribute to reducing carbonate stocks, reef growth and therefore diminish the ability of Brazilian reefs to act in response to rising seas.

10.3 Local and Regional Impacts on the Brazilian Coast

10.3.1 Urban, Touristic, and Industrial Development

Rapid and unplanned urban, touristic, and industrial development along the Brazilian coast threaten especially the nearshore reefs (Leão et al. 2019; Magris et al. 2018, 2020). The Brazilian population on the coastal zone is estimated at 111.2 million and this number intends to increase over time, along with urban and economic growth (Soares et al. 2017). When these activities are carried out indiscriminately and with inappropriate use of natural resources, the reefs suffer multiple impacts such as the loss of reef richness, community shifts (i.e., decline in Phaeophyceae and a substantial increase of Chlorophyta) and large declines of calcareous algal cover (Portugal et al. 2016; Magris et al. 2018).

The changing flow of water and urban runoff, due to the dredging of the flooded areas and the waterproofing of the surface, besides the widespread use of septic tanks and cesspits, drive the transfer of sediments and other polluting and nutrients materials to the groundwater (Costa Jr et al. 2000) such as the dunes, seagrass beds, and estuaries that are interconnected to many Brazilian reefs. The coral reefs in the Northeastern and Eastern marine ecoregions (Leão et al. 2016) are under intense urban pressure (Fig. 10.2) due to the fact that 10 of 11 states' capitals of Northeastern Brazil are on the coast. In these ecoregions, some areas have experienced unplanned urbanization, socio-environmental problems such as lack of basic sanitation, pollution, overcrowding of beaches, and illegal land occupation (Soares et al. 2021a). This causes high densities of macroalgae and heterotrophic organisms growing on the reefs due to high nutrient concentrations (nitrate, nitrite, ammonia, phosphate and silicate) and increased primary productivity (Soares et al. 2021a). This eutrophication reduces the coral growth and favors other organisms on the reef composition (Costa Jr et al. 2000).

Adding to the urban sprawl, touristic activities bring together actions that impact Brazilian reefs, both directly and indirectly. Those that cause physical damage directly to the reefs are diving activities (leisure or fishing), trampling, removal of organisms and anchoring (Leão et al. 2019). On the Abrolhos reefs, one of the most visited diving destinations on the SW Atlantic, receive, approximately, 1300 scuba divers annually and it was estimated that they touch on corals 74,529 times causing damage 9,464 times and raising sediment onto corals 10,647 times per year (Giglio et al. 2016).

The Coral Coast MPA (Fig. 10.2) has been under a growing process of degradation due to tourism impacts and easy access. On the Porto de Galinhas beach, usually boats take people to walk on the reef and practice snorkeling that results in a higher percentage of bare areas due to human trampling and anchoring. These activities can limit coral growth (e.g., *Millepora alcicornis*), causing colony death and overgrowth of other non-calcified organisms, increasing sedimentation rates and decreasing water transparency (Woodland and Hooper 1977). Moreover, on Maragogi beach (Fig. 10.1), artificial feeding of fish by tourists was responsible for influencing on reef fish communities, increasing mobile invertebrate feeders and aggregations of omnivores (Feitosa et al. 2012).

10.3.2 Nutrient Inputs

Brazilian coral reefs are significantly influenced by river discharge and continental runoff (Castro and Pires 2001; Moura et al. 2016; Soares et al. 2021a). The large number of hydrographic basins and wetlands contribute with nutrient inputs to the primary production of algae and mixotrophic corals. Among the main sources of nutrients are the urban population clusters, agriculture, industry, livestock and aquaculture farms (Wassmann and Olli 2005) such the shrimp farms along the Northeastern Brazilian ecoregion (Fig. 10.1). Untreated sewage discharge is consolidated as the main promoter of pollution. According to the National Sanitation Information System, only ~ 49.1% of the sewage generated in Brazil receives treatment before being released into bodies of water.

The main sources of nutrients are phosphorus and nitrogen compounds providing anoxic environments in low-circulation Brazilian areas (e.g., tropical bays), impairing the development of marine organisms and coral mortality. Nearshore reefs have highest levels of water-column productivity and benthic algal cover (Aued et al. 2018). Concentrations of chlorophyll observed in the inshore region has a higher variation compared to oceanic regions. In Cabo Frio (Rio de Janeiro State) the range was 0.12–1.41 Lg/L on the coast waters, and had values 0.3 Lg/L in the oceanic region (Metzler et al. 1997), chlorophyll-a in the coastal region of Alagoas State had high spatial variation (0.03–18.23 mg m^{-3}) highly influenced by the hydrological regime of rivers and wind-induced coastal currents, while on the deepest zone the range was between 0.376 and 0.721 mg m^{-3}, also this similar variation was found between nearshore and offshore corals reefs of Porto Seguro Bay (Bahia

State) (Costa Jr et al. 2002). In this regard, these reefs are impacted due to urban and agricultural pollution, changes in the watercourse and changes in soil occupation in coastal areas, which can promote eutrophication with the accelerated growth of filamentous algae that compete with corals for space (Moura et al. 2015).

10.3.3 Disasters Caused by Human Action

There are human-caused disasters that place additional acute pressures on these Brazilian reefs. In this regard, marine oil and gas activities are important and common which brings high risks to important reef ecosystems such as Abrolhos Bank and the Great Amazon Reef System—GARS (Fig. 10.1). Human-made disasters involving oil spills are common, such as the oil spill in Guanabara Bay in Rio de Janeiro in 2000 (Lyra et al. 2008). Almost 20 years after this spill, in 2019/2020 the most extensive and severe oil-related disaster occurred on the Brazilian coast (Soares et al. 2020a, b). About 5,000–12,000 m^3 of Venezuelan oil were spilled, reaching more than 990 beaches along the 3,000 km of tropical coast, mainly the Brazilian Northeast (Soares et al. 2020a, b). This oil spill strongly affected more than 55 MPAs, reaching important coral reef areas such as the Coral Coast and the Abrolhos (Fig. 10.2).

Generally, upon arrival at the coast, the oil floats at the air-water interface, possibly over the corals, and contaminates them during tidal changes as they become more "exposed" at low tide (Haapkyla et al. 2007). Another means of contamination is by the action of waves that can disperse the oil droplets and subsequently the potential for contact with the corals and reef organisms (Haapkyla et al. 2007). And, finally, long-term contamination due to the oil incorporated in the sediments, having a strong impact on the reef biota with the cyclic release of toxic components of the sediment oil (Haapkyla et al. 2007).

Polycyclic Aromatic Hydrocarbons (PAHs) comprise an important fraction of most petroleum fuels because they are more bioavailable and toxic to reef organisms such as corals such as the 16 PAHs found in this extensive disaster in the Brazilian coast (Soares et al. 2021b). PAHs are part of the group of compounds known as Persistent Organic Pollutants (POPs), which have characteristics of greater persistence, low biodegradability and high lipophilicity, requiring greater attention from society (Jones and Voogt 1999). Recent analysis indicates that this extensive oil spill on the Brazilian coast has affected coral larvae in reef areas (Campelo et al. 2021). This demonstrates that even after larvae escape from contaminated seawater, the impact of chemical stressors remains.

Another human-caused disaster was the rupture of the Fundão Dam on the Rio Doce in November 2015 (Coimbra et al. 2020). Around 50 million m^3 of iron ore tailings were spilled, characterizing one of the greatest mining tragedies in the world (Coimbra et al. 2020). The toxic sludge has traveled more than 650 km through the Rio Doce basin (Coimbra et al. 2020). After 16 days, the toxic waste reached the sea and, by January 2016, about 7000 km^2 of total marine area was affected (Hatje et al. 2017). In July 2016, the plume of toxic sediments and trace elements migrated north, posing sub-lethal coral contamination in the Abrolhos reefs (Fig. 10.1) (Coimbra et al. 2020).

Evangelista and Valeriano (2017) analyzed the skeletons of two species, *Mussismilia hartii* and *Siderastrea siderea*, comparing the concentrations of metals before and after the arrival of the plume of sediments. The results showed a significant increase in the incorporation of Zinc (Zn) and Copper (Cu) for both corals, in addition to the significant presence of other heavy metals such as Phosphorus (P), Arsenic (As) and Barium (Ba) in the Abrolhos Marine Park (Evangelista and Valeriano 2017). A decrease in the growth rate and an increase in the degradation rate caused by organisms such as mollusks and sponges in the coral *Siderastrea siderea* (Evangelista and Valeriano 2017) were also observed. Coimbra et al. (2020) estimated the particulate matter in suspension before, during and after the dam rupture. The evaluation of the Abrolhos Bank revealed an upward trend in the concentration of suspended particulate matter between 2015 and 2019 (Coimbra et al. 2020). These sediments decrease pelagic light penetration, leading to a decline in zooplankton production (Coimbra et al. 2020). This mining-related contamination, together with other local and global stressors, contribute to the reduction of the resilience of the Brazilian reefs.

10.3.4 Marine Debris

Brazil is the largest producer of plastic in Latin America and pollutes the sea with 325,000 tons of waste per year (Iwanicki 2020). Therefore, the impact of macro- and microplastics and fishing gears on Brazilian reefs are still poorly understood, although has already been detected in many environments (Costa and Barletta 2015; Adelir-Alves et al. 2016; Andrades et al. 2018; Link et al. 2019; Monteiro et al. 2020). Moreover, numerous taxa have been observed in contact with marine litter, including fishes, crustaceans, cephalopods, and echinoderms, threatening reef ecological processes (de Carvalho-Souza et al. 2018).

As suspension feeders, corals are vulnerable to microplastic contact and ingestion, and many researchers suggest that microplastics also cause several impacts, particularly to stress-sensitive species (Reichert et al. 2018; Soares et al. 2020c). The presence of marine debris can suffocate polyps and deprive colonies of light and oxygen facilitating the spread of diseases (Lamb et al. 2018). Directly, microplastic can become adhered to colonies (Martin et al. 2019), or be ingested by polyps (Reichert et al. 2018; Tang et al. 2021) since corals probably lack a selection mechanism to allow the polyps to discern between food items and microplastics (Savinelli et al. 2020). One of the indirect effects is related to the contact of the colonies with pathogenic microorganisms (e.g., viruses and bacteria) present in the biofilm of microplastic particles, which can cause coral diseases. A summary of the main negative effects of microplastics in reef-building corals is listed in Fig. 10.6.

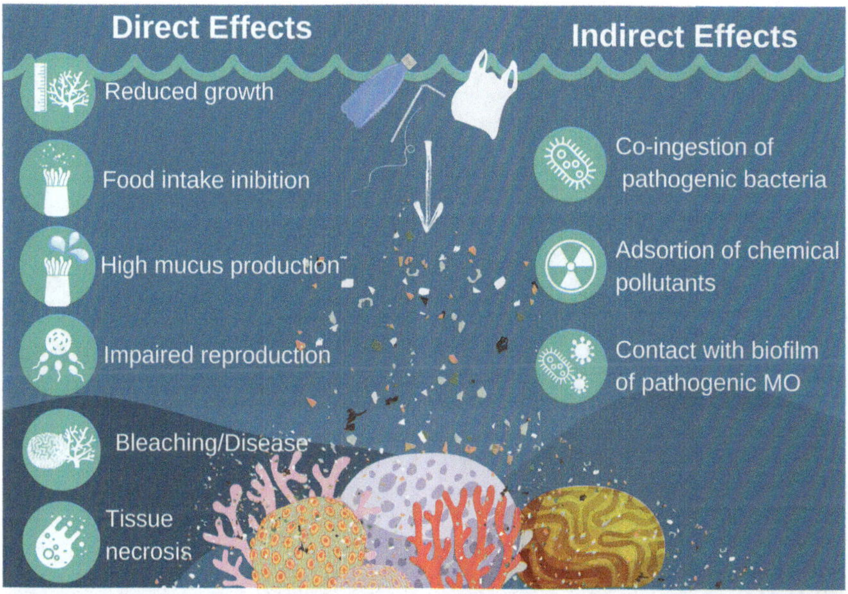

Fig. 10.6 Negative effects of microplastics in corals include direct and indirect effects (Reichert et al. 2018, 2019; Martin et al. 2019; Soares et al. 2020c)

Besides the impacts of plastic waste, the lost fishing gears such as gillnets, trammel nets, longlines, traps gear, and pots continue to deliberately species in the reefs, with no selectivity, causing the death of the megafauna, such as dolphins, turtles, sharks, bonefish, sea birds (Fig. 10.7) (Gilman 2016; Shester and Micheli 2011).

In Brazil, studies showing ghost fish are still scarce, which may reflect the lack of public initiatives to mitigate the impact, since most of the time, in situ expeditions using scuba are necessary (Link et al. 2019). However, unfortunately, it is already known that there is a record of the death of organisms in Brazilian protected areas (Santos et al. 2012; Soares et al. 2011). In the south coast of Bahia state, recent research interviewed artisanal fishers. Most of the respondents (90%) reported having found ghost nets in their fishing areas (Barbosa-Filho et al. 2020).

Ghost fishing is a silent threat that is affecting the Brazilian reefs (Fig. 10.7). Thinking about a solution for this impact is still quite challenging, however, initiatives such as removal get nets from the sea, use of less durable materials, or even biodegradable ones, in the making of fishing gear (Gilman 2016), or even the monitoring of nets and traps using satellite technologies can be an alternative for monitor and recover these devices in case they are lost. Public policies associated with the participation of the local population contribute together to the mitigation of this problem (Barbosa-Filho et al. 2020).

Fig. 10.7 No selecting ghost-fishing nets and its impacts on the seafloor. (**a–c**) Nurse shark (*Ginglymostoma cirratum*), a fish, and a crab entangled in net, respectively; (**d**) Abandoned fishing net covering the reef substrate (Ceará coast, NE Brazil). (Photos: Marcus Davis Andrade Braga)

10.3.5 Fisheries

The impact of overfishing predates all other anthropogenic disturbances in coastal ecosystems (Jackson et al. 2001). Specifically in reef environments, this problem gains greater magnitude, with around 55% of the world's reefs suffering from the effects of overfishing (Burke et al. 2011) which, in addition to the problems in the reef ecosystem, directly affects about 28% small-scale fishers globally (Teh et al. 2013).

Similar to what is occurring on a global scale, commercial, recreational and artisanal fisheries have been causing negative ecological effects on many reef-associated species throughout the Brazilian Biogeographical Province (Eggertsen et al., 2024; Floeter et al. 2006; Pinheiro et al. 2018). Although reef fishing catches a wide variety of taxa, in many cases, the fishing effort and, consequently, most of the biomass production, have been directed to large carnivores (Costa et al. 2003), such as lutjanids (e.g., *Lutjanus analis*; *Lutjanus jocu*; *Lutjanus synagris*; *Ocyurus chrysurus*; *Lutjanus purpureus*; *Lutjanus vivanus*), epinephelids (e.g., *Mycteroperca bonaci* (Fig. 10.8a)), scombrids (e.g., *Scomberomorus brasiliensis*; *Scomberomorus cavalla*), and carangids (e.g., *Seriola dumerili*). However, with the fall in the production of the target species due to overfishing (e.g., Fonteles-Filho 2007; Frédou et al. 2009), part of the fishing effort has been directed to species until then less explored and/or that belong to different trophic levels, for example, representatives of the Labridae (Scarinae) (Fig. 10.8b, c), Acanthuridae (Fig. 10.8d), Mullidae, and Haemulidae families (Eggertsen et al., 2024; Cunha et al. 2012; Melo et al. 2020).

Faced with this new reality, some species of nominally herbivores, such as *Scarus trispinosus*, due to their vulnerability to overexploitation (Freitas et al. 2019), already show signs of population decline in different reef environments along the Brazilian coast (Bender et al. 2014; Roos et al. 2020). As a result of overfishing, essential ecological functions, such as predation and herbivory, may already be seriously compromised in many Brazilian reefs.

Despite being basic information for management, a good part of the historical catch data has deficiencies, including low taxonomic resolution (Freire et al. 2015).

Fig. 10.8 Some examples of species fished in Brazilian reef environments. (**a**) *Mycteroperca bonaci*, one of the main species of epinephelids caught in northeastern Brazil. (**b**) *Scarus zelindae*; (**c**) *Sparisoma frondosum* and (**d**) *Acanthurus chirurgus*, examples of nominally herbivorous (sensu Cordeiro et al., 2016) currently exploited by fisheries, including for export purposes (Cunha et al. 2012). All photos were taken at Pedra da Risca do Meio Marine State Park, Ceará, Northeastern Brazil. (Photos: J. E. P. Freitas)

In recent years, between 2014 and 2020, Brazil has not officially reported capture and aquaculture data to be included in global statistics (FAO 2020). The lack of data makes it difficult to manage fishing activity and evaluate the impacts on Brazilian reefs, where characteristics such as multi-species and multi-gear fisheries, decentralized landings and socioeconomic importance make this management an even more challenging task.

Marine Protected Areas (MPAs) with distinct levels of protection (no-take or multiple-use) have been used to assist fisheries management in different locations on the coast (Moura et al. 2009). Some studies point to a greater biomass of reef fish within these MPAs (Francini-Filho and Moura 2008; Morais et al. 2017). Nevertheless, depending on the level of degradation of the reef environment, it would be necessary an average of 35 years or more, given protection from fishing, to recover your fish biomass (MacNeil et al. 2015). Unfortunately, most Brazilian Federal MPAs have serious management difficulties, with about 63% without management plans (Gerhardinger et al. 2011; Mills et al. 2020).

Although extremely important, the delimitation of MPAs alone does not seem to be sufficient to promote the recovery of some overexploited species and, in these cases, fisheries management with a focus on the species, in addition to MPAs, can help the recovery of these populations (Roos et al. 2020). Despite their peculiarities, the Brazilian reef environments represent a valuable opportunity for an integrated management, as long as fishing data are generated (Eggertsen et al., 2024; Floeter et al. 2006; Magris et al. 2020).

10.3.6 Invasive Species

Along the Brazilian coast, an update revision indicated at least 138 non-indigenous species, of which 19 were considered invasive, including chordates (N = 2), arthropods (N = 1), annelids (N = 2), molluscs (N = 3), cnidarians (N = 7), chlorophytes (N = 1), ochrophytas (N = 1), and myxozoans (N = 2) (Teixeira and Creed 2020). Among those, the most widespreads are the Malacostraca *Charybdis (Charybdis) hellerii*, the Bivalvia *Isognomon bicolor*, the lionfish *Pterois volitans*, and the Anthozoa corals *Tubastraea coccinea* and *Tubastraea tagusensis*, all spreading thousands of kilometers along the Brazilian coast.

Charybdis hellerii is surported to have been introduced in the Western Atlantic coast by ballast waters (Tavares and Mendonça Jr 1996), whereas *I. bicolor* and *Tubastraea* spp. encrusting oil and gas platforms (Oliveira and Creed 2008; Creed et al. 2017). The lionfish invasion in Brazil is already in the consolidation stage, with > 600 adult individuals recorded so far (2020–2024) along 4,000 km of coastline. These lionfish records cover a broad depth range (1–110 m depth), fourteen marine protected areas, nine Brazilian states and multiple coastal habitats (i.e., mangroves, shallow-water and mesophotic reefs, seagrass beds, artificial reefs, and sandbanks), indicating a successful invasion process in Brazilian waters (Soares et al., 2023). Among these invasive species, the large detrimental effects from

Tubastraea spp. invasions appear to be unprecedented. Sun (or orange-cup) corals *Tubastraea* spp. were first recorded in Brazil during the late 1980s (Castro and Pires 2001), and identified to species level in 2004 (de Paula and Creed 2004). Molecular studies have shown that besides being responsible for their first introduction, the transport of contaminated artificial structures can be blamed for facilitating the dispersion of these invasive scleractinian corals along the coast (Capel et al. 2019). Nowadays, nearly 30 years after their first records, *Tubastraea* spp. has dominated large areas of both natural and artificial substrates from Ceará to Santa Catarina states (Creed et al. 2017; Soares et al. 2016), causing negative impacts to local communities by homogenization, competition, changes in seascape and, therefore, modifications in the ecosystem functioning (Mantelatto et al. 2011; Creed 2006; Capel et al. 2020).

Laboratory experiments and *in situ* observations have demonstrated antagonistic relationships among *T. coccinea* and/or *T. tagusensis* and native corals such as *Siderastrea stellata*, *Madracis decactis* (Miranda et al. 2016), the endemic *Mussismilia hispida* (Creed 2006; Santos et al. 2013) (Fig. 10.3), and the Zoantharia *Palythoa caribaeorum* (Luz and Kitahara 2017). Additionally, to the direct competition, *Tubastraea* can inhibit the settlement of several benthic species by producing allelopathic compounds (Lages et al. 2010). Combining the aforementioned biological capabilities to their gregarious settlement behavior, *Tubastraea* can saturate the available hard substrate, ultimately leading to the homogenization of the benthic fauna in some areas such as in Búzios Island, southeastern Brazil (see Mantelatto et al. 2011). Localities that have high densities of *Tubastraea* harbor not only a less diverse community of sessile invertebrates but also a poorer community of vagile ones (Silva et al. 2019), which, altogether, are purported to significantly decrease the foraging of itinerant herbivore fishes (Miranda et al. 2018). The dominance of *Tubastraea* also triggers a more profound effect in the seascape than previously thought, with potential changes of the macro and meiofauna communities from the adjacent soft sediments (Capel et al. 2020). Management actions aiming to control and reduce the invasive population of *Tubastraea* spp. are scarce and punctual. Such efforts have been focusing mainly on protected areas (e.g., Tamoios Ecological Station, Tupinambás Ecological Station, Alcatrazes Wildlife Refuge, Laje de Santos Marine State Park, and Arvoredo Marine Biological Reserve). Nevertheless, effective control must regulate and control the occurrence of new and secondary introductions through contaminated vectors, the last known to have a significant role in dispersing the invasive corals throughout the Brazilian coast (Capel et al. 2019).

In the context of *Tubastraea* spp. invasion and climate change in the Brazilian coast (see Sect. 10.2.1 of this book), the effects of the higher water temperatures will probably result in an even worse scenario. Temperatures that have been shown to severely affect the native coral fauna with bleaching and death (Banha et al. 2019; Duarte et al. 2020) apparently do not affect the azooxanthellate invasive corals such as *Tubastraea* spp. Thus, the synergetic effect of high sea surface temperature and competition against invasive species impose a considerable challenge for native corals and reef species on the Brazilian marine ecosystems.

10.4 Conclusions and Final Remarks

In conclusion, Southwestern Atlantic reefs are unique due an amazing combination of endemism, low functional redundancy, rare species, dominance of massive, depth-specialists and stress-tolerant/weedy corals, and also occurrence in turbid waters with significant nutrient content and sediment resuspension. Brazilian reefs present a large number of phase-shift events and indicators of reef decline (e.g., decreasing richness, biomass and loss of architectural complexity) which indicates significant degradation in the supply of geo-ecological functions such as reef framework production and sediment generation, the maintenance of reef habitat complexity and reef growth potential. Moreover, the disproportionate distribution of rare fish species (Araújo et al. 2020) and decreased biomass reveals a higher vulnerability of Brazilian reef communities to impacts and stochastic density fluctuations.

The large Brazilian's marginal reef nearshore area targets seem to be in fact more resistant to global changes impacts as current heatwaves (Mies et al. 2020; Soares 2020; Soares et al. 2021a). In the meantime, a recent study has demonstrated that these same reefs could be more susceptible to local impact and also to phase shift them reefs developed near optimal conditions (Cruz et al. 2018) which provide evidence against the BRRH. Meanwhile, in the years 2019, 2020, and 2024, mass mortality of coral species *Mussismilia hartii* and especially *Millepora alcicornis* were recorded in distinct Brazilians reefs after marine heatwaves (Duarte et al. 2020; Soares et al. 2021a). Because of impacts, two kinds of phase shift were recorded on large reef areas, the dominance of macroalgae and the dominance of zoanthid (Cruz et al. 2018). The macroalgae phase shift is a wider problem occurring in 29% of studied reefs and could be related to presence of ports, urban surface and nutrient runoff. While zoanthid dominance reaches 17% of studied reefs and could be associated with urban surface, proximity with urban areas with medium and large human population and dredged ports. Both together suggest that Brazilian marginal reefs are more endangered (Cruz et al. 2018).

Some reef areas (e.g., Abrolhos Bank and Vitória-Trindade Seamount Chain) (Hoegh-Guldberg et al. 2017; Mies et al. 2020) may be potential short-term ecological refuge in the upcoming years (or decades) for few stress-tolerant species but need special attention for their conservation especially the reduction of local and regional impacts. Overall, all the Brazilian reefs are highly threatened by a plethora of human pressures until the end of this century, have limited potential to provide climate-change refugia, and need urgent attention to reduce and mitigate impacts at local and global scale.

We remember the importance of maintaining pathways of larval connectivity in the Brazilian reef areas and MPAs, reducing local stressors (e.g., environmental education, sewage systems in the Brazilian cities, national plans to fight against invasive species, fishing regulations), and also protecting especially the endemism hotspots. Moreover, the shallow-water and mesophotic reefs are ecologically distinct and need to be protected together to ensure the conservation of alpha and beta South Atlantic reef diversity (Morais and Santos 2018; Soares et al. 2019b).

Acknowledgements We wish to acknowledge the editor of this book chapter Prof. Ruy Kenji Papa de Kikuchi for comments on our draft, which significantly improved this manuscript. MOS thanks the financial support provided by the Conselho Nacional de Desenvolvimento Científico e Tecnológico—CNPq (Grant 313518/2020-3), FUNCAP Chief Scientist Program, CAPES-PRINT Program, CAPES-COFECUB, CAPES-AVH (Alexander Von Humboldt Foundation), PELD Costa Semiárida do Brasil (442337/2020-5), and INCT AmbTropic (National Institute of Science and Technology for the Tropical Marine Environment, CNPq 465634/2014-1). ALA Gurgel and BD Lopes thank CNPq scholarships. MVK thanks the financial support provided by São Paulo Research Foundation (FAPESP 2021/06866-6). MOS, CEPT and MVK thank CNPq Research Productivity Fellowships.

References

Adelir-Alves J, Rocha GRA, Souza TF et al (2016) Abandoned, lost or otherwise discarded fishing gears in rocky reefs of southern Brazil. Braz J Oceanogr 64:427–434

Andrades R, Santos RG, Joyeux J, Chelazzi D, Cincinelli A, Giarrizzo T (2018) Marine debris in Trindade Island, a remote island of the South Atlantic. Mar Pollut Bull 137:180–184

Angulo RJ, Lessa GC, Souza MC (2006) A critical review of mid- to late Holocene sea-level fluctuations on the eastern Brazilian coastline. Quat Sci Rev 25:486–506. https://doi.org/10.1016/j.quascirev.2005.03.008

Araújo ME, Mattos FMG, Melo FPL et al (2020) Diversity patterns of reef fish along the Brazilian tropical coast. Mar Environ Res 160:105038

Aued AW, Smith F, Quimbayo JP et al (2018) Large-scale patterns of benthic marine communities in Brazilian Province. PLoS One 13:e0198452

Aubrey D.G., Emery K.O., Uchupi E. (1988) Changing coastal levels of South America and the Caribbean region from tide-gauge records Tectonophysics 154(3-4) 269-284 10.1016/0040-1951(88)90108-4

Banha TNS, Capel KCC, Kitahara MV et al (2019) Low coral mortality during the most intense bleaching event ever recorded in subtropical Southwestern Atlantic reefs. Coral Reefs 39:515–521

Barbosa-Filho MLV, Seminara CI, Tavares DC et al (2020) Artisanal fisher perceptions on ghost nets in a tropical South Atlantic marine biodiversity hotspot: challenges to traditional fishing culture and implications for conservation strategies. Ocean Coast Manag 192:105189

Bender MG, Machado GR, Silva PJA et al (2014) Local ecological knowledge and scientific data reveal overexploitation by multigear artisanal fisheries in the Southwestern Atlantic. PLoS One 9(10):e110332

Belem, A. L. Impacto das mudanças climáticas globais no risco de inundações em zonas costeiras. In: JORNADAS INTERNACIONALES SOBRE GESTIÓN DEL RIESGO DE INUNDACIONES Y DESLIZAMIENTOS DE LADERAS, 2007

Bigarella JJ, Andrade GO (1964) Considerações sobre a estratigrafia dos sedimentos cenozóicos em Pernambuco (Grupo Barreiras). Arq Inst Ciênc Terra 2:2–14

Bigarella JJ, Mousinho MRM, Silva JX (1965) Processes and environments of the Brazilian quaternary. In: 10th INQUA congress. Caderno Especial, Rio de Janeiro, p 156

Bittencourt ACSP, Martin L, Vilas Boas GS, Flexor JM (1979) Quaternary marine formations of the coast of the state of Bahia (Brazil). In: Suguio K et al (eds) Proceedings of the international symposium on coastal evolution in the quaternary. Brazilian National Working Group for the IGCP, São Paulo, pp 232–253

Bleuel J, Pennino MG, Longo G (2021) Coral distribution and bleaching vulnerability areas in Southwestern Atlantic under ocean warming. Sci Rep 11:12833

Burke L, Reytar K, Spalding M, Perry A (2011) Reefs at risk revisited. World Resources Institute, Washington, DC

Cai W-J, Feely RA, Testa JM et al (2021) Natural and anthropogenic drivers of acidification in large estuaries. Annu Rev Mar Sci 13:23–55

Camp EF, Schoepf V, Mumby PJ et al (2018) The future of coral reefs subject to rapid climate change: lessons from natural extreme environments. Front Mar Sci 5:1–15

Campelo RPS, Lima CDM, Santana CS, Silva AJ, Neumann-Leitão S, Ferreira BP, Soares MO, Júnior MM, Melo PAMC (2021) Oil spills: the invisible impact on the base of tropical marine food webs. Mar Pollut Bull 167:112281

Capel KCC, Creed JC, Kitahara MV et al (2019) Multiple introductions and secondary dispersion of *Tubastraea* spp. in the Southwestern Atlantic. Sci Rep 9:13978

Capel KCC, Creed JC, Kitahara MV (2020) Invasive corals trigger seascape changes in the Southwestern Atlantic. Bull Mar Sci 96:217–218

Castro CB, Pires DO (2001) Brazilian coral reefs: what we already know and what is still missing. Bull Mar Sci 69:357–371

Coimbra KTO, Alcântara E, de Souza Filho CR (2020) Possible contamination of the Abrolhos reefs by Fundao dam tailings, Brazil—new constraints based on satellite data. Sci Total Environ 733:138101

Cordeiro CAMM, Mendes TC, Harborne AR, Ferreira CEL (2016) Spatial distribution of nominally herbivorous fishes across environmental gradients on Brazilian rocky reefs Journal of Fish Biology 89(1):939–958. https://doi.org/10.1111/jfb.2016.89.issue-1

Costa MF, Barletta M (2015) Microplastics in coastal and marine environments of the western tropical and sub-tropical Atlantic Ocean. Environ Sci Process Impacts 11:1868–1879

Costa OS Jr, Leão ZMAN, Nimmo M, Attrill MJ (2000) Nutrification impacts on coral reefs from northern Bahia, Brazil. Hydrobiologia 440:307–315

Costa OS Jr, Attrill MJ, Pedrini AG et al (2002) Benthic macroalgal distribution in coastal and offshore reefs at Porto Seguro Bay, Brazilian Discovery coast. In: Proceedings of the 9th international coral reef symposium, vol 1. Ministry for the Environment, Jakarta, pp 499–508

Costa PAS, Braga AC, Rocha LOF (2003) Reef fisheries in Porto Seguro, eastern Brazilian coast. Fish Res 60:577–583

Cotovicz LC, Knoppers BA, Brandini N et al (2018) Aragonite saturation state in a tropical coastal embayment dominated by phytoplankton blooms (Guanabara Bay – Brazil). Mar Pollut Bull 129:729–739

Cotovicz LC, Chielle R, Marins RV (2020a) Air-sea CO2 flux in an equatorial continental shelf dominated by coral reefs (Southwestern Atlantic Ocean). Cont Shelf Res 204:104175

Cotovicz LC, Vidal LO, de Rezende CE et al (2020b) Carbon dioxide sources and sinks in the delta of the Paraíba do Sul River (Southeastern Brazil) modulated by carbonate thermodynamics, gas exchange and ecosystem metabolism during estuarine mixin. Mar Chem 226:103869

Creed JC (2006) Two invasive alien azooxanthellate corals, *Tubastraea coccinea* and *Tubastraea tagusensis*, dominate the native zooxanthellate *Mussismilia hispida* in Brazil. Coral Reefs 25:350

Creed JC, Fenner D, Sammarco PW et al (2017) The invasion of the azooxanthellate coral *Tubastraea* (Scleractinia: Dendrophylliidae) throughout the world: history, pathways and vectors. Biol Invasions 19:283–305

Cruz ICS, Waters LG, Kikuchi RKP et al (2018) Marginal coral reefs show high susceptibility to phase shift. Mar Pollut Bull 135:551–561

Cunha FEA, Carvalho RAA, Araújo ME (2012) Exportation of reef fish for human consumption: long-term analysis using data from Rio Grande do Norte, Brazil. Bol Inst Pesca 38:369–378

de Carvalho-Souza GF, Tinôco ML, Medeiros DV et al (2018) Marine litter disrupts ecological processes in reef systems. Mar Pollut Bull 133:464–471

de Paula AF, Creed JC (2004) Two species of the coral *Tubastraea* (Cnidaria, Scleractinia) in Brazil: a case of accidental introduction. Bull Mar Sci 74(1):175–183

Doney SC, Fabry VJ, Feely R, Kleypas JA (2009) Ocean acidification: the other CO2 problem. Annu Rev Mar Sci 1:169–192

Doney SC, Busch DS, Cooley SR, Kroeker KJ (2020) The impacts of ocean acidification on marine ecosystems and reliant human communities. Annu Rev Environ Resour 45:83–112

Dove SG, Brown KT, Van Den Heuvel A et al (2020) Ocean warming and acidification uncouple calcification from calcifier biomass which accelerates coral reef decline. Commun Earth Environ 1:55

Duarte GAS, Villela HDM, Deocleciano M et al (2020) Heat waves are a major threat to turbid coral reefs in Brazil. Front Mar Sci 7:179

de Mesquita, A. R. (2003). Sea-Level Variations Along the Brazilian Coast: A Short Review. Journal of Coastal Research, 21–31. http://www.jstor.org/stable/40928745

Elliff CI, Silva IR (2017) Coral reefs as the first line of defense: shoreline protection in face of climate change. Mar Environ Res 127:148–154

Evangelista H, Valeriano C (2017) Sumário das Análises de Isótopos Radiogênicos de Sr e Nd em sedimentos da foz do rio Doce-ES e do Parque Nacional dos Abrolhos-BA (antes e depois do rompimento da barragem de rejeitos da Samarco em Mariana-MG)

FAO (2020) The state of world fisheries and aquaculture 2020. Sustainability in action. FAO, Rome

Feitosa CV, Chaves LDCT, Ferreira BP, De Araujo ME (2012) Recreational fish feeding inside Brazilian MPAs: impacts on reef fish community structure. J Mar Biol Assoc U K 92:1525–1533

Floeter SR, Halpern BS, Ferreira CEL (2006) Effects of fishing and protection on Brazilian reef fishes. Biol Conserv 128(3):391–402

Fonteles-Filho AA (2007) Síntese sobre o pargo (*Lutjanus purpureus*). In: Haimovici M (ed) A Prospecção Pesqueira e Abundância de Estoques Marinhos No Brasil Nas Décadas de 1960 a 1990: Levantamento de Dados e Avaliação Crítica. Ministério do Meio Ambiente, Brasília, pp 249–255

Francini-Filho RB, Moura RL (2008) Evidence for spillover of reef fishes from a no-take marine reserve: an evaluation using the before-after control-impact (BACI) approach. Fish Res 93:346–356

Francini-Filho R, Asp NE, Siegle E et al (2018) Perspectives on the great Amazon reef: extension, biodiversity, and threats. Front Mar Sci 5:142

Frédou T, Ferreira BP, Letourneur Y (2009) Assessing the stocks of the primary snappers caught in Northeastern Brazilian reef systems. 1: Traditional modelling approaches. Fish Res 99:90–96

Freire KMF, Aragão JAN, Araújo ARR et al (2015) Revisiting Brazilian catch data for Brazilian marine waters 1950-2010. In: Freire KMF, Pauly D (eds) Fisheries catch reconstructions for Brazil's mainland and oceanic islands. The Fisheries Centre – University of British Columbia, Vancouver, pp 3–29

Freitas MO, Previero M, Leite JR et al (2019) Age, growth, reproduction and management of Southwestern Atlantic's largest and endangered herbivorous reef fish, *Scarus trispinosus* Valenciennes, 1840. PeerJ 7:e7459

Gattuso J-P, Magnan A, Billé R et al (2015) Contrasting futures for ocean and society from different anthropogenic CO_2 emissions scenarios. Science 349(6243):aac4722

Gerhardinger LC, Godoy EAS, Jones PJS et al (2011) Marine protected dramas: the flaws of the Brazilian national system of marine protected areas. Environ Manag 47:630–643

Giglio VJ, Luiz OJ, Schiavetti A (2016) Recreational diver behavior and contacts with benthic organisms in the Abrolhos National Marine Park, Brazil. Environ Manag 57:637–648

Gilman E (2016) Biodegradable fishing gear: part of the solution to ghost fishing and marine pollution. Anim Conserv 19:320–321

Harari J, Camargo R (1994) Simulação da propagação das nove principais componentes de maré na plataforma sudeste brasileira através de modelo numérico hidrodinâmico. Boletim do Instituto Oceanográfico, São Paulo 42(1):35–54

Haapkyla J, Ramade F, Salvat B (2007) Oil pollution on coral reefs: a review of the state of knowledge and management needs. Vie et Milieu 57:91–107

Hatje V, Pedreira RMA, Rezende CE et al (2017) The environmental impacts of one of the largest tailing dam failures worldwide. Sci Rep 7:10706

Hoegh-Guldberg TP, Kerry JT, Álvarez-Noriega M et al (2017) Global warming and recurrent mass bleaching of corals. Nature 543:373–377

Holbrook NJ, Gupta AS, Oliver EC et al (2020) Keeping pace with marine heatwaves. Nat Rev Earth Environ 1:482–493

Hughes TP, Barnes ML, Bellwood DR et al (2017) Coral reefs in the Anthropocene. Nature 546:82–90

IPCC, 2019: Summary for Policymakers. In: IPCC Special Report on the Ocean and Cryosphere in a Changing Climate [H.-O. Pörtner, D.C. Roberts, V. Masson-Delmotte, P. Zhai, M. Tignor, E. Poloczanska, K. Mintenbeck, A. Alegría, M. Nicolai, A. Okem, J. Petzold, B. Rama, N.M. Weyer (eds.)]. Cambridge University Press, Cambridge, UK and New York, NY, USA, pp. 3-35. https://doi.org/10.1017/9781009157964.001

Irion G, Morais JO, Bungenstock F (2012) Holocene and Pleistocene sea-level indicators at the coast of Jericoacoara, Ceará, NE Brazil. Quat Res 77:251–257

Iwanicki L (2020) Um oceano livre de plástico: desafios para reduzir a poluição marinha no Brasil/ Lara Iwanicki, Ademilson Zamboni, 1st edn. Oceana Brasil, Brasília

Jackson JBC, Kirby MX, Berger WH et al (2001) Historical overfishing and the recent collapse of coastal ecosystems. Science 293:629–638

Jones KC, Voogt P (1999) Persistent organic pollutants (POPs): state of science. Environ Pollut 100:209–221

Klein AHF, Short D (2016) Brazilian beach systems: introduction. In: Short AD, Klein AHdF (eds) Brazilian beach systems, Coastal research library, vol 17. Springer, Cham, pp 1–35

Kleypas JA, Yates KK (2009) Coral reefs and ocean acidification. Oceanography 22:108–117

Kline DI, Teneva L, Okamoto DK et al (2019) Living coral tissue slows skeletal dissolution related to ocean acidification. Nat Ecol Evol 3:1438–1444

Lages BG, Fleury BG, Pinto AC, Creed JC (2010) Chemical defenses against generalist fish predators and fouling organisms in two invasive ahermatypic corals in the genus *Tubastraea*. Mar Ecol 31:473–482

Lamb JB, Willis BL, Fiorenza EA et al (2018) Plastic waste associated with disease on coral reefs. Science 359:460–462

Leão ZMAN, Kikuchi RKP, Ferreira BP et al (2016) Brazilian coral reefs in a period of global change: a synthesis. Braz J Oceanogr 64:97–116

Leão ZMAN, Kikuchi RKP, Oliveira MDM (2019) The coral reef province of Brazil. In: World seas: an environmental evaluation, vol 1. Elsevier, pp 813–833

Link J, Segal B, Casarini LM (2019) Abandoned, lost or otherwise discarded fishing gear in Brazil: a review. Perspect Ecol Conserv 17:1–8

Linda, Eggertsen André L., Luza César A. M. M., Cordeiro Cristian, Dambros Carlos E. L., Ferreira Sergio R., Floeter Ronaldo, B. Francini-Filho Kátia M. F., Freire Maria A., Gasalla Tommaso, Giarrizzo Vinicius J., Giglio Natalia, Hanazaki Priscila F. M., Lopes Guilherme O., Longo Osmar J., Luiz Rafael A., Magris Thiago C., Mendes Hudson T., Pinheiro Juan P., Quimbayo José Amorim, Reis-Filho Daniele A., Vila-Nova Mariana G., Bender (2024) Complexities of reef fisheries in Brazil: a retrospective and functional approach Reviews in Fish Biology and Fisheries 34(1) 511-538 10.1007/s11160-023-09826-y

Loiola M, Cruz ICS, Lisboa DS et al (2019) Structure of marginal coral reef assemblages under different turbidity regime. Mar Environ Res 147:138–148

Losada IJ, Reguero BG, Méndez FJ et al (2013) Long-term changes in sea-level components in Latin America and the Caribbean. Glob Planet Chang 104:34–50

Luz BLP, Kitahara MV (2017) Could the invasive scleractinians *Tubastraea coccinea* and *T. tagusensis* replace the dominant zoantharian *Palythoa caribaeorum* in the Brazilian subtidal? Coral Reefs 36:875

Lyra AP, Lyrio R, Cavalcanti P et al (2008) Avaliação dos impactos ambientais dos derrames de óleo sobre a biota e a coluna dágua da Baía de Guanabara, RJ: monitoramento ambiental e identificação de fontes poluidoras na bacia hidrográfica

MacNeil MA, Graham NAJ, Cinner JE et al (2015) Recovery potential of the world's coral reef fishes. Nature 520:341–344

Magris RA, Grech A, Pressey RL (2018) Cumulative human impacts on coral reefs: assessing risk and management implications for Brazilian coral reefs. Diversity 10(2):26

Magris RA, Costa MDP, Ferreira CEL et al (2020) A blueprint for securing Brazil's marine biodiversity and supporting the achievement of global conservation goals. Divers Distrib 27:198–215

Mantelatto MC, Mourão GG, Migotto AE, Creed JC (2011) Range expansion of the invasive corals *Tubastraea coccinea* and *Tubastraea tagusensis* in the Southwest Atlantic. Coral Reefs 30:397

Martin C, Corona E, Mahadik GA, Duarte CM (2019) Adhesion to coral surface as a potential sink for marine microplastics. Environ Pollut 255:113281

Melo CC, Soares APC, Pelage L et al (2020) Haemulidae distribution patterns along the Northeastern Brazilian continental shelf and size at first maturity of the most abundant species. Reg Stud Mar Sci 35:101226

Metzler PM, Glibert PM, Gaeta SA et al (1997) New and regenerated production in the South Atlantic off Brazil. Deep-Sea Res I Oceanogr Res Pap 44:363–384

Mies M, Francini-Filho RB, Zilberberg C et al (2020) South Atlantic coral reefs are major global warming refugia and less susceptible to bleaching. Front Mar Sci 7:514

Mills M, Magris RA, Fuentes MMPB et al (2020) Opportunities to close the gap between science and practice for marine protected areas in Brazil. Perspect Ecol Conserv 18:161–168

Miranda RJ, Cruz ICS, Barros F (2016) Effects of the alien coral *Tubastraea tagusensis* on native coral assemblages in a Southwestern Atlantic coral reef. Mar Biol 163:45

Miranda RJ, Nunes AJCC, Mariano-Neto E et al (2018) Do invasive corals alter coral reef processes? An empirical approach evaluating reef fish trophic interactions. Mar Environ Res 138:19–27

Monteiro RCP, Ivar do Sul JA, Costa MF (2020) Small microplastics on beaches of Fernando de Noronha Island, Tropical Atlantic Ocean. Ocean Coastal Res 68:e20235

Morais J, Santos BA (2018) Limited potential of deep reefs to serve as refuges for tropical Southwestern Atlantic corals. Ecosphere 9:e02281

Morais RA, Ferreira CEL, Floeter SR (2017) Spatial patterns of fish standing biomass across Brazilian reefs, Southwestern Atlantic. J Fish Biol 91:1642–1667

Moura RL, Minte-Vera CV, Curado IB et al (2009) Challenges and prospects of fisheries co-management under a marine extractive reserve framework in northeastern Brazil. Coast Manag 37:617–632

Moura NSV, Moran EF, Strohaecker TM, Kunst AV (2015) Urbanization in the coastal zone: local and regional processes and environmental changes – the case of the northern coast of the state of Rio Grande do Sul, Brazil. Cienc Nat 37:594–612

Moura RL, Amado-Filho GM, Moraes FC et al (2016) An extensive reef system at the Amazon River mouth. Sci Adv 2:1501252

NOAA Coral Reef Watch. 2018, updated daily. NOAA Coral Reef Watch Version 3.1 Daily Global 5km Satellite Coral Bleaching Degree Heating Week Product, Jun. 3, 2013-Jun. 2, 2014. College Park, Maryland, USA: NOAA Coral Reef Watch. Data set accessed 2020-09-01 at ftp://ftp.star.nesdis.noaa.gov/pub/sod/mecb/crw/data/5km/v3.1/nc/v1.0/daily/dhw/.

Oliveira AES, Creed JC (2008) Mollusca, Bivalvia, *Isognomon bicolor* (C. B. Adams 1845): distribution extension. Check List 4:386–388

Oliver ECJ, Donat MG, Burrows MT et al (2018) Longer and more frequent marine heatwaves over the past century. Nat Commun 9:1–12

Oliver ECJ, Benthuysen JA, Darmaraki S et al (2020) Marine heatwaves. Annu Rev Mar Sci 13:313–342

PBMC (2016) Impacto, vulnerabilidade e adaptação das cidades costeiras brasileiras às mudanças climáticas: Relatório Especial do Painel Brasileiro de Mudanças Climáticas

Perry CT, Alvarez-Filip L (2018) Changing geo-ecological functions of coral reefs in the Anthropocene. Funct Ecol 33:976–988

Perry CT, Alvarez-Filip L, Graham NA et al (2018) Loss of coral reef growth capacity to track future increases in sea level. Nature 558:396–340

Pinheiro HT, Rocha LA, Macieira RM et al (2018) South-western Atlantic reef fishes: zoogeographical patterns and ecological drivers reveal a secondary biodiversity centre in the Atlantic Ocean. Divers Distrib 24:951–965

Pirazzoli, P. A. (1986). Secular Trends of Relative Sea-Level (RSL) Changes Indicated by Tide-Gauge Records. Journal of Coastal Research, 1–26. http://www.jstor.org/stable/44863318

Portugal AB, Carvalho FL, Carneiro PBM, Rossi S et al (2016) Increased anthropogenic pressure decreases species richness in tropical intertidal reefs. Mar Environ Res 120:44–54

Reichert J, Schellenberg J, Schubert P, Wilke T (2018) Responses of reef building corals to microplastic exposure. Environ Pollut 237:955–960

Rodrigues RR, Taschetto AS, Gupta AS, Foltz GR (2019) Common cause for severe droughts in South America and marine heatwaves in the South Atlantic. Nat Geosci 12:620–626

Roos NC, Longo GO, Pennino MG et al (2020) Protecting nursery areas without fisheries management is not enough to conserve the most endangered parrotfish of the Atlantic Ocean. Sci Rep 10:1–10

Santos AJB, Bellini C, Bortolon LF, Coluchi R (2012) Ghost nets haunt the olive ridley turtle (Lepidochelys olivacea) near the Brazilian Islands of Fernando de Noronha and Atol das Rocas. Herpetol Rev 43:245–246

Santos LAH, Ribeiro FV, Creed JC (2013) Antagonism between invasive pest corals *Tubastraea* spp. and the native reef-builder *Mussismilia hispida* in the southwest Atlantic. J Exp Mar Biol Ecol 449:69–76

Savinelli B, Fernandez TV, Galasso NM et al (2020) Microplastics impair the feeding performance of a Mediterranean habitat-forming coral. Mar Environ Res 155:104887

Schoepf V, Baumann JH, Barshis DJ, Browne NK, Camp EF, Comeau S, Cornwall CE, Guzmán HM, Riegl B, Rodolfo-Metalpa R, Sommer B (2023) Corals at the edge of environmental limits: A new conceptual framework to re-define marginal and extreme coral communities Science of The Total Environment 884163688-10.1016/j.scitotenv.2023.163688

Shester GG, Micheli F (2011) Conservation challenges for small-scale fisheries: bycatch and habitat impacts of traps and gillnets. Biol Conserv 144:1673–1681

Silva R, Vinagre C, Kitahara MV et al (2019) Sun coral invasion of shallow rocky reefs: effects on mobile invertebrate assemblages in Southeastern Brazil. Biol Invasions 21:1339–1350

Silva, G. N. Variação de longo período do nível médio do mar: causas, consequências e metodologia de análise. 1992. Tese (Mestrado em Engenharia Oceânica) – COPPE/UFRJ, Rio de Janeiro.

Soares MO , Pereira PHC, Feitosa CV, Maggioni R , Rocha RS, Bezerra LEA, Duarte OS , Paiva SV, Noleto-Filho E, Silva MQM, Csapo-Thomaz M, Garcia TM, Arruda Júnior JPV , Cottens KF, Vinicius B, Araújo R, Eirado CB, Santos LPS , Guimarães TCS , Carlos Targino CH., Tommaso Giarrizzo (2023) Lessons from the invasion front: Integration of research and management of the lionfish invasion in Brazil Journal of Environmental Management 340117954-10.1016/j.jenvman.2023.117954

Soares MO (2020) Marginal reef paradox: a possible refuge from environmental changes? Ocean Coast Manag 185:105063

Soares MO, Paiva CC, Freitas JEP, Lotufo TMC (2011) Gestão de unidades de conservação marinhas: o caso do Parque Estadual Marinho da Pedra da Risca do Meio, NE – Brasil. Rev Gestão Costeira Integr 11:257–268

Soares MO, Davis M, de Macêdo-Carneiro PB (2016) Northward range expansion of the invasive coral (*Tubastraea tagusensis*) in the Southwestern Atlantic. Mar Biodivers 48:1651–1654

Soares MO, Teixeira CEP, Ferreira SMC et al (2019a) Thermal stress and tropical reefs: mass coral bleaching in a stable temperature environment? Mar Biodivers 49:2921–2929

Soares MO, Tavares TCL, Carneiro PBM (2019b) Mesophotic ecosystems: distribution, impacts and conservation in the South Atlantic. Divers Distrib 25:255–268
Soares MO, Teixeira CEP, Bezerra LEA et al (2020a) Brazil oil spill response: time for coordination. Science 367(6474):155
Soares MO, Teixeira CEP, Bezerrra LEA et al (2020b) Oil spill in South Atlantic (Brazil): environmental and governmental disaster. Mar Policy 115:103879
Soares MO, Matos E, Lucas C et al (2020c) Microplastics in corals: Na emergent threat. Mar Pollut Bull 161:111810
Soares MO, Rossi S, Gurgel AR, Lucas CC, Tavares TCL, Diniz B, Feitosa CV, Rabelo EF, Pereira PHC, Kikuchi RKP, Leão ZMAN, Cruz ICS, Carneiro PBM, Alvarez-Filip L (2021a) Impacts of a changing environment on marginal coral reefs in the Tropical Southwestern Atlantic. Ocean Coast Manag 210:105692. https://doi.org/10.1016/j.ocecoaman.2021.105692
Soares EC, Bispo MD, Vasconcelos VC, Soletti JI, Carvalho SHV, Oliveira MJ, Santos MC, Freire ES, Nogueira ASP, Cunha FAS, Sandes RDD, Santos RAR, Neta MTSL, Narain N, Garcia CAB, Costa SSL, Santos JCC (2021b) Oil impact on the enviroment and aquatic organisms on the coasts of the states of Alagoas and Sergipe, Brazil – a preliminary evaluation. Mar Pollut Bull 171:112723. https://doi.org/10.1016/j.marpolbul.2021.112723
Soares MO, Cruz ICS, Santos BA et al (2021c) Marginal reefs in the Anthropocene: they are not Noah's Ark. In: Rossi S, Bramanti L (eds) Perspectives on the marine animal forests of the world, 1st edn. Springer Nature Switzerland AG: Springer International Publishing, Cham, pp 1–572
Spalding MD, Fox HE, Allen GR et al (2007) Marine ecoregions of the world: a bioregionalization of coastal and shelf areas. Bioscience 57:573–583
Suguio K (2010) Geologia do quaternário e mudanças ambientais, 2nd edn. Oficina de Textos, São Paulo, p 408
Suguio K, Martin L, Bittencourt ACSP, Dominguez JML, Flexor JM, Azevedo AEG (1985) Flutuações do nível relativo do mar durante o Quaternário Superior ao longo do litoral brasileiro e suas implicações na sedimentação costeira. Rev Bras Geosci 15:273–286
Tang J, Wu Z, Wan L et al (2021) Differential enrichment and physiological impacts of ingested microplastics in scleractinian corals in situ. J Hazard Mater 404(B):124205
Tavares M, Mendonça JB Jr (1996) *Charybdis hellerii* (A. Milne Edwards, 1867) (Brachyura: Portunidae), eighth nonindigenious marine decapod from Brazil. Crustac Res 25:151–157
Teh LSL, Teh LCL, Sumaila UR (2013) A global estimate of the number of coral reef fishers. PLoS One 8(6):e65397
Teixeira LMP, Creed JC (2020) A decade on: an updated assessment of the status of marine non-indigenous species in Brazil. Aquat Invasions 15:30–43
Teixeira CD, Leitão RLL, Ribeiro FV et al (2019) Sustained mass coral bleaching (2016–2017) in Brazilian turbid-zone reefs: taxonomic, cross-shelf and habitat-related trends. Coral Reefs 38:801–813
Van Woesik R, Houk P, Isechal AL et al (2012) Climate-change refugia in the sheltered bays of Palau: analogs of future reefs. Ecol Evol 2:2474–2484
Wassmann P, Olli K (2005) Integrated approaches to drainage basin nutrient inputs and coastal eutrophication: an introduction. University of Tromsø, Norway
Woodland DJ, Hooper NA (1977) The effect of human trampling on coral reefs. Biol Conserv 11:1–4
Ximenes Neto AR (2020) Indicadores Geológico-Geomorfológicos de Variação do Nível Relativo do Mar em Sistemas de Natureza Mista (Siliciclástico-Carbonático): Costa do Ceará, NE Brasil. Tese (Doutorado) – Curso de Geografia, Programa de Pós-Graduação em Geografia, Universidade Estadual do Ceará, Fortaleza. 330 f

Chapter 11
Tourism, Environmental Education, and Aquarium Trade

Liana F. Mendes, Janaina F. Calado, Caroline V. Feitosa, Livio M. Gurjão, Maria Teresa J. Gouveia, Ligia M. Rocha, and Leopoldo M. Barreto

Abstract Coral reefs are places of remarkable scenic beauty, with colorful biodiversity and associated with clear and warm waters. Visiting a reef system usually carries a sense of adventure and emotion for people. As such, it also combines this excitement with much-needed environmental education actions. This chapter connects three human activities in reef areas on the Brazilian tropical coast: tourism, environmental education, and aquarium trade. These are sources of employment, with foreign direct investment and gross domestic product. The economy of coastal states depends on these revenues, which are linked to recreational activities and the quality of the beaches. However, poorly planned tourism causes severe damage to these delicate reef environments, so educational activities must be designed to contribute to the conservation and preservation of the environment. After all, nobody wants to visit a degraded setting. Some possible solutions include defining good

L. F. Mendes (✉)
Department of Ecology, Federal University of Rio Grande do Norte (UFRN), Natal, Rio Grande do Norte, Brazil

J. F. Calado
State University of Amapa (UEAP), Macapa, Amapá, Brazil

C. V. Feitosa
Institute of Marine Sciences, Federal University of Ceará (LABOMAR-UFC), Fortaleza, Ceará, Brazil

L. M. Gurjão
Brazilian Institute of the Environment and Renewable Natural Resources (IBAMA-SUPES-CE), Fortaleza, Ceará, Brazil

M. T. J. Gouveia
Coral Vivo Project, Environmental Education and Public Policies (2007-2023), Porto Seguro, Bahia, Brazil

L. M. Rocha
Oceanica NGO – Research, Education and Conservation, Parnamirim, Rio Grande do Norte, Brazil

L. M. Barreto
Federal University of Recôncavo da Bahia (UFRB), Cruz das Almas, Bahia, Brazil

R. K. P. Kikuchi et al. (eds.), *Brazilian Coral Reefs*, Brazilian Marine Biodiversity,
https://doi.org/10.1007/978-3-031-59152-5_11

indicators to monitor the capacity to support tourism and incentives for ecotourism, including environmental education practices and sustainable fishkeeping. With these approaches, visitors will have the opportunity to learn about different aspects of aquatic ecosystems and their species, to draw analogies between artificial and natural environments, and to raise awareness of the importance of balance for the survival of living beings.

Keywords Tourism impact · Environmental education in reefs · Marine ornamental species · Brazilian reefs

11.1 Introduction

Coral reefs are places of remarkable scenic beauty, exhibiting intense polychromatic biodiversity associated with clear and warm waters. These ecosystems have been the target of tourist activities that often lead to environmental education actions. Besides enchanting thousands of people daily, visiting a reef system also involves a sense of adventure and emotion, which are perfect ingredients to raise awareness and educate. These areas are home to countless showy and charming species, such as fish and invertebrates, which arouse the interest of aquarists. This chapter connects these three human activities in reef areas on the Brazilian tropical coast: tourism, environmental education, and aquarium trade.

11.2 Tourism

Scenic quality is the target of search and interest for tourism, and this characteristic has become an eligibility criterion (Costa-Chamas and Schmidt 2011), where reefs fit perfectly into this rule. Throughout the Americas, mainly in the Caribbean, tourism represents one of the main sources of employment, with foreign direct investment and gross domestic product—GDP (UNWTO 2017). In this way, the economy of coastal states has become dependent on these revenues, which are linked to recreational activities that, in turn, depend on the quality of their beaches. Therefore, the coast comprises an environmental and socioeconomic complex that interact and are interdependent (Araújo and Da Costa 2008).

Tourism generates revenues in the billions (WTO 2004) and is an economic power, representing 10% of the world's GDP, with 30% of services exports and 10% of jobs worldwide. In 2015, it was considered the third most important source of revenue from exports worldwide (World Tourism Organization and United Nations Development Programme 2017). In Brazil in 2016, tourism generated more than seven million jobs, according to the WTTC (World Travel and Tourism Council 2016). Reefs provide important goods and services, with an estimated US$ 375 billion (Melo et al. 2005), benefiting more than 500 million people in tropical countries

(Hughes et al. 2018). A recent survey in some Brazilian reefs (e.g., in Pernambuco, Alagoas, and Bahia) revealed that this ecosystem generates up to R$ 167 billion in coastal protection and tourism services (Bumbeer et al. 2023). Considering reef distribution over 3000 km along the Brazilian coast and the results obtained in situ by the survey, the amount was extrapolated to all the cities that have coral reefs, considering the size, infrastructure, and population of each one (Bumbeer et al. 2023). In the famous Brazilian oceanic archipelago of Fernando de Noronha, known for the beauty and conservation of its reef ecosystem, tourism generates US$ 73.8 million annually. Most of this revenue refers to services provided to tourists, which includes services offered to divers attracted by the quality of the reefs. Divers are responsible for 50.4% (US$ 36.4 million) of total revenue, including accommodation, food, transportation, souvenirs, and diving (Pires et al. 2016).

The valuable natural heritage is of fundamental importance for nations with reefs in their coastal regions due to the potential for high numbers of tourists. They are distributed in warm water regions around the world, concentrated in shallow tropical regions (between 20°N and 20°S), usually close to a coastline with beautiful beaches. Most reef environments in Brazil do not differ from this pattern. They are distributed between the states of Maranhão and Bahia (Ferreira et al. 2006), corresponding to an extensive coastal strip that receives an expressive volume of tourists throughout the year.

Brazil's reefs offer different opportunities for visitation and tourist experience, as they are distributed along this extensive coastline and are diverse in composition, climatic, and environmental characteristics (Laborel-Deguen et al. 2019; Zilberberg et al. 2016). Although they are often dominated by macroalgae (Figueiredo et al. 2008; Feitosa and Ferreira 2014), some exhibit expressive coral coverage, such as the reefs of Bahia (Leão et al. 2008), others have a significant representation of sponges, such as the reefs of Rio Grande do Norte (Mendes and Grimaldi 2020), with some made up of calcareous algae and vermetid mollusks, such as the Atol das Rocas (Grossman et al. 2012). The reefs of the Brazilian tropical region are diversified in terms of composition and have a considerable abundance of herbivorous fish, such as surgeons (Acanturids), damselfishes (Pomacentrids), and parrotfishes (Labrids), in addition to those that eat invertebrates as the grunts (Haemulids) (Zilberberg et al. 2016; Ferreira et al. 2004). In addition, although reef fish communities along the coast share common and dominant species, at low latitudes, reefs are diverse in rare species, revealing greater vulnerability to the impacts of human communities in these regions (De Araújo et al. 2020).

For-profit nautical tours conducted in reef areas began in the 1970s and have become more intense and frequent over the years. Tourist companies that transport customers to visit the reefs along the Brazilian coast offer contemplative snorkeling dives (e.g., Parrachos de Pirangi and Perobas—RN, Maragogi—AL, Taipu de Fora—BA) and in many locations there is also *scuba* diving (e.g., Maracajaú—RN, Porto de Galinhas—PE, Fernando de Noronha—PE, Maragogi—AL, Abrolhos—BA) (Fig. 11.1). This last activity is less common than snorkeling in Brazil, although the reef ecosystem is an important destination for the practice of scuba diving. One

Fig. 11.1 Tourists led by guide diver in the Area of Environmental Protection of Coral Reefs (APARC), Maracajaú, state of Rio Grande do Norte. (© Bruno César)

of the fastest-growing recreational sports (Hasler and Ott 2008), *Scuba* diving tourism in Brazil is more recent, starting in the 1980s (de Queiroz Neto 2012).

Diving tourism can be an activity that promotes economic gains, with a joint focus on education and social transformation. Thus, more important than the definition of the diver's impact on the reef environment, it is important to understand the relationship between human beings and the environment to show that ecotourism has the opportunity to promote changes in this relationship. Such behavioral change can reverberate, including practicing fishkeeping,[1] a worldwide hobby.

Diving is considered a low-impact activity and a viable economic alternative for the local community in relation to fishing (Tapsuwan and Asafu-Adjaye 2008). Characteristics such as climate, currents, fresh water supply, and regional geography are variables that influence the composition and human accessibility to these environments. In some tourist locations, reef visitation is monitored, including established quotas for the numbers of people, such as marine protected areas (e.g., Maracajaú reefs—Coral Reef APA, Fernando de Noronha archipelago and reefs of Maragogi—APA Costa dos Corais). However, visitation in most reef areas still lacks control (e.g., Pirangi reefs—RN) without inspection or monitoring.

As for visitation, some reefs are less frequented, probably due to the difficult access and adverse oceanographic and climatic conditions, characteristics of the northern coast of the Northeast, which impose more elaborate visitation logistics, examples seen in the reefs located in the States of Maranhão (e.g., Manuel Luiz Parcel), Ceará (e.g., Pedra da Risca do Meio Marine State Park) and the north coast of Rio Grande do Norte (e.g., Urcas). On the other hand, reefs located on the eastern coast, in more accessible and pleasant places, are the target of a large number of

[1] Fishkeeping is a popular hobby practiced by aquarists concerned with keeping ornamental aquatic organisms in a home aquarium or garden pond.

visitors, such as the already mentioned Maracajaú (RN), Porto de Galinhas (PE), Maragogi (AL) and reefs located in the State of Bahia.

National public policies aimed at Ecological Tourism were initiated in the 1990s, and thus, environmental issues gained greater visibility and regulation. Ecotourism, considered a new tourism segment, is practiced sustainably, helping to conserve natural and cultural heritage and encouraging environmental awareness. All of this is facilitated by the interaction with the environment and the local population, promoting even better environmental awareness for those previously unaware of the importance of nature in sustainability (Ardoin et al. 2015).

Marine ecotourism occurs in the reef and rocky shore ecosystems, which have favorable scenic landscapes, such as in Brazil's Northeast, Southeast, and South regions. According to the National Tourism Plan 2018–2022 (Brasil – MT 2018), tourist leisure travel has contributed to its growth in recent years, including generating jobs and income. One example is the Fernando de Noronha National Marine Park and the Marine Protected Area Costa dos Corais (APACC), which feature lush reefs and is among the ten most visited Conservation Units (CU) in the country (Duarte 2017).

At APACC, artisanal fisheries and community tourism professionals were trained to promote sustainable tourism (Jangadeiros da Rota Ecológica Project). This project strengthened community-based associations, increasing the generation of jobs and income for local populations, improving service provision, and the participation of the local community in the zoning process recommended by the APACC Management Plan. In this region, community-based tourism occurs through raft rides on the reefs of each location (Pereira et al. 2017; Pinheiro et al. 2016). Also, in this protected marine area, the prefectures of some municipalities linked to institutes of environmental management have encouraged professionals working in tourism through training courses (e.g., Maragogi on the sustainable tourism route). The training involves divers, underwater photographers, and conductors of nautical tours (B.R. Pinheiro, personal communication).

However, it should be noted, and with great concern, that there is currently a consensus among the scientific community that poorly planned tourism causes severe damage to reef environments (e.g., breaking the three-dimensional structure of the reef, changing the structure of the community, among others) and artisanal fishers (Melo et al. 2021). On the other hand, when the tourist activity is planned and linked to educational activities, this type of tourism contributes to the conservation of the environment through environmental education,[2] which allows sensitization, integration, and economic alternatives for local communities. Examples are observed in marine protected areas on the Brazilian coast (Melo et al. 2005; Giglio et al. 2015, 2016). Even so, the growing demand for tourism, coupled with its practice without controlling and monitoring environmental impacts, is worrisome. If

[2] According to the National Environmental Education Policy (Brasil 1999), Environmental Education (EE) promotes processes by which the individual and the community build social values, knowledge, skills, and attitudes aimed at environmental conservation and sustainability in its most diverse aspects, such as social and economic.

measures are not taken based on sustainability principles, these activities will produce troubling results. After all, *who wants to visit a degraded setting?* Impacts that deteriorate reefs are due to a combination of natural and anthropogenic factors, including tourism (Van't Hof 2001).

The impacts generated by unregulated tourism in reef environments include the damage caused by the logistics activity, such as traffic and anchoring of vessels (Creed and Amado-Filho 1999; Giglio et al. 2017), the installation of marine structures, and coastal leisure accommodations for visitors, such as resorts, hotels, inns, restaurants, and water parks (Gladstone et al. 2013). Another group of direct-action damage concern is divers, both professionals and amateurs, who tramp, collect, touch, and break reef organisms, in addition to disturbing sediments, feeding fish, disturbing fauna, polluting the water with garbage and sunscreen residues, and the leakage of fuel and/or oil from vessels (Augustowski and Francine 2002; Pedrini et al. 2007; Lamb et al. 2014; Downs et al. 2015).

In general, the impacts of tourism on coastal areas are still poorly understood. It is important to discuss and understand the functional aspects of the ecosystem and resilience of Brazilian reefs compared to other coral reefs. Most of the tropical coastal reefs of the South Atlantic are composed of sandstone rock formations (old beach lines) with coral reefs and deposits of encrusted calcareous algae, which share the space with high dominance of macroalgae and other non-coralline encrusting organisms (Castro and Pires 2001; Zilberberg et al. 2016; Laborel-Deguen et al. 2019; Mendes and Grimaldi 2020). Much of the impact studies associated with tourism activity on coral reefs in the world use corals as indicators, and thus, considering the particularities of Brazilian reefs, it is necessary to highlight specific indicators in addition to the implementation of long-term monitoring. For example, organisms living over the substrate (epibenthic) are generally more susceptible to the impact caused by human trampling, especially sessile organisms, in addition to stony corals, such as zoanthids, sponges, algae, and small associated organisms, such as polychaetes and crustaceans. The reduction in the abundance of these organisms can harm the food chain, causing ecological imbalance since small invertebrates make up a relevant part of the base of the food chain in reef environments (Barboza and Leite 2020).

11.2.1 Impacts of Tourist Activities on Brazilian Reefs

A systematic survey of academic studies on the influence of tourism in Brazilian reef environments was done concerning the period between 1996 and 2020 (Calado et al. 2021). Eight biological groups were identified and evaluated on the reefs that responded to tourism use in different states of Brazil, considering aspects such as wealth, abundance, and density, referring to algae, foraminifera, benthic animals (e.g., mollusks, echinoderms, polychaetes), hard and soft corals, fish and marine mammals. In addition to the impacts on the biological groups mentioned, some research has also addressed environmental impacts in an integrated manner, such as assessments of changes in water quality and other abiotic variables, ecosystem services, integrated coastal management, environmental perception/education, and the behavior of divers on reefs (Fig. 11.2).

It is known that disorderly tourism can make changes in the structure of the reef environment evident, arousing interest in understanding these impacts by the scientific community (Giglio et al. 2017). In addition, the social pressure arising from the need to organize tourism on the reefs can be another factor that drives research on the topic. Companies, diving professionals, fishers, and communities closely related to the reefs interact in this space and demand answers about the support capacity of tourist areas and possible changes and losses that tourism may or may not cause (Silva

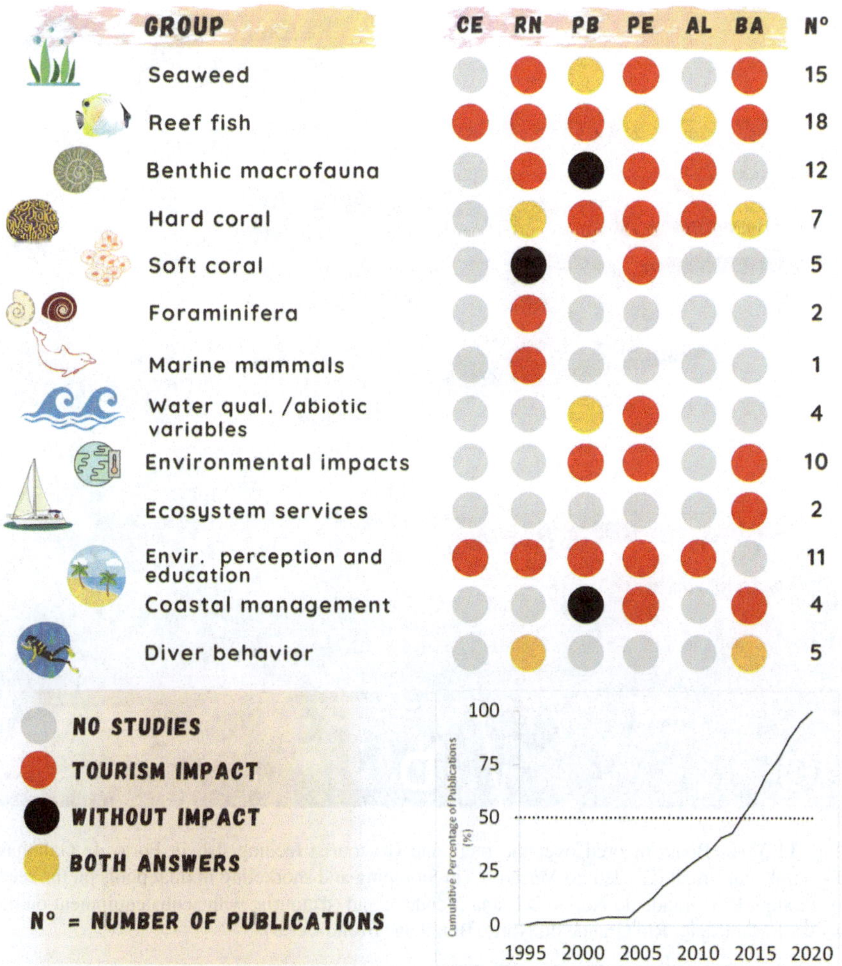

Fig. 11.2 Diving impacts on tropical reefs in Brazilian states. CE—Ceará, RN—Rio Grande do Norte, PB—Paraíba, PE—Pernambuco, AL—Alagoas, BA—Bahia (Calado et al. 2021)

et al. 2009; IDEMA 2012). Finally, the growth in the number of publications in Brazil follows the growing interest in this topic in the rest of the world, especially in reefs in the Caribbean, Red Sea, and Great Barrier Reef in Australia (Lamb et al. 2014).

Eighty-one publications were identified, and there has been an increase in the number of publications in the last 5 years. Such studies come from the gray literature[3] (37%) and publications in national (32%) and international (30.8%) scientific journals. It is noted that almost 50% of all studies carried out in Brazil are concentrated in the last 5 years (2015–2020). This fact reflects a possible relationship with the increased disordered tourism in reef areas off its coast (Rowe and Santos 2016) and, perhaps, an increase in newly trained diving researchers (Fig. 11.3). The fact that many publications come from gray literature reflects the micro-regional

Fig. 11.3 (**a**) Boats moored over the reefs and (**b**) tourist-feeding fish in Porto de Galinhas, Pernambuco, Brazil (© Cláudio Macedo); (**c**) Stomping and snorkeling in tidal pools on the reefs of Pirangi, Rio Grande do Norte (© Liana Mendes); and (**d**) tourist with scuba equipment on the reefs of Maracajaú, Rio Grande do Norte, Brazil. (© Bruno César)

[3] Gray literature refers to unconventional and non-commercial, semi-published publications, difficult to find in traditional distribution channels, with ineffective bibliographic control, often not included in bibliographies and catalogs.

approach to research and the country's lack of incentive for scientific journals, which fluctuates according to Brazil's political environment.

Studies incorporating multiple levels of ecological interactions in the analysis of the impacts of tourism on reefs are needed, highlighting, for example, how species can respond to damage when subjected to intense visitation (Renfro and Chadwick 2017). Systematic and long-term studies seeking to understand ecological patterns and processes are also necessary in Brazilian reef ecosystems (Ferreira and Maida 2006).

It is necessary to define which indicators of diving impacts are appropriate, to implement their monitoring, identify changes caused by tourist activity, and, as soon as possible, propose measures to mitigate the effects before damages are irreversible. Impact bioindicators, alternatives to corals, such as foraminifera, can be used to analyze the health of reef areas to warn about the level of anthropogenic stresses related to tourism (Eichler and de Moura 2020). Another study conducted in Rio Grande do Norte showed that benthic foraminifera around the reefs had low levels of diversity and environmental stability, with dominance of opportunistic species and absence of symbiotic foraminifera with algae. This scenario denotes the decline of fauna (Eichler et al. 2020). Other animals, like fish, can also be used as indicators of tourist activities since the behavior of some species can be directly influenced by human presence (Feitosa et al. 2002) (Fig. 11.3).

Another important activity is mass tourism, with a high potential to generate degrading socio-environmental impacts in communities where it occurs (Araújo and Carvalho 2013). This type of tourism, promoted by tour packages, has been taking place in Brazilian reef environments, making it possible for large numbers of people to visit certain areas in a short period. It is very important to pay special attention to the reefs that receive this type of visitation in Brazil (e.g., Maracajaú—RN and Porto de Galinhas—PE), as these are more susceptible to impacts due to the very nature of the activity. Some characteristic tourist destinations culminate in companies dominating the local community with low prices, attracting tourists with less purchasing power, and finally generating an economic, social, and environmental impact in the target community (Butler 2006).

The definition of the support capacity for tourism use is an excellent starting point for establishing initial planning and management measures on a reef (Dixon et al. 1993; Hawkins et al. 2005). However, it is essential to continuously monitor the ecosystem and visitor's profiles to develop strategies to reduce the damage caused by visitation (Hammerton 2017). It is documented that males and photographers without diving experience are the most significant cause of injury to the reef. This relatively small number of people cause much more of this type of impact, most often accidentally and at the beginning of the dive (Rouphael and Inglis 2001; Zakai and Chadwick-Furman 2002; Giglio et al. 2016).

The central management strategies for reducing such impacts indicate the appropriate use of educational briefings (Rouphael and Hanafy 2007; Luna et al. 2009; Giglio et al. 2017), monitoring and intervention by leading divers (Barker and Roberts 2004), establishing a limited number of divers per point (Dixon et al. 1993; Hawkins and Roberts 1997) and creating underwater trails with appropriate lectures

(Plathong et al. 2000; Rhormens et al. 2017). Near areas of intense tourist activity, the levels of biodiversity were even lower. In addition to the rotation of the sacrifice area, it is suggested to delimit a protected area for conservation as a recovery source (Eichler et al. 2020).

Marine Protected Areas (MPAs) are expected to host sustainable tourism, considering efficient management committed to controlling the activities performed within its limits (Luna et al. 2009; Augustowski and Francine 2002; Green and Donnelly 2003). However, the lure of immediate profit from mass tourism, as well as the obstacles in the management of MPAs, especially in developing countries, make the practice of sustainable tourism infrequent in Brazil. In unprotected areas, concern for reefs occurs only when some notable change in the community structure is perceived. On the other hand, research conducted at diving points within MPAs guided management measures aimed at preventing or minimizing the impacts caused by tourism, such as the reefs of Maracajaú—RN (State Environmental Protection Area for Coral Reefs) and the Abrolhos National Marine Park—BA (Silva et al. 2009; Giglio et al. 2016).

One solution that could minimize tourism's impact on reef environments is ecotourism. Although this activity is little explored in the Northeast of Brazil, this segment grows proportionally more throughout the world, with great socioeconomic potential added to the conservation of natural assets (EMBRATUR 2016), fundamental pillars of sustainability. This activity is an alternative to mass visitation and is responsible for a series of impacts (Gössling 1999). Ecotourism sustainably uses natural and cultural heritage, encouraging the conservation of the environment and promoting the well-being of the populations involved.

An example of the economic potential of ecotourism is observation diving. Research shows that the maintenance of certain species in the reefs (e.g., groupers, sharks) is more profitable than their exploitation by fishing, whether for consumption or ornamental. For sharks, several authors cite the importance of maintaining their populations in diving destinations, such as the Caribbean (Cline 2008), Micronesia (Vianna et al. 2010), South Africa (Dicken and Hosking 2009; Dicken 2014), the Maldives (Cagua et al. 2014) and Australia (Catlin et al. 2010). For example, while the global revenue generated by shark fishing was $630 million annually, the first global economic assessment of shark diving was valued at $314 million/year (Cisneros-Montemayor et al. 2013). These two activities have different spatial patterns in their economic impact, as fishing is distributed throughout the world, while shark tourism is restricted to places where sharks are present and the diving industry is well developed (Pires et al. 2016). Fernando de Noronha archipelago is a good example of shark tourism.

A range of pro-environmental attitudes and behaviors can be modified by experience with nature tourism (Ardoin et al. 2015). Responsible tourism, like ecotourism, in addition to generating foreign exchange, can incorporate and disseminate local knowledge while providing socio-environmental development. In this way, it is possible to stimulate the much-desired sustainable tourism, which has an intimate and inseparable relationship with Environmental Education, also mirroring a conscious practice of fishkeeping, which will be mentioned later.

In summary, the promotion of tourism on the reefs requires strategies and policies. As part of the strategies, it is suggested that the tourism sector be advised of the relevance, establishing regulations for the activity in the management plans of protected areas, as well as qualifying professionals in the sector (e.g., consistent briefings and guidelines). Policies need to include conservation and restoration actions, as well as boundaries and user plans for the areas visited.

11.3 Environmental Education

Despite the growing number of tourists and actors involved in visiting reef areas, the marine world is still largely unknown. For example, the importance of the oceans as a source of food and energy, as a climate regulator on Earth, and as a provider of economic, social, and aesthetic services for humanity remains ignored by most people (Santoro et al. 2018). In Brazil, ignorance of the richness of ecosystems and biodiversity associated with reef formations is also a bottleneck for sustainable development. Even with the immense Brazilian coast and its numerous reef formations that are important tourist attractions, most of society does not *see* the magnificence or importance of reef environments.

In general, only social groups that interact directly with the sea, such as fishers who depend on it for survival, researchers, and professionals in the marine tourism and aquarium industry, have information about the rich biodiversity and the intricate relationships that exist with their environment. Tourism and recreational dives have brought the population closer to these environments and disseminated information about them. It is worth remembering that the opportunities for direct experiences like these allow the establishment of bonds of affective memory that transform nature into something that needs to be taken care of and respected now and in the future. As Richard Louv (2016) said: "Nature protection does not depend only on the organizational strength of conservation institutions; it also depends on the quality of the relationship between young people and nature—on how young people connect with nature if at all connect."

In this perspective, several studies were undertaken, such as investigating attitudes related to the conservation of reef environments in the Fernando de Noronha Archipelago, including students through varied activities, with field visits and the implementation of a Sea Museum at the school (do Amaral et al. 2014); research involving the educational process in reef formations, based on field visits and works such as the preparation of models, posters and interviews (Oliveira et al. 2014); or studies pointing out the high level of misinformation about coastal environments, both among residents and tourists, especially regarding the structure, biodiversity and ecology of reefs ecosystems (Silva et al. 2013).

In general, reefs need attention in the field of education, not only to promote knowledge but also to its value for society. So, *how do we conduct environmental education for the conservation of reef ecosystems?* Considering that people already recognize forests as high biodiversity systems, an initial approach can be comparing reef systems to forests. Both have immensity, scope, beauty, and the provision of ecosystem and environmental services in common. Thus, social recognition of the existence of environmental degradation in terrestrial and coastal ecosystems, such as disordered land occupation and pollution in rivers, allows educational processes to relate the connectivity of these environments to the marine, in the case of reef environments. In this way, social ignorance about marine biodiversity requires that educational actions be planned and carried out under different pedagogical strategies, aiming to meet both social diversities and promote dialogue between the productive and environmental conservation sectors. The promotion of oceanic culture is one of UNESCO's proposals for the United Nations Decade of Ocean Science for Sustainable Development (UNESCO 2020).

The inclusion of reefs as an object of awareness-raising and environmental education practices is recent. Literature about these theme in coastal areas in Brazil is restricted. A review up to 2010 indicated 32 publications (Pedrini 2010). In a survey carried out by ReBentos (Coastal Benthic Habitats Monitoring Network) (Berchez et al. 2016), of the ten initiatives in marine environment, only two directly addressed reef formations: one involving field classes for students (de Oliveira and Correia 2013) and the other for the creation of virtual games (Ghilardi-Lopes et al. 2013). Although most of the projects have restricted focus and time-lapse, few are developed on a long-term basis and will be discussed below.

11.3.1 Coral Vivo Project

During almost two decades, the Coral Vivo Project developed its pedagogical strategy over the pillar of conservation based on the socio-environmental sustainability of coralline habitats. This aspect sets it apart from other initiatives involving environmental education. With nationwide operations, the focus area for the application of educational interventions is the Costa do Descobrimento (Bahia),[4] which aims at different audiences in that region. The panorama of challenges and possibilities (described above) constitutes the educational trajectory of the Coral Vivo Project, identifying and practicing a permanent dialogue with diverse social groups to achieve reef conservation with socio-environmental sustainability. Thus, the strategies are in line with the teachings of Paulo Freire, who points out that all knowledge is relevant, coupled with the assumptions of Morin's (2011) complex thinking about the need to "Teach earthly identity", as he states:

> The pertinent knowledge must face complexity. *Complexus* means what was woven together; in fact, there is complexity when different elements are inseparable from the whole (such as economic, political, sociological, psychological, affective, and mythological). There is an interdependent, interactive, and interretroactive fabric between the object of knowledge and its context, the parts and the whole, the whole and the parts, the parts among themselves.

[4]Costa do Descobrimento is one of the 27 Identity Territories in the state of Bahia, where it is located almost on the border with the southern end of the Brazilian Northeast.

The project follows the Brazilian public education and policies, such as the National Environmental Education Policy (Brasil 1999), which indicates that promotes processes by which the individual and the community build social values, knowledge, skills, and attitudes aimed at environmental conservation and sustainability in its most diverse aspects, such as social and economic. However, the Coral Vivo Project's EE strategy involves a critical and transformative approach, giving it certain protagonism in the area of Marine EE. It is worth highlighting the offer of continuing education courses for teachers, environmental educators, and tourism professionals from the Costa do Descobrimento (Fig. 11.4). Additionally, there is continuous training with teams of BIOMAR, a network of projects sponsored by Petrobras,[5] namely Albatroz, Humpback Whale, Coral Vivo, Golfinho Rotador, and TAMAR. The covered contents, linked to the previously identified demands for knowledge, contributed to a paradigm shift in educational activities, constituting a new educational "way of doing" aimed at the conservation of species and marine environments, incorporating theoretical and policy fundamentals of EE.

The actions performed in schools, through technical and financial support for pedagogical projects, contextualize concepts such as environmental complexity, connectivity, and socio-environmental and socio-economic relations. They incorporate scientific information on coastal and marine environments and bridge the gap between these contents in instructional materials available to the public. Such content was materialized in publications, videos, and books, distributed during the training, and is available on the project page (available at https://coralvivo.org.br/pesquisa-e-educacao/publicacoes). The most relevant material produced regarding EE are Education for the Conservation of Reefs, Teacher Training Manual in EE (Gouveia 2008, 2011); Youth Training Course: A new generation of the Youth Collective of Costa do Descobrimento (Gouveia 2015); Marine and Coastal Ecosystems for Educators (Raineri et al. 2016); Sustainable Tourism in Reefs and

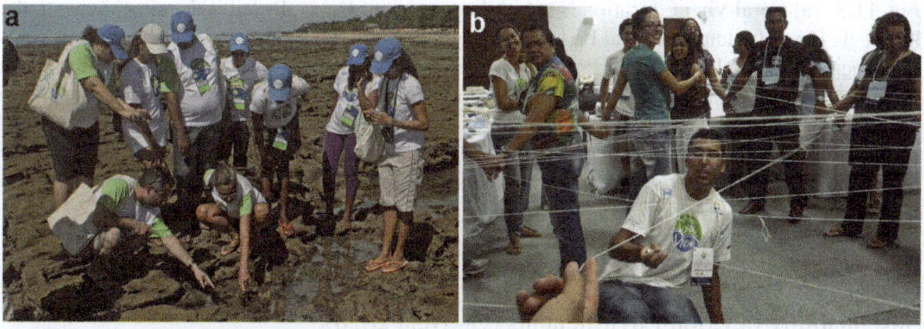

Fig. 11.4 (**a**) Class in tidal pools (© Coral Vivo) and (**b**) dynamics of socio-environmental relations in the Teacher Training Course in EE, Costa do Descobrimento, state of Bahia, Brazil (© Clovis Castro)

[5]The Brazilian oil company - PETROBRAS - sponsors conservation projects in the country through its socio environmental Program.

Fig. 11.5 (**a**) Coral Vivo Education Network in the state of Bahia, Brazil: Educational visit to the Recife de Fora Municipal Natural Park, Porto Seguro (© Clovis Castro); (**b**) Fishers labeling reefs in the activity of the Pedagogical Project "Fishing and Marine Life in the Coraline Environments of Santa Cruz Cabrália" (© Matheus Deocleciano); (**c**) Educational visit to the ice factory of the Colony of Fishers Z-51 during the Pedagogical Project "School and Fisheries Meetings" (© Matheus Deocleciano); (**d**) Collection and classification of debris on the beach by students in the pedagogical project "Fighting marine debris with Citizen Science". (© Clovis Castro)

Coral Environments (Gouveia 2016); and EE in Basic Education: Conservation and Sustainability of Coral Environments (Gouveia 2020).

The educational actions developed in partnership with schools constitute the Coral Vivo Education Network. It also includes financial and technical support for the development of pedagogical projects (Fig. 11.5), as well as attendance of undergraduate students from University Extension Programs. For young people, Coral Vivo incorporated three innovations. The first one was to create the "Costa do

Fig. 11.6 Presentation of the Spoken Map (Social Cartography) of Arraial D'Ajuda, Bahia, Brazil, by the young Heyon Carrasco in the Youth Training Course A new generation of the Youth Collective of Costa do Descobrimento. (© Clovis Castro)

Descobrimento Youth Collective" based on the program "Youth Collective of the Environment", created by the Environmental Ministry. The second was to integrate into the collective the young people from different south coastal localities in the state of Bahia that interact in several actions of the Project (Fig. 11.6). Third, connect them to the youth network ("Jovem Mar Collective") made by the young people involved in the project as members of the BIOMAR Network, as already mentioned.

It should also be noted that the Coral Vivo Project has direct participation in the federal public policy of the Conscious Conduct in Reef Environments program, developed by the Ministry of the Environment since 2001 (Brasil – MMA 2009, 2018). This program produced leaflets, booklets, videos, posters, and stickers, informing the population about the importance of these environments and the conduct that must be followed when visiting a coral reef. The program aims to educate tourists and residents of coastal communities about the importance of preservation when carrying out any activity in these areas.

11.3.2 Environmental Education Projects in Brazilian Reef Areas

Another project that stands out for creating educational processes aimed at valuing and understanding reef formations on the Brazilian coast is the Ponta de Pirangi Project, carried out by Oceânica—Research, Education and Conservation (@ oceanica.osc). The project focuses on biodiversity conservation and coastal planning. It counts on the active participation of coastal communities and other stakeholders in the state of Rio Grande do Norte.

Considering coastal areas as a territory shared by several social actors where socio-environmental conflicts exist, marine conservation should involve several stakeholders. Therefore, since 2010, the Ponta de Pirangi Project has involved fishers, teachers, professors, students, environmental managers, nautical entrepreneurs, leaders, and local organizations in conservation and planning action. Among the awareness and education activities, there are events such as Ocean Day (Fig. 11.7), beach and reef cleaning campaigns (Fig. 11.8), ongoing training for public school teachers, development of workshops, and support for council members and committees involved in coastal management. In addition, educational materials such as books and posters valuing the reef biodiversity (Rocha et al. 2016; Rocha and Bonilha 2020; Souza et al. 2016; Suassuna and Mendes 2020) were published and distributed in schools, universities, public and commercial institutions along the coast.

Although first-hand experience is important, it is not always possible to go to the beach, dive in the reefs, or participate in educational activities in coastal areas. In these cases, aquariums have the extraordinary role of creating marine microcosms that can be observed, even when far from the sea. The invention of aquariums emerged in the nineteenth century. It became fundamental to better understand and disseminate the marine world (Brunner 2005). Created as a tool for research, aquariums became strategic areas for education and entertainment in the twentieth century (Salgado and Marandino 2014).

In aquariums, visitors can learn about aspects of different aquatic ecosystems and their species, draw analogies between an artificial and natural ecosystem, and raise awareness of the importance of balance for the survival of living beings (Scopel

Fig. 11.7 During the Ocean Day in 2019, Oceânica organized an event for the protection of Búzios Beach, a nursery of sea turtles. Nísia Floresta, State of Rio Grande do Norte, Brazil. (© Tiago Lima)

Fig. 11.8 During a cleanup campaign at Búzios Beach (2019), the cradle had a warning question: "What if this beach was the cradle for your species?" Nísia Floresta, Rio Grande do Norte, Brazil. (© Tiago Lima)

et al. 2019). The exhibitions created from tanks, texts, and luminous panels are the main forms of communication in these environments, indicating the intention that the spectator, when visiting them, can produce meanings, learn, and, eventually, change behaviors (Salgado and Marandino 2014). More recently, the potential of fishkeeping as a practice that promotes environmental awareness is highlighted, which "contributes to the formation of a human being more concerned with environmental causes and the sustainability of the environment in which he is inserted, in addition to teaching science content, motivating for participation and understanding, and, equally, for the critical discussion of the existing social model" (Barreto 2020, p. 17).

Brazilian public aquariums are visited by hundreds of thousands of people annually. Starting activities in 2016, the Rio de Janeiro Public Aquarium is the largest marine aquarium in South America, with an annual visitation of one million people (https://www.aquariomarinhodorio.com.br/o-aquario). In the state of São Paulo, the Municipal Aquarium of Santos opened in 1945 and receives 500,000 annual visitors (Salgado and Marandino 2014), while the Ubatuba Aquarium opened in 1996 and receives about 150,000 visitors/year (Gallo and Barbosa 2010). Thus, aquariums are not only important tools to sensitize people about the importance of marine ecosystems in environmental, social, and economic sustainability but are also an economic activity with significant revenues.

11.4 Aquarium Trade

Home tanks or public aquariums, besides enabling people who live a thousand miles away from the sea to be close to species they would hardly meet in the wild, allow the observation of the fascinating color of reef fish and their intriguing interspecific relationships. Such observations make people aware of how intricate marine ecosystems are and how direct implications to one species can indirectly impact many others.

Fig. 11.9 Some native aquarium species found in the Parque Estadual Marinho da Pedra da Risca do Meio reefs, Ceará state, Brazil—Queen angelfish (*Holacanthus ciliaris*) (**a**), French angelfish (*Pomacanthus paru*) (**b**) and a Barber goby (*Elacatinus figaro*) cleaning around the eye of a Green moray (*Gymnothorax funebris*) (**c**) (© Eduardo Freitas), and a striped cleaner shrimp (*Lysmata grabhami*) cleaning a brown chromis (*Chromis multilineata*) in Saint Peter and Saint Paul's Archipelago, Brazil (**d**). (© Luiz Rocha)

For instance, by looking at marine angelfish in artificial habitats, people can not only appreciate their notable beauty (Fig. 11.9a, b) but also observe their feeding behavior and then infer how those fish perform vital ecological roles in the coral reef food web (Hill 1998; Andréa et al. 2007; Konow and Bellwood 2011). Invertebrate grazers and surgeonfish, mostly used to control algal growth in home tanks, play a corresponding role in the wild, maintaining the health of coral reefs by means of high rates of herbivory to suppress macroalgae that compete with cnidarians for settling substrate (Rhyne et al. 2009; Comeros-Raynal et al. 2012; Bonaldo and Hay 2014). Interspecific associations, recorded in captivity, make it possible to understand that the maintenance of reef ecosystems' health depends on this symbiotic relationship in 'cleaning stations' (Fig. 11.9c, d) (Campos and Sá-Oliveira 2011; Espino and Riera 2017; Sampaio et al. 2017), which benefit not only aquarium fishes but also other commercially important species recognized as 'clients' (Monteiro-Neto et al. 2003). In this sense, both home tanks and public aquariums increase people's empathy towards the importance of coral reefs conservation and represent exceptional opportunities to promote environmental education (Wood 2001; Barreto 2020).

Here, a question emerges: *do we preserve what we don't know?* In this regard, we highlight the practice of fishkeeping, which is known to have great potential for raising awareness about the aquatic biotope and its relationships, having an important mission related to environmental education (Gallo and Barbosa 2010). Moreover, concerning economic and social aspects, fishkeeping represents a huge market worldwide, generating jobs for both fishers and producers, as well as increasing revenues for producing states and countries. In 2014, for example, Brazil, at its peak, exported over $13 million of live fish (Trademap 2020).

Additionally, many public aquariums and zoos are renowned institutions that make significant contributions to the conservation and preservation of wildlife around the world (WAZA 2005), producing relevant scientific studies in many areas. In Brazil, as there is an extensive maritime coast with contrasting features, those institutions can exhibit parts of different ecosystems in distinct tanks, which can be used as educational tools to promote environmental awareness. Furthermore, some public aquariums and zoos allow physical interaction between guests and marine life, and although controversial (e.g., animal stress from interaction with humans or stingray barbs clipping to avoid accidents), touch-tanks have become increasingly popular to stir up people's interest upon different species and even to stimulate empathy and reduce discomfort or fear of certain organisms (Kearns et al. 2017; Biasetti et al. 2020). From the ca. 22 existing public aquariums and oceanariums in Brazil (Holanda et al. 2015), touch-tanks can be found in at least four of them: Aquário Natal (Rio Grande do Norte state), Oceanário de Aracaju (Sergipe state), AquaRio (Rio de Janeiro state) and Aquário de Ubatuba (São Paulo state), besides TAMAR Project (Bahia state) (Fig. 11.10).

The welfare of reef species maintained in captivity demands careful control of the physicochemical parameters that regulate water quality and, consequently, guarantee marine life's health. Thus, by looking closely at these artificial environments, aquarium owners and visitors of public aquariums can understand the importance of

Fig. 11.10 People interacting with a stingray in a touch-tank at TAMAR Project, Bahia state, Brazil. (© Leopoldo Barreto)

water parameters, such as alkalinity, calcium carbonate, and temperature control. Visitors can also monitor coral growth and infer how ocean acidification and global warming prevent the development of reef environments and promote coral bleaching. Therefore, such microcosms also represent fascinating tools to reinforce the importance of coral reef conservation (Penning et al. 2009).

On the other hand, since the majority of marine aquarium species are wild-caught, damaging practices elicited controversies regarding the sustainability of reef species collection, as over-harvesting is among the most serious causes of coral reef degradation around the world (Wabnitz et al. 2003; Bellwood et al. 2004; Olivotto et al. 2011; Rhyne et al. 2014). In some reported cases, fish population depletion was recorded, social conflicts emerged, management techniques had to be adopted, and rigid regulations were implemented (Tissot et al. 2004, 2009, 2010; Ortiz and Tissot 2012; Stevenson and Tissot 2013; Dawes 2018). Nevertheless, if extractions are performed on a sustainable basis, aquarium trade can be a source of income for vulnerable people, supporting jobs and contributing to coral reef conservation where other options for generating revenues are limited (Wabnitz et al. 2003; Rhyne et al. 2014).

In addition, it is important to consider that sustainable ornamental fisheries are driven to target species (there is no bycatch), and harvesting activities aim at good-featured specimens. Thus, since exploited individuals will be kept alive for many years in captivity (King 2019), best collecting practices shall be used to guarantee animal welfare. Indeed, if these harvesting practices are also performed respecting science-based governmental norms, conciliating social, economic, and environmental interests, the sustainable collection of reef species is a feasible possibility (e.g., Piaba Project—Brazilian freshwater environments and LINI—Indonesian marine ecosystems).

In that sense, Brazilian studies have reported concern about the sustainable exploitation of reefs to supply both national and international aquarium trade (Gasparini et al. 2005; Nottingham et al. 2005a). Conversely, reef species collection is less harmful to Brazilian marine ecosystems than other anthropogenic activities, especially when compared to the illegal use of bottom-set nets for lobster fishery and shrimp trawling (Nottingham et al. 2005b), besides being an important source of income for some families that depend exclusively on ornamental harvesting to survive (Nottingham et al. 2005a). Shrimp trawls may present a 1:2.5 kg bycatch ratio (Lira et al. 2019), which means that for each portion of produced shrimp two and a half fold, that same amount of accompanying fauna is captured and usually rejected. On the other hand, reef species harvesting can aggregate value to bycatch organisms that are discharged for human consumption. As an example, small individuals of lookdown fish, *Selene vomer*, are commonly rejected when captured in beach-seine fisheries in Brazil, whereas the species is sold for more than $100 to final marine tank clients through e-commerce websites. So, if ornamental reef species collection is regulated on a sustainable basis, not only can local communities be economically benefited, but natural resources can also be better exploited.

The former administrative norms that regulated the collection and commerce of marine ornamental fish in Brazil were IN IBAMA 14/2004 and IN IBAMA 56/2004, and then IN IBAMA 202/2008, which was likewise expected to be periodically reviewed. However, the latter remained for more than a decade without significant improvements, bringing fragilities to the rules applied to the dynamic exploitation chain (Sampaio and Ostrensky 2013). Additionally, frequent changes in the Brazilian fishing policy made the management of the marine aquarium trade even worse (Gurjão 2018). Thus, in recent years, the Brazilian Association of Aquarium Shops (Associação Brasileira de Lojas de Aquariofilia—ABLA) committed efforts with the current government to modify the existing rules by means of two other new norms: IN MAPA 10/2020, which was subsequently repealed by Portaria SAP/MAPA 17/2021. Although aquarium traders celebrated advances toward debureaucratization promoted by the most recent legislation changes, the issue is controversial, and environmental regressions have been claimed by other social sectors (Ferraz et al. 2020).

It is noteworthy that Brazil figures as a major supplier of the global marine ornamental species market (Bruckner 2005; Rhyne et al. 2012, 2017), but there is a lack of updated information on the exports (Gurjão and Lotufo 2018a), besides a paucity of systematic control of the Brazilian domestic commerce, which hinders the accurate assessment of market size and volumes traded (Sampaio and Ostrensky 2013). However, it is known that approximately 370 native species are wild-harvested to supply the Brazilian marine aquarium trade (Gasparini et al. 2005; Gurjão and

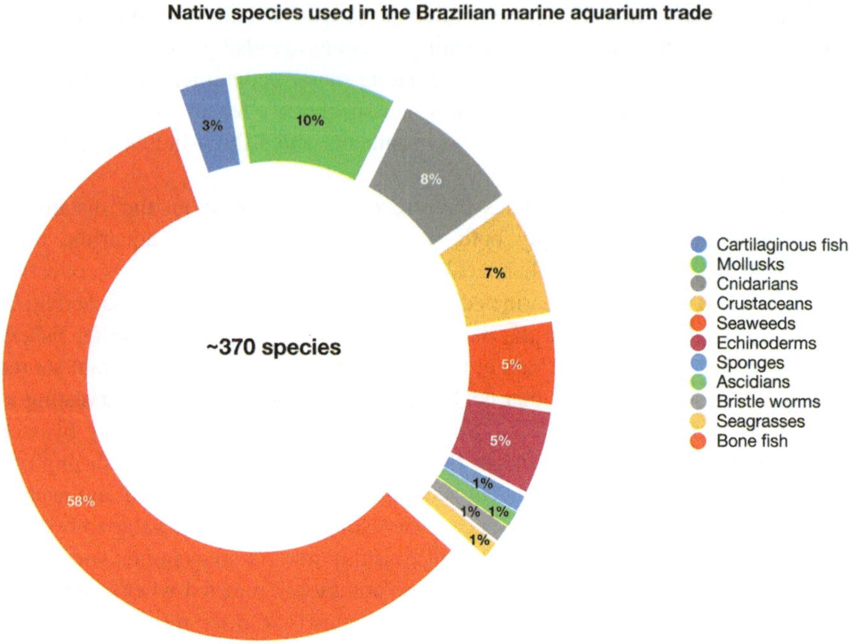

Fig. 11.11 Species are wild-harvested to supply the Brazilian marine aquarium trade

Lotufo 2018b) (Fig. 11.11), and the great majority of them are collected in reef ecosystems whereas just a few others come from estuaries (e.g., seahorses, batfish, frogfish and morays—Nottingham et al. 2005a).

It is important to highlight that nowadays, people can keep "sustainable" marine tanks that do not depend on the collection of native species to be set since a wide range of captive-bred reef species is currently available in the ornamental market, reducing collecting pressure over populations of targeted species (Calado et al. 2003; Pomeroy et al. 2006; Olivotto et al. 2011; Murray and Watson 2014). The well-succeed production of clownfish (Amphiprioninae) in captivity around the world is a wonderful example of how aquacultured species can reduce harvesting pressure over native populations, particularly considering the increasing demand for those fish exacerbated by the motion picture 'Finding Nemo' (Militz and Foale 2017). In the wild, clownfish larvae travel almost 250 miles to find anemones, without which they cannot live, and 90% are killed by natural predators in this period (Sekharan 2018).

In Brazil, the great majority of the traded corals are non-indigenous home-propagated or large-scale farmed species. In the same way, exotic mollusks (e.g., *Berghia stephanieae*) and a broad range of non-native fish and their varieties (e.g., gobies, blennies, dottybacks, cardinals, damsels and especially clownfish) have also been produced by aquaculture companies. Unfortunately, captive breeding of Brazilian native marine organisms is restricted to a few species produced on commercial scale, such as *Elacatinus figaro* and, in the past, *Hippocampus reidi* (Hora and Joyeux 2009; IBAMA 2009; Gurjão et al. 2017). However, there are several attempts to develop technology aiming at the production of other reef species (e.g., Centropyge aurantonotus, Gramma brasiliensis, and *Opistognathus* sp.) (Gonçalves 2017; Mendonça et al. 2020). Therefore, more efforts to develop the aquaculture of Brazilian native species are needed, particularly by combining both private initiatives and public research institutions.

Regarding the usage of non-indigenous species (NIS) in marine tanks, it is important to note the risk of introduction of those species into native reef ecosystems. NIS bioinvasion can cause loss of biodiversity and severe damage to the structure of communities, besides dissemination of diseases and pathogens in the wild (Semmens et al. 2004; Whittington and Chong 2007; Green et al. 2012; Evangelista et al. 2016; Mannino et al. 2019). One of the sources of NIS introduction into reef environments is aquarium dumping, when tank owners, upon giving up their hobby, are reluctant to sacrifice or even sell their pets and choose to release them into natural ecosystems (Padilla and Williams 2004; Walters et al. 2011; Diaz et al. 2012; Holmberg et al. 2015).

Until May 2019, 138 NIS were recorded in Brazilian marine ecosystems (Teixeira and Creed 2020) and at least five of them werefound in reef environments. The most likely explanation for these introductions are intentional releases by aquarium owners (Falcãoand Széchy 2005; Lopes et al. 2009; Mantelatto et al. 2018). Therefore, special attention should be paid to octocorals Sarcotheliasp. and Briareum hamrum, that are covering reef surfaces and displacing native species in Bahia state (Fig. 11.12a, b) (Menezes etal., 2021) and the recurrent records of lionfish, *Pterois sp.,* in Brazilian reefs (Fig. 11.12c) (Soares et al., 2022; Soares et al., 2023).

The latter species was initially reported in the country on the Arraial do Cabo reefs, Rio de Janeiro state (Ferreira et al. 2015), followed by another record in this same region 14 months later and then, more recently, by recurrent reports off Northern and Northeasthern Brazilian coasts, respectively, under the Amazon plume and in Fernando de Noronha archipelago (Luiz et al., 2021), and along shallow-water coastal environments (Soares et al., 2022). It is noteworthy that the arrival of lionfish in Brazilian reefs as a result of the spread caused by previous releases of the species in the US (Schofield, 2010; Hixon et al., 2016) is expected to cause one of the most negative ecological impacts among all marine invasions, whose consequences will be worse than that in the Caribbean reefs (Albins and Hixon, 2008; Luiz et al., 2013; Ferreira et al., 2015). Predictive models showed that lionfish can invade the whole Brazilian coast and expand its range as far as Argentina (Evangelista et al., 2016) and, since the species feed on a wide variety of economic and ecologically important reef organisms (Eddy et al., 2016), trophic models demonstrated that reef fish would be the most affected species (Bumbeer et al., 2017). Hence, lionfish invasion in Brazil demands urgent management measures as functional eradication (Luiz et al., 2021) and inter-institutional integrated approaches (Soares et al., 2023).

Thus, end consumers must be educated not only about the potential risks involving the unsustainable removal of species from coral reefs but also about the damage

Fig. 11.12 Non-indigenous soft corals (*Sarcothelia* sp.) (**a**) and *Briareum hamrum* (**b**) dominating the substrate and displacing native species in Baía de Todos os Santos reefs, Bahia state, Brazil (© Claudio Sampaio); Lionfish (**c**) in Jijoca de Jericoacoara, Ceará state, Brazil. (© Tommaso Giarrizzo)

associated with aquarium dumping and bioinvasion (Azevedo-Santos et al. 2015), highlighting the role of tank owners in the conservation of coral reefs.

In a comprehensive view of the marine aquarium trade, we can observe its connection with tourism. At the same time, underwater activities stimulate curiosity and passion for the ocean and may encourage people to have marine tanks as a piece of the ocean at home. Therefore, poorly practiced aquarium trade can negatively impact reef environments either through the collection of species or the release of exotic organisms into the wild, as well as poorly practiced tourism, like trampling on corals and fish feeding, indirectly harms the marine aquarium market.

The connection between aquarium trade and environmental education (EE) is also notorious, as mentioned throughout the chapter and highlighted here as mutually related. EE may use the practice of fishkeeping as a tool for promoting knowledge about many different environmental issues.

In this way, tourism, environmental education, and aquarium trade in reef environments are activities that emerge in a sea of possibilities beyond the appreciation of the scenic beauty of these habitats when interconnected by the principles of sustainability. When properly managed, these fabulous ecosystems become a powerful tool with the potential to modify human behavior. Through direct contact with marine beings, the three activities approached in this chapter, tourism, environmental education, and aquarium trade, can actively interact to promote awareness of the maintenance of these rich natural systems. These activities are developed in key sites along the Brazilian tropical seas (Fig. 11.13).

11.5 Final Remarks

Scenic quality, particularly beaches and reefs, is a crucial factor attracting tourists, contributing significantly to employment, foreign direct investment, and the GDP of coastal states. The economic dependency on tourism revenues, linked to recreational activities and beach quality, underscores the intertwined nature of environmental and socioeconomic factors. Brazil's reefs, rich in biodiversity, offer diverse opportunities for tourism, including scuba diving and snorkeling. The economic benefits of diving tourism are important to consider, especially if practiced as a low-impact activity with the potential for economic gains and increase in environmental awareness. However, concerns arise regarding unregulated tourism causing damage to reef environments, impacting the structure of the ecosystem, and disturbing marine life. There is a need for sustainable tourism practices, with a focus on ecotourism, marine protected areas, and responsible tourism. These sustainable practices must be coordinated with monitoring and management strategies to mitigate the negative impacts of tourism on reef ecosystems. Ecotourism is a viable alternative, especially for observing marine life, like sharks, and can generate economic benefits surpassing those of traditional fishing practices.

Environmental education is a strategy to cope with a lack of awareness and knowledge about the marine world, particularly reef environments, among the general population, despite their increasing importance for food, energy, climate regulation, and economic services. Only specific groups, like fishers, researchers, and those in marine tourism, are well-informed about the biodiversity and intricate relationships within these ecosystems. Tourism and recreational dives are identified as avenues that bring people closer to reef environments, fostering an understanding of their significance. There is a need to bridge the gap in knowledge, developing studies and initiatives like field visits, the establishment of a Sea Museum, and educational processes comparing reef systems to forests. It is important to promote oceanic culture, aligning with UNESCO's goals for sustainable development. Despite the recent inclusion of reefs in educational practices, there is a call for more comprehensive and long-term initiatives to raise awareness and foster a connection with these vital ecosystems.

The Coral Vivo Project, spanning nearly two decades, distinguishes itself in environmental education by emphasizing socio-environmental sustainability in coral environments. Operating nationwide, with a focus on Costa do Descobrimento (Bahia), the project employs a pedagogical strategy rooted in a dialogue with diverse social groups, aligning with Paulo Freire's critical and transformative approach.

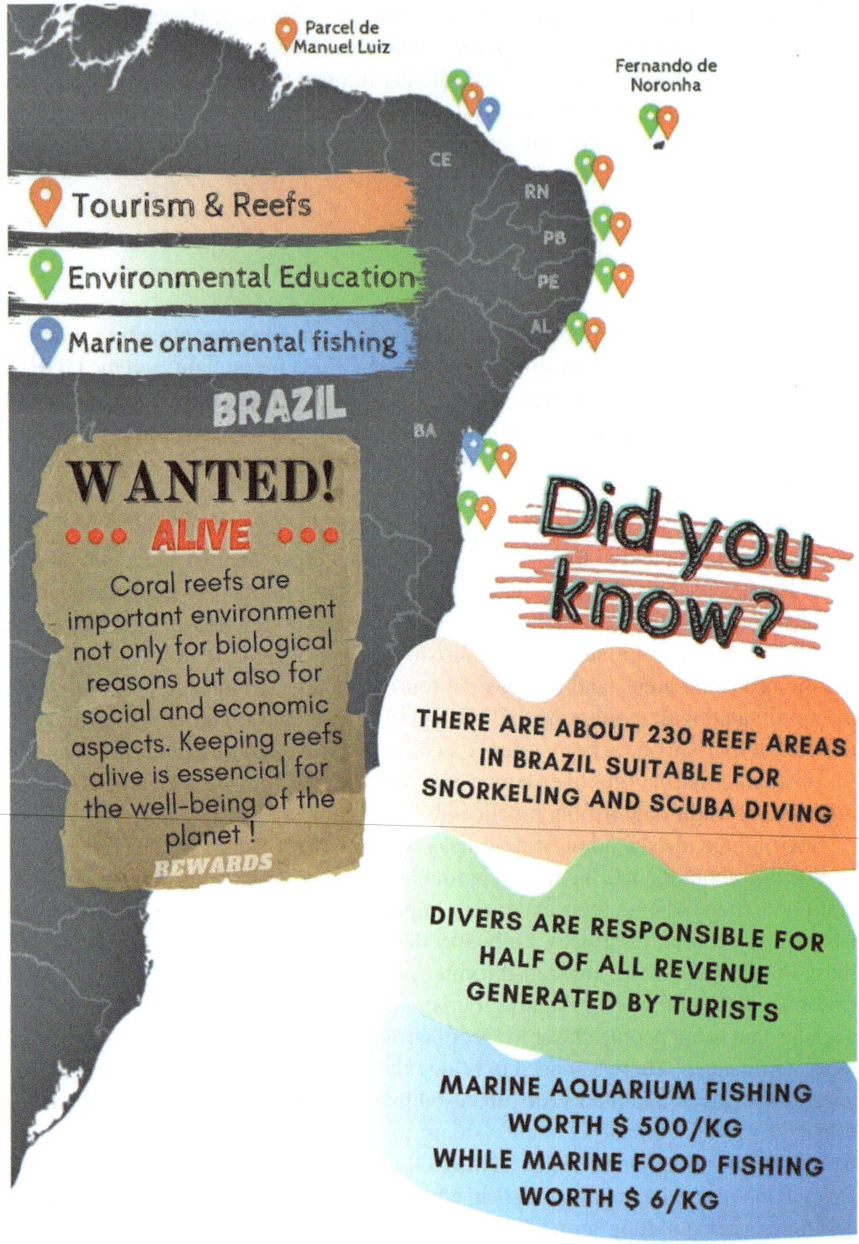

Fig. 11.13 Graphic synthesis of tourism, environmental education, and marine ornamental activities in sites described in this chapter

Embracing Morin's complex thinking, the project aims to "teach earthly identity" by intricately weaving different elements into the educational fabric. It adheres to Brazilian policies, offering continuing education courses and contributing to the Conscious Conduct in Reef Environments program. The project's publications, videos, and books, distributed during training, embody a paradigm shift in educational activities, fostering a new "way of doing" for species and marine environment conservation. The Coral Vivo Education Network extends its impact through partnerships with schools, training programs, and active participation in federal policies, collectively promoting awareness and preservation of coral reefs. Similarly, the Ponta de Pirangi Project in Rio Grande do Norte engages coastal communities and stakeholders, employing diverse awareness and education activities, workshops, and publications to promote biodiversity conservation and coastal planning. Both projects recognize the crucial role of education in marine conservation and aim to inspire a sense of responsibility and connection to coral reef ecosystems.

Finally, the impact of the aquarium trade is multifaceted, involving both positive and negative aspects. Home tanks and public aquariums offer a unique opportunity for people far from the sea to connect with marine life, fostering awareness of intricate ecosystems and interspecific relationships. Observations in artificial habitats, such as those of marine angelfish, not only showcase their beauty but also provide insights into their ecological roles in coral reef food webs. While aquariums contribute to education, economic benefits, and scientific research, the trade raises concerns about sustainability. Wild-caught species, essential for the trade, may contribute to coral reef degradation, prompting debates on regulations, ethical practices, and the need for sustainable alternatives like captive breeding. The interconnectedness of aquarium trade, tourism, and environmental education highlights the potential for positive change through sustainable practices and awareness campaigns.

Acknowledgements The authors wish to thank Coral Vivo Project, NGO Oceanica—Ponta de Pirangi Project, Parrachos Turism and Maracajú Divers for the support received along their working experience. The authors also thank Clóvis Castro, Matheus Deocleciano, Tiago Lima, Eduardo Freitas, Luiz Rocha, Claudio Sampaio, Tommaso Giarrizzo, Cláudio Macedo and Bruno Cesar Antunes for granting permission to use their photographs.

References

Albins MA, Hixon, MA (2008) Invasive Indo-Pacific lionfish Pterois volitans reduce recruitment of Atlantic coral-reef fishes. Mar Ecol Prog Ser 367:233–238

Andréa BR, Batista D, Sampaio CLS, Muricy G (2007) Spongivory by juvenile angelfish (Pomacanthidae) in Salvador, Bahia State, Brazil. In: Custodio MR, Lôbo-Hajdu G, Hajdu E, Muricy G (eds) Porifera research: biodiversity, innovation and sustainability. Museu Nacional, Rio de Janeiro, pp 131–137

Araújo MCB, Da Costa MF (2008) Environmental quality indicators for recreational beaches classification. J Coast Res 24:1439–1449. West Palm Beach (Florida), ISSN 0749-0208

Araújo LM, Carvalho RC (2013) O turismo de massa em debate: a importância de sua análise para o planejamento turístico do estado de Alagoas, Brasil. Proceeding of the 10th Seminar of the National Association for Research and Postgraduate Studies in Tourism (pp. 1–20). Caxias do Sul, RS

Ardoin NM, Wheaton M, Bowers AW, Hunt CA, Durham WH (2015) Nature-based tourism's impact on environmental knowledge, attitudes, and behavior: a review and analysis of the literature and potential future research. J Sustain Tour 23:838–858. https://doi.org/10.1080/09669582.2015.1024258

Augustowski M, Francine RJ (2002) O Mergulho Recreacional como Ferramenta para o Turismo Sustentável em Unidades de Conservação Marinhas. In: Proceedings of the 3th Brazilian Congress of Conservation Units, Fortaleza, Brasil, 22–26

Azevedo-Santos VM, Pelicice FM et al (2015) How to avoid fish introductions in Brazil: education and information as alternatives. J Nat Conserv 13:123–132. https://doi.org/10.1016/j.ncon.2015.06.002

Barboza ARP, Leite TS (2020) Impactos ambientais do turismo intensivo na área recifal: efeitos do pisoteio em organismos bentônicos nos recifes rasos de Pirangi. In: Rocha LM, Bonilha LEC (eds) APA Recifes de Pirangi: proposta de criação de área protegida costeiro-marinha no Rio Grande do Norte. ONG Oceânica, Parnamirim, pp 131, 195 p–132

Barker NHL, Roberts CM (2004) Scuba diver behaviour and the management of diving impacts on coral reefs. Biol Conserv 120:481–489

Barreto LM (2020) Potencialidades do aquarismo como prática promotora da consciência ambiental. Acta Fish Aquat Resour 8:14–17. https://doi.org/10.46732/actafish.2020.8.1.14-17

Bellwood DR, Hughes TP, Folke C, Nystro M (2004) Confronting the coral reef crisis. Nature 429:827–833. https://doi.org/10.1038/nature02691

Berchez FAS, Ghilardi-Lopes NP et al (2016) Marine and coastal environmental education in the context of global climate changes-synthesis and subsidies for ReBentos (Coastal Benthic Habitats Monitoring Network). Braz J Oceanogr 64:137–156

Biasetti P, Florio D, Gili C, Mori B (2020) The ethical assessmente of touch pools in aquariums by means of the ethical matrix. J Agric Environ Ethics 33:337–353. https://doi.org/10.1007/s10806-020-09823-2

Bonaldo RM, Hay ME (2014) Seaweed-coral interactions: variance in seaweed allelopathy, coral susceptibility, and potential effects on coral resilience. PLoS One 9(1):e85786. https://doi.org/10.1371/journal.pone.0085786

Brasil (1999) Política Nacional de Educação Ambiental, Lei 9795. Diário Oficial da República Federativa do Brasil, Brasília, DF, 27 abr.

Brasil – MMA (2009) Conduta consciente em ambientes recifais/Gerência de Biodiversidade Aquática e Recursos Pesqueiros. Ministério do Meio Ambiente/SBF, Brasília

Brasil – MMA (2018) Conduta consciente em ambientes recifais – manual para multiplicadores. Ministério do Meio Ambiente, Secretaria da Biodiversidade, Secretaria de Recursos Hídricos e Qualidade Ambiental, MMA, Brasília

Brasil – MT (2018) Plano Nacional de Turismo 2018–2022: mais emprego e renda para o Brasil. Ministério do Turismo. Available in: http://www.turismo.gov.br/images/mtur-pnt-web2.pdf

Bruckner AW (2005) The importance of the marine ornamental reef fish trade in the wider Caribbean. Rev Biol Trop 53(1):127–138. https://doi.org/10.15517/rbt.v53i1.26627

Brunner B (2005) The ocean at home: an illustrated history of the aquarium. Princeton University Press, New York

Bumbeer J, da Rocha RM, Bornatowski H, Robert MDC, Ainsworth C (2017) Predicting impacts of lionfish (Pterois volitans) invasion in a coastal ecosystem of southern Brazil. Biol Invasions 20:1257–1274

Bumbeer J, Jardeweski C, Alves A, Andrade M, Rosado M, Fernandes M (2023) Oceano sem mistérios Desvendando os recifes de corais. Available in: https://www.fundacaogrupoboticario.org.br/pt/Biblioteca/OsM_outubro-3_leve(31-10-23).pdf

Butler RW (2006) The tourism area life cycle: conceptual and theoretical issues. Channel View Publications, Clevedon

Cagua EF, Collins N, Hancock J, Rees R (2014) Whale shark economics: a valuation of wildlife tourism in South Ari Atoll, Maldives. PeerJ 2:e515

Calado R, Lin J, Rhyne AL, Araújo R, Narciso L (2003) Marine ornamental decapods – popular, pricey, and poorly studied. J Crustac Biol 23(4):963–973. https://doi.org/10.1651/C-2409

Calado JF, Moreira ALP, Mendes LF (2021) O que sabemos sobre os impactos ambientais do turismo nos recifes tropicais do Brasil? Revista Brasileira de Pesquisa em Turismo 16:e-2420

Campos CEC, Sá-Oliveira JC (2011) Atividade de limpeza e clientes de *Elacatinus figaro* (Pisces: Gobiidae) nos recifes de coral dos Parrachos de Muriú, Nordeste do Brasil. Biota Neotrop 11(1):47–52. https://doi.org/10.1590/S1676-06032011000100004

Catlin J, Jones T, Norman B, Wood D (2010) Consolidation in a wildlife tourism industry: the changing impact of whale shark tourist expenditure in the Ningaloo coastal region. Int J Tour Res 12:134–148

Castro CB, Pires DO (2001) Brazilian coral reefs: what we already know and what is still missing. Bull Mar Sci 69(2):357–371

Cisneros-Montemayor AM, Barnes-Mauthe M et al (2013) Global economic value of shark ecotourism: implications for conservation. Oryx 47:381–388

Cline W (2008) Shark diving overview for the islands of the Bahamas. Economy Report of the Bahamas Ministry of Tourism. Bahama Dive Association

Comeros-Raynal MT, Choat JH, Polidoro BA et al (2012) The likelihood of extinction of iconic and dominant herbivores and detritivores of coral reefs: the parrotfishes and surgeonfishes. PLoS One 7(7):e39825. https://doi.org/10.1371/journal.pone.0039825

Costa-Chamas CC, Schmidt A (2011) Ecotourism and heritage conservation. In: Micallef, A. (ed.), MCRR3-2010. Conference proceedings. J Coast Res 61:234–241

Creed JC, Amado-Filho GM (1999) Disturbance and recovery of the macroflora of a seagrass (Halodule wrightii, Ascherson) meadow in the Abrolhos marine National Park, Brazil: an experimental evaluation of anchor damage. J Exp Mar Biol Ecol 235:285–306

Dawes J (2018) Editorial. Ornam Fish Int 88:3–4

De Araújo ME, Mattos FMG, Lopes FP, Melo FPL, Chaves LCT, Feitosa CV, Lippi DL, Hackradt FCF, Hackradt CW, Nunes JLS, Leão ZMAN, Kikuchi RKP, Ferreira Junior AV, Pereira PHC, Macedo CHR, Sampaio CLS, Feitosa JLL (2020) Diversity patterns of reef fish along the Brazilian tropical coast. Mar Environ Res 160:105038. https://doi.org/10.1016/j.marenvres.2020.105038

de Oliveira APL, Correia MD (2013) Aula de campo como mecanismo facilitador do ensino-aprendizagem sobre os ecossistemas recifais em Alagoas. Alexandria 6(2):163–190

de Queiroz Neto AC (2012) Dive tourism – um mergulho conceitual. In: Proceedings of the 7th Mercosur tourism research seminar. Caxias do Sul, RS

Diaz S, Smith JR, Zaleski SF, Murray SN (2012) Effectiveness of the California state ban on the sale of *Caulerpa* species in aquarium retail stores in Southern California. Environ Manag 50:89–96. https://doi.org/10.1007/s00267-012-9860-3

Dicken ML (2014) Socio-economic aspects of the Sodwana Bay SCUBA diving industry, with a specific focus on sharks. Afr J Mar Sci 36:39–47

Dicken ML, Hosking SG (2009) Socio-economic aspects of the tiger shark diving industry within the Aliwal Shoal marine protected area, South Africa. Afr J Mar Sci 31:227–232

Dixon JA, Scura LF, Van't Hof T (1993) Meeting ecological and economics goals: marine parks in the Caribbean. Ambio 22:117–125

do Amaral FMD, de Araújo Vieira MDF et al (2014) The role of environmental education in changing school students' perceptions of and attitudes toward coral reefs in the Fernando de Noronha Archipelago, Brazil. Revista de Gestão Costeira Integrada/J Integr Coast Zone Manag 14:581–590

Downs CA, Kramarsky-Winter E, Segal R, Fauth J, Knutson S, Bronstein O, Ciner FR, Jeger R, Lichtenfeld Y, Woodley CM, Pennington P, Cadenas K, Kushmaro A, Loya Y (2015) Toxicopathological effects of the sunscreen UV filter, oxybenzone (benzophenone-3), on coral planulae and cultured primary cells and its environmental contamination in Hawaii and the U.S. Virgin Islands. Arch Environ Contam Toxicol 70:265–288

Duarte C (2017) ICMBio divulga dados de visitação em UCs. Available in: http://www.icmbio.gov.br/portal/ultimas-noticias/20-geral/8711-cresce-numero-de-visitantes-nos-parques-nacionais. Accessed 22 Jan 2021

Eddy C, Pitt J, Morris Jr JA, Smith S, Goodbody-Gringley G, Bernal D (2016) Diet of invasive lionfish (Pterois volitans and P. miles) in Bermuda. Mar Ecol Prog Ser 558:193–206

Eichler PPB, de Moura DS (2020) Symbiont-bearing foraminifera as health proxy in coral reefs in the equatorial margin of Brazil. Environ Sci Pollut Res 27:13637–13661. https://doi.org/10.1007/s11356-019-07483-y

Eichler PPB, Gupta BS, Vital H, Barker CP (2020) Uso de microrganismos indicadores em recifes de corais brasileiros e sua resposta às perturbações ambientais ao longo do tempo. In: Rocha LM, Bonilha LEC (eds) APA Recifes de Pirangi: proposta de criação de área protegida costeiro-marinha no Rio Grande do Norte. ONG Oceânica, Parnamirim, pp 133–134, 195 p. ISBN: 9786586069617

Espino F, Riera R (2017) The cleaning mutualism between the shrimp, *Lysmata grabhami* (Malacostraca: Hippolytidae) and fish, with description of a new interaction with *Mycteroperca fusca* (Actinopterygii: Serranidae) from the Eastern Atlantic Ocean. Rev Acad Canar Cienc 29:49–60

Evangelista PH, Young NE, Schofield PJ, Jarnevich CS (2016) Modeling suitable habitat of invasive red lionfish *Pterois volitans* (Linnaeus, 1758) in North and South America's coastal waters. Aquat Invasions 11:313–326. https://doi.org/10.3391/ai.2016.11.3.09

Falcão C, de Széchy MTM (2005) Changes in shallow phytobenthic assemblages in southeastern Brazil, following the replacement of *Sargassum vulgare* (Phaeophyta) by *Caulerpa scalpelliformis* (Chlorophyta). Bot Mar 48:208–217. https://doi.org/10.1515/BOT.2005.033

Feitosa JLL, Ferreira BP (2014) Distribution and feeding patterns of juvenile parrotfish on algal-dominated coral reefs. Mar Ecol 36:1–13. https://doi.org/10.1111/maec.12154

Feitosa CV, Pimenta DAS, Araújo ME (2002) Ictiofauna recifal dos Parrachos de Maracajaú (RN) na área dos flutuantes: Inventário e estrutura da comunidade. Arq Cien Mar Fortaleza 35:39–50

Ferraz JD, Geller IV, Casimiro AC, Yabu MH, Garcia DA, Jarduli LR, Vitule JRS, Occhi TVT, Carneiro LO, Faria L, Cochak C, Sampaio FDF, Magalhães ALB, Lima-Júnior DP, Orsi ML (2020) Instrução Normativa n° 10, de 17 de abril de 2020, para peixes ornamentais: a urgência em reduzir as distâncias entre as esferas legislativa, científica e popular em aquariofilia. Boletim – Sociedade de Ictiologia de Londrina 132:21–38

Ferreira BP, Maida M (2006) Monitoramento dos recifes de coral do Brasil. MMA, Brasília, 250 p

Ferreira CEL, Floerter SR, Gasparini JL, Ferreira BP, Joyeux JC (2004) Trophic structure patterns of Brazilian reef fishes: a latitudinal comparison. J Biogeogr 31:1093–1106

Ferreira BP, Messias LT, Maida M (2006) The environmental municipal councils as an instrument in coastal integrated management: the Área de Proteção Ambiental Costa dos Corais (AL/PE) experience. J Coast Res 2:1003–1100

Ferreira CEL, Luiz OJ, Floeter SR, Lucena MB, Barbosa MC, Rocha CR, Rocha LA (2015) First record of invasive lionfish (*Pterois volitans*) for the Brazilian coast. PLoS One 10:e0123002. https://doi.org/10.1371/journal.pone.0123002

Figueiredo MAO, Horta PA, Pedrini AG, Nunes JMC (2008) Benthic marine algae of the coral reefs of Brazil: a literature review. Oecol Bras 12:258–269

Gallo H, Barbosa CB (2010) Educação Ambiental em aquários de visitação pública: a experiência do aquário de Ubatuba. In: Pedrini AG (ed) Educação Ambiental marinha e costeira no Brasil. EDUERJ, Rio de Janeiro, pp 227–242

Gasparini JL, Floeter SR, Ferreira CEL, Sazima I (2005) Marine ornamental trade in Brazil. Biodivers Conserv 14:2883–2899. https://doi.org/10.1007/s10531-004-0222-1

Ghilardi-Lopes NP, Braga JC et al (2013) Environmental education through an online game about global environmental changes and their effects on coastal and marine ecosystems. In: Proceedings of XII Simpósio Brasileiro de Jogos e Entretenimento Digital, pp 469–474

Giglio VJ, Luiz OJ, Schiavetti A (2015) Marine life preferences and perceptions among recreational divers in Brazilian coral reefs. Tour Manag 51:49–57

Giglio VJ, Luiz OJ, Schiavetti A (2016) Recreational diver behavior and contacts with benthic organisms in the Abrolhos National Marine Park, Brazil. Environ Manag 57:637–648. https://doi.org/10.1007/s00267-015-0628-4

Giglio VJ, Ternes MLF, Mendes TC, Cordeiro CAMM, Ferreira CEL (2017) Anchoring damages to benthic organisms in a subtropical scuba dive hotspot. J Coast Conserv 21:311–316

Gladstone W, Curley B, Shokri MR (2013) Environmental impacts of tourism in the Gulf and the Red Sea. Mar Pollut Bull 72:375–388

Gonçalves ELT (2017) Ictiopatologia marinha no Brasil: um estudo prospectivo. Doctoral thesis, Universidade Federal de Santa Catarina, Florianópolis

Gössling S (1999) Ecotourism: a means to safeguard biodiversity and ecosystem functions? Ecol Econ 29:303–320

Gouveia MTJ (Org) (2008) Educação para conservação de recifes e ambientes coralíneos. Manual de capacitação do professor em educação ambiental. Projeto Coral Vivo, Rio de Janeiro

Gouveia, MTJ (2011) Educação para Conservação de Recifes e Ambientes Coralineos: Manual de Capacitação do Professor em Educação Ambiental. Rio de Janeiro: Projeto Coral Vivo/ Associação Amigos do Museu Nacional

Gouveia MTJ (Org) (2015) Educação para conservação de recifes e ambientes coralíneos. Manual de formação de jovens: uma nova geração do Coletivo Jovem da Costa do Descobrimento. SAMN, Rio de Janeiro

Gouveia MTJ (Org) (2016) Manual do turismo sustentável em recifes e ambientes coralíneos. SAMN, Rio de Janeiro

Gouveia MTJ (Org) (2020) Educação ambiental na educação básica: conservação e sustentabilidade de ambientes coralíneos. Projeto Coral Vivo, Rio de Janeiro

Green E, Donnelly R (2003) Recreational SCUBA diving in Caribbean marine protected areas: do the users pay? Ambio 32:65–69

Green SJ, Akins JL, Maljković A, Côté IM (2012) Invasive lionfish drive Atlantic coral reef fish declines. PLoS One 7:e32596. https://doi.org/10.1371/journal.pone.0032596

Grossman A, Granville M, Matheus Z (2012) Atol das Rocas 3°51′S 33°48′W. Bei Editora, 228 p

Gurjão LM (2018) Legislação brasileira aplicada ao aquarismo marinho: a torre de babel ornamental. Arq Ciênc Mar 51(1):130–139. https://doi.org/10.32360/acmar.V51i1.19328

Gurjão LM, Lotufo TMC (2018a) The importance of Ceará state to the Brazilian marine ornamental fish exports. Arq Ciênc Mar 51:75–92. https://doi.org/10.32360/acmar.v51i1.33973

Gurjão LM, Lotufo TMC (2018b) Native species exploited by marine aquarium trade in Brazil. Biota Neotrop 18:e20170387. https://doi.org/10.1590/1676-0611-bn-2017-0387

Gurjão LM, Barros GML, Lopes DP, Machado DAN, Lotufo TMC (2017) Illegal trade of aquarium species through the Brazilian postal service in Ceará State. Mar Freshw Res 69:178–185. https://doi.org/10.1071/MF16257

Hammerton Z (2017) Determining the variables that influence SCUBA diving impacts in eastern Australian marine parks. Ocean Coast Manag 142:1–9

Hasler H, Ott JA (2008) Diving down the reefs? Intensive diving tourism threatens the reefs of the northern Red Sea. Mar Pollut Bull 56:1788–1794

Hawkins JP, Roberts CM (1997) Estimating the carrying capacity of coral reefs for scuba diving. In: Proceedings of the 8th international coral reef symposium, Panama, vol 2, pp 1923–1926

Hawkins JP, Roberts CM, Kooistra D, Buchan K, White S (2005) Sustainability of scuba diving tourism on coral reefs of Saba. Coast Manag 33:373–387

Hill MS (1998) Spongivory on Caribbean reefs releases corals from competition with sponges. Oecologia 17:143–150. https://doi.org/10.1007/s004420050642

Hixon MA, Green SJ, Albins MA, Akins JL, Morris Jr JA (2016) Lionfish: a major marine invasion. Mar Ecol Prog Ser 558:161–165

Holanda PC, Furtado-Neto MAA, Cabral NRAJ (2015) An evaluation of public aquariums in São Paulo (Brazil) in light of the "global aquarium strategy for conservation and sustainability". Arq Ciênc Mar 48:9–19

Holmberg RJ, Tlusty MF, Futoma E, Kaufman L, Morris JA, Rhyne AL (2015) The 800-pound grouper in the room: asymptotic body size and invasiveness of marine aquarium fishes. Mar Policy 53:7–12. https://doi.org/10.1016/j.marpol.2014.10.024

Hora MSC, Joyeux JC (2009) Closing the reproductive cycle: growth of the seahorse *Hippocampus reidi* (Teleostei, Syngnathidae) from birth to adulthood under experimental conditions. Aquaculture 292:37–41. https://doi.org/10.1016/j.aquaculture.2009.03.023

Hughes TP, Anderson KD, Connolly SR, Heron SF, Kerry JT et al (2018) Spatial and temporal patterns of mass bleaching of corals in the Anthropocene. Science 359:80–83

Instituto Brasileiro de Turismo (EMBRATUR) (2016) Segmento do turismo: marcos conceituais. Ministério do Turismo, Brasília, 56 p

Instituto Brasileiro do Meio Ambiente e dos Recursos Naturais Renováveis (IBAMA) (2009) Proposta de Plano de Gestão para o Uso Sustentável dos Cavalos-Marinhos do Brasil. Ministério do Meio Ambiente, Brasília. Disponível em: http://www.ibama.gov.br/phocadownload/biodiversidade/biodiversidade-aquatica/gestao-pesqueira/publicacoes/2011-proposta-plano-de-gestao-uso-sustentavel-cavalos-marinhos.pdf. Accessed 26 Jan 2021

Instituto de Desenvolvimento Sustentável e Meio Ambiente (IDEMA) (2012) Plano de Manejo da Área de Proteção Ambiental dos Recifes de Corais: resumo executivo. Disponível em: http://adcon.rn.gov.br/ACERVO/idema/DOC/DOC000000000034185.PDF. Accessed 22 Oct 2021

Kearns PJ, Bowen JL, Tlusty MF (2017) The skin microbiome of cow-nose rays (*Rhinoptera bonasus*) in an aquarium touch-tank exhibit. Zoo Biol 36:226–230. https://doi.org/10.1002/zoo.21362

King TA (2019) Wild caught ornamental fish: a perspective from the UK ornamental aquatic industry on the sustainability of aquatic organisms and livelihoods. J Fish Biol 94:925–936. https://doi.org/10.1111/jfb.13900

Konow N, Bellwood DR (2011) Evolution of high trophic diversity based on limited functional disparity in the feeding apparatus of marine angelfishes (f. Pomacanthidae). PLoS One 6:e24113. https://doi.org/10.1371/journal.pone.0024113

Laborel-Deguen F, Castro CB, Nunes F, Pires DO (2019) Recifes Brasileiros: o Legado de Laborel. Museu Nacional, Rio de Janeiro, 376 p

Lamb JB, True JD, Piromvaragorn S, Willis BL (2014) Scuba diving damage and intensity of tourist activities increases coral disease prevalence. Biol Conserv 178:88–96

Leão ZMAN, Oliveira MDM, Kikuchi RKP (2008) Os recifes de coral da APA Ponta da Baleia, Bahia. OLAM Ciênc Tecnol 8:287–315

Lira AS, Silva-Júnior CAB, Eduardo LN, Viana AP, Lucena-Frédou F, Frédou T (2019) Ichthyofauna bycatch of the artisanal fishery of penaeid shrimps in Pernambuco, northeastern Brazil. Bol Inst Pesca 45:435. https://doi.org/10.20950/1678-2305.2019.45.1.435

Lopes RM, Cunha DR (eds) (2009) Informe sobre as espécies exóticas invasoras marinhas no Brasil – Biodiversidade 33. Ministério do Meio Ambiente, Brasília

Louv R (2016) A última criança na natureza: resgatando nossas crianças do transtorno do déficit de natureza. Aquariana, São Paulo

Luiz OJ, Floeter SR, Rocha LA, Ferreira CE (2013) Perspectives for the lionfish invasion in the South Atlantic: are Brazilian reefs protected by the currents? Mar Ecol Prog Ser 485:1–7

Luiz OJ, Santos WC, Marceniuk AP, Rocha LA, Floeter SR, Buck CE, Klautau AGCM, Ferreira CE (2021) Multiple lionfish (*Pterois* spp.) new occurrences along the Brazilian coast confirm the invasion pathway into the Southwestern Atlantic. Biol Invasions 23:3013–3019. https://doi.org/10.1007/s10530-021-02575-8

Luna B, Pérez CV, Sánchez-Lizaso JL (2009) Benthic impacts of recreational divers in a Mediterranean marine protected area. ICES J Mar Sci 66:517–523

Mannino AM, Cicero F, Toccaceli M, Pinna M, Balistreri P (2019) Distribution of *Caulerpa taxifolia* var. *distichophylla* (Sonder) Verlaque, Huisman & Procaccini in the Mediterranean Sea. J Nat Conserv 37:17–29. https://doi.org/10.3897/natureconservation.37.33079

Mantelatto MC, Silva AG, Santos-Louzada T, McFadden CS, Creed JC (2018) Invasion of aquarium origin soft corals on a tropical rocky reef in the southwest Atlantic, Brazil. Mar Pollut Bull 130:84–94. https://doi.org/10.1016/j.marpolbul.2018.03.014

Melo RDS, Crispim MC, De Lima ER (2005) O turismo em ambientes recifais: em busca da transição para a sustentabilidade. Caderno Virtual de Turismo 5:34–42

Melo PW, Mattos SMG, Araújo ME (2021) Disminución de los pescadores artesanales en Rio Formoso, Pernambuco: Una contextualización histórica en Brasil. In: Narchi NE, Beitl CM (eds) América profunda: visiones y convergencias en la oceanografía social del continente. El Colegio de Michoacán, México, 36 p

Mendes LF, Grimaldi GG (2020) A biodiversidade e as formações recifais. cap 3. In: Rocha LM, Bonilha LEC (eds) APA Recifes de Pirangi: proposta de criação de área protegida costeiro-marinha no Rio Grande do Norte. ONG Oceânica, Parnamirim, pp 29–42, 195 p. ISBN: 9786586069617

Mendonça RC, Ikebata SP, Araújo-Silva SL, Manhães JVA, Tsuzuki MY (2020) Thermal influence on the embryonic development and hatching rate of the flameback pygmy angelfish *Centropyge aurantonotus* eggs. Zygote 28:80–82. https://doi.org/10.1017/S096719941900056X

Menezes NM, McFadden CS, Miranda RJ, Nunes JACC, Lolis L, Barros F, Sampaio CLS, Pinto TK (2021) New non-native ornamental octocorals threatening a South-west Atlantic reef. J Mar Biol Assoc U K 101(6):911–917

Militz TA, Foale S (2017) The "Nemo effect": perception and reality of Finding Nemo's impact on the marine aquarium fisheries. Fish Fish 18:596–606. https://doi.org/10.1111/faf.12202

Monteiro-Neto C, Cunha FEA, Nottingham MC, Araújo ME, Rosa IL, Barros GML (2003) Analysis of the marine ornamental fish trade at Ceará State, Northeast Brazil. Biodivers Conserv 12:1287–1295. https://doi.org/10.1023/A:1023096023733

Morin E (2011) Os sete saberes necessários à educação do futuro. Cortez/Unesco, São Paulo

Murray JM, Watson GJ (2014) A critical assessment of marine aquarist biodiversity data and commercial aquaculture: identifying gaps in culture initiatives to inform local fisheries managers. PLoS One 9(9):e105982. https://doi.org/10.1371/journal.pone.0105982

Nottingham MC, Barros GML, Araújo ME, Rosa IML, Ferreira BP, Mello TRR (2005a) O ordenamento da explotação de peixes ornamentais marinhos no Brasil. Bol Técn Cient CEPENE 13:75–107

Nottingham MC, Barreto LM et al (2005b) A explotação de peixes ornamentais marinhos no estado do Ceará, Brasil: capturas e manutenção nas empresas e exportação. Bol Técn Cient CEPENE 13:53–73

Oliveira APL, Correia MD, Sovierzoski HH (2014) Análise da Exposição sobre as Atividades das Aulas de Campo nos Ecossistemas Recifais. Rev Educ Ambient Ação 49:1–16

Olivotto I, Planas M, Simões N, Holt GJ, Avella MA, Calado R (2011) Advances in breeding and rearing marine ornamentals. J World Aquacult Soc 42:135–166. https://doi.org/10.1111/j.1749-7345.2011.00453.x

Ortiz DM, Tissot BN (2012) Evaluating ontogenetic patterns of habitat use by reef fish in relation to the effectiveness of marine protected areas in West Hawaii. J Exp Mar Biol Ecol 432:83–93. https://doi.org/10.1016/j.jembe.2012.06.005

Padilla DK, Williams SL (2004) Beyond ballast water: aquarium and ornamental trades as sources of invasive species in aquatic ecosystems. Front Ecol Environ 2:131–138

Pedrini AG (Org) (2010) Educação Ambiental Marinha e Costeira no Brasil. Editora da Universidade do Estado do Rio de Janeiro, Rio de Janeiro

Pedrini AG, Costa C, Newton T, Maneschy FSA, Silva VG, Berchez F, Spelta L, Ghilardi NP, Robim MJ (2007) Efeitos ambientais da visitação de turistas em áreas protegidas marinhas. Estudo de caso na Piscina Natural Marinha, Parque Estadual da Ilha Anchieta, Ubatuba, São Paulo, Brasil. OLAM Ciênc Tecnol 7:678–696

Penning M, Reid G et al (2009) Turning the tide: a global aquarium strategy for conservation and sustainability. World Association of Zoos and Aquariums, Berna

Pereira PHC, Pinheiro BR, Coelho-Jr C, Magalhães K, Souza C, Almeida E, Santos J, Normande IC (2017) Coral reef conservation on the largest Brazilian MPA – linking hybrid governance and social-ecological resilience. In: The 10th Indo-Pacific fish conference, Tahiti, p 467

Pinheiro BR, Coelho-Jr C, Pereira PHC, Greenhut JC, Souza CN, Normande IC, Santos JU, Almeida EM (2016) Mapas temáticos elaborados por jangadeiros e pescadores da Rota Ecológica. In: Seminário de Pesquisa APA Costa dos Corais: prioridades para a conservação da biodiversidade, I, 2016, Tamandaré, p 35

Pires NM, Garla RC, Carvalho AR (2016) The economic role of sharks in a major ecotourism archipelago in the western South Atlantic. Mar Policy 72:31–39

Plathong S, Inglis GJ, Huber ME (2000) Effects of self-guided snorkeling trails in a tropical marine park. Conserv Biol 14:1821–1830

Pomeroy RS, Parks JE, Balboa CM (2006) Farming the reef: is aquaculture a solution for reducing fishing pressure on coral reefs? Mar Policy 30:111–130. https://doi.org/10.1016/j.marpol.2004.09.001

Raineri C, Fernandes L, Gouveia MTJ et al (2016) Manual de ecossistemas marinhos e costeiros para educadores. Editora Comunnicar, São Paulo

Renfro B, Chadwick NE (2017) Benthic community structure on coral reefs exposed to intensive recreational snorkeling. PLoS One 12:1–22

Rhormens MS, Pedrini AG, Ghilardi-Lopes NP (2017) Implementation feasibility of a marine ecotourism product on the reef environments of the marine protected areas of Tinhare and Boipeba Islands (Cairu, Bahia, Brazil). Ocean Coast Manag 139:1–11. https://doi.org/10.1016/j.ocecoaman.2017.01.022

Rhyne A, Rotjan R, Bruckner A, Tlusty M (2009) Crawling to collapse: ecologically unsound ornamental invertebrate fisheries. PLoS One 4:e8413. https://doi.org/10.1371/journal.pone.0008413

Rhyne AL, Tlusty MF, Kaufman L (2012) Long-term trends of coral imports into the United States indicate future opportunities for ecosystem and societal benefits. Conserv Lett 5:478–485. https://doi.org/10.1111/j.1755-263X.2012.00265.x

Rhyne AL, Tlusty MF, Kaufman L (2014) Is sustainable exploitation of coral reefs possible? A view from the standpoint of the marine aquarium trade. Curr Opin Environ Sustain 7:101–107. https://doi.org/10.1016/j.cosust.2013.12.001

Rhyne AL, Tlusty MF, Szczebak JT, Holmberg RJ (2017) Expanding our understanding of the trade in marine aquarium animals. PeerJ 5:e2949. https://doi.org/10.7717/peerj.2949

Rocha LM, Bonilha LEC (2020) APA Recifes de Pirangi: proposta de criação de área protegida costeiro-marinha no Rio Grande do Norte/Coordenadores Ligia Moreira da Rocha, Luiz Eduardo Carvalho Bonilha. Oceânica, Parnamirim

Rocha LM, Grimaldi GDG, Batista JLDV et al (2016) Vivendo com o mar: um olhar para a conservação do litoral sul potiguar – Projeto Ponta de Pirangi. Oceânica, Parnamirim

Rouphael AB, Hanafy M (2007) An Alternative Management Framework to Limit the Impact of SCUBA Divers on Coral Assemblages. J Sustain Tour 15(1):91–103. https://doi.org/10.2167/jost611.0

Rouphael AB, Inglis GJ (2001) "Take only photographs and leave only footprints"?: an experimental study of the impacts of underwater photographrs on corals reefs dives sites. Biol Conserv 100:281–287

Rowe RYG, Santos GEO (2016) Turismo de mergulho: análise do comportamento de viagem dos mergulhadores brasileiros. Caderno Virtual de Turismo 16:61–75

Salgado MDM, Marandino M (2014) O mar no museu: um olhar sobre a educação nos aquários. Hist Cienc Saude Manguinhos 21:867–882

Sampaio FDF, Ostrensky A (2013) Brazilian environmental legislation as a tool to conserve marine ornamental fish. Mar Policy 42:280–285. https://doi.org/10.1016/j.marpol.2013.03.009

Sampaio CL, Loiola M, Colman LP et al (2017) Cryptobenthic fish as clients of french angelfish *Pomacanthus paru* (Pomacanthidae) during cleaning behaviour. Mar Biodivers Rec 10:8. https://doi.org/10.1186/s41200-017-0109-y

Santoro F et al (2018) Cultura oceánica para todos: kit pedagógico. UNESCO Office Venice and Regional Bureau for Science and Culture in Europe (Italy), Intergovernmental Oceanographic Commission

Schofield PJ (2010) Update on geographic spread of invasive lionfishes (Pterois volitans [Linnaeus, 1758] and P. miles [Bennett, 1828]) in the Western North Atlantic Ocean, Caribbean Sea and Gulf of Mexico. Aquat Invasions 5:S117–S122

Scopel JM, Cavalli GL, da Costa FM et al (2019) UCS Aquarium e a sensibilização ambiental para a preservação dos ecossistemas aquáticos. Sci Ind 7:7–40

Sekharan MN (2018) Conservation and management of Indian ornamental fishes. Ornam Fish Int 86:20–22

Semmens BX, Buhle ER, Salomon AK, Pattengill-Semmen CV (2004) A hotspot of nonnative marine fishes: evidence for the aquarium trade as an invasion pathway. Mar Ecol Prog Ser 266:239–244. https://doi.org/10.3354/meps266239

Silva CB, Ferreira RG, Amaral RF (2009) Análise da imagem que turistas e comunidade local têm da Área de Proteção Ambiental dos Recifes de Corais – APARC (RN). Caderno Virtual de Turismo 9:85–102

Silva L, Correia M, Sovierzoski H (2013) Percepção Ambiental Sobre os Ecossistemas Recifais em duas diferentes Áreas do Litoral Nordeste do Brasil. Rev Educ Ambient Ação 45:1–13

Soares, M. D. O., Feitosa, C. V., Garcia, T. M., Cottens, K. F., Vinicius, B., Paiva, S. V., Duarte, O. S., Gurjão, L. M., Silva, G. D. V., Maia, R. C., Previatto, D. M., Carneiro, P. B. M., Cunha, E., Amâncio, A. C., Sampaio, C. L. S., Ferreira, C. E. F., Pereira, P. H. C., Rocha, L. A., Tavares, T. C. L. & Giarrizzo, T. (2022). Lionfish on the loose: Pterois invade shallow habitats in the tropical southwestern Atlantic. Front Mar Sci 9 (956848).

Soares MO, Pereira PH, Feitosa CV et al. (2023) Lessons from the invasion front: Integration of research and management of the lionfish invasion in Brazil. Environ Manag 340:117954

Souza IM, Mendes LF, Rocha LM, Grimaldi GDG (2016) A vida marinha no litoral sul Potiguar. Oceânica, Parnamirim

Stevenson TC, Tissot BN (2013) Evaluating marine protected areas for managing marine resource conflict in Hawaii. Mar Policy 39:215–223. https://doi.org/10.1016/j.marpol.2012.11.003

Suassuna F, Mendes LF (2020) Mar doce lar. Ilustrações Luiza Laffitte; adaptação do CD ao livro ilustrado Ligia Moreira da Rocha, Jessica de Paiva Bezerra, Joane Luiza Dantas Vieira Batista, 2nd edn. Oceânica, Parnamirim

Tapsuwan S, Asafu-Adjaye J (2008) Estimating the economic benefit of SCUBA diving in the Similan Islands, Thailand. Coast Manag 36:431–442

Teixeira LMP, Creed JC (2020) A decade on: an updated assessment of the status of marine non-indigenous species in Brazil. Aquat Invasions 15:30–43. https://doi.org/10.3391/ai.2020.15.1.03

Tissot BN, Walsh WJ, Hallacher LE (2004) Evaluating effectiveness of a marine protected area network in West Hawai'i to increase productivity of an aquarium fishery. Pac Sci 58:175–188. https://doi.org/10.1353/psc.2004.0024

Tissot BN, Walsh WJ, Hixon MA (2009) Hawaiian islands marine ecosystem case study: ecosystem- and community-based management in Hawai'i. Coast Manag 37:255–273. https://doi.org/10.1080/08920750902851096

Tissot BN, Best BA, Borneman EH et al (2010) How US ocean policy and market power can reform the coral reef wildlife trade. Mar Policy 34:1385–1388. https://doi.org/10.1016/j.marpol.2010.06.002

Trademap (2020) List of importing markets for a product exported by Brazil: 0301 Live fish. Geneva, Switzerland. Disponível em: https://www.trademap.org/Index.aspx. Accessed 26 Jan 2021

UNESCO (Organização das Nações Unidas para a Educação, Ciência e Cultura) (2020) Cultura Oceânica para todos – kit pedagógico, Manuais e Guias da COI, vol 80. UNESCO

UNWTO (World Tourism Organization and United Nations Development Programme) (2017) Tourism and the Sustainable Development Goals – journey to 2030. UNWTO, Madrid. https://doi.org/10.18111/9789284419401

Van't Hof T (2001) Tourism impacts on coral reefs: increasing awareness in the tourism sector. UNEP and Tour Operators Initiative. Funded by the French Ministry of the Environment (MATE)

Vianna GMS, Meekan MG, Pannell D, Marsh S, Meeuwig LJ (2010) Wanted dead or alive? The relative value of reef sharks as a fishery and an ecotourism asset in Palau. Australian Institute of Marine Science and University of Western Australia, Perth

Wabnitz C, Taylor M, Green E, Razak T (2003) From ocean to aquarium. UNEP-WCMC, Cambridge

Walters L, Odom R, Zaleski S (2011). The aquarium hobby industry and invasive species: has anything changed?. Front Ecol Environ 9(4):206–207

WAZA (World Association of Zoos and Aquariums) (2005) Construindo um futuro para a vida selvagem – Estratégia Mundial dos Zoos e Aquários para a conservação. WAZA Executive Office, Berna

Whittington RJ, Chong R (2007) Global trade in ornamental fish from an Australian perspective: the case for revised import risk analysis and management strategies. Prev Vet Med 81:92–116. https://doi.org/10.1016/j.prevetmed.2007.04.007

Wood E (2001) Collection of coral reef fish for aquaria: global trade, conservation issues and management strategies. Marine Conservation Society, Ross-on-Wye

World Tourism Organization (2004) Tourism and poverty alleviation recomendations for action. World Tourism Organization, Madrid

World Tourism Organization and United Nations Development Programme (2017) Tourism and the Sustainable Development Goals – journey to 2030, highlights. UNWTO, Madrid

World Travel & Tourism (2016) World Travel & Tourism Council: travel & tourism economic impact 2016 Londres. WTTC, Inglaterra

Zakai D, Chadwick-Furman NE (2002) Impacts of intensive recreational diving on reef corals at Eilat, northern Red Sea. Biol Conserv 105:179–187

Zilberberg C, Abrantes DP, Marques JA (eds) (2016) Conhecendo os recifes brasileiros: rede de pesquisas Coral vivo. Museu Nacional, Universidade Federal do Rio de Janeiro

Chapter 12
Conservation and Management of Reef Ecosystems

Rafael A. Magris, Harildon M. Ferreira, Juan P. Quimbayo, and Leandro Godoy

Abstract This chapter focuses on discussing the main strategies for the conservation and management of Brazilian reefs. Marine Protected Areas (MPAs) are one of the key components of reef conservation. Although the existing system of MPAs covers 15.3% of the total reef area, the reality of this figure is misleading because only a fraction of reef habitats is covered by no-take MPAs (3.6%), and several reef types have no coverage by fully-protected MPAs. Discrepancies between the amount of coverage by all MPAs and what is covered by no-take ones were stronger on the Northeastern continental shelf and decreased towards oceanic, deep reef areas, thus confirming the strong protection bias. Existing system of MPAs also does not meet the minimum requirements to adequately protect threatened species, particularly the endemic ones. The predominance of multiple-use MPAs highlights critical gaps between MPA coverage and protection levels afforded. The conservation effectiveness of Brazilian MPAs is variable and the difference between MPA coverage and protection level reached up to 94%. The extensive amount of reef area lying within priority areas for further protection highlights the need of expanding and strengthening the existing system of MPAs. We also identified a general lack of long-term monitoring programs running over no-take MPAs and the absence of more specific studies developing recovery tools designed for coral reefs.

Keywords Marine protected areas · Marine reserves · Conservation effectiveness · Ecological monitoring · Restoration · Brazil

---

R. A. Magris (✉) · H. M. Ferreira
Instituto Chico Mendes de Conservação da Biodiversidade (ICMBIO), Federal District, Brasilia, Brazil

J. P. Quimbayo
Centro de Biologia Marinha (CEBIMAR), Universidade de São Paulo, São Sebastião, São Paulo, Brazil

Department of Biology, University of Miami, Miami, USA

L. Godoy
Universidade Federal do Rio Grande do Sul, Porto Alegre, Rio Grande do Sul, Brazil

R. K. P. Kikuchi et al. (eds.), *Brazilian Coral Reefs*, Brazilian Marine Biodiversity,
https://doi.org/10.1007/978-3-031-59152-5_12

12.1 Introduction

Coral reefs are highly vulnerable ecosystems globally and are in the frontline of exposure to multiple human pressures (Hughes et al. 2017). As in other parts of the world, the escalating magnitude of ecological changes in Brazilian reefs necessitates increased efforts for the conservation and management of these unique and iconic ecosystems (Leão and Kikuchi 2005; Leão et al. 2003; Magris et al. 2018; Soares et al. 2021). Conservation and management of coral reefs are imperative and have gained attention particularly for their potential to maintain coral reef functionality and associated ecosystem services (Bellwood et al. 2012; Hughes et al. 2010). A key tool for mitigating threats, conserving biodiversity, and managing fisheries is the designation of marine protected areas (MPAs) (Watson et al. 2014). As the extent of MPAs on reef ecosystems increases (Mora et al. 2006), the effective implementation of MPAs is a key important step to ensure this strategy delivers the desired conservation outcomes (Marinesque et al. 2012).

However, existing MPAs might be limited in their contribution to safeguard and recover biodiversity due to their poor design and management (Magris et al. 2020; Mouillot et al. 2016). There is thus a need of identifying future conservation priorities for the expansion and strengthening of existing MPAs. Yet, there are major constraints on the establishment of future MPAs and this strategy might be too slow for biodiversity facing imminent decline (Venter et al. 2014). Subsequent monitoring programs are then necessary to examine sudden changes in the ecosystem or inform about threats that operate across a wide range of temporal and spatial scales (e.g., climate change, invasive species). In addition, MPAs are surrounded by areas facing high levels of threats and threatened species face imminent extinction beyond the borders of protected areas. Arguably, MPAs alone are inadequate to address the current biodiversity crisis in coral reef ecosystems (Anthony et al. 2017; Duarte et al. 2020). Developments in management strategies to assist the recovery of damaged reefs are also needed for those lacking any formal protection or having protection that fails to combat declines in biodiversity. Taken together, these complementary conservation strategies are essential to be effective in maintaining the integrity of coral reefs for future generations.

The goal of this chapter is to provide an overview of the five main conservation and management strategies for reef ecosystems at a regional scale (Brazil): establishment of MPAs, implementation of MPAs, identification of future conservation priorities, development of monitoring programs, and the use of restoration tools. Finally, we finish with an overall synthesis and highlight some key challenges for ways forward.

12.2 Establishment of Marine Protected Areas

Coral reef conservation and management often includes the designation of marine protected areas (MPAs), which aim to limit or prohibit human activities within their boundaries and offer protection to the habitats and species. The use of MPAs, and especially no-take areas[1] (i.e., those offering strict protection, where extractive activities are not allowed), has became popular among scientists, policymakers, and conservationist as the most effective tool to ensure greater species biodiversity and biomass than nearby exploited areas (Di Lorenzo et al. 2020; Ferreira et al. 2022; Giakoumi et al. 2017; Lester et al. 2009; Sciberras et al. 2015). Although the extent of MPAs in Brazil increased substantially from 1.5% to 24.5% in March 2018 (Magris and Pressey 2018), the area coverage by no-take MPAs remained low (i.e., only 2.5% of the Brazilian waters to date). A major part of the MPA system encompasses multiple-use MPAs[2]—covering about 21% of the Brazilian waters. This management category provides partial protection, where various extractive activities associated with multiple users are permitted but are regulated, spatially and temporally, to a greater extent than areas outside these MPAs. The remainder of the MPA system includes extractive reserves,[3] which aim to combine biodiversity conservation with sustainable use of natural resources, and are co-managed by traditional, small-scale communities (i.e., about 1%). However, all these coverage estimates do not consider the absolute areas of habitats within MPAs and therefore it is difficult to ascertain whether the full breadth of biodiversity is represented within them.

Although in the last few decades the science of MPAs has advanced towards optimizing their spatial design for conserving biodiversity in the long term (Álvarez-Romero et al. 2018; Kukkala and Moilanen 2013), the Brazilian national system of MPAs has been developed more based on opportunity than on strategic planning, without taking into account standard conservation planning principles such as biodiversity representation (Magris et al. 2020). The adoption of this planning principle has been recommended by several international policies and agreements (Claudet et al. 2020; Rees et al. 2018). To be representative, MPAs should sample a set of conservation features (i.e., habitats or species), ideally offering greater protections to those features that are more threatened. Because it is challenging in practice to obtain detailed inventories of species, biodiversity representation is often applied to habitat types, which are defined by biotic and/or abiotic characteristics and assumed to be surrogates for distinctive biological assemblages, or selected species. The

[1] No-take MPAs corresponds to the categories I-III according to the International Union for Conservation of Nature (IUCN) and to the category named "*strict (or integral) protection*" according to the National System of Protected Areas (SNUC—acronym in Portuguese).

[2] Multiple-use MPAs corresponds to the IUCN category VI and are also labelled as "Environmental Protection Areas" and "Areas of Relevant Ecological Interest" within the category of "*sustainable use*" by SNUC.

[3] Extractive reserves correspond to the IUCN category V and are also within the category of "*sustainable use*" by SNUC.

challenge of designating representative MPAs increases when performing conservation assessments that require fine-scale habitat classifications across large geographical scales and comprehensive data on the distribution of biodiversity (Magris et al. 2017). Additionally, biodiversity representation can be achieved by an uneven protection among the conservation features within an area (Jantke et al. 2018), thus analysing only the percentage of area under protection risks misjudging the performance of MPA systems.

Evaluating biodiversity representation and identifying gaps in MPA coverage involves gathering spatial information on features of conservation interest and MPA boundaries, and performing a spatial overlap between them. We collated spatial information on reef habitats (i.e., coral reefs and associated habitats) and related threatened species, data that had already been assembled and made available by Magris et al. (2020). Reef habitats were first classified into two classes: shallow-water and mesophotic reefs. While mesophotic reefs could be further classified into subtypes based on the ecoregion they are located in (*sensu* Spalding et al. 2007; N = 3 classes), shallow-water reefs were subdivided depending upon specific combinations of ecoregions, substrate, and geomorphologic types. This hierarchical classification resulted in the following reef classes: shallow-water neritic coral reefs (N = 10), rocky shores (N = 1), and shallow-water oceanic reefs (N = 1). This full range of reef types are important to structure biodiversity patterns across a large geographical scale (Araújo et al. 2020; Aued et al. 2018).

Threatened species were selected based on their inclusion in the National Plan for the Conservation of Brazilian Reefs (ICMBio 2020) and categorized by the Brazilian Red List as Critically Endangered, Endangered, or Vulnerable, including fish (N = 37) and invertebrate species (N = 15). To explore the distribution of coverage by all MPA categories across reef habitats, we thus provided additional measures such as mean and median cover by each management category, as well as the equality of protection (Barr et al. 2011). Equality of protection ranges from 0% (completely inequitable) to 100% (a perfectly equitable distribution of protection).

Currently, 15.3% of reef area in Brazil has been covered by 61 MPAs under the different levels of government (federal, state, and municipal; Fig. 12.1a): 29 assigned as no-take MPAs, covering 3.6% of the total reef area; 27 assigned as multiple-use MPAs, covering 10.9% of reef area; and 5 assigned as extractive reserves, covering 0.8% of reef area. This percentage of the total habitat area covered by MPAs is similar to the ones found in the Coral Triangle (i.e. 17.8%; (White et al. 2014)). Mapped shallow-water and mesophotic reefs from all ecoregions have been to some extent covered by MPAs (Fig. 12.2a). Although there were discrepancies between habitats in terms of MPA coverage, all shallow-water reef types have been also covered by MPAs of any management category (Fig. 12.2b). Total coverage varied from 0.73% (i.e., mesophotic reefs in the Amazon continental shelf) to 100% (i.e., offshore banks in the Northeastern continental shelf and shallow-water, mushroom-shaped pinnacles in the Eastern continental shelf). Yet, these two habitats with the largest coverage by MPAs had only between 0% and 3.7% of coverage by no-take ones, respectively. Patch reefs in the Northeastern continental shelf also had no coverage by no-take MPAs. Other reef habitats that had very little coverage

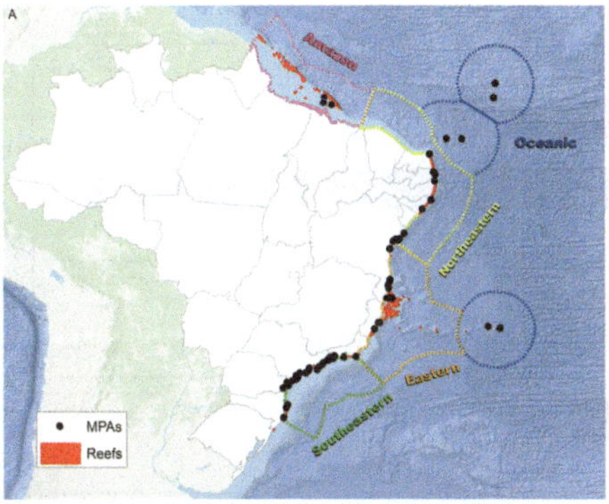

B

MPA coverage (%)	No-take coverage (%)	Protection level[1] (%)	Protection level[2] (%)
1.23	1.23	0.41	0.63
35.3	29.85	28.3	16.18
76.40	0.71	33.84	16.20
43.61	8.75	20.84	20.40
39.4	3.99	20.61	15.35

[1] Protection level calculated according to the management effectiveness of the MPAs
[2] Protection level calculated according to the ecological effectiveness of the MPAs

Fig. 12.1 Brazilian reefs and the location of marine protected areas (shown as the centroids to facilitate visualization) (**a**). Coverage and protection levels of reefs at each ecoregion (Amazon shelf, Oceanic, Northeastern shelf, Eastern shelf, and Southeastern shelf) by MPAs (**b**). Coverage values considered the whole MPA system (including MPAs from all management categories) and no-take MPAs only. Protection levels were presented considering the management and ecological effectiveness of MPAs. See also Chap. 1 for an adapted ecoregion classification

by no-take areas were nearshore banks in the Northeastern shelf and rocky shores in the Southeastern continental shelf, with 1% and 3.9% coverage, respectively. Conversely, four reef types towards remote, deeper waters had more coverage by no-take MPAs than other categories offering partial protection. They were shallow-water reefs bordering islands and mesophotic reefs in oceanic ecoregions (i.e., Fernando de Noronha Island and Atoll das Rocas, São Pedro and São Paulo Islands and Trindade and Martin Vaz Islands); offshore banks in the Amazon continental shelf; and offshore banks in the Easter continental shelf. This pattern demonstrates that, despite large spatial extent of MPAs protecting Brazilian reefs, there are coverage bias towards more remote, deeper areas, and in relation to MPA management category.

Mean and median coverage of reef habitats by no-take MPAs were 18% and 4.6%, respectively (Fig. 12.3a, b). Nevertheless, mean and median coverage of reef habitats by multiple-use areas increased to 33.4% and 15.4%, respectively. Because high coverage levels by extractive reserves was generally lacking, extractive reserves had the highest protection equality, followed by multiple-use MPAs, then no-take MPAs (Fig. 12.3c). While the national system of MPAs in Brazil qualitatively represents almost all reef habitats, unevenness and bias are evident, particularly towards higher coverage of reef habitats by MPAs that allows some forms of human use (i.e., multiple-use ones) and lower coverage by MPAs offering strict protection (i.e., no-take ones). Although there is some uncertainty in the estimates of MPA coverage due to unmapped features, these findings nonetheless reflect important protection gaps and the biased distribution of existing MPAs. Globally, in assessments considering multiple marine habitats, the emerging trend is also to create a biased MPA system in terms of distribution of different management categories (e.g., Brooks et al. 2020; Devillers et al. 2015).

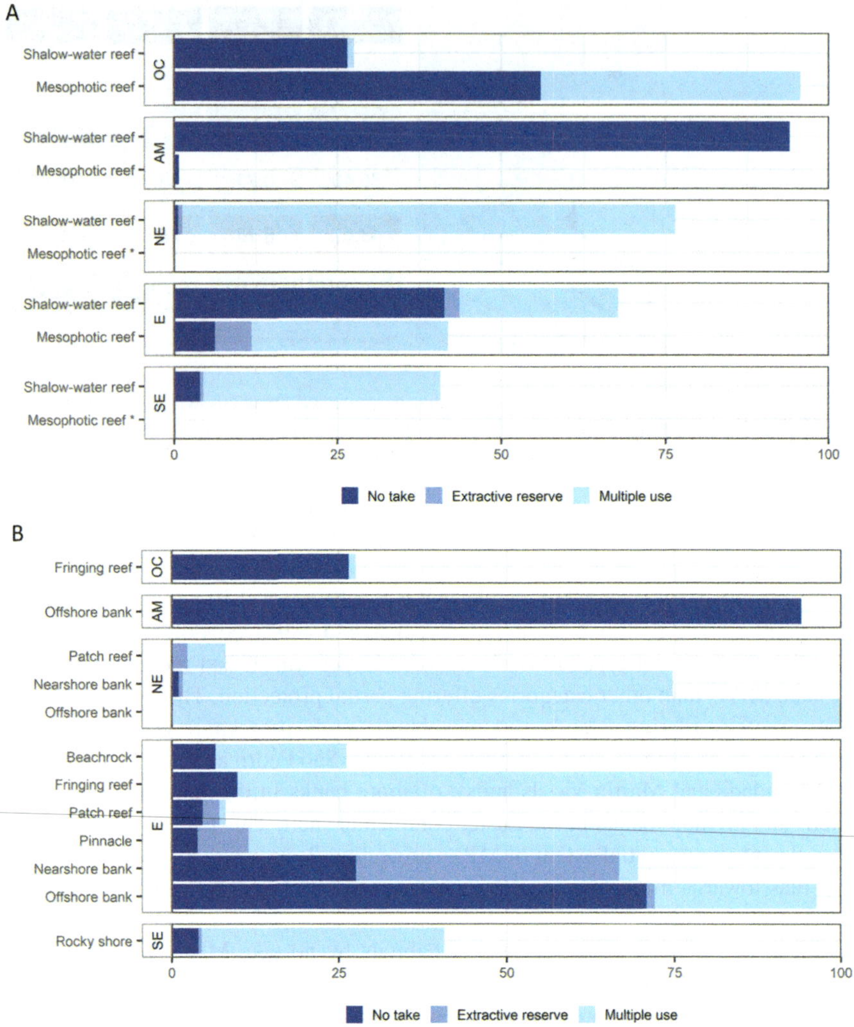

Fig. 12.2 Percentage cover of reef habitats by the three MPA management categories (No-take, Extractive reserves, and Multiple-use MPAs) within each ecoregion in Brazil. In Panel (**a**) (top), reefs were classified into shallow-water and mesophotic reefs. In Panel (**b**) (bottom), shallow-water coral reefs were further classified into 12 classes. Abbreviations for ecoregions: OC—Oceanic and encompasses large tracts of the ocean surrounding Fernando de Noronha Islands, Atoll das Rocas, São Pedro and São Paulo Islands and Trindade and Martin Vaz Islands; AM = Amazon Shelf; NE = Northeastern Shelf; E = Eastern Shelf; and SE = Southeastern Shelf. *Mesophotic reefs at the ecoregions NE and SE have been unmapped

Seven out of the top ten MPAs in terms of reef coverage were multiple-use areas, such as the Environmental Protection Area Ponta da Baleia, Environmental Protection Area Litoral Norte de Sao Paulo, and Environmental Protection Area dos Recifes de Corais (Table 12.1). The first aforementioned multiple-use MPA was

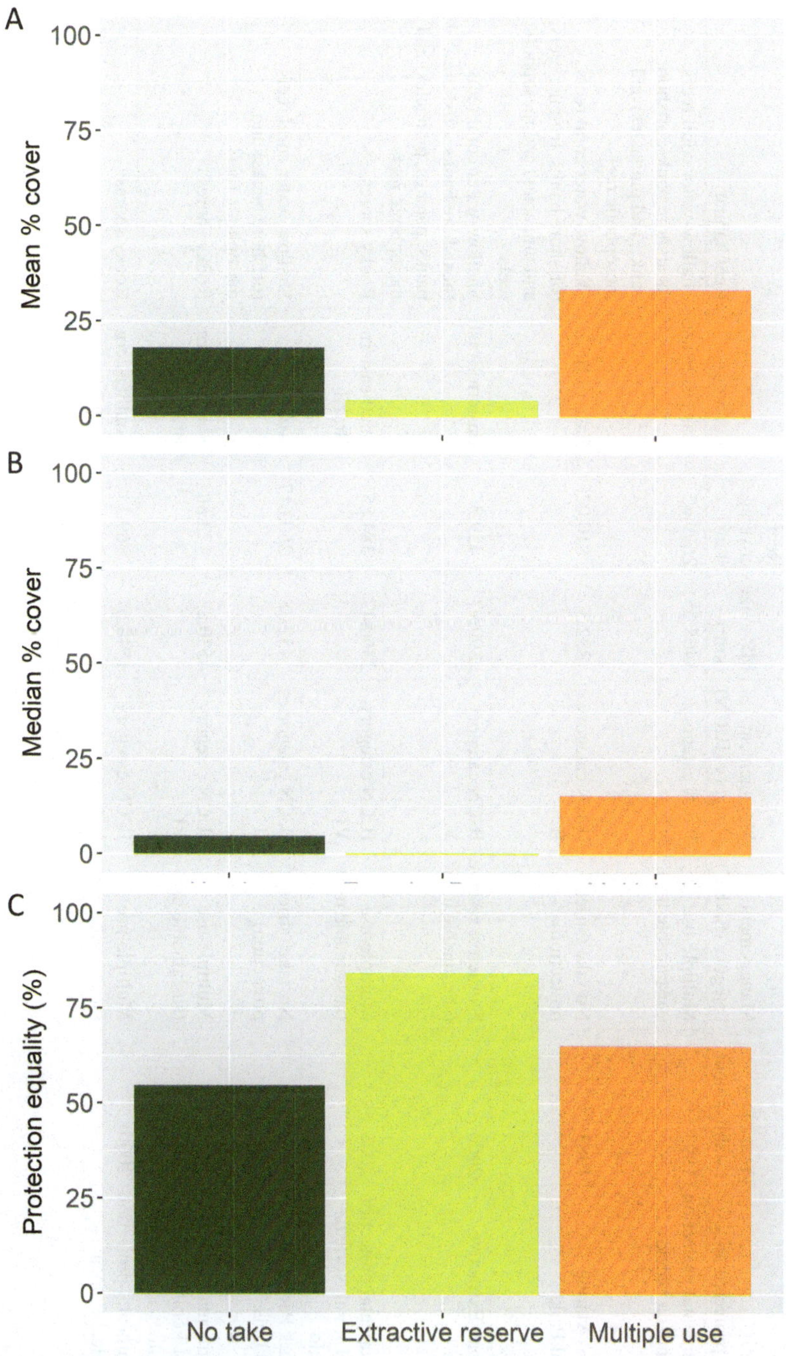

Fig. 12.3 Mean (**a**), median coverage (**b**) and protection equality (**c**) by MPA management categories among reef habitats in Brazil

Table 12.1 Main marine protected areas with reefs along the Brazilian coast and their characteristics. Ecoregions according to Spalding et al. (2007)

Marine protected area	Year of establishment	Management category (SNUC)	Management category (IUCN)	MPA size (km^2)	Reef coverage (km^2)	Ecoregion	Reef habitat
Environmental protection area Ponta da Baleia	1993	Multiple-use (sustainable use)	IUCN category VI	3465.35	2699.65	Eastern Brazil	Shallow-water coral reefs (nearshore banks, offshore banks, and pinnacles) and mesophotic reefs
Abrolhos Marine National Park	1983	No-take (strict protection)	IUCN category II	882.5	810.02	Eastern Brazil	Shallow-water coral reefs (offshore banks, fringing reefs, and pinnacles) and mesophotic reefs
Corumbau Extractive Reserve	2000	Extractive reserve (sustainable use)	IUCN category V	899.97	459.3	Eastern Brazil	Shallow-water coral reefs (nearshore banks, offshore banks, patches, pinnacles) and mesophotic reefs
Environmental protection area of Litoral Norte de Sao Paulo	2008	Multiple-use (sustainable use)	IUCN category VI	2360.47	384.85	Southeastern Brazil	Rocky shores
State Park Parcel de Manoel Luis	1991	No-take (strict protection)	IUCN categories II	452.38	260.617	Amazon	Shallow-water coral reefs (offshore banks) and mesophotic reefs
Environmental protection area of Litoral Centro de Sao Paulo	2008	Multiple-use (sustainable use)	IUCN category VI	4530.83	122.86	Southeastern Brazil	Rocky shores
Environmental protection area Baleia Franca	2000	Multiple-use (sustainable use)	IUCN category VI	1548.67	69.1	Southeastern Brazil	Rocky shores

Environmental protection area dos Recifes de Corais	2001	Multiple-use (sustainable use)	IUCN category VI	1363.44	57.2	Northeastern Brazil	Shallow-water coral reefs (nearshore banks, offshore banks, and patches)
Environmental protection area Caraiva/Trancoso	1993	Multiple-use (sustainable use)	IUCN category VI	319.0	49.3	Eastern Brazil	Shallow-water coral reef (patches) and mesophotic reefs
Environmental protection area Costa dos Corais	1997	Multiple-use (sustainable use)	IUCN category VI	4135.63	43.4	Northeastern Brazil	Shallow-water coral reefs (nearshore banks)

See Chap. 1 for an adapted ecoregion classification

designated in 1993 and covers the largest coral reef area in Brazil, encompassing the region's most important estuary and nearly all reefs forming the coastal arc of the Abrolhos Bank. Brazil's first no-take MPA (i.e., Abrolhos Marine National Park) was proclaimed in 1983 to protect an offshore parcel around the Abrolhos archipelago and a nearshore bank on the coastal arc of the Abrolhos Bank. The MPA contains at least 20 species of coral, including six species that are endemic to Brazil (Francini-Filho et al. 2013), and one species (*Mussismilia braziliensis*) that is endemic to the Abrolhos region alone (Leão and Kikuchi 2005; Leão et al. 2003). Its international recognition as an important site for conservation has gained attention because of its importance for the species and ecosystems it hosts (Mission Blue 2020; Blue Park 2020; Ramsar 2014), beyond containing the most developed reef in the Southwestern Atlantic (Loiola et al. 2019).

Analysis of the overlap of the geographic range of threatened species demonstrated that, if considering all MPAs, just over half of all species (N = 27, 51% of total species) could be considered well represented (percentage of coverage >10%; Fig. 12.4). Mean and median coverage by all MPAs were 21.8% and 9.7%, respectively. However, this was mainly due to the presence of several, large multiple-use MPAs that afforded high coverage to many species, especially fishes. Mean and median coverage by this MPA category was 15.3% and 7.2%, respectively. When we considered the degree of species coverage by no-take MPAs, we found that only five (9.6%) of threatened species were well represented (>10%). Moreover, none of the invertebrate species were considered well represented by no-take MPAs. Mean and median coverage by no-take MPAs were only 5.5% and 0.69%, respectively.

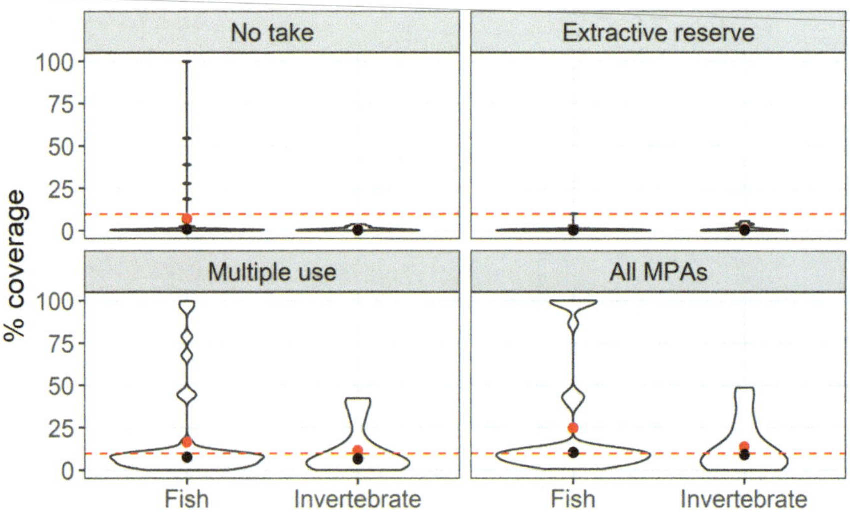

Fig. 12.4 Percentage coverage of threatened fish and invertebrate species by Brazilian MPAs, using four subsets of MPAs (no take, extractive reserves, multiple use and all MPAs). The red dot indicates the mean and the black dot the median

Using this 10% threshold, we found 47 gap species (92%), 52% of which (N = 25) that were considered gap species for all MPAs and no-take ones. Threatened species considered poorly represented by the existing MPA system included the fire coral *Millepora laboreli* (0.15% coverage by no-take and all MPAs), the sandbar shark *Carcharhinus plumbeus* (3.98% coverage by all MPAs and 0.33% by no-take MPAs), and the dusky grouper *Epinephelus marginatus* (6.79% coverage by all MPAs and 0.52% by no-take MPAs). Twenty threatened species that are also endemic to Brazilian waters fall well below the 10% critical coverage threshold by no-take MPAs (e.g., *Mussismilia braziliensis*, *Scarus trispinosus*). This coverage target has been specifically claimed to maintain the persistence of wide-ranging species (>250,000 km^2) (Venter et al. 2014) but has not proven to be enough to sustain small-ranged species (such as endemic ones) given their ecological requirements (Table 12.2).

12.3 Conservation Effectiveness of Marine Protected Areas

Although the number and coverage of MPAs have increased substantially, percentages of regions under formal protection are not reliable measures of conservation success (Roberts et al. 2018). MPA implementation lags in many developing countries, where conservation funding is particularly limited (Gill et al. 2017) and there is a strong competition between multiple users with conflicting interests (McClanahan et al. 2009). Understanding how effective and well-implemented MPAs are require information on comparative studies of the extent to which conservation initiatives improve ecological conditions (Lester et al. 2009) and the degree to which management practices follow scientific recommendations (Mills et al. 2020). MPAs that have adequate human and financial capacity and incorporate good practices into decision making processes can become effective in management terms because they are more likely to accomplish their objectives. This, in turn, can enhance ecosystem functioning, and increase the biomass and size of individual species (Bejarano et al. 2019). The ecological effectiveness of MPAs is an important end goal, with influence on the level of local support for conservation (Bennett et al. 2019). Ensuring that MPAs are effectively managed and associated with positive ecological impacts are thus important for social acceptance and improved decision making.

Evaluation of MPA coverage can be used not only to assess representation of reef biodiversity, but also, to assess the extent to which conservation features are in fact protected within a MPA system—i.e., their genuine contribution to habitats and species. For protection analyses, such assessments require information on the relative effectiveness of MPAs in achieving biodiversity outcomes. The conservation effectiveness of MPAs for protecting coral reefs in Brazil has been demonstrated by previous studies focusing on the ecological (e.g., Francini-Filho and Moura 2008) and management (e.g., Brandão et al. 2017) benefits of individual MPAs. Nevertheless, there has been no synthesis nor systematic assessment of the

Table 12.2 Gap species considered to have a very low coverage (<10%) of their distribution by all MPAs and no-take ones

Taxonomic group	Species name	Gap species —all MPAs	Gap species —no take	Endemic to Brazilian waters
Invertebrate	*Aliger costatus*	10.29	1.09	×
Invertebrate	*Condylactis gigantea*	13.58	2.25	×
Invertebrate	*Coscinasterias tenuispina*	9.15	1.26	×
Invertebrate	*Eustrombus goliath*	13.66	0.93	✓
Invertebrate	*Halichondria (Halichondria) cebimarensis*	0.25	0	✓
Invertebrate	*Halichondria (Halichondria) tenebrica*	30.16	0	✓
Invertebrate	*Latrunculia (Biannulata) janeirensis*	0.32	0.32	✓
Invertebrate	*Linckia guildingi*	4.8	0.38	×
Invertebrate	*Lytechinus variegatus*	4.29	0.35	×
Invertebrate	*Millepora laboreli*	0.15	0.15	✓
Invertebrate	*Mussismilia braziliensis*	23.98	2.69	✓
Invertebrate	*Mussismilia harttii*	48.6	3.72	✓
Invertebrate	*Oreaster reticulatus*	6.88	0.56	×
Invertebrate	*Petaloconchus myrakeenae*	0	0	✓
Invertebrate	*Synaptula secreta*	42.25	0	✓
Fish	*Carcharhinus perezi*	3.8	0.43	×
Fish	*Carcharhinus plumbeus*	3.98	0.33	×
Fish	*Cerdale fasciata*	43.29	0	✓
Fish	*Choranthias salmopunctatus*	100	0.27	✓
Fish	*Elacatinus figaro*	10.49	0.73	✓
Fish	*Epinephelus itajara*	5.11	0.42	×
Fish	*Epinephelus marginatus*	6.79	0.52	×
Fish	*Epinephelus morio*	8.96	0.65	×
Fish	*Ginglymostoma cirratum*	4.93	0.46	×
Fish	*Halichoeres rubrovirens*	86.49	7.49	✓
Fish	*Hippocampus erectus*	11.85	0.82	×
Fish	*Hippocampus patagonicus*	3.92	1.27	×
Fish	*Hippocampus reidi*	10.4	0.72	×
Fish	*Hyporthodus nigritus*	8.39	0.71	×
Fish	*Hyporthodus niveatus*	9.06	0.68	×
Fish	*Lutjanus cyanopterus*	5.23	0.43	×
Fish	*Lutjanus purpureus*	10.78	0.72	×
Fish	*Malacoctenus brunoi*	11.54	2.5	✓
Fish	*Manta birostris*	3.19	0.29	×
Fish	*Micrognathus erugatus*	98.99	2.31	✓
Fish	*Microspathodon chrysurus*	97.7	8.73	×
Fish	*Mobula tarapacana*	3.6	0.29	×
Fish	*Mycteroperca bonaci*	8.26	0.8	×
Fish	*Mycteroperca interstitialis*	7.3	0.61	×
Fish	*Negaprion brevirostris*	10.89	0.88	×
Fish	*Polyprion americanus*	0.5	0.18	×
Fish	*Scarus trispinosus*	11.64	0.96	✓
Fish	*Scarus zelindae*	5.83	0.48	✓
Fish	*Sparisoma axillare*	13.89	1.32	✓
Fish	*Sparisoma frondosum*	12.42	1.18	×
Fish	*Sparisoma rocha*	1.53	0.33	✓
Fish	*Stegastes trindadensis*	11.54	2.5	✓

Gray cells highlight values that were below the minimum base value of 10%. The checkmark symbol indicates that the gap species is endemic to Brazilian waters; the cross-mark symbol indicates that the gap species is not endemic to Brazilian waters conservation effectiveness of MPAs from different management categories across MPA systems (but see Oliveira Júnior et al. 2016, Ferreira et al. 2022). We gathered information on empirical estimates of both the management and ecological effectiveness of each MPA containing reefs in Brazil to examine how well these habitats are protected. Therefore, we measured progress made towards protecting Brazilian reefs in two ways. First, we evaluated MPA management effectiveness, which describes the level of implementation of MPAs based on indicators such as MPA regulations, administrative capacity, level of threats faced by MPAs, management plan suitability, etc. To empirically estimate the potential management effectiveness of MPAs, we derived a metric from the dataset held by the national agency for biodiversity conservation (http://samge.icmbio.gov.br/). Second, we evaluated MPA ecological effectiveness, which was measured by ecological studies comparing measures of biodiversity in MPAs and adjacent unprotected areas. To empirically estimate the potential ecological effectiveness of MPAs, we derive a metric of effectiveness based on a meta-analysis of MPA effects on biodiversity in Brazil (Ferreira et al. 2022). These effectiveness measures indicate the additional benefit of biodiversity conservation beyond unprotected areas.

To examine the extent to which MPAs are implemented, we gathered the most recent overall score for the management effectiveness of each MPA and normalised it to the reference range of zero to one (value of 0 represented the least well-managed MPA and value of 1 considered the most well managed MPA across all MPAs considered here). The management effectiveness scores reflect the perceptions of the MPA staff regarding the appropriateness of management resources and actions towards the achievement of the MPA conservation objectives. For those MPAs that had not undergone the evaluation (N = 22), we used an overall score representing the mean value for that management category across all Brazilian MPAs. We found that no-take MPAs had a mean effectiveness score of 0.61, extractive reserves had a mean effectiveness score of 0.51, and multiple-use MPAs had a mean effectiveness score of 0.44 (Fig. 12.5). Abrolhos Marine National Park had a very high score (0.8) and could therefore be considered one of the most effective in protecting coral reefs from the management perspective. Conversely, a multiple-use MPA on the Eastern continental shelf (i.e., Environmental Protection Area Trancoso/Caraiva) was considered the least effective, with a very low score (0.1).

To examine the ecological effectiveness of MPAs, we conducted a meta-analysis of scientific literature documenting and comparing the effects of no-take MPAs to partially protected areas such as extractive reserves and multiple-use MPAs, and partially protected areas to no protection (i.e., to fished areas). We used data that were collected by Ferreira et al. (2022), which collated 424 empirical observations from 19 studies for 18 MPAs along the Brazilian coast. We reanalysed their data by quantifying MPA-level overall effects on biodiversity. If no entry was found for a specific MPA, we used the mean effect size for that management category across all studied MPAs. Following the procedure for the management effectiveness score, we normalised the mean effect size for each MPA to the reference range of zero to one (value of 0 represented the least effective MPA in ecological terms and value of 1 considered the most ecologically-effective MPA across all MPAs considered here).

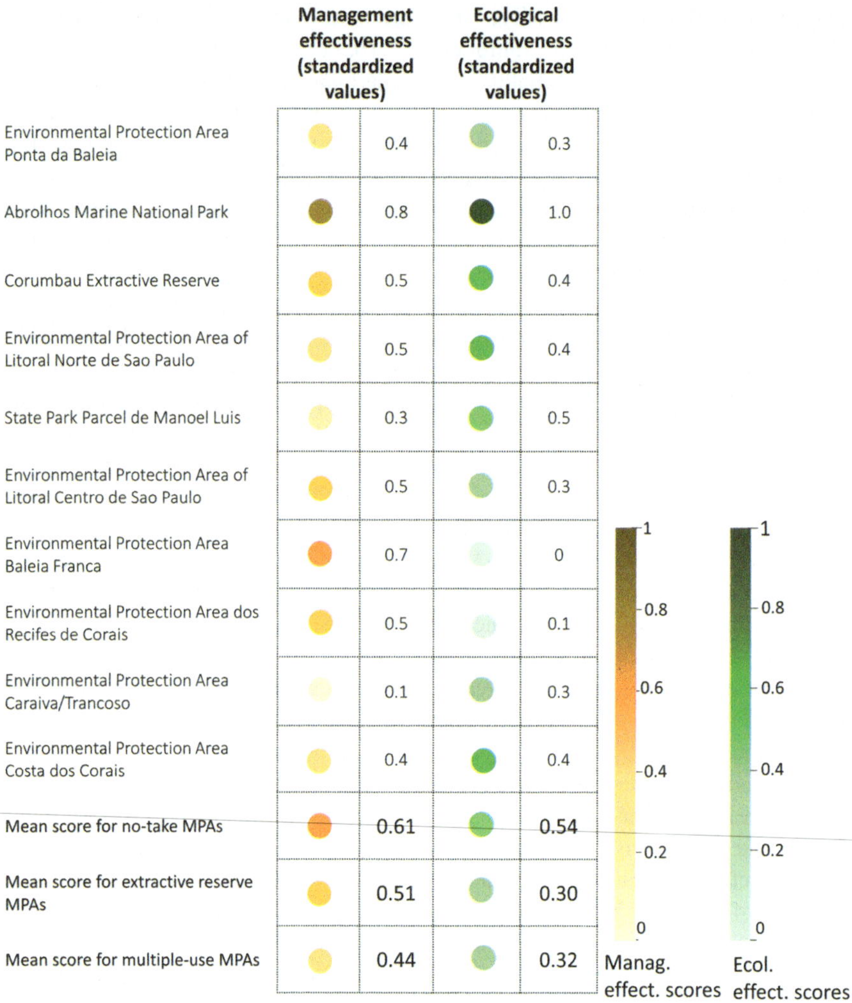

Fig. 12.5 Management and ecological effectiveness of MPAs in Brazil. Effectiveness values range from zero (MPA considered not effective) to 1 (MPA considered fully effective). The report results are the individual scores for the top 10 MPAs in terms of reef area in Brazil and the mean values across all MPAs within each management category

Similarly to the management effectiveness assessment, we found that no-take MPAs were effective with a mean score of 0.54 (Fig. 12.5). Multiple-use MPAs scored much lower, with an overall value of 0.32, and extractive reserves scored an average of 0.30. We found a lower score for extractive reserves, compared to multiple-use MPAs, but this value should be considered with caution given that all available biodiversity data was derived from only three MPAs belonging to this management category. As more information becomes available, further investigations of the ecological impacts of this type of MPA are required. The Abrolhos National Park was also considered the most effective in this assessment, reaching the highest score (i.e., 1.0), followed by State Park Parcel de Manoel Luis with an overall score of 0.5.

Using these MPA-specific effectiveness scores, we conducted two additional analyses of protection gaps considering the estimates of relative effectiveness of MPAs for protecting reef habitats in Brazil. We estimated the total amount of reef area protected by MPAs from a given management category as follows:

$$PL_{feature} = \sum_{i=j}^{k} CV_{feature} \times Ef_i$$

Where $PL_{feature}$ is the protection level of a given feature (i.e., reef habitat), i is the MPA within a specific management category (no take, extractive reserve, or multiple-use), $CV_{feature}$ is the coverage level in the corresponding MPA for the given feature, and Ef_i is the effectiveness score for the specific MPA and calculated in line with each of the two methods described above. By doing so, we quantified the total amount of reef habitats protected by all MPAs within each management category.

Overall protection of reef habitats within MPAs varied from 0.2% to 56% when management effectiveness was considered (mean: 12.5%; median: 4.2%; Fig. 12.6) and from 0.1% to 70.8% (mean: 10.8%; median: 4.28%; Fig. 12.7) when ecological effectiveness was included in the assessments. The protection level across reef types showed a similar pattern with MPA coverage analysis: more remote, offshore reefs were better protected than shallow-water reefs. Offshore banks on the Eastern shelf (Fig. 12.6a) and mesophotic reefs in oceanic ecoregions (Fig. 12.6b) were considered best protected (i.e., 66.1% and 67.8%, respectively), when incorporating MPA management effectiveness. Similarly, offshore banks in the Amazon ecoregion (Fig. 12.7a) and mesophotic reefs in oceanic ecoregions (Fig. 12.7b) were considered best protected (i.e., 48.4% and 38%, respectively), when incorporating MPA ecological effectiveness. Our results demonstrated that reefs habitats that were prevalent in multiple-use MPAs suffered the strongest reductions in the total amount of reef are protected in relation to the total coverage by MPAs. That was the case of offshore and nearshore banks on the Northeastern continental shelf, and fringing reefs and pinnacles on the Eastern continental shelf. The incorporation of MPA effectiveness into the assessments reduced the protection level by up to 75% for the management effectiveness and 94% for the ecological effectiveness. As expected, mesophotic reefs in the Amazon shelf (0.2–0.4%), patch reefs along the Northeastern shelf (0.8–3.8%), as well as along the Eastern coast (3.5–5.2%) remained with very low protection levels.

Finally, we calculated the total area of MPA coverage, area of no-take MPAs, and protection levels derived from both measures of conservation effectiveness as a percentage of total reef habitat within each ecoregion (Fig. 12.1b). While reefs at the Northeastern shelf had the largest total coverage by MPAs (~76%), the coverage level by no-take MPAs was extremely low (0.71%) in this ecoregion, and protection levels were somewhat lower (16–33%, depending on the effectiveness metric used). All values were uniformly low for the Amazon shelf ecoregion. Overall, the percentage of areas effectively protected in ecological terms was lower than their counterparts in each ecoregion.

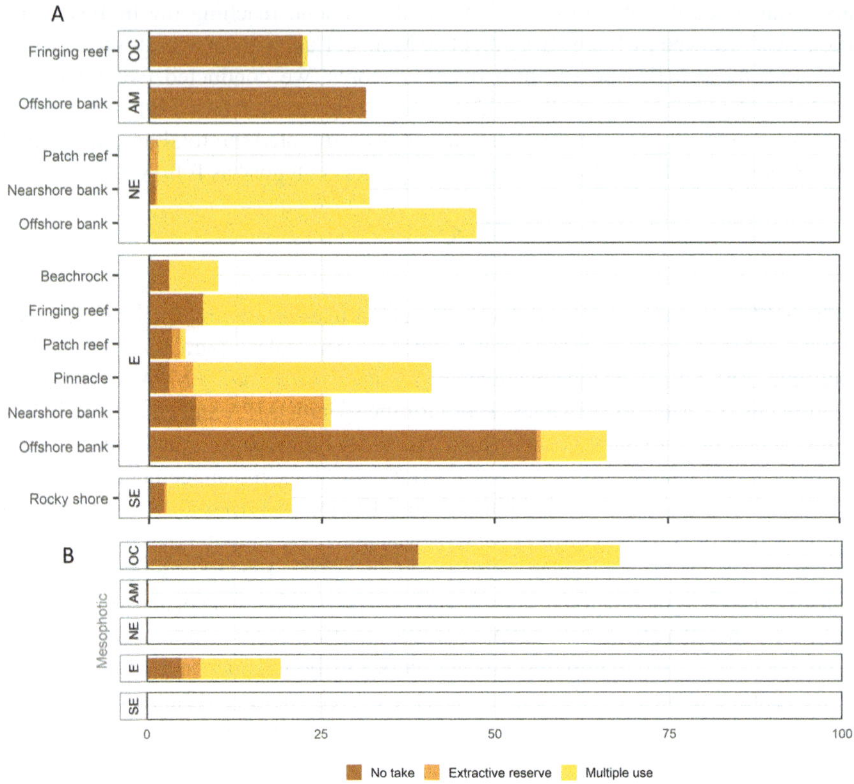

Fig. 12.6 Protection levels afforded by MPAs from different management categories considering the management effectiveness scores for shallow-water (**a**) and mesophotic (**b**) reefs

Identifying gaps in MPA protection and assessing MPA implementation are vital to developing systematic and effective protected area networks. The current extent, distribution, and effectiveness of MPAs for reef habitats fail to constitute a comprehensive national system of MPAs that is required to achieve legitimate and equitable representation of reef biodiversity. There is clear evidence from the scientific literature that multiple-use MPAs have fewer ecological benefits (Ferreira et al. 2022) and the results presented here thus are of concern regarding the long-run capacity of Brazilian MPAs to sustain reef biodiversity. In addition to a systematic expansion of existing MPAs in the future, the establishment of core no-take zones within multiple-use MPAs might be an adaptive-management approach to increase the effectiveness of the current system of MPAs. The conservation assessments presented here also suggested that the general link between good management performance and positive biodiversity outcomes of MPAs (Gill et al. 2017) might not always be obvious. This was because some MPAs scored high for ecological effectiveness assessment while scored low in assessments using management effectiveness (e.g., State Park Parcel Manoel Luis and see Cordeiro et al. 2021).

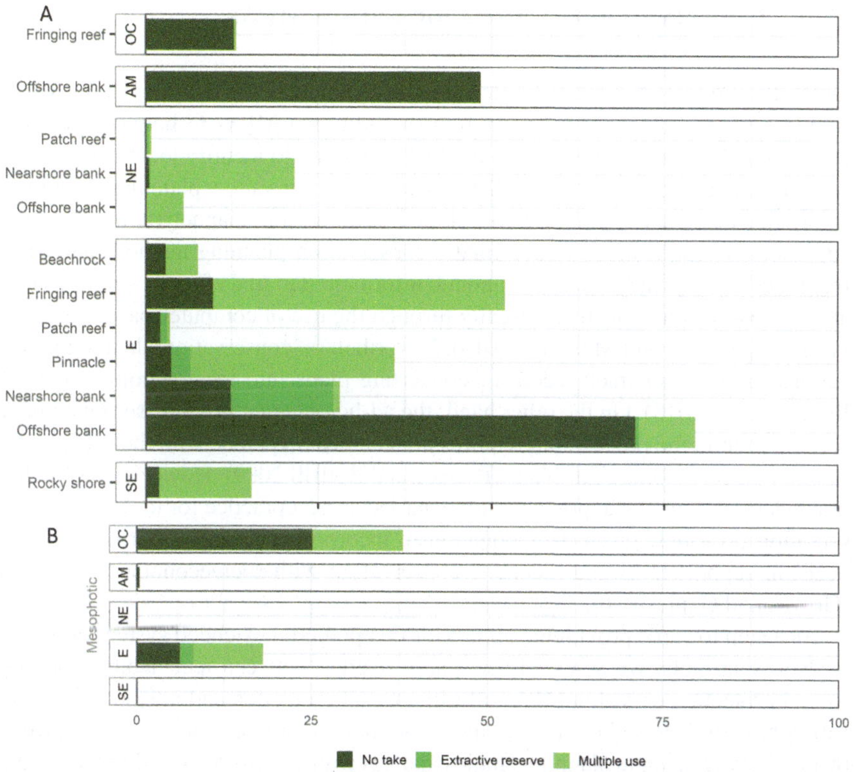

Fig. 12.7 Protection levels afforded by MPAs from different management categories considering the ecological effectiveness scores for shallow-water (**a**) and mesophotic (**b**) reefs

While there are several methods intended to measure MPA effectiveness (e.g., Pajaro et al. 2010; Pomeroy et al. 2005), most of these tools emphasise only the management process (Pyhälä et al. 2019), neglecting any assessment of the state of biodiversity. Although the framework presented here uses management and ecological measures of MPA success, the approach to incorporating management effectiveness rely upon qualitative or semi-qualitative judgments, which might provide inaccurate indications of management achievements. On the other side, the ecological effectiveness assessments were based on quantitative comparisons of biodiversity indicators between protected and unprotected sites. Meta-analytic approaches can be helpful in the evaluation of ecological effectiveness of MPAs in the absence of baseline and monitoring data over appropriate temporal scales, but more detailed and rigorous techniques will be needed to develop better estimates in the future.

12.4 Conservation Priorities for Future MPAs

Given the protection gaps presented here, the implementation of existing MPAs should be supplemented by designation of additional MPAs to help mitigate the continuing decline of reef biodiversity. MPA planning to identify locations for conservation actions can be undertaken following conservation planning principles (Margules and Pressey 2000) or developed in an ad-hoc manner (non-systematic approaches). Applications of systematic conservation planning methods and tools integrates ecological and socio-economic information to find efficient solutions that attain conservation objectives, usually through the use of computer-based software systems (Kukkala and Moilanen 2013). Nevertheless, conservation planning applications have a poor track record of converting plans into local actions (Álvarez-Romero et al. 2018). On the other hand, the ad-hoc manner might ease stakeholder participation in the process and requires no technical expertise with the use of software when developing draft plans (Pressey and Bottrill 2009). However, despite the drawbacks, systematic approaches are regarded as best practice for identifying conservation priorities, given their potential to incorporate ecological principles into decision support tools and reconcile conservation with socioeconomic objectives (Groves and Game 2016).

As a step towards providing support for this potential extended system of MPAs on Brazilian reefs, we assessed the spatial overlap between reef habitats lacking formal protection or within MPAs offering partial protection and integrated conservation priorities. Based on an overlay of the individual maps identifying priorities by previous systematic planning studies and government efforts, we identified areas of spatial coincidence (hereafter "integrated conservation priorities"). To date, only two systematic planning studies have provided a national scale analysis of where MPAs should be located (Magris et al. 2020; Vilar et al. 2020). Government efforts to develop marine plans intending to create new MPAs include: (i) the prioritisation process led by the Brazilian Ministry of the Environment, which uses a systematic planning approach; and (ii) proposed MPA boundaries identified through a non-systematic MPA planning process led by the national agency for biodiversity conservation (ICMBio) in collaboration with multiple stakeholders. We calculated the proportion of each reef habitat that falls within the boundaries of conservation priorities at each level of convergence (1 through 4; Fig. 12.8).

Of the four MPA planning exercises undertaken by previous studies and government efforts, priority areas covered most of the reef habitats present in Brazilian waters, but none of the reefs present in oceanic ecoregions were consistently identified as having priority for marine conservation because they are mostly within existing MPAs. While none of the reefs were identified by the four plans, small portions of six reef types were selected by three plans simultaneously. They were patch reefs and offshore banks on the Northeastern shelf; patch reefs, offshore banks, pinnacles, and mesophotic reefs on the Eastern shelf. Furthermore, mesophotic reefs in the Amazon ecoregion, offshore banks on the Northeastern shelf, and pinnacles on the Eastern shelf were entirely, or almost entirely, covered by at least one plan.

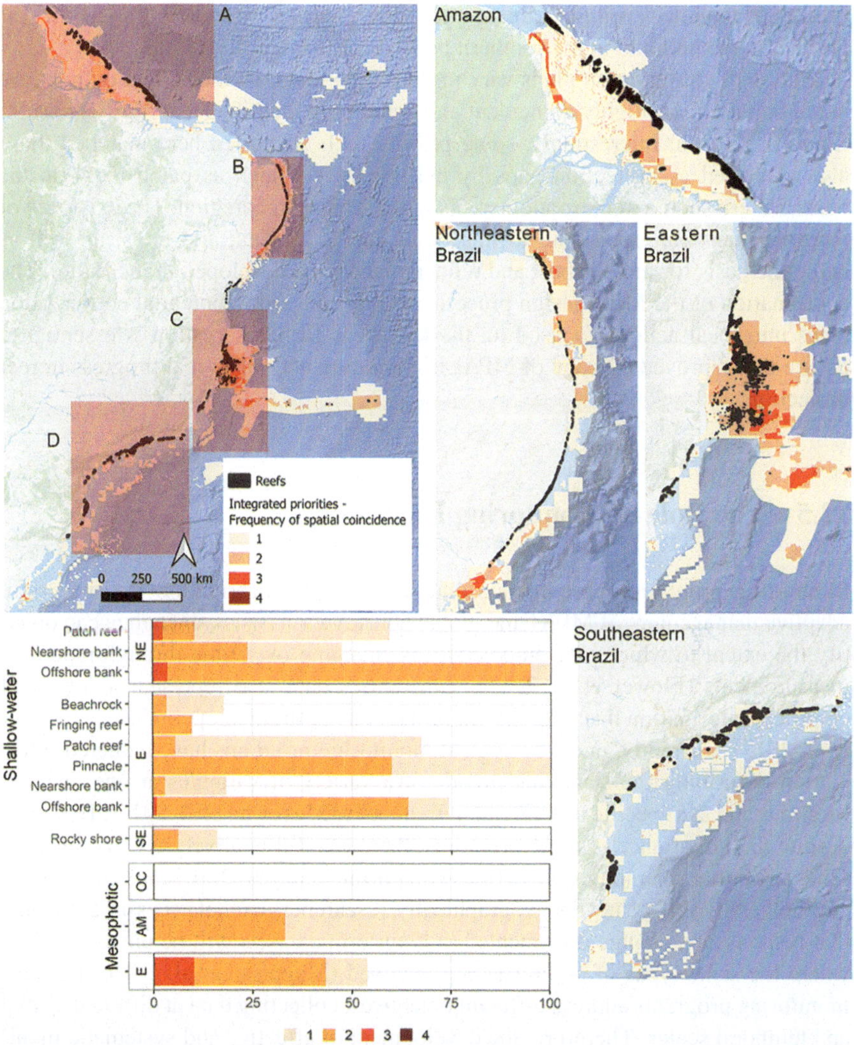

Fig. 12.8 Spatial coincidence between priority areas identified by MPA planning exercises and reef habitats in Brazil. For the frequency analysis, darker colours represent areas identified as priorities in more conservation plans. Panels on the right depict zoomed views of ecoregions with reef habitats as priorities by at least one plan (A: Amazon; B: Northeastern Brazil; C: Eastern Brazil; D: Southeastern Brazil). Bar plot on the left indicates the percentage of reef habitat included in each class of frequency of spatial coincidence across plans, from one (light orange) to 4 (dark red). Reef habitats that had no overlap with any single plan were omitted in the bar plot. Ecoregions defined according to Spalding et al. (2007). See Chap. 1 for an adapted ecoregion classification

Overall, integrated priority areas were concentrated in coastal areas of the Northeastern shelf and in the southern portions of the Abrolhos Bank.

Although conservation plans vary in their input data, methods, and outputs, we noted that reefs are highly coincident and commonly identified as priority areas for future protection. Interestingly, some priority reefs identified here matched those also identified as protection gaps by existing MPAs such as patch reefs on the Northeastern shelf and mesophotic reefs in the Amazon ecoregion. The existence of multiple plans can contribute to improving methods, data, and tools, as well as being the basis for more robust and refined plans to be developed in the future. The coordination of the MPA design process towards converting potential conservation areas into local actions focused on the key areas for conservation is essential to ensure the improved efficacy of MPAs in the future and make real progress in reef protection.

12.5 The Role of Monitoring Programs

Monitoring programs represent a valuable source of information for successful adaptive management of MPAs and global conservation issues since they can quantify the extent to which reef biodiversity is changing over time through repetitive measurements (Flower et al. 2017). Time series data also represent a source of empirical information that help inform projections of future reef conditions (Flower et al. 2017). Currently, many ecological monitoring programs have gathered information describing the states and trends of a variety of variables at one or more localities and at a specific measurement frequency (Magurran et al. 2010). However, despite the importance of these programs, there are still relatively few biodiversity time-series that span long time periods (Magurran et al. 2010). Because the drivers of biodiversity shifts can vary geographically, researchers cannot necessarily assume that responses to change documented in long-term datasets will be universal, even where the same types of organisms are involved (Magurran et al. 2010). Besides, monitoring programs address different objectives, collecting data at different spatial and temporal scales. Therefore, the development of effective and systematic monitoring frameworks is critical, accounting for the capacity and governance of each country (Pereira et al. 2013).

Monitoring programs are particularly useful to ensure successful MPA management and track progress towards the achievement of their conservation goals (Mills et al. 2020). However, some scientists have voiced their concerns about the scientific credibility of MPA monitoring mainly because of their inappropriate design and statistical analysis techniques. In fact, MPA literature contains very little discussion about good monitoring approaches, particularly when compared to other fields in marine ecology. According to Lindenmayer and Likens (2010), a "good monitoring program" contains four key characteristics, considered essential to make them effective: (1) Key questions posed, which is associated with the use of appropriate and relevant biological indicators in relation to the conservation issues and the monitoring program hypotheses; (2) Good design, which requires to include both spatial and temporal variation of the features of interest and the appropriate choice of data collection methods; (3) Data presentation, analysis and data storage, which must be oriented to address the key questions posed for the monitoring whereas following global standard patterns (e.g., Darwin Core); and (4) Reporting of results, which must follow the appropriate style depending on the targeted audience (e.g., scientists, managers, and members of the public).

In Brazil, long-term monitoring programs on coral reefs started in 1998 in the Abrolhos Marine National Park, and since then, other 12 monitoring programs have been implemented on the Brazilian coast. According to a compilation made from reports and scientific workshops to describe the strengths and weaknesses of ecological monitoring programs in general, we observed that existing programs focusing on Brazilian reefs cover different lengths of time. Most of the studies focused on the seasonality of changes (44%, N = 70), whereas studies tracking interannual variability (i.e., between 2 and 4 years) represented only 28%, and those considering long-term variability (i.e., spanning longer than 5 years) represented 29% of the total (Fig. 12.9a). Moreover, we observed that most programs focused on benthic and coral communities, followed by reef fishes, and sea urchins, respectively. We also identified that the most frequently used metrics used as indicators of ecosystem change were cover benthic community (e.g., sponges, algae, and invertebrates), for interannual and long-term programs, and fish abundance and biomass, specifically for those programs focused on seasonal variability. The use of genetic metrics were very few and recorded only in seasonal programs (Fig. 12.9b).

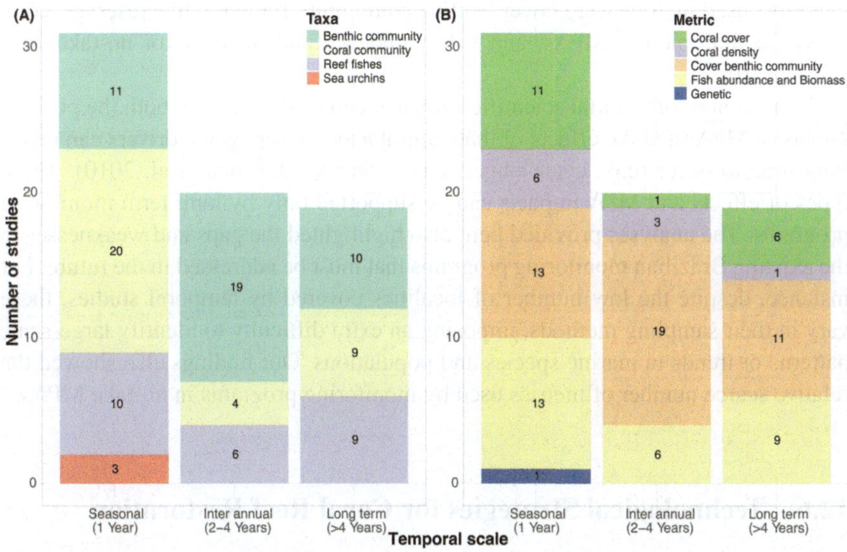

Fig. 12.9 Number of studies performing ecological monitoring of reef ecosystems over different temporal scales and considering different (**a**) taxonomic groups; and (**b**) metric used. The number represented the percentual from the total number the studies

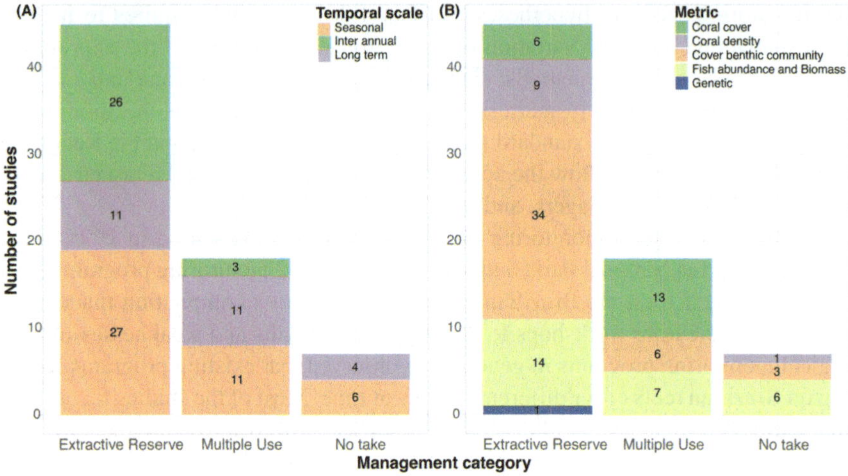

Fig. 12.10 Number of studies performing ecological monitoring of reef ecosystems within marine protected areas (classified according to their management categories), depending on their temporal scales (**a**) and metric used (**b**). The number represents the percentual from the total number the studies observed in each category

In addition, we observed that most of the monitoring programs collected data within extractive reserves (N = 35, 64%), followed by multiple-use MPAs (26%; N = 18), and no-take MPAs (10%; N = 7; Fig. 12.10a). The few studies focused on no-take MPAs were mostly focused on seasonal variability. Additionally, we found that most used metrics were cover benthic community for extractive reserves, coral cover for multiple-use MPAs, and fish abundance and biomass for no-take ones (Fig. 12.10).

There is now substantial scientific evidence demonstrating that both the positive effects of MPAs and the effects of that natural and anthropogenic drivers can take a long time to occur (e.g., even longer than a decade; Babcock et al. 2010). These types of effects and MPA impacts can be supported only by long-term monitoring programs. The analyses provided here also highlighted the gaps and weaknesses in the existing Brazilian monitoring programs that must be addressed in the future. For instance, despite the low number of localities covered by temporal studies, these vary in their sampling methods, imposing an extra difficulty to identify large-scale patterns or trends in marine species and populations. Our findings also showed the relative scarce number of metrics used by monitoring programs in no-take MPAs.

12.6 Technological Strategies for Coral Reef Restoration

Although *in situ* conservation practices such as MPAs help to reduce local impacts, global drivers of impact such as ocean warming and acidification do not respect legal boundaries. If not contained, such impacts will continue to cause population declines at ever-increasing rates. Large-scale, mass mortalities without recovery lead to declines in the genetic diversity of coral populations and increases the likelihood of species extinction (Laikre et al. 2010). In the face of a global climate crisis, researchers worldwide have been looking for strategies that can somehow guarantee the survival and assist the recovery of damaged reefs (Boström-Einarsson et al. 2020).

Over the last years, several methods have been developed and applied with varying levels of success. Here, the term restoration describes a suite of interventions aimed at improving reef structure, ecosystem function and increasing populations of key species in areas that have been damaged (see Hein et al. 2020). Some reef restoration strategies were initially adapted from methods used in the terrestrial ecosystems, as the concept of transplantation and coral gardening developed in the 1990s (Rinkevich 1995). When using the direct transplantation strategy, coral colonies or fragments are planted on the reef without an intermediate nursery phase. On the other hand, gardening strategies use fragmented corals, grown in nurseries, which are planted back on the reef. These nurseries can be *in situ* (in the ocean) or *ex situ* (kept in aquaria).

Another strategy that has been applied together with coral gardening and transplantation is the so-called microfragmentation (Page et al. 2018). The idea of microfragmentation is to break a colony into small pieces, with less than 10 polyps each. It accelerates the growth rate of corals by up to 50 times when compared to non-fragmented colonies. While this research is still incipient in Brazil (only one institution currently develops studies in this topic—UFPE), several institutions around the world have practiced microfragmentation such as: SECORE (Mexico), Bonaire Reef Renewal Foundation (Caribbean), Coral Restoration Foundation (United States), Ocean Quest (Thailand), Tropical Research and Conservation Center (Malaysia), and Gili Eco Trust (Indonesia). These strategies were initially applied to branching species such as *Acropora cervicornis* (staghorn coral) and *Acropora palmata* (elkhorn coral), and are showing some success also in massive corals such as *Orbicella faveolata* (mountainous star coral) and *Montastraea cavernosa* (great star coral). Whether keeping the fragmented colonies in laboratory, where they have a controlled environment, or directly in the ocean, these techniques enable mass production of corals. However, since they are clones, fragmentation does not introduce individuals genetically more resistant to climate change. As a result, these techniques alone are extremely fragile for conservation, since climatic events can continue to decimate reefs over large spatial scales. However, they are alternatives to quickly recover degraded reef environments, helping to save a little more time in the battle against extinction.

The term 'assisted evolution' refers to a range of approaches that involve active intervention to accelerate the rate of naturally occurring evolutionary processes. These approaches aim to enhance the resilience of corals to predicted future ocean scenarios of high temperature and acidification by improving certain attributes such as temperature tolerance and growth. Among the assisted evolution processes are stress conditioning, assisted gene flow, hybridization, modifications of algal symbiont communities and other processes that include manipulations of microbes such as bacteria and studies of gene function employing novel techniques like RNAi and CRISPR/Cas9 gene editing using transfection and microinjection (van Oppen et al. 2015).

Cryopreservation is another strategy that has been explored by scientists. Cryopreserving means maintaining viable living systems (cells, tissues) stored at low temperatures. These temperatures are so low (usually at −196 °C) that paralyze any biological activity, including aging, degradation of genetic material (DNA) and cell death. It is considered one of the most effective tools in the conservation of endangered wild species. Only three research groups in the world have managed to develop protocols for cryopreservation of coral gametes. One of them is coordinated by Dr. Mary Hagedorn (Smithsonian Institution, Hawaii) and the other by Dr. Sujune Tsai (Mingdao University, Taiwan). Currently, these protocols are successfully used to freeze sperm from 31 species, three of which are located in Hawaii, three in the Caribbean, four in French Polynesia and 21 in the Great Barrier Reef (Hagedorn et al. 2019). In addition, methods for freezing eggs and coral larvae are also being tested. The Coral Vivo Research Network is the first and only one in the entire South Atlantic Ocean to test freezing protocols for gametes of two endemic brain coral species (*Mussismilia harttii* and *Mussismilia hispida*). This technology allows gametes to be frozen and stored indefinitely and, when desired, this material can be thawed and used to give life to new corals, which in the future may assist in the restoration of degraded reefs. The assisted evolution and cryopreservation strategies have been applied to species with sexual reproduction, especially broadcast-spawning corals, thus allowing the use of assisted fertilization to generate new corals as well was to achieve restoration goals.

On-land (*ex situ*) solutions aim to preserve corals outside their natural habitat, with the benefit that such maintained animals are not subject to marine heatwaves or other natural events that might destroy them (see Zoccola et al. 2020). *Ex situ* culturing builds on the conservation work already carried out by aquaria and zoos for endangered populations of vertebrates and invertebrates. This approach employs a network of aquaria with the benefits of (1) conserving coral species, (2) generating coral material without depleting wild stocks for research into areas such as assisted evolution, and (3) providing a multispecies test bed for the performance and interaction of manipulated corals prior to *ex situ* culturing and field testing.

Coral reef restoration is not designed to reduce climate impacts, but rather, it is intended as a complementary tool to support natural recovery following disturbance in high-value areas. Given the many uncertainties associated with different climate scenarios, the key challenge is to design coral restoration efforts such that the realities of climate change are embedded in the choice of goals and methods. It is not an 'either or' situation; multiple actions need to be implemented concurrently as part of a larger integrated management strategy in order to provide coral reefs with the greatest hope for the future.

12.7 Final Remaks: Future Challenges

Although we have outlined some important aspects of reef conservation and management, emerging frontiers that have potential to improve sustainability of Brazilian reefs include: (i) developing management tools for off-MPA conservation (e.g., recovery tools, fisheries and water quality policies); (ii) improving knowledge on the relative effectiveness of different management actions to protecting reef biodiversity; (iii) better integrating ecological modelling and social variables into MPA design; (iv) addressing impediments to the adoption of strategic planning in real-world contexts; and (v) running long-term monitoring programs that follow best-practice principles and are aligned to meet conservation needs. Brazilian reefs are in dire need of effective protection and other conservation strategies, and thus we need to adequately secure this staggering biodiversity.

References

Álvarez-Romero JG, Mills M, Adams VM, Gurney GG, Pressey RL, Weeks R, Storlie CJ (2018) Research advances and gaps in marine planning: towards a global database in systematic conservation planning. Biol Conserv 227:369–382. https://doi.org/10.1016/j.biocon.2018.06.027

Anthony K, Bay LK, Costanza R, Firn J, Gunn J, Harrison P, Walshe T (2017) New interventions are needed to save coral reefs. Nat Ecol Evol 1:1420–1422. https://doi.org/10.1038/s41559-017-0313-5

Araújo ME, Mattos FMG, Melo FPL, Chaves LCT, Feitosa V, Lippi DL, Félix Ha FC, Ha CW, Nunes JLS, Leão ZMAN, Kikuchi RKP, Ferreira Junior AV, Pereira PHC, Macedo HR, Sampaio CLS, Feitosa JLL (2020) Diversity patterns of reef fish along the Brazilian tropical coast. Mar Environ Res 160:105038. https://doi.org/10.1016/j.marenvres.2020.105038

Aued AW, Smith F, Quimbayo JP, Cândido DV, Longo GO, Ferreira CEL, Witman JD, Floeter SR, Segal B (2018) Large-scale patterns of benthic marine communities in the Brazilian province. PLoS One 13:1–15. https://doi.org/10.1371/journal.pone.0198452

Babcock RC, Shears NT, Alcala AC, Barrett NS, Edgar GJ, Lafferty KD, McClanahan TR, Russ GR (2010) Decadal trends in marine reserves reveal differential rates of change in direct and indirect effects. Proc Natl Acad Sci USA 107:18256–18261

Barr LM, Pressey RL, Fuller RA, Segan DB, McDonald-Madden E, Possingham HP (2011) A new way to measure the world's protected area coverage. PLoS One 6(9):e24707. https://doi.org/10.1371/journal.pone.0024707

Bejarano S, Pardede S, Campbell SJ, Hoey AS, Ferse SCA (2019) Herbivorous fish rise as a destructive fishing practice falls in an Indonesian marine national park. Ecol Appl 29(8):e01981. https://doi.org/10.1002/eap.1981

Bellwood DR, Hoey AS, Hughes TP (2012) Human activity selectively impacts the ecosystem roles of parrotfishes on coral reefs. Proc R Soc B Biol Sci 279(1733):1621–1629. https://doi.org/10.1098/rspb.2011.1906

Bennett NJ, Di Franco A, Calò A, Nethery E, Niccolini F, Milazzo M, Guidetti P (2019) Local support for conservation is associated with perceptions of good governance, social impacts, and ecological effectiveness. Conserv Lett 12:1–10. https://doi.org/10.1111/conl.12640

Blue Park (2020) Marine Conservation Institute. Retrieved December (2020) from https://marine--conservation.org/blueparks

Boström-Einarsson L, Babcock RC, Bayraktarov E, Ceccarelli D, Cook N, Ferse SCA, Hancock B, Harrison P, Hein M, Shaver E, Smith A, Suggett D, Stewart-Sinclair PJ, Vardi T, McLeod IM (2020) Coral restoration – a systematic review of current methods, successes, failures and future directions. PLoS One 15(1):e0226631. https://doi.org/10.1371/journal.pone.0226631

Brandão CDS, Malta A, Schiavetti A (2017) Temporal assessment of the management effectiveness of reef environments: the role of marine protected areas in Brazil. Ocean Coast Manag 142:111–121. https://doi.org/10.1016/j.ocecoaman.2017.03.015

Brooks CM, Chown SL, Douglass LL, Raymond BP, Shaw JD, Sylvester ZT, Torrens CL (2020) Progress towards a representative network of Southern Ocean protected areas. PLoS One 15(4):e0231361. https://doi.org/10.1371/journal.pone.0231361

Claudet J, Bopp L, Cheung WWL, Devillers R, Escobar-Briones E, Haugan P, Gaill F (2020) A roadmap for using the UN decade of ocean science for sustainable development in support of science, policy, and action. One Earth 2:34–42. https://doi.org/10.1016/j.oneear.2019.10.012

Cordeiro CAMM, Quimbayo JP, Nunes JACC, Nunes LT, Sissini MN, Sampaio CLS, Morais RA, Horta PA, Aued AW, Carraro JL, Hajdu E, Rocha LA, Segal B, Floeter RS (2021) Conservation status of the southernmost reef of the amazon reef system: the Parcel de Manuel Luís. Coral Reefs 40:165–185. https://doi.org/10.1007/s00338-020-02026

Devillers R, Pressey RL, Grech A, Kittinger JN, Edgar GJ, Ward T, Watson R (2015) Reinventing residual reserves in the sea: are we favouring ease of establishment over need for protection? Aquat Conserv Mar Freshwat Ecosyst 25:480–504. https://doi.org/10.1002/aqc.2445

Di Lorenzo M, Guidetti P, Di Franco A, Calò A, Claudet J (2020) Assessing spillover from marine protected areas and its drivers: a meta-analytical approach. Fish Fish 21:1–10. https://doi.org/10.1111/faf.12469

Duarte CM, Agusti S, Barbier E, Britten GL, Castilla JC, Gattuso JP, Worm B (2020) Rebuilding marine life. Nature 580(7801):39–51. https://doi.org/10.1038/s41586-020-2146-7

Ferreira HM, Magris RA, Floeter SR, Ferreira CEL (2022) Drivers of ecological effectiveness of marine protected areas: a meta-analytic approach from the Southwestern Atlantic Ocean (Brazil). J Environ Manag 301:113889

Flower J, Ortiz JC, Chollett I, Abdullah S, Castro-Sanguino C, Hock K, Lam V, Mumby PJ (2017) Interpreting coral reef monitoring data: a guide for improved management decisions. Ecol Indic 72:848–869

Francini-Filho RB, Moura RL (2008) Dynamics of fish assemblages on coral reefs subjected to differentmanagement regimes in the Abrolhos Bank, eastern Brazil. Aquat Conserv Mar Freshwat Ecosyst 18:1166–1179

Francini-Filho RB, Coni EOC, Meirelles PM, Amado-Filho GM, Thompson FL, Pereira-Filho GH, Moura RL (2013) Dynamics of coral reef benthic assemblages of the Abrolhos Bank, eastern Brazil: inferences on natural and anthropogenic drivers. PLoS One 8:1–12. https://doi.org/10.1371/journal.pone.0054260

Giakoumi S, Scianna C, Plass-Johnson J, Micheli F, Grorud-Colvert K, Thiriet P, Guidetti P (2017) Ecological effects of full and partial protection in the crowded Mediterranean Sea: a regional meta-analysis. Sci Rep 7:1–12. https://doi.org/10.1038/s41598-017-08850-w

Gill DA, Mascia MB, Ahmadia GN, Glew L, Lester SE, Barnes M, Fox HE (2017) Capacity shortfalls hinder the performance of marine protected areas globally. Nature 543:665–669. https://doi.org/10.1038/nature21708

Groves CR, Game ET (2016) Conservation planning: informed decisions for a healthier planet. Roberts and Company Publishers, Colorado

Hagedorn M, Spindler R, Daly J (2019) Cryopreservation as a tool for reef restoration: 2019. In: Reproductive sciences in animal conservation. Springer, Cham, pp 489–505. https://doi.org/10.1007/978-3-030-23633-5_16

Hein MY, McLeod IM, Shaver EC, Vardi T, Pioch S, Boström-Einarsson L, Ahmed M, Grimsditch G (2020) Coral reef restoration as a strategy to improve ecosystem services – a guide to coral restoration methods. United Nations Environment Program, Nairobi

Hughes TP, Graham NAJ, Jackson JBC, Mumby PJ, Steneck RS (2010) Rising to the challenge of sustaining coral reef resilience. Trends Ecol Evol 25:633–642. https://doi.org/10.1016/j.tree.2010.07.011

Hughes TP, Barnes ML, Bellwood DR, Cinner JE, Cumming GS, Jackson JBC, Scheffer M (2017) Coral reefs in the Anthropocene. Nature 546:82–90. https://doi.org/10.1038/nature22901

ICMBio (2020) National plan for the conservation of Brazilian reefs (WWW document). https://www.icmbio.gov.br/portal/faunabrasileira/plano-de-acao-nacional-lista/3620-plano-de-acao-nacional-para-conservacao-dos-recifes-de-corais. Accessed 10 Dec 2020

Jantke K, Jones KR, Allan JR, Chauvenet ALM, Watson JEM, Possingham HP (2018) Poor ecological representation by an expensive reserve system: evaluating 35 years of marine protected area expansion. Conserv Lett 11:1–8. https://doi.org/10.1111/conl.12584

Kukkala AS, Moilanen A (2013) Core concepts of spatial prioritisation in systematic conservation planning. Biol Rev 88:443–464. https://doi.org/10.1111/brv.12008

Laikre L, Allendorf FW, Aroner LC, Baker CS, Gregovich DP, Hansen MM, Jackson JA, Kendall KC, McKelvey K, Neel MC, Olivieri I, Ryman N, Schwartz MK, Bull RS, Stetz JB, Tallmon DA, Taylor BL, Vojta CD, Waller DM, Waples RS (2010) Neglect of genetic diversity in implementation of the convention on biological diversity. Conserv Biol 24:86–88. https://doi.org/10.1111/j.1523-1739.2009.01425.x

Leão ZMAN, Kikuchi RKP (2005) A relic coral fauna threatened by global changes and human activities, Eastern Brazil. Mar Pollut Bull 51:599–611. https://doi.org/10.1016/j.marpolbul.2005.04.024

Leão ZMAN, Kikuchi RKP, Testa V (2003) Corals and coral reefs of Brazil. In: Latin American coral reefs. Elsevier, Amsterdam, pp 9–52. https://doi.org/10.1016/B978-044451388-5/50003-5

Lester SE, Halpern BS, Grorud-Colvert K, Lubchenco J, Ruttenberg BI, Gaines SD, Warner RR (2009) Biological effects within no-take marine reserves: a global synthesis. Mar Ecol Prog Ser 384:33–46. https://doi.org/10.3354/meps08029

Lindenmayer DB, Likens GE (2010) The science and application of ecological monitoring. Biol Conserv 143:1317–1328

Loiola M, Cruz ICS, Lisboa DS, Mariano-Neto E, Leão ZMAN, Oliveira MDM, Kikuchi RKP (2019) Structure of marginal coral reef assemblages under different turbidity regime. Mar Environ Res 147:138–148. https://doi.org/10.1016/j.marenvres.2019.03.013

Magris RA, Pressey RL (2018) Marine protected areas: just for show? Science 360(6390):723–724. https://doi.org/10.1126/science.aat6215

Magris RA, Pressey RL, Mills M, Vila-Nova DA, Floeter S (2017) Integrated conservation planning for coral reefs: designing conservation zones for multiple conservation objectives in spatial prioritisation. Global Ecol Conserv 11:53–68. https://doi.org/10.1016/j.gecco.2017.05.002

Magris RA, Grech A, Pressey RL (2018) Cumulative human impacts on coral reefs: assessing risk and management implications for Brazilian coral reefs. Diversity 10:26. https://doi.org/10.3390/d10020026

Magris RA, Costa MDP, Ferreira CEL, Vilar CC, Joyeux JC, Creed JC, Floeter SR (2020) A blueprint for securing Brazil's marine biodiversity and supporting the achievement of global conservation goals. Divers Distrib 27:198–215. https://doi.org/10.1111/ddi.13183

Magurran AE, Baillie SR, Buckland ST, Dick JMP, Elston DA, Scott EM, Smith RI, Somerfield PJ, Watt AD (2010) Long-term datasets in biodiversity research and monitoring: assessing change in ecological communities through time. Trends Ecol Evol 25:574–582

Margules CR, Pressey RL (2000) Systematic conservation planning. Nature 405:243–253. https://doi.org/10.1038/35012251

Marinesque S, Kaplan DM, Rodwell LD (2012) Global implementation of marine protected areas: is the developing world being left behind? Mar Policy 36:727–737. https://doi.org/10.1016/j.marpol.2011.10.010

McClanahan TR, Cinner JE, Graham NAJ, Daw TM, Maina J, Stead SM, Polunin NVC (2009) Identifying reefs of hope and hopeful actions: contextualizing environmental, ecological, and social parameters to respond effectively to climate change. Conserv Biol 23:662–671. https://doi.org/10.1111/j.1523-1739.2008.01154.x

Mills M, Magris RA, Fuentes MMPB, Bonaldo R, Herbst DF, Lima MCS, Freitas RR (2020) Opportunities to close the gap between science and practice for marine protected areas in Brazil. Perspect Ecol Conserv 18:161–168. https://doi.org/10.1016/j.pecon.2020.05.002

Mission Blue (2020) Hope Spots. Mission Blue. Retrieved December 2020 from https://mission--blue.org/hope-spots/

Mora C, Andréfouët S, Costello MJ, Kranenburg C, Rollo A, Veron J, Myers RA (2006) Coral reefs and the global network of marine protected areas. Science 312:1750–1751. https://doi.org/10.1126/science.1125295

Mouillot D, Parravicini V, Bellwood DR, Leprieur F, Huang D, Cowman PF, Guilhaumon F (2016) Global marine protected areas do not secure the evolutionary history of tropical corals and fishes. Nat Commun 7:1–8. https://doi.org/10.1038/ncomms10359

Oliveira Júnior JGC, Ladle RJ, Correia R, Batista VS (2016) Measuring what matters – identifying indicators of success for Brazilian marine protected areas. Mar Policy 74:91–98. https://doi.org/10.1016/j.marpol.2016.09.018

Page CA, Muller EM, Vaughan DE (2018) Microfragmenting for the successful restoration of slow growing massive corals. Ecol Eng 123:86–94. https://doi.org/10.1016/j.ecoleng.2018.08.017

Pajaro MG, Mulrennan ME, Alder J, Vincent ACJ (2010) Developing MPA effectiveness indicators: comparison within and across stakeholder groups and communities. Coast Manag 38:122–143. https://doi.org/10.1080/08920751003633094

Pereira HM, Bruford MW, Brummitt N, Butchart SHM, Cardoso AC, Coops NC, Dulloo E (2013) Essential biodiversity variables. Science 339:277–278

Pomeroy RS, Watson LM, Parks JE, Cid GA (2005) How is your MPA doing? A methodology for evaluating the management effectiveness of marine protected areas. Ocean Coast Manag 48:485–502. https://doi.org/10.1016/j.ocecoaman.2005.05.004

Pressey RL, Bottrill MC (2009) Approaches to landscape- and seascape-scale conservation planning: convergence, contrasts and challenges. Oryx 43:464–475. https://doi.org/10.1017/S0030605309990500

Pyhälä A, Eklund J, McBride MF, Rakotoarijaona MA, Cabeza M (2019) Managers' perceptions of protected area outcomes in Madagascar highlight the need for species monitoring and knowledge transfer. Conserv Sci Pract 1:1–9. https://doi.org/10.1111/csp2.6

Ramsar (2014) Ramsar Sites Information Services. Abrolhos Marine National Park (WWW document). https://rsis.ramsar.org/ris/1902. Acessed Dec 2020

Rees SE, Foster NL, Langmead O, Pittman S, Johnson DE (2018) Defining the qualitative elements of Aichi Biodiversity Target 11 with regard to the marine and coastal environment in order to strengthen global efforts for marine biodiversity conservation outlined in the United Nations Sustainable Development Goal 14. Mar Policy 93:241–250. https://doi.org/10.1016/j.marpol.2017.05.016

Rinkevich B (1995) Restoration strategies for coral reefs damaged by recreational activities: the use of sexual and asexual recruits. Restor Ecol 3:241–251. https://doi.org/10.1111/j.1526-100X.1995.tb00091.x

Roberts KE, Valkan RS, Cook CN (2018) Measuring progress in marine protection: a new set of metrics to evaluate the strength of marine protected area networks. Biol Conserv 219:20–27. https://doi.org/10.1016/j.biocon.2018.01.004

Sciberras M, Jenkins SR, Mant R, Kaiser MJ, Hawkins SJ, Pullin AS (2015) Evaluating the relative conservation value of fully and partially protected marine areas. Fish Fish 16:58–77. https://doi.org/10.1111/faf.12044

Soares MO, Rossi S, Gurgela AR, Lucasa CC, Tavaresa TCL, Diniza B, Feitosa CV, Rabelod EF, Pereira PHP, Kikuchi RKP, Leao ZMAN, Cruz ICS, Carneiro PBN, Alvarez-Filip L (2021) Impacts of a changing environment on marginal coral reefs in the Tropical Southwestern Atlantic. Ocean Coast Manag 210:105692. https://doi.org/10.1016/j.ocecoaman.2021.105692

Spalding MD, Fox HE, Allen GR et al (2007) Marine ecoregions of the world: a bioregionalization of coastal and shelf areas. Bioscience 57:573–583. https://doi.org/10.1641/B570707

van Oppen MJ, Oliver JK, Putnam HM, Gates RD (2015) Building coral reef resilience through assisted evolution. Proc Natl Acad Sci 112:2307–2313. https://doi.org/10.1073/pnas.1422301112

Venter O, Fuller RA, Segan DB, Carwardine J, Brooks T, Butchart SHM, Watson JEM (2014) Targeting global protected area expansion for imperiled biodiversity. PLoS Biol 12:e1001891. https://doi.org/10.1371/journal.pbio.1001891

Vilar CC, Magris RA, Loyola R, Joyeux JC (2020) Strengthening the synergies among global biodiversity targets to reconcile conservation and socio-economic demands. Aquat Conserv Mar Freshwat Ecosyst 30:497–513. https://doi.org/10.1002/aqc.3269

Watson JEM, Dudley N, Segan DB, Hockings M (2014) The performance and potential of protected areas. Nature 515:67–73. https://doi.org/10.1038/nature13947

White AT, Aliño PM, Cros A, Fatan NA, Green AL, Teoh SJ, Wen W (2014) Marine protected areas in the coral triangle: progress, issues, and options. Coast Manag 42:87–106. https://doi.org/10.1080/08920753.2014.878177

Zoccola D, Ounais N, Barthelemy D, Calcagno R, Gaill F, Henard S, Hoegh-Guldberg O, Janse M, Jaubert J, Putnam H, Salvat B, Voolstral CR, Allemand D (2020) The World Coral Conservatory (WCC): a Noah's ark for corals to support survival of reef ecosystems. PLoS Biol 18(9):e3000823. https://doi.org/10.1371/journal.pbio.3000823

Correction To: Brazilian Coral Reefs

Ruy K. P. Kikuchi, Zelinda M. A. N. Leão,
Maria Elisabeth de Araújo, and Tito M. C. Lotufo

Correction to:
R. K. P. Kikuchi et al. (eds.), *Brazilian Coral Reefs*, Brazilian Marine Biodiversity, https://doi.org/10.1007/978-3-031-59152-5

The original version of the book was inadvertently published with incorrect spelling of the authors' names. This has now been amended throughout the book and approved by the author.

The updated version of this book can be found at
https://doi.org/10.1007/978-3-031-59152-5

R. K. P. Kikuchi et al. (eds.), *Brazilian Coral Reefs*, Brazilian Marine Biodiversity,
https://doi.org/10.1007/978-3-031-59152-5_13

Index

R. K. P. Kikuchi et al. (eds.), *Brazilian Coral Reefs*, Brazilian Marine Biodiversity, https://doi.org/10.1007/978-3-031-59152-5

The manufacturer's authorised representative in the EU is Springer Nature Customer Service Centre GmbH, Europaplatz 3, 69115 Heidelberg, Germany. If you have any concerns regarding our products, please contact ProductSafety@springernature.com

Printed and bound by CPI Group (UK) Ltd, Croydon, CR0 4YY

28/09/2026

02224056-0001